GALAXY EVOLUTION:
Connecting the Distant Universe with the Local Fossil Record

Proceedings of a Colloquium held at the Observatoire de Paris-Meudon
from 21–25 September, 1998

Edited by

M. SPITE

DASGAL, Observatoire de Paris Meudon, France

Reprinted from *Astrophysics and Space Science*
Volume 265, Nos. 1–4, 1999

SPRINGER SCIENCE+BUSINESS MEDIA, B.V.

Library of Congress Cataloging-in-Publication Data

Galaxy evolution : connecting the distant universe with the local
  fossil record / edited by M. Spite.
      p.   cm.
    Includes index.
    ISBN 978-0-7923-5943-2   ISBN 978-94-011-4213-7 (eBook)
    DOI 10.1007/978-94-011-4213-7
    1. Galaxies--Evolution Congresses. 2. Nucleosynthesis Congresses.
  I. Spite, Monique, 1939-
  QB857.5.E96G35   1999
  523.1'12--dc21
                                                    99-42392

ISBN 978-0-7923-5943-2
_____

Printed on acid-free paper

TABLE OF CONTENTS

IV – THE OLD DISK POPULATION

V – THE BULGES

VI – EVOLUTION OF THE MILKY WAY

PREFACE

This book contains the Proceedings of the first 'Rencontres de l'Observatoire' which was held at the Meudon campus of the Paris Observatory on September 21–25, 1998.

The subject addressed was 'Galaxy evolution: Connecting the distant Universe with the local fossil record'. This colloquium was a project devised by a group of astronomers who, in France but also in Europe, in the United States, in South America, and even in Japan, have worked for a long time with Roger and Giusa Cayrel. Roger and Giusa had a tremendous influence on us all. They taught us in their discrete way, the care of the observations and the rigor in the interpretation. Besides the presentations of Jun Jugaku and René Racine and the presentation of Pierre Couturier of the story of the Canada-France-Hawaii Telescope (presentation which unfortunately does not appear in this book) several papers refer to the role of Roger and Giusa in astronomy.

Their contribution to the study of the galactic evolution has been fundamental. Our Milky Way is the galaxy which can be studied in most details. From the study of our Galaxy's old population, one can understand what our Milky Way was like very soon after its birth, at the same stage where we now observe the very distant and thus very 'young' galaxies.

The organizers wanted to bring together with Roger and Giusa Cayrel experts of the old Galactic Population, of the distant galaxies and of the galactic evolution. The meeting was attended by 153 participants from 22 different countries.

A wealth of basic questions were discussed: Stellar Nucleosynthesis and the primordial formation of the elements, Galactic halos, the structure of the Galactic disc, Bulges, and generally speaking, the chemical and dynamical evolution of galaxies. To highlight the role of Roger in the design and the construction of the 3.6 m telescope in Hawaii, a session was devoted to the perspectives opening up with the recent development of a new generation of very large telescopes. Thanks to the enthousiasm of most participants this meeting will certainly favor new collaborations and a better use of these new telescopes.

It is a pleasure to thank here Ludowijk Woltjer who accepted the difficult task of drawing the conclusions of the meeting, and who did it so brightly.

The organizers would like to warmly thank Chantal Balkowski, who was at the time of the meeting vice-director of Paris-Observatory, and who initiated these annual 'Rencontres de l'Observatoire'. She greatly helped us organizing this first venue, and even found generous sponsors.

 Astrophysics and Space Science is the original source of publication of this article. It is recommended that this article is cited as: *Astrophysics and Space Science* **265:** xi–xiv, 1999.
© 1999 *Kluwer Academic Publishers.*

We want to thank also
- Michel Combes then director of Paris Observatory, who accepted to open the 'Cassini Room', the historical meridian room of Paris Observatory for the welcome party.
- Paul Felenbok for his permanent help and the organization of a memorable conference dinner at the 'Palais du Sénat'.
- Jacqueline Pluet, the Secretary of the colloquium and Christiane Adam the Secretary of the scientific council of Paris Observatory.
- And also Eliane Aulard, Claire Bentolila, Suzanne Berton, Pascal Hammès, Monique Michel, Lucia Picard, Nicole Romain, Mireille Petit, et Jean-Jacques Poisot who helped efficiently in many details of the local organization.

We also wish to express our acknowledgement to
- The 'Ministère de l'Education Nationale et de la Recherche'
- The 'Ministères des Affaires étrangères'
- The Canada-France Hawaii telescope Corporation
- The 'CNRS – SDU'
- The 'GdR Galaxies' et 'Structure interne'
- The 'département de Physique des étoiles et des Galaxies (DASGAL)'
- The 'Institut d'Astrophysique de Paris'
- The 'Bureau des Longitudes'
whose financial help made possible these 'Rencontres de l'Observatoire – 1998', and we thank as well the Société Générale for its participation.

François and Monique Spite
May 1999

AVANT-PROPOS

Ce volume contient les Actes des premières 'Rencontres de l'Observatoire' qui ont eu lieu à l'Observatoire de Paris sur le campus de Meudon du 21 au 25 Septembre 1998.

Le thème choisi était 'L'évolution galactique : des étoiles vieilles à l'Univers lointain'. Ce colloque a été voulu par tout un petit groupe d'astronomes qui tant en France qu'en Europe, aux Etats-Unis, en Amérique du Sud, et même au Japon ont travaillé depuis de longues années avec Roger et Giusa Cayrel.

Roger et Giusa ont eu une énorme influence sur tous leurs 'collaborateurs', ils nous ont appris dans la discrétion, le soin dans l'observation et la rigueur dans l'interprétation. Outre les exposés introductifs de Jun Jugaku et René Racine et l'évocation par Pierre Couturier de l'histoire du Télescope Canada-France Hawaii (qui malheureusement n'est pas publiée ici) plusieurs articles font explicitement référence au rôle joué par Roger et Giusa Cayrel en astronomie.

Leur contribution à l'étude de l'évolution galactique a été fondamentale. Or, de toutes les galaxies, notre Galaxie est celle qui peut être étudiée le plus en détail. En particulier l'étude de la vieille population galactique permet de comprendre comment était notre Galaxie peu après sa naissance, à une phase d'évolution où sont observées aujourd'hui les galaxies très lointaines et donc très 'jeunes'.

Nous avons voulu rassembler autour de Roger et Giusa des spécialistes de la vieille population Galactique, des galaxies lointaines jeunes et de l'évolution des galaxies. Ce colloque a réuni 153 participants venant de 22 pays différents.

Y furent étudiés: La nucléosynthèse stellaire et la formation primordiale des éléments, les halos galactiques, la structure du disque de notre Galaxie, les bulbes, et de façon générale l'évolution des galaxies tant du point de vue chimique que du point de vue dynamique. En écho au dynamisme montré par les Cayrel en maintes occasions (on se souvient du rôle de Roger dans la conception et la construction du télescope de 3,6m d'Hawaii) une session a été consacrée aux perspectives ouvertes par la mise en service des Très Grands Télescopes. Grâce à l'enthousiasme des participants cette conférence favorisera certainement de nouvelles et enrichissantes collaborations.

C'est un plaisir de remercier ici Ludowijk Woltjer qui a tiré avec tant de brio les conclusions de ce colloque.

Nous remercions aussi tout particulièrement Chantal Balkowski qui, alors directeur adjoint de l'Observatoire de Paris, a voulu ces Rencontres annuelles de l'Observatoire et nous a beaucoup aidé pour leur première organisation, allant jusqu'à nous chercher des 'mécènes' et à nous débloquer des crédits dans l'urgence.

Nous remercions encore:

- Michel Combes directeur de l'Observatoire qui, notamment, a mis à notre disposition pour le Cocktail la salle 'Cassini', salle méridienne historique de l'Observatoire de Paris.
- Paul Felenbok qui nous a beaucoup aidé et nous a organisé un banquet mémorable au palais du Sénat.
- Jacqueline Pluet la secrétaire du colloque, et Christiane Adam la secrétaire du conseil scientifique.
- tous ceux qui à l'observatoire nous ont assistés dans l'organisation matérielle du colloque: Eliane Aulard, Claire Bentolila, Suzanne Berton, Pascal Hammès, Monique Michel, Lucia Picard, Nicole Romain, Mireille Petit, et Jean-Jacques Poisot.

Ces 'Rencontres de l'Observatoire - 1998' n'auraient pas été possibles sans le soutien financier de plusieurs institutions :
- Le Ministère de l'Education Nationale et de la Recherche
- Le Ministères des Affaires étrangères
- La Société du Telescope Canada-France Hawaii
- Le CNRS - SDU
- Les GdR 'Galaxies' et 'Structure interne'
- Le département de Physique des étoiles et des Galaxies (DASGAL)
- L'Institut d'Astrophysique de Paris
- Le Bureau des Longitudes
Nous remercions également la Société Générale pour sa participation.

François et Monique Spite
Mai 1999

Roger and Giusa Cayrel

GALAXY EVOLUTION : CONNECTING THE DISTANT UNIVERSE WITH THE LOCAL FOSSIL RECORD

ÉVOLUTION GALACTIQUE: DES ÉTOILES VIELLES Á L'UNIVERS LOINTAIN

Comité scientifique d'Organisation

Spite F.	(DASGAL, Observatoire de Paris)
Andersen J.	(Niels Bohr Institute, Denmark)
Barbuy B.	(IAGUSP, Sao Paulo, Brazil)
Beers T.	(Michigan State University, USA)
Boesgaard A.	(Hawaii University USA)
Chmielevski Y.	(Observatoire de Genève, Suisse)
Friel E.	(NSF Arlington, USA)
Gratton R.	(Osservatorio Astronomico di Padova, Italy)
Gustafsson B.	(Astron. Obs. Upssala, Sweden)
Kunth D.	(Institut d'Astrophysique de Paris)
Lelièvre G.	(DASGAL, Observatoire de Paris)
Rich M.	(Columbia Univ., NY, USA)
Turon C.	(DASGAL, Observatoire de Paris)

Comité local d'Organisation

Spite M.	(DASGAL, Observatoire de Paris)
Baglin A.	(DESPA, Observatoire de Paris)
Briot D.	(DASGAL, Observatoire de Paris)
Crifo F.	(DASGAL, Observatoire de Paris)
Felenbok P.	(DAEC, Observatoire de Paris)
Jablonka P.	(DAEC, Observatoire de Paris)
Adam C.	(DAG, Observatoire de Paris)
Bentolila C.	(DASGAL, Observatoire de Paris)
Pluet J.	(DASGAL, Observatoire de Paris)

 Astrophysics and Space Science is the original source of publication of this article. It is recommended that this article is cited as: *Astrophysics and Space Science* **265:** xv–xvi, 1999.
© 1999 *Kluwer Academic Publishers.*

Conference photo

Welcome party

LIST OF PARTICIPANTS

Christophe Alard   `alard@iap.fr`
Georges Alecian   `alecian@obspm.fr`
Christine Allen   `allen@venus.astroscu.unam.mx`
Danielle Alloin   `dalloin@eso.org`
Johannes Andersen   `ja@astro.ku.dk`
Nobuo Arimoto   `arimoto@mtk.ioa.s.u-tokyo.ac.jp`
David Arnett   `dave@bohr.as.Arizona.EDU`
Jean Audouze   `jean.audouze@palais-decouverte.fr`
Marc Azzopardi   `azzopardi@OBMARA.CNRS-MRS.FR`
Suchitra Balachandran   `suchitra@astro.umd.edu`
Beatriz Barbuy   `barbuy@iagusp.usp.br`
J.Philippe Beaulieu   `beaulieu@astro.rug.nl`
Tim Beers   `beers@pa.msu.edu`
Peter Berczik   `berczik@mao.kiev.ua`
Jan Bernkopf   `jbe@usm.uni-muenchen.de`
Olivier Bienaymé   `Bienayme@astro.u-strasbg.fr`
Samuel Boissier   `boissier@iap.fr`
Piercarlo Bonifacio   `bonifaci@oat.ts.astro.it`
Angela Bragaglia   `angela@astbo3.bo.astro.it`
Gustavo Bruzual   `bruzual@cida.ve`
Margaret Burbidge   `mburbidge@ucsd.edu`
Carla Cacciari   `cacciari@bo.astro.it`
Vittoria Caloi   `caloi@victoria.ias.rm.cnr.it`
Giovani Carraro   `carraro@pd.astro.it`
Eugenio Carretta   `carretta@pd.astro.it`
Michel Cassé   `casse@iap.fr`
Fiorella Castelli   `castelli@astrts.oat.ts.astro.it`
Francesco A. Catalano   `Fcatalano@alpha4.ct.astro.it`
Robert Cavallo   `rob@astro.umd.edu`
Giusa Cayrel   `Giusa.Cayrel@obspm.fr`
Roger Cayrel   `Roger.Cayrel@obspm.fr`
Sandra Chapelon   `sandra@astrsp-mrs.fr`
Corinne Charbonnel   `Corinne.Charbonnel@obs-mip.fr`
Cristina Chiappini   `cristina@andromeda.iagusp.usp.br`
Norbert Christlieb   `nchristlieb@hs.uni-hamburg.de`
Suzanne Collin   `Suzy@obspm.fr`
Francoise Combes   `bottaro@obspm.fr`
Christopher Corbally SJ   `corbally@as.arizona.edu`
Roberto D.D. Costa   `roberto@iagusp.usp.br`

 Astrophysics and Space Science is the original source of publication of this article. It is recommended that this article is cited as: *Astrophysics and Space Science* **265**: xix–xxii, 1999. © 1999 *Kluwer Academic Publishers*.

Pierre Couturier	pierrec@cfht.hawaii.edu
Françoise Crifo	Francoise.Crifo@obspm.fr
Bertrand Dauphole	dauphole@observ.u-bordeaux.fr
Sébastien Derriere	derriere@astro.u-strasbg.fr
Nicole Feautrier	Nicole.Feautrier@obspm.fr
Paul Felenbok	paul.felenbok@OBSPM.FR
Sofia Feltzing,	sofia@astro.lu.se
Chris Flynn	cflynn@astro.utu.fi
Eileen Friel	efriel@nsf.gov
Jay Frogel	frogel@galileo.mps.ohio-state.edu
Burkhard Fuchs	fuchs@ari.uni-heidelberg.de
Klaus Fuhrmann	fuhrmann@usm.uni-muenchen.de
Laura Fullton	Laura.Fullton@obs.unige.ch
Ortwin Gerhard	gerhard@astro.unibas.ch
Raffaele Gratton	gratton@pd.astro.it
Michel Grenon	michel.grenon@obs.unige.ch
Frank Grundahl	Frank.Grundahl@hia.nrc.ca
Bruno Guiderdoni	guider@iap.fr
Sara Heap	hrsheap@stars.gsfc.nasa.gov
Pr. Gerhard Hensler	hensler@astrophysik.uni-kiel.de
Vanessa Hill	vhill@eso.org
Thais Idiart	thais@iagusp.usp.br
Garik Israelian	gil@iac.es
Inese Ivans	iivans@astro.as.utexas.edu
Pascale Jablonka	jablonka@gin.obspm.fr
Hartmut Jahreiss	hartmut@ari.uni-heidelberg.de
Jun Jugaku	jugakujn@cc.nao.ac.jp
Torgny Karlsson	torgny.karlsson@astro.uu.se
David Katz	David.Katz@obspm.fr
Valentina Klochkova	valenta@sao.ru
Tadayuki Kodama	kodama@ast.cam.ac.uk
Eira Kotoneva	eianko@astro.utu.fi
Attay Kovetz	attay@etoile.tau.ac.il
Ben Zion Kozlowsky	bzk@pair.gsfc.nasa.gov
Robert Kraft	kraft@ucolick.org
Daniel Kunth	kunth@iap.fr
Erwan Lastennet	E.Lastennet@qmw.ac.uk
Yveline Lebreton	Yveline.Lebreton@obspm.fr
Thibault Lejeune	lejeune@astro.unibas.ch
James Lequeux	aanda@obspm.fr
Rainer Luetticke	luett@astro.ruhr-uni-bochum.de
Walter Maciel	maciel@iagusp.usp.br
Pierre Magain	magain@astro.ulg.ac.be

Steven Majewski	srm4n@didjeridu.astro.virginia.edu
Maria Lucia Malagnini	malagnini@ts.astro.it
Jean-Michel Martin	Jean-Michel.Martin@obspm.fr
Louis Martinet	martinet@obs.unige.ch
Francesca Matteucci	francesc@newton.sissa.it
Andrew McWilliam	andy@ociw.edu
Claude Megessier	Claude.Megessier@obspm.fr
Laurent Meillon	Laurent.Meillon@obspm.fr
Poul E. Nissen	pen@obs.aau.dk
Arlette Noels	noels@astro.ulg.ac.be
Ken'ichi Nomoto	nomoto@astron.s.u-tokyo.ac.jp
Birgitta Nordström	birgitta@astro.ku.dk
John E. Norris	jen@mso.anu.edu.au
Edward Olszewski	edo@as.arizona.edu
Sergio Ortolani	ortolani@pd.astro.it
Goran Östlin	ostlin@iap.fr
Bernard Pagel	bejp@star.cpes.susx.ac.uk
Juliennne Palasi	Julienne.Palasi@obspm.fr
Ivan Panchenko	ivan@sai.msu.su
Vladimir Panchuk	panchuk@sao.ru
Laura Pasinetti	laura.pasinetti@mi.infn.it
Yakiv Pavlenko	yp@mao.kiev.ua
Ruth Peterson	peterson@ucolick.org
Patrick Petitjean	petitjean@iap.fr
Stefanie Phleps	phleps@mpia-hd.mpg.de
Cathy Pilachowski	catyp@noao.edu
Françoise Praderie	Francoise.Praderie@obspm.fr
Nicolas Prantzos	prantzos@iap.fr
Francesca Primas	fprimas@eso.org
Ren é Racine	racine@astro.umontreal.ca
Reuven Ramaty	ramaty@pair.GSFC.nasa.gov
Rafael Rebolo	rrl@ll.iac.es
Johannes Reetz	reetz@usm.uni-muenchen.de
Hubert Reeves	reeves@club-internet.fr
R. Michael Rich	rmr@ciel.astro.ucla.edu
Noël Robichon	robichon@strw.leidenuniv.nl
Annie Robin	robin@obs-besancon.fr
Brigitte Rocca-Volmerange	rocca@iap.fr
Helio Rocha-Pinto	helio@iagusp.usp.br
Frederic Royer	Frederic.Royer@obspm.fr
Wallace Sargent	wws@astro.caltech.edu
Evry Schatzman	schatzman@obspm.fr
Ricardo Schiavon	ripisc@andromeda.iagusp.usp.br

Regina Schulte-Ladbeck	rsl@phyast.pitt.edu
William J. Schuster	schuster@bufadora.astrosen.unam.mx
Uwe Schwarzkopf	schwarz@astro.ruhr-uni-bochum.de
Maartje Sevenster	msevenst@mso.anu.edu.au
Olga Sil'chenko	olga@sai.msu.su
Joseph Silk	silk@pac2.berkeley.edu
Christopher Sneden	chris@astro.as.utexas.edu
Jesper Sommer-Larsen	jslarsen@tac.dk
George Sonneborn	sonneborn@stars.gsfc.nasa.gov
Caroline Soubiran	soubiran@observ.u-bordeaux.fr
François Spite	Francois.Spite@obspm.fr
Monique Spite	Monique.Spite@obspm.fr
Grazina Tautvaišienė	taut@itpa.lt
Eline Tolstoy	etolstoy@eso.org
Francoise Tran Minh	tranmin@obspm.fr
James Truran	truran@nova.uchicago.edu
Takuji Tsujimoto	tsuji@misty.mtk.nao.ac.jp
Catherine Turon	Catherine.turon@obspm.fr
David Valls-Gabaud	dvg@astro.u-strasbg.fr
Claude van't Veer-Menneret	claude.vantveer@obspm.fr
Elisabeth Vangioni-Flam	flam@iap.fr
Olivier Varenne	varenne@newb6.u-strasbg.fr
Sylvie Vauclair	svcr@srvdec.obs-mip.fr
Mira Véron	mira@obs-hp.fr
Gerard Wlerick	gerard.wlerick@obspm.fr
Lodewijk Woltjer	woltjer@obs-hp.fr
Jean-Paul Zahn	Jean-Paul.Zahn@obspm.fr
Claude Zeippen	zeippen@obspm.fr

I – INTRODUCTION

THE PUBLICATIONS AND CITATIONS OF GIUSA AND ROGER CAYREL

JUN JUGAKU

2-29-3 Sakuragaoka, Tama-shi, Tokyo 206-0013, Japan

Abstract. Giusa and Roger Cayrel have published more than 170 papers and have been cited in more than 2173 times in the main journals. A total of 104 researchers have collaborated with them. These and other statistics are shown.

1. Introduction

We are aware of and grateful for Giusa's and Roger's influence on astronomy and astrophysics in the past half century. While important things may not be judged by quantification, it might be instructive to see what the standard objective criteria reveal about their work and to highlight the papers that have received the most citations.

2. Papers

Giusa's first publication appeared in 1948 (de Strobel 1948). The title of the paper is 'Determinazione della latitudine dell' Osservatorio astrofisico di Asiago', and is quite different from her main theme thereafter. Roger's first paper was published in 1951 (Cayrel, 1951). The paper entitled as 'Contribution à l'étude de l'équilibre radiatif (Influence de la zone convective)' represents a subject in which he still has a strong interest as evident in his recent papers. Since then, they have written more than 170 papers before 1998. Nineteen of them have been written together, often with their colleagues. A total of 104 researchers have written papers with Giusa and/or Roger. The distribution of these 170 papers is shown in Table I. In the category of 'other', we find many papers presented in conferences. In particular, we note their many contributions to Transactions, Highlights of Astronomy, Symposia and Colloquia of IAU.

Astrophysics and Space Science is the original source of publication of this article. It is recommended that this article is cited as: *Astrophysics and Space Science* **265**: 3–6, 1999.
© 1999 *Kluwer Academic Publishers.*

TABLE I

Distribution of papers

Journal	Number of papers
Astron. Astrophys.	34
Ann. d'Astrophys.	12
Mem. Astron. Soc. Ital.	12
Astrophys. J.	6
Comptes Rendus (Paris)	5
Publ. Astron. Soc. Pacific	2
Z. Astrophys.	1
Other	98
(including IAU publications	37)

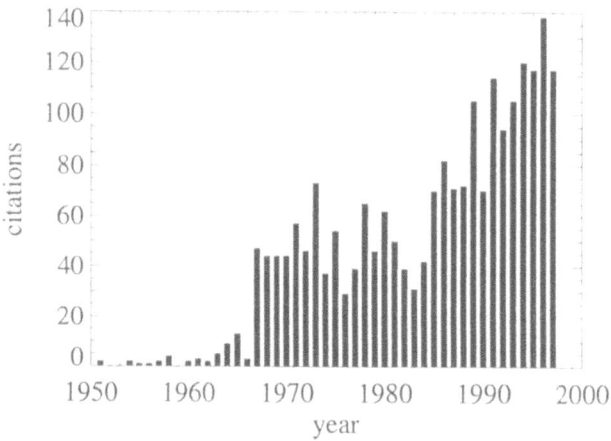

Figure 1. Yearly citation number of papers

3. Citations

Although Science Citation Index is by no means a perfect source, it gives us a reasonable impression and a lower bound on the number of times papers have been cited in the main journals. Yearly citation number of Cayrels' papers is shown in Figure 1. It increases from 0 in 1950 to 137 in 1996 and 117 in 1997. The accumulated total number is 2173.

Their top ten papers judged by total number of citations before 1998 are shown in Table II.

Since four fairly recent papers appear in Table II, it is not surprising that top ten ranking changes somewhat when we look at number of citations per year (Table III). Perhaps we can make a modest prediction from inspection of Table III

TABLE II

Top ten papers judged by total number of citations before 1998

Citations	Paper	Topic
137	Cayrel de Strobel *et al.* (1985)	Catalogue
136	Cayrel *et al.* (1985)	Hyades (heavy elements)
134	Cayrel and Jugaku (1963)	Model atmospheres
131	Cayrel and Cayrel (1963)	Epsilon Virginis
115	Cayrel *et al.* (1984)	Hyades (Li)
97	Cayrel de Strobel *et al.* (1992)	Catalogue
96	Perrin *et al.* (1977)	H-R diagram
66	Cayrel (1968)	Population II
59	Cayrel de Strobel (1966)	K stars
56	Cayrel de Strobel *et al.* (1980)	Catalogue

TABLE III

Top ten papers judged by citations per year (total citations /number of years after publication)

Citations yr^{-1}	Paper	Topic
16.2	Cayrel de Strobel *et al.* (1992)	Catalogue
10.5	Cayrel de Strobel *et al.* (1985)	Catalogue
10.5	Cayrel *et al.* (1985)	Hyades (heavy elements)
8.2	Cayrel *et al.* (1984)	Hyades (Li)
4.8	Perrin *et al.* (1977)	H-R diagram
3.8	Cayrel and Jugaku (1963)	Model atmospheres
3.7	Cayrel and Cayrel (1963)	Epsilon Virginis
3.5	Cayrel de Strobel and Bentolila (1983)	F,G,K field stars
3.3	Cayrel (1986)	Population III
3.1	Cayrel de Strobel *et al.* (1980)	Catalogue

that the latest 1996 edition of [Fe/H] catalogue (Cayrel de Strobel *et al.*, 1997) will appear in this list soon.

References

Cayrel, G. and Cayrel, R.: 1963, *Astrophys. J.* **137**, 431.
Cayrel, R.: 1951, *Ann. d'Astrophys.* **14**, 1.

Cayrel, R.: 1968, *Astrophys. J.* **151**, 997.

Cayrel, R.: 1986, *Astron. Astrophys.* **168**, 81.

Cayrel, R. and Jugaku, J.: 1963, *Ann. d'Astrophys.* **26**, 495.

Cayrel, R., *et al.*: 1984, *Astrophys. J.* **283**, 205.

Cayrel, R., *et al.*: 1985, *Astron. Astrophys.* **146**, 249.

Cayrel de Strobel, G.: 1966, *Ann. d'Astrophys.* **29**, 413.

Cayrel de Strobel, G. and Bentolila, C.: 1983, *Astron. Astrophys.* **119**, 1.

Cayrel de Strobel, G., *et al.*: 1980, *Astron. Astrophys. Suppl.* **41**, 405.

Cayrel de Strobel, G., *et al.*: 1985, *Astron. Astrophys. Suppl.* **59**, 145.

Cayrel de Strobel, G., *et al.*: 1992, *Astron. Astrophys. Suppl.* **95**, 273.

Cayrel de Strobel, G., *et al.*: 1997, *Astron. Astrophys. Suppl.* **124**, 299.

de Strobel, G.: 1948, *Mem. Astron. Soc. Ital.* **19**, 103.

Perrin, M.-N., *et al.*: 1977, *Astron. Astrophys.* **54**, 779.

THE CANADA-FRANCE-HAWAII TELESCOPE AND THE COSMIC CONNECTION

RENÉ RACINE

Départements de physique, Université de Montréal, Université Laval, Canada

1. Introduction

When François and Monique invited me to speak on the Canada-France-Hawaii Telescope (CFHT) contributions to our understanding of the 'Cosmic Connection' revealed by galaxy evolution, I worried that this far reaching topic would make it nearly impossible to *exclude* anything done at the CFHT over the twenty years elapsed since its inauguration. I was much reassured when they suggested that I could speculate on future contributions, about which nothing needs to be excluded since nothing is of course known. I am thus at liberty to reminisce about the gestation of the CFHT, brush a broad outline of its multifaceted and important scientific contributions, and indulge into visions where it can still take its community of users.

This *Rencontre* celebrates the two persons who have arguably played the most pivotal roles in the birth of the CFHT. I shall briefly and candidly tell my perception of the impacts Roger and Guisa have had throughout the pioneering era of this project and boast of the unique qualities of the facility they made possible. I will illustrate, with publication statistics, the breadth of the advances the telescope has made possible over the last 20 years. Finally, I shall speculate over the future challenges and possible glories of the next 20 years, in total ignorance of what, if anything, will reside on the extraordinary site presently occupied by the CFHT.

2. The Distinctive Characters of Roger and Guisa

The factual episodes which led to the first international agreement to install a 4-m class telescope on Mauna Kea are well known. I have added my version of this story (Racine, 1981) to numerous others in the literature, including even a brief one in *Le Petit Larousse Illustré*.

A detail of this story is worth noting in connection with this *Rencontre de l'Observatoire*. The suggestion that the large telescope being planned in France in the late 60's, which became the CFHT, should be located on Mauna Kea was

Astrophysics and Space Science is the original source of publication of this article. It is recommended that this article is cited as: *Astrophysics and Space Science* **265**: 7–13, 1999.

first made by a young University of Hawaii solar astronomer on a visit to Meudon. It would appear that John Jefferies' vision was enthusiastically shared by Roger Cayrel who carried the ball from then on.

The distinctive characters of people like Roger and Guisa, who made the CFHT happen, played a crucial role in the success of the enterprise but have been less publicised in press. My first personal contact with the project, the Cayrels, and indeed French astronomy, was at the initial meeting of the Scientific Advisory Committee in Meudon's 'Salle des Communs' in 1974, under the chairmanship of regretted Pierre Charvin. I was amazed! Project Director Roger Cayrel, and associate Graham Odgers, paraded before the committee a series of engineers proposing strange ideas to suspicious astronomers demanding even stranger features. A bewildering array of instruments were proposed and summarily adopted by budding scientists whose imagination knew of no frontiers. Guisa sporadically erupted into the room, reminding Roger in a no subtle tone of things he was not likely to have forgotten. Roger, imperturbable, was taking in stride the growing pile of projects, injecting here and there a subdued but firm cautionary comment, trying to appease the more ebullient members of his Project Office while retiring with them to the *Wahinés*, the compound of temporary buildings recuperated from past war-time glories in the South Pacific and which housed the Project Office. Roger's marvellous capabilities as a team builder and leader shone brilliantly on those occasions. For years to come, in Meudon, La Rochelle and Waimea, Roger's talents were to be the pillar over which the growth of the CFHT could thrive.

During the Waimea Project Office era (1977–1980), when the CFHT took shape at the summit of Mauna Kea, the project faced a number difficult political problems on Big Island. When I took over from Roger as CFHT Director in July 1980, it became immediately clear that the Cayrel's warm and conciliatory personalities, and their generous involvement in the local community, had been crucial to the resolution of these thorny issues. Their going-away Aloha Oe' luau in Honokaa has been a most heartfelt occasion. Guisa handed my wife Claudine an extraordinarily long list of local people and suggested we invite them all without delay, together with the CFHT staff, to our house warming party, which we did. We never regretted it, not did the CFHT I reckon, despite administrator Claude Bethoud's well meaning frown upon such festivities, to which he always politely accepted the invitation.

Barely a day of our four-year stay in Hawaii passed by without someone in the street, at the shopping mall or elsewhere stopping us to enquire 'How's Roger and Guisa? So nice people yeah!' We might have felt somewhat ignored, were it not for their addressing Claudine as 'Madame Canada-France' and also asking her, with a smile, how we were doing.

3. The Distinctive Character of the CFHT

A strength of the CFHT, which it shares with other facilities on Mauna Kea, is well known: 4200 m of altitude. Standing tall above the worst 40 percent of the atmosphere, telescopes at the summit of Mauna Kea benefits from unsurpassed natural image quality and outstanding transparency in the infrared. Considerable efforts have been devoted to allow the telescope to reap the full benefits of the site, with remarkable success.

Of equally important impact has been the often very imaginative and always high performance instruments available at the telescope. Especially well know to people at this meeting is the f/8 coudé spectrograph and its 1875 Reticon detector which served Guisa Cayrel, the Spite and many others so well over so many years. Thanks to the combination of site, telescope and instrumentation excellence, astronomers at the CFHT were able to probe, in space and in wavelength, deeper, sharper, at higher signal-to-noise (SNR) and with greater precision that at any other facilities of its kind. And the imagination of the users did put this potential to good use. A survey of the publications based on CFHT data *(www.cfht.hawaii.edu/Science/ Publications/)*, shows the immense breadth of their contributions to nearly all topics in astronomy, from high-redshift QSOs to studies of solar system objects, including the (eclipsed!) Sun itself. (The Moon has also been observed but, to my knowledge, only as a calibration source).

4. The First 20 Years

The Cosmic Connection between the distant Universe and the various local fossils of evolution has been largely discussed at this conference in terms of the chemical abundance of the elements spewed out by the Big Bang and ensuing stars, and their distribution in galaxies and old stellar systems such as halo stars and globular clusters. I suppose this is a fruitful route for inquiries. Precious clues on the earliest conditions in the Universe have indeed come from the epochal studies of Li abundance in halo dwarfs carried out at the early CFHT by François and Monique Spite and by Ann Boesgaard. These were the very first coudé runs after the commissioning of the Reticon detector in late 1980, together with the start of the precision radial velocity program of Gordon Walker and Bruce Campbell. That program had nothing to do with the Cosmic Connection: it merely searched for unfashionable extra-solar planets, ahead of its time by a couple of decades, but had been the primary science driver for the Reticon detector. Detailed spectroscopic studies of abundance continued at the CFHT coudé for many years, notwithstanding Giusa's vocal concerns that infrared astronomy threatened to squeeze all coudé (bright time) spectroscopy and real abundance studies out of the dome. Indeed I recall CFHT Director Bob McLaren joking, ten years after the commissioning of the CFHT, that the main problem with the coudé spectrograph was that the periodic

TABLE I

Topics	1980–1989	1990–1997
QSOs and High-z Galaxies	30%	15%
Gravitational Lenses	1%	10%
Local ($z < 0.05$) Galaxies	15%	25%
Globular Clusters and GC Systems	10%	10%
Milky Way Stars and ISM	35%	25%
Solar System Objects	4%	10%
Instruments and Techniques	5%	5%
TOTAL	100%	100%

table of the elements was too long! This is illustrated by the following CFHT papers: '*Lithium*, Age, and Metallicity in Open Clusters' (Boesgaard, 1991);

'*Lithium* in Old Binary Stars' (Spite *et al.*, 1994);
'Galactic Evolution of '*Beryllium*'' (Boesgaard *et al.*, 1993);
'*Beryllium* in *Lithium*-Deficient F and G Stars' (Stephens *et al.*, 1997);
'*Oxygen* Abundance in Metal-Poor Dwarfs ...' (Spite and Spite, 1991);
'... *C/H* and *C/Fe* in F Dwarfs ...' (Friel and Boesgaard, 1990);
'The Abundance of *Li, Al, Si and Fe*' (Burkhart and Coupry, 1991);
'Determination of the *[Th/Eu]* Ratio in Halo Stars' (Spite and Spite, 1993);

But unravelling the Cosmic Connection is also a matter of looking back in time, far in space whose extent must be measured, searching for early objects and early structures, and tracking the evolution of their kinematics and composition as $(1+z)$ decreases. It is also a matter of geometrically, photometrically and spectroscopically probing the unseen material along the line of sight by the signature it imprints on the photon streams as they travel toward the telescope: dark matter, gravitational deflectors, absorbers, L_α clouds. With this added point of view, much of what has been done at the CFHT over the first 20 years of it life has to do with unravelling the Cosmic Connection.

Some 90 papers based on CFHT material are published each year. The relative frequencies of somewhat arbitrary broad classes of topics covered by all papers published between 1980 and 1997 are listed in Table I. These figures reveal some interesting facts and trends. By my biased criterion that work on all but the last two topics, is directly related to the Cosmic Connection, some 90 percent of the CFHT work qualifies as such. I would even be willing to concede that some fine abundance analysis in the atmosphere of the Jovian planets have considerable bearing on the issue and raise that 90 percent estimate somewhat.

There has been over time some significant evolution of the topics' relative weight. Contrary to what one might have expected perhaps, fewer papers have appeared more recently on the most distant objects. This is explained by a transition from studies of individual sources, which led to numerous papers, to very demanding broad surveys such as the Canada-France-Redshift Survey (Lilly *et al.*, 1995), the Canadian Network for Observational Cosmology project (Yee *et al.*, 1996) and studies of cluster lensing (Miralda-Escudé and Fort, 1993; Surdej *et al.*, 1993; Yee *et al.*, 1993). Indeed, the studies of gravitational lenses, including arclets, and weak shear caused by dark matter in galaxy clusters, has become a mainstay of research at the CFHT. The facility is eminently well suited for this work, thanks to the excellent images the telescope delivers to faint limits over wide fields. The first gravitational lens (0957 + 62) was discovered by D. Weedman in 1980 at the CFHT. So was the geometry of the much studied quadruple image 2237 + 0305 (Yee, 1985) and the confirmation of the first cluster gravitational arcs, in A370 (Mellier *et al.*, 1991; Fort *et al.*, 1992). Such objects should help to resolve the too long standing dark matter issue, probably the most important and certainly the least understood link in the Cosmic Connection.

The relative decline in Milky Way studies and accompanying increase in nearby galaxy research displayed above, seem to be genuinely related. The capabilities of multi-object instruments and more efficient detectors make it increasingly feasible to explore in details external galaxies. Astronomers are thus able to carry out much needed comparative studies from which evolutionary trends can be discovered and more easily linked to galaxy morphology and dynamics than is possible within an horizon limited to the Milky Way.

5. The Next 20 Years

5.1. CHALLENGES AND PLANS

Plan for the long-range future of the CFHT are at this time, and to put it mildly, in a state of flux. Nothing much will change at the facility over the next few years as wide field CCD and near-IR imagers currently under development fully exploit the present telescope capabilities for morphological studies of large scale structures, weak lensing and general survey work. Spectroscopically, the facility cannot remain at the forefront and attract previous limelight's as a host of 8–10 m telescopes acquire first light. Nor can it remain competitive in a field it now completely dominates, high resolution imaging with adaptive optics: the D^4 advantage of the larger telescopes in this regime is simply overwhelming.

The CFHT community has been struggling to define new long range directions for the facility for some years now. There seems to emerge a consensus that 'bigger is better and needed'. Plans to replace the 3.6-m telescope with an 8-m instrument have been put forth. The CFHT building, after some remodelling, could accommodate such a telescope. It is argued in places that wide field imaging with exquisite image quality would make this a uniquely powerful research tool. Others feel, in view of the plethora of 6 to 10-m telescopes that will be in operation by 2005, that *much* bigger is needed. The CFHT partners are already involved in demanding 8-m projects: the ESO VLT and the Gemini telescopes. This leaves them rather limited financial and, more importantly, human resources to tackle much larger projects at this time.

5.2. An option for coudé spectroscopy

A promising and affordable approach to enhance the CFHT's spectroscopic capabilities, and further elucidate the Cosmic Connection through abundance analysis, would be to feed its excellent cross-dispersed coudé spectrograph with the photons collected by the 8-m primary of the Gemini telescope located 200 m away. Suitable trades with Gemini users for that connection would have to be made on the basis of scientific balance. Wide field imaging, at which the CFHT will be unsurpassed for some years, and which Gemini will lack, should be an enticing enough commodity. It would remain to be seen if, on both side of the trade, those who relish high dispersion spectroscopy can convince Time Allocation Committees of the priority of their quest over that of their colleagues whose primary pleasures rest at the CFHT prime focus or Gemini Cassegrain focus.

Should all that come to pass, chemical abundance analysis, stellar kinematics and atmosphere studies at the CFHT would soar to new heights. Relatively faint but crucially important stars such as halo RR Lyrae and luminous stars in nearby galaxies could be investigated to the limits imposed by microturbulence. Detailed abundance data would become available throughout the Local Group and beyond. The conditions and isotopic ratios in the interstellar medium could be determined with unprecedented precision. Such results would certainly shed much new light on the Cosmic Connection.

5.3. The dark side

I shall conclude on a dark note, of which all are well aware. We know of only a few percent of the content of the Universe. All objects of our studies are drowned in a sea of dark matter whose nature completely escapes us. We can hope that its cross sections for interactions are totally negligible over cosmic distances and column densities, except for what we interpret as its revealing gravitational field. But hope is not a scientific attitude; admission of ignorance rather is. Until the problem of dark matter which has been with us for 65 year can be put to rest, we must face the possibility that much of our ideas about cosmic connections are grossly ill-founded and must humbly admit that faith in our models is presumptuous.

Mapping dark matter distributions, by solving the potential equation with velocity and image data as boundary conditions, has reached new levels of sophistication in recent years. The richest data base is coming from deep, wide field imaging of distant objects lensed by galaxies and clusters of galaxies. Weak shear observations are especially precious. There, the gravitational lensing effects, and the solution they allow, extend over enormous angular extent. But these effects are barely detectable in the presence of image spread caused by atmosphere, telescope and instrument. And they require enormous arrays of high performance detectors to be mapped with reasonable efficiency. Thanks to the excellence of the images it can deliver over wide fields of view, and the coming of giant visible and near IR panoramic imagers such as MegaCam and WIRCAM, the CFHT will lead this field and make tremendous contributions. It may not solve the problem of the *nature* of dark matter but it will certainly provide detailed maps of its distribution which should point the way to what it is made off.

Until we have the answer to this question, nothing of what we might uncover in the future should surprise us.

References

Boesgaard, A.M.: 1991, *Astrophys. J.* **370**, L95.
Boesgaard, A.M. and King, J.R.: 1993, *Astron. J.* **106**, 2309.
Burkhart, C. and Coupry, M.F.: 1991, *Astron. Astrophys.* **249**, 205.
François, P., Spite, M. and Spite, F.: 1993, *Astron. Astrophys.* **274**, 821.
Fort, B., Le Fèvre, O., Hammer, F. and Cailloux, M.: 1992, *Astrophys. J.* **399**, L125.
Friel, E.D. and Boesgaard, A.M.: 1990, *Astrophys. J.* **351**, 480.
Lilly, S.J., Le Fèvre, O., Crampton, D., Hammer, F. and Tresse, L.: 1995, *Astrophys. J.* **455**, 50.
Mellier, Y, Fort, B., Soucail, G., Mathez, G. and Cailloux, M.: 1991, *Astrophys. J.* **380**, 334.
Miralda-Escudé, J. and Fort, B.: 1993, *Astrophys. J.* **417**, L5.
Racine, R.: 1981, *JRASC* **75**, 6.
Spite, M., Pasquini, L. and Spite, F.: 1994, *Astrophys. J.* **290**, 217.
Spite, M. and Spite, F.: 1991, *Astron. Astrophys.* **252**, 689.
Stephens, A., Boesgaard, A.M., King, J.R. and Deliyannis, C.P.: 1997, *Astrophys. J.* **491**, 339.
Surdej, J., Claeskens, J.F., Crampton, D., Filippenko, A.V., Hutsemékers, D., Magain, P., Pirenne, B., Vanderriest, C. and Yee, H.K.C.: 1993, *Astron. J.* **105**, 2064.
Yee, H.K.C.: 1985, *Astron. J.* **90**, 1082.
Yee, H.K.C., Ellingson, E. and Carlberg, R.G.: 1996, *Astrophys. J. Suppl.* **102**, 269.
Yee, H.K.C., Filippenko, A.V. and Tang, D.: 1993, *Astron. J.* **105**, 7.

II – THE PREGALACTIC EVOLUTION AND THE NUCLEOSYNTHESIS OF THE ELEMENTS

PRIMORDIAL NUCLEOSYNTHESIS

HUBERT REEVES

Service d'Astrophysique, CEA, Saclay, France

1. Introduction

It is with great pleasure that I have accepted to contribute to these 'Rencontres de l'Observatoire' in honor of Giusa and Roger Cayrel. They are amongst the first astronomers I have met here in Paris when I arrived more than 30 years ago. (I already knew the astronomical papers of Roger during my previous stay at The Institute for Space Studies in New York). I will always remember their warm family welcome. With their charming little girls (Françoise, Marguerite, Laura) climbing on my shoulders: I found it difficult to follow the conversations.

Since then, I have had many scientific conversations with Giusa and Roger. I have often profited from their expertise on many subjects of common interests. In particular Roger was my reference when I wanted to have an opinion on recently published observations. His verdict, competent and prudent, was always of great value to me. Roger is a rare example of an astronomer equally competent in theoretical and observationnal matters who can pronounce sound judgments on both fields. All through these years Giusa has played a most important role in the life of young astronomers in Paris and Meudon. Her presence at seminars and thesis would always inject a particular vitality to the discussions. She would always bring forward the really important issues from behind the screen of (often!) too abstract terminology. She always defended young students against occasionally unkind superiors. Vivid in my memory is a thesis at the end of which a jury member adressed rather harsh remarks to the student. She raised up and said: 'your blame is undeserved. Furthermore I saw you sleeping throughout the presentation!' She is aptly called 'la mamma de l'astronomie française'. Everyone loves her .... .

2. Helium and Carbon Isotopic Ratios

And now to my subject. Jean Audouze is reviewing the Big Bang Nucleosynthesis (BBN) as far as D and Li are concerned. Here I will concentrate on two relatively recent developments, one on ^{3}He and one on neutrino masses.

 Astrophysics and Space Science is the original source of publication of this article. It is recommended that this article is cited as: *Astrophysics and Space Science* **265**: 17–21, 1999.
© 1999 *Kluwer Academic Publishers.*

It is often of interest to reinsert a subject in its historical context. For two reasons. First because one gains a broader perspective which often opens the way to new reflexions and sometimes avoids repeating the same mistakes. And second, because this often gives to the author the pleasure of recalling previous contributions and collaborations. In 1969, Johannes Geiss and his colleagues in Bern measured the ^{3}He/^{4}He ratio in stellar wind from an aluminum sheet brought back from the moon (Geiss and Gloeckler, 1998). Since the solar convective zone in hot enough to burn D+p = ^{3}He it was clear that this solar wind contained the sum of the primordial D and ^{3}He of the protosolar nebula. But was it possible that during the past history of this zonal material, some ^{3}He had been destroyed in ^{4}He?

The presence of ^{9}Be in the solar photosphere at essentially the meteoritic (i.e. protosolar) value argued against this possibility: in the SCZ (surface convective zone) the ^{9}Be+p reaction occurs at lower temperature than the ^{3}He burning reactions. Hence any ^{3}He depletion would have been accompanied by an even stronger ^{9}Be depletion. This does not seem to be the case. Recent solar photospheric Be (Balachandran and Bell, 1997, 1998) measurements have confirmed that the depletion of this element is less than 20%.... . Thus the solar wind ^{3}He can be used as a good estimate of D+^{3}He in the ISM at solar birth.

For many years the observed sum of (D+^{3}He) was considered as a safe lower limit to the Big Bang Nucleosynthesis (BBN) value. The initial argument was that, as in the case of the SCZ, D would have burned in ^{3}He while some ^{3}He could have been generated in stars. Later on, more sophisticated calculations were made (Yang *et al.*, 1987) to estimate the stellar contributed ^{3}He and hence, by subtraction, evaluate the BBN yields. Unfortunately ^{9}Be could not be of much use here since this nucleus is not generated in stellar interiors. The small depletion of ^{9}Be accompanying any ^{3}He burning would represent only a very marginal effect on the corresponding stellar surface. It is interesting to note the similarity between this ^{9}Be situation and the present carbon isotopic ratio situation in relation with ^{3}He, largely discussed in recent litterature.

The problem raised by the observations of ^{3}He is illustrated in Figure 6 of Rood *et al.*, (1998). The observed ratios of (^{3}He/H) are = 1 to 3 10^{-5}, independant of age (the same between the protosolar nebula and today) and also of galactocentric distances. Furthermore these values are well in the range expected from BBN yields in the baryonic density range determined by D, ^{4}He and ^{7}Li observations. In other words it appears that the ^{3}He stellar contribution is at best comparable to the BBN contribution. This quite at odd with standard stellar model computations as described in Iben (1974). An ^{3}He build-up occurs during the Main Sequence at intermediate depth layers where the temperature is high enough to induce the ^{3}He+D reactions but not the ^{3}He+^{3}He or the ^{3}He+^{4}He reactions. Ratios of ^{3}He/^{4}He $\approx 10^{-2}$ can be reached there. In stars with $M < 2M_\odot$ this ^{3}He surplus is expected to be carried to the surface by later deep convection motions (dredge-up) during the red giant branch ascension and to be observable in planetary nebulae. A few cases have indeed been reported (Rood *et al.*, 1998) such as NGC 3242, a planetary nebula with ^{3}He/H $\approx 10^{-3}$.

The later expulsion of these atoms are expected to have enriched the ISM. In view of the low mass of the corresponding stars this process is not expected to be important in the first billion years of the Galaxy. Nevertheless according to galactic evolution models (Tosi, 1998; Vangoni-Flam *et al.*, 1994) the present ^{3}He/^{4}He ratio should by now have reached values of $\approx 10^{-4}$, much larger than the BBN early contribution. Where is this stellar ^{3}He?

It is of interest to couple this problem with another problem of stellar evolution in relation with the ^{12}C/^{13}C ratio. Theoretically, as a star ascends the RG branch, the surface convective zone deepens and a mixing occurs between the stellar outer layers (the so-called 'first dredge-up'). The resulting surface carbon isotopic ratio is expected to lie around 20–30 (Palla *et al.*, 1998). Charbonnel and do Nascimiento (1998) have recently investigated a large sample of red giants and shown an interesting correlation between their luminosity and their ^{12}C/^{13}C ratio. Stars with high luminosity (Mv < 0) show a ratio appreciably lower than 15, while stars with lower luminosity show higher ratios (15 to 50). Their Figure 3 illustrates the observational situation. This fact implies that the surface layers of the brighter stars must have been brought to temperature high enough to complete the full CNO cycle, thereby burning ^{3}He into ^{4}He. This, in turn, implies an extra dredge-up episode plausibly taking place after the 'Red Giant Bump'.

The present value of ^{12}C/^{13}C in ISM is ≈ 50, appreciably larger than the protosolar value of 90. One question comes to mind: what is the contribution of this new episode to the ISM value of ^{13}C? A simple estimate can be obtained in the following way. From Tosi (1998), we find that, in absence of this extra dredge-up, the present value of ^{3}He/H should be $= 2 \times 10^{-4}$. Using the ^{3}He/H ratio of 10^{-3} observed in the PN NGC 3242 and using ^{4}He/H $= 0.1$ we require that the PN contribution to ISM be Δ^4He/H ≈ 0.02, quite reasonable in view of the increase from Y = 0.23 in BBN to Y = 0.3 in present ISM. Assuming an increase of Δ^{12}C/^{4}He $= 10^{-3}$ in PNe and using the CNO equilibrium ratio ^{12}C/^{13}C $= 4$, one gets Δ^{13}C/H $= 5 \times 10^{-6}$ while the 'cosmic' ^{13}C/H $\approx 10^{-6}$ showing a possibly important decrease in the carbon isotopic ratio. Notwithstanding these speculations, we note that the success of BBN has allowed a neat interpretation of the ^{3}He data: they are consistent with no important stellar contribution. In turn this inference has the important implications for stellar and galactic evolution discussed in the previous paragraphs ... .

3. Neutrino Mass Observations

After many years of false rumors on the determinations of the masses of neutrinos, good news have come from the Superkamiokande neutrino observatory in Japan (Normile, 1988). Previously we had only the lower limits:

$M(\nu_e) < 10$ eV, $M(\nu_\mu) < 24$ keV, $M(\nu_\tau) < 300$ MeV.

The neutrinos under investigation are generated by the bombardment of high energy cosmic rays on the upper terrestrial atmosphere. An approximately equal number of ν_e and ν_μ are produced in this region (the instrument is insensitive to the ν_τ). Comparing the fluxes coming from above the lab (Zenith) and below (Nadir) a clear deficit of ν_μ was observed with respect of the ν_e. The standard quantum mechanical explanation for this phenomena is called 'neutrino oscillation'. It implies that during the travel throughout the earth, about one half of the ν_μ were transmuted in (plausibly) ν_τ, thus escaping detection. According to quantum physics this phenomena can only occur if one (at least) of the two particles (the ν_μ or the ν_τ) has a non-zero mass. Unfortunately the theory only gives a rather cryptic information on the physics of the situation. It measures the difference between the square of the two masses

$$M^2 (\nu_i) - M^2 (\nu_j) \approx 0.01 \text{ eV}^2 \text{ where } i \text{ and } j \text{ can be } \mu \text{ or } \tau.$$

This result requires that at least one has a mass of at least 0.1 eV. There is also the possibility that both have larger masses, implying however a very near (rather unlikely) equality between the two masses.

This important observation has several cosmological implications.

First, according to BBN there shoud still remain to day a neutrino fossil radiation (at ≈ 1.8 K) similar to the photon 3K background radiation. With the lower mass limit of 0.1 eV for one neutrino species, this corresponds to a lower cosmic neutrino density of ≈ 0.004 in terms of the critical density (required to eventually halt the expansion). Interestingly this is quite comparable to the luminous density in the form of stars and gases. It is quite striking to note that this evanescent species, predicted on purely theoretical basis by Wolfgang Pauli in 1932 and experimentally observed (with great pains . . .) in 1954 by Reines and Cowan, is comparable in mass density to all the stars in our skies. Nowadays however such a statement does not sound so strange in view of the fact that more than 90% of the cosmic matter is still of a nature unknown to us. We live indeed in a mysterious universe.

Second, using the previous lower limits on $M(\nu_e)$ and $M(\nu_\mu)$ it appears clearly that the ν_τ must have been relativistic at BBN ($M(\nu_\tau) \ll 0.14$ MeV). This determination is of importance for the calculations of the yields of the light elements. The ordinarily made assumption of three relativistic species of neutrinos in the standard BBN is thus confirmed.

Third, it is well known that mass distribution of structures in the universe can only be theoretically fitted by a mixture of ingredients. Amongst these, two varieties of so-called 'dark matter' (i.e. particles with coupling constants much lower than electromagnetic): one cold CDM (non relativistic at recombination) and one hot HDM (relativistic at recombination) are required. This last component could well be composed of neutrinos.The usually quoted best number to reproduce the distribution of small structures is $\Omega_{HDM} \approx 0.2$.

Where do we stand in the respect? Again the SuperKamiokande data has shown a lower limit of Ω_ν of 0.002. This result is qualitatively of great interest since it goes in the right direction (massless neutrinos would give only $\Omega_\nu \approx 0.0005$). It remains to be seen if the observations of the tau-neutrinos, by the Sudbury observatory, will give a higher value to its mass. Or if the cosmological modelist will bargain for a lower HDM value. One thing is sure: neutrinos are rapidly gaining importance in cosmology.

References

Balachandran, S. and Bell, R.: 1997, *Astron. Astrophys. Suppl.* **191**, 7408B.

Balachandran, S. and Bell, R.: 1998, *Nature* **392**, 791.

Charbonnel, C.: 1998, Proceeding of ISSI Symposium, *Space Sci. Rev.* **84**, 199.

Charbonnel, C. and do Nascimento, J.D., Jr.: 1998, *Astron. Astrophys.* **336**, 915.

Normile, D.: 1988, *Science* **220**, 1690.

Geiss, J. and Gloekler, G.: 1998, Proceedings of ISSI Symposium, *Space Sci. Rev.* **84**, 239.

Iben, I., Jr.: 1967, *Astrophys. J.* **143**, 642.

Palla, F., Galli, D., Bachiller, R. and Perez Gutierrez, M.: 1998, Proceeding of ISSI Symposium, *Space Sci. Rev.* **84**, 177.

Rood, R.T., Bania, T.M., Balser, D.S. and Wilson, T.L.: 1998, Proceeding of ISSI Symposium, *Space Sci. Rev.* **84**, 185.

Tosi, M.: 1998, Proceeding of ISSI Symposium, *Space Sci. Rev.* **84**, 207.

Vangioni-Flam, E., Olive., K.A. and Prantzos, N.: 1998, *Astrophys. J.* **148**, 3.

Yang, J., Turner, M., Steigman, G., Schramm, D.N. and Olive, K.: 1984, *Astrophys. J.* **281**, 493.

DEUTERIUM, LITHIUM AND THE DENSITY OF THE UNIVERSE

JEAN AUDOUZE

Institut d'Astrophysique de Paris, Paris, France

Abstract. The main outcome of the primordial nucleosynthesis is the ability to account for the abundances of D, ^{3}He, ^{4}He and ^{7}Li with the proper choice of the nuclear density parameter Ω_B. The relative advantages/disadvantages of D and ^{7}Li as the proper 'baryometer' are discussed. In favour of D, the main arguments are the relative simplicity of the formation/destruction schema, but this is challenged by the large uncertainties on the choice of its actual 'primordial' abundance and on the galactic evolution scheme. In favour of ^{7}Li there are the confirmation of the so called 'Spite plateau' and the observation of ^{6}Li at the surface of at least one (may be two) Population II stars, but the paucity of such stars such as the possibility of scenarios in which the ^{7}Li abundance could be affected even in these stars cannot be overlooked.

1. Introduction

'Primordial' nucleosynthesis which is one of the physical consequences of the classical Big Bang schemes is considered as the process allowing the formation of D, ^{3}He, ^{4}He and ^{7}Li. Moreover the standard (simplest) models lead to the determination of the baryonic density parameter Ω_B and the number of neutrino (relativistic particle) flavors, N_ν. As shown by many authors including our friend the late David N. Schramm (see e.g. Schramm and Turner, 1998), $N_\nu = 3$ (which is consistent with the Grand Unification Theory) and $\Omega_B \approx 1\%$. The constraint on N_ν comes mainly from the comparison between the observations and the computations relative to ^{4}He while D and ^{7}Li are the nuclei which are the most sensitive to the choice of Ω_B. The purpose of this short presentation is to attempt to evaluate what is the more precise (less controversial) value for Ω_B: in other words what is the more reliable baryometer D or ^{7}Li?

2. The Relevance of the ^{7}Li Abundance

Two arguments speak in favour of the choice of ^{7}Li. They are (i) the confirmation of the 'Spite Plateau' and (ii) the observation of the presence of ^{6}Li at the surface of pop II stars such as HD 84937 and possibly also BD +26 3578.

All the existing observations of Li abundances at the surface of (very old) Pop. II stars confirm the 'Spite plateau' behaviour i.e. the constancy of this abundance

 Astrophysics and Space Science is the original source of publication of this article. It is recommended that this article is cited as: *Astrophysics and Space Science* **265**: 23–27, 1999.
© 1999 *Kluwer Academic Publishers*.

concerning these stars with $T_{eff} > 5700$ K. Although many ^{7}Li depletion mechanisms have been considered in various papers (such as overshooting, turbulence, shear rotation, microscopic diffusion, mass loss, meridional circulation, gravity waves, 'plumes' ...) there is this striking convergence of the ^{7}Li observed abundance recently reassessed e.g. by Molaro (1999) at the value ^{7}Li/H $\approx 1 - 2\ 10^{-10}$ corresponding to $N_\nu = 3$ ($\Omega_B \approx 1\%$ for $H_0 = 70$ km s^{-1} Mpc^{-1}). This is indeed the first and may be more solid argument favouring the choice of ^{7}Li as the proper baryometer. This statement is reinforced by the point made by Spite et al. (1996) who clearly showed that the dispersion of the Li abundance in the 'plateau' almost disappears when one considers an homogeneous set of observations.

(ii) Recent and growing interest has been focused on the search of ^{6}Li at the surface of some of these Pop. II stars: it has been known for many years that ^{6}Li is more fragile against thermonuclear destruction than ^{7}Li (see e.g. Lemoine et al., 1996). In that respect it is interesting to call attention on the recent contribution by Cayrel et al. (1999) who showed that there is a clear evidence in favour of $N(^6Li)/N(^7Li) \approx 0.07$ in the atmosphere of HD 84937.

By contrast to these two arguments supporting the choice of ^{7}Li as the proper 'baryometer', one cannot overlook the fact that a few low Z stars seem to show a relatively higher ^{7}Li/H abundance than that of the 'Spite plateau'. As noticed e.g. by Deliyannis (1998) the dispersion of the Li/H abundance inside the plateau is still real and more importantly one needs to find more '^{6}Li rich' Population II stars to be sure that this first observation reported by Cayrel et al. (1999) is effective: as said in a french proverb: 'the occurrence of a single bird does not mean the return of the spring!'.

3. What About Deuterium

Dave Schramm again in most of his recent public presentations was advocating that deuterium is the ultimate baryometer: D can only be destroyed and not formed during the course of the galactic (stellar) evolution. Moreover its abundance is a steep decreasing function of Ω_B. A recent (excellent) book of Prantzos et al. (1998) assembles the basic contributions providing most of our present knowledge regarding its observed abundance in various astrophysical sites (see also Vidal-Madjar et al., 1998). To sum up the presently available information: in the local interstellar medium, D/H $\approx 10^{-5}$ (with conspicuous large variations from one line of sight to the other as recalled by Vidal-Madjar et al., 1998); in the solar system $(4.5 \times 10^9$ years old) material, D/H $\approx (2.6 \pm 0.6 \pm 0.3) \times 10^{-5}$ (Geiss and Gloeckler, 1998). When one makes use of the lines of sight in the direction of large redshift quasars, one expects to find (D/H) ratios closer to the origin ('primordial'?) value. But the present situation is still quite confusing since very large variations have been found among the different presently studied lines of sight (Table I).

TABLE I

Values of D/H

QSO name	Z	D/H value	Comments
0014 + 813	2.8 − 3.3	0 − 2.104	The 'controversial' QSO
1009 + 2956	2.504	$(4 − 5) \, 10^{-5}$	The two Tytler QSOs
1937 − 1009	3.572	$(4 − 5) \, 10^{-5}$	
0454 − 220	0.535	$(2 \pm 0.5) \, 104$	The two Webb *et al.*
1718 + 4807	0.7	$< 1.3 \, 0^{-5}$	QSOs

This presently large uncertainty on the D/H abundances deduced from the present analyses of the QSO lines of sight is indeed an argument against the choice of D as the proper baryometer although one can try to make a choice from arguments based on galactic evolution schemes.

Two situations have been considered (i) the possibility of a relatively low (D/H) $\approx 3 − 5 \, 10^{-5}$ abundance, as advocated by D. Tytler in many occasions. In that case $D_{primordial}/D_{present} \approx 2$, which means a relatively low astration rate, such as those used in the models of galactic evolution built up by Tosi (1996) and Timmes *et al.* (1995). In that case $Y_p \approx 0.245$ and $^{7}Li/H \approx 4 \, 10^{-10}$, excluding in principle the presence of ^{6}Li in the Pop. II atmospheres.

(ii) $D/H_{primordial} \approx 1 − 2 \, 10^{-4}$, which is the choice e.g. of L. Cowie, corresponds to $D_{primordial}/D_{present} \approx 10$. In that case the astration rate is large. Several models of galactic evolution developed e.g. by Scully *et al.* (1997) include effects such as varying star formation rates, outflows, reduced ^{3}He formation yields and are able to account for such a large decrease of D over the evolution of the galaxy (stellar $\pm$ interstellar) material. This large $D/H_{primordial}$ abundance is also consistent with $Y_p \approx 0.23$ and $^{7}Li/H \approx 1.6 \, 10^{-10}$ (possible presence of ^{6}Li).

At that point, one can choose between relatively large Ω_B (0.01–0.08) values corresponding to low (D/H $\approx 3 \, 10^{-5}$) D values or smaller Ω_B values ($7 \, 10^{-3} − 2 \, 10^{-2}$) in accordance to a high (D/H $\approx 1 − 2 \, 10^{-4}$) D value.

In an attempt to distinguish between these two possibilities (which normally should exclude each other) Cassé *et al.* (1998) have proposed to use the available information on the overall galactic luminosities distributions at varying redshifts in an attempt to choose between galactic models assuming low astration rates (leading to the prediction of relatively low primordial D values) and those built up with large astration rates. The Cassé *et al.* (1998) comparison clearly favours the latter hypothesis (large astration rates and big primordial D values) on the grounds of an analysis of the overall galactic luminosity evolution. One would like to conclude from it that D/H is near 10^{-4} which is consistent with Li/H $\approx 10^{-10}$, the presence of

^{6}Li in the atmosphere of some Population II stars, and Ω_B restricted to 1–2%. But the relation between the luminosity distribution and the rate of star formation can also be affected by the choice of the cosmological constant Λ (see e.g. Cassé *et al.*, 1998). Moreover the D abundances can also be affected during phases subsequent to the Big Bang nucleosynthesis occurrence. As shown by Cassé *et al.* (1999), high energy photons can trigger either the formation or the destruction of D by partial photodisintegration processes ($\gamma + {}^4$He $\to$ D or $\gamma + $D $\to$ p, n) which may occur in various QSO vicinities.

4. Last But Non Final Remarks

From this quick overview, it is still difficult to choose between ^{7}Li and D as the ultimate baryometer, although there is a consistency between, the possible confirmation of the presence of ^{6}Li in Pop. II star, a low ^{7}Li/H value corresponding to the Spite Plateau, a large (D/H $\approx 10^{-4}$) value consequence of large rates of astration, themselves in agreement with the overall luminosity redshift distribution, and ending up by predicting a relatively low value for Y_p (≈ 0.23) and Ω_B($\approx 1-2\%$).

But we have pointed out all the weaknesses faced by such a consistency quest: not yet proved presence of ^{6}Li – uncertainty on the actual primordial D/H value – complications coming from the subsequent D alteration processes .... . To sum up, one could adopt either the pessimistic point of view according which neither ^{7}Li nor D can be considered as the proper baryometer because of all these remaining and irreducible uncertainties; by contrast, one could be optimistic by noting that in all cases Ω_B is only a few percent which may be sufficient to demonstrate ultimately that the expansion of the Universe will continue for ever.

Acknowledgements

All these discussions have been inspired by the work, the efforts and the inspiration provided by Giusa and Roger Cayrel, to whom I am pleased to convey my respectful affection here. I am very grateful to François and Monique Spite who not only have made the most striking contributions in that field but also have been kind enough to invite me to present these views and have been extremely patient in the course of the writing of these notes.

References

Cassé, M., Olive, K.A., Vangioni-Flam, E. and Audouze, J.: 1998, *New Astronomy* **3**, 259.
Cassé, M., Roland, J. and Vangioni-Flam, E.: 1999, to be published.

Cayrel, R., Spite, M., Spite, F., Vangioni-Flam, E., Cassé, M. and Audouze, J.: 1999, *Astron. Astrophys.*, in press.

Deliyannis, C.P.: 1999, in: M. Prantzos (ed.), *Nuclei in the cosmos* (to be published).

Geiss, J. and Gloeckler, G.: 1998, in: Prantzos *et al.* (loc. cit.), p. 275.

Lemoine, M., Schramm, D.N., Truran, J.W. and Copi, C.J.: 1997, *Astrophys. J.* **478**, 554.

Molaro, P.: 1999 (to be published).

Prantzos, N., Tosi, M. and von Steiger, R. (eds.): 1998, *Primordial nuclei and their galactic evolution*, Kluwer Academic Publishers, Dordrecht.

Schramm, D.N. and Turner, M.S.: 1998, *Rev. Mod. Phys.* **70**, 303.

Scully, S.T., Cassé, M., Olive, K.A. and Vangioni-Flam, E.: 1997, *Astrophys. J.* **476**, 521.

Spite, M., François P., Nissen, P.E. and Spite F.: 1996, *Astron. Astrophys.* **307**, 172.

Timmes, F.X., Truran, J.W., Lauroesch, J.T. and York, D.G.: 1997, *Astrophys. J.* **476**, 464.

Tosi, M.: 1996 in: *From Stars to Galaxies, ASP Conference Series* **98**, 299.

Vidal-Madjar, A., Ferlet, R. and Lemoine, M.: 1998, in: Prantzos *et al.* (loc. cit.), p. 297.

THE FIRST SUPERNOVAE
A Challenge to Astrophysicists

DAVID ARNETT

Steward Observatory, University of Arizona, Tucson, AZ, USA

Abstract. What were the first supernova like, and what nuclei did they synthesize? What limits our ability to accurately simulate their behavior? These and related questions are critically reviewed.

1. Introduction

Within a Big Bang cosmology there must be an epoch at which the first stars formed. The massive stars of this generation produced the first heavy elements ($Z > 5$, i.e., elements carbon and above), which can now be observed in the most metal-poor stars and plasma. Theory of this nucleosynthesis is deeply entwined with the assumptions used to model the evolution of these stars.

One of the first indications of early stellar nucleosynthesis was the 'alpha effect' (Greenstein, 1970; Wallerstein, 1968), in which the abundance of the 'alpha elements' such as (O, Mg, Si, S, Ca and Ti) is high relative to Fe in the most metal-poor stars. These metal-poor stars were of sufficiently low mass that their ages could be long enough for them to have been formed at early cosmological time, that is, at an epoch now being (or soon to be) revealed in distant objects at high redshift. Because of their low mass, these stars are unlikely sources for the production of the elements observed in their spectra. Consequently we assume that their abundance anomalies were acquired when they formed from previous nucleosynthesis in their more massive siblings, and that these same processes may also be discerned in high redshift observations.

As a consequence of neutrino cooling, massive stars produce large yields of alpha elements; this is because their presupernova structure has massive layers of unburned fuel which survives the explosion (Arnett, 1978; Woosley and Weaver, 1995; Arnett, 1996; Thielemann, Nomoto and Hashimoto, 1996). The early overabundance of alpha elements relative to iron could result in a delay in the production of Fe. Type Ia supernovae are thought to be the explosions of white dwarf stars, probably in a binary system which transfers mass. Explosions of white dwarfs tend to turn highly degenerate matter into ^{56}Ni, which decays through Co to Fe, and are prolific producers of Fe in most scenarios (Thielemann, Hashimoto and Nomoto, 1990; Arnett, 1996). Thus a delay in the onset of supernova of Type Ia relative

Astrophysics and Space Science is the original source of publication of this article. It is recommended that this article is cited as: *Astrophysics and Space Science* **265**: 29–35, 1999.
© 1999 *Kluwer Academic Publishers.*

to other supernovae (which result from the core collapse of more massive stars) would give an 'alpha effect' much like that observed.

This general picture gained considerable ground when it was realized (Truran, 1981) that a similar effect in the abundances of s-process and r-process elements could also be seen in the most extreme metal-poor stars, with an early production of r-process from the explosion of massive stars, followed by s-process from more slowly evolving intermediate mass stars.

These ideas are appealing, and likely to be correct in general outline. The challenge lies in the details. In particular, can we use our theoretical notions to make quantitative connection with present and future observations?

How reliable are yield estimates which are based on one dimensional simulations? Significant success in reproducing the observed abundance pattern in the solar system has been obtained (Arnett, 1978; Woosley and Weaver, 1995; Thielemann, Nomoto and Hashimoto, 1996; Chieffi, Limongi and Straniero, 1998). The simulations of (Woosley and Weaver, 1995), (Thielemann, Nomoto and Hashimoto, 1996), and (Chieffi, Limongi and Straniero, 1998) represent outstanding computational efforts. However, equal success results from using a 'typical zone' integration (Arnett, 1996). The differences between this approach, and the more detailed numerical work (Woosley and Weaver, 1995; Thielemann, Nomoto and Hashimoto, 1996; Chieffi, Limongi and Straniero, 1998) is no larger than the differences found *between* these three numerical efforts themselves. This implies that the differences between specific detailed models are of the same order as computational uncertainties. Only by understanding the reasons for these differences will we be able to proceed to understand what the observations may imply. Beyond the purely computational uncertainties, there are several problems of physical formulation which will be discussed below.

Estimates of bulk yields (Arnett, 1978) are more robust than estimates of yields of rarer nuclei. For example, in the CNO group, estimates of nitrogen production are sensitive to mixing of the star prior to explosion; see recent work on extremely metal-poor massive stars in (Arnett, 1999), and references therein. A similar sensitivity occurs for ratios of odd-Z to even-Z elements below Fe, as was suggested long ago (Arnett, 1971).

2. Direct Computation of Presupernovae

Throughout almost all of their lives, stars are incredibly stable systems. Having well defined structure means that the amount of matter at a given temperature is well defined too. This, along with the length of exposure, defines the amount of nuclear burning that can occur.

Stars are almost perfect spheres, a fact which has made simulations of their evolution much easier. This perfection is broken by rotation and by mixing currents. This mixing can severely modify the simple picture just presented by affecting exposure times; matter can flow in and out of hot regions. Historically, stellar evolutionary simulations have dealt with mixing as a homogenizing process governed by a mixing-length theory of convection. It is supposed that mixing occurs at very subsonic velocities but still far faster than evolutionary time scales, thus allowing a sort of 'random walk' of convective blobs through the convectively unstable region. While perhaps adequate for slowly evolving stars, this approach is incorrect for the late stages of evolution of massive stars, a stage which recently has been directly simulated (Bazan and Arnett, 1998).

Several general features appear, which are not included in the conventional picture. The motion is only slightly subsonic, so that pressure waves (sound waves) carry a significant part of the energy flux, as do the mass motions. The convection is intermittent in space and time; the flame is not a uniform spherical shell, but an ensemble of hot spots, flashing on and off. This means the mean burning temperature is higher, shifting the nucleosynthesis yields. The nature of mixing is macroscopic not microscopic. In the evolutionary time available, only a few stirring motions occur. Mixing occurs as blobs of matter are successively processed by similar conditions, not by microscopic mixing of processed nuclei. The extent of mixing is not so well defined as in the one dimensional models; some matter mixes from one burning shell to another.

Prof. Schatzman (this conference) has proposed that gravity waves are significant in stellar mixing. The simulations show dramatic evidence for such waves; they contribute to both the mixing and the energy flux (Bazan and Arnett, 1998).

As yet such simulations have only be done in two dimensions, and have a duration limited to less than half the evolutionary time of the oxygen shell, the penultimate stage prior to core collapse. Such simulations strain the computer resources available. Further modifications of our understanding may result when these limitations are removed. Because we do not yet understand the full implications of these simulations, we are not yet able to provide reliable predictions of the detailed yields. This is clearly an unsatisfactory situation. However, because the gross behavior is similar to the one dimensional models, we expect the bulk yields may be little changed.

3. Other Three-dimensional Effects

There are a number of other effects which are expected to modify the behavior of stars prior to supernova explosion. One of the oldest problems is that of rotation (Tassoul, 1978). Simple approaches to incorporate rotation in stellar evolution codes result in stellar cores spinning near breakup velocity after helium burning (Kippenhahn, Meyer-Hoffmeister and Thomas, 1970; Endal and Sofia, 1976). See (Langer *et al.*, 1998) for a recent example of this approach. However, magneto-hydrodynamic theory (Parker, 1979) suggests that magnetic fields would resist such strong shear. Such winding might strengthen the fields to the point at which their buoyancy would cause them to rise (Parker, 1974). Because they are entrained in the plasma, the matter would rise too. This would both give additional mixing, and transport angular momentum. It appears that treatment of rotation in a star is incomplete without consideration of the MHD effects (stellar matter is a plasma after all ...).

Recent work (Kudritski, 1999) promises to put stellar mass loss on a firm theoretical foundation, and one in good agreement with observations. As yet the theory is limited to hot stars, that is, those with effective temperatures high enough for hydrogen to be well ionized. Mass loss in massive stars affects the yields by (1) removing matter from further processing, and (2) modifying the stellar state by changing its mass. As Langer has long emphasized (see (Langer *et al.*, 1998) and references therein), the effects of rotation and mass loss on the yields of massive stars are profound. Direct simulations (S. Asida, private communication), using the VULCAN hydrodynamic code of E. Livne in its implicit mode, suggest that rotation with hydrodynamic flow will reveal suprises as striking as did the direct simulation of oxygen burning.

We do not yet have observations which determine what are the true progenitors of type Ia supernovae, although there is little argument that they are most likely members of an interacting binary system. We are unable to compute the merging scenario of type Ia supernovae; adequate resolution and long secular time scales are not yet possible. To what extent does accreted matter stir the surface of the accretor? How do mass loss and angular momentum loss vary with accretion? How does accretion vary with time in dynamic systems? How do the dynamics of accretion disks modify the secular evolution of the system? These questions involve hydrodynamic flows with 3D geometry and time dependence, and are a severe challenge. Although the type Ia supernovae are not likely to be a major part of the population of the first supernovae, their nature is crucial to unravelling the puzzle of how the abundance distribution evolves.

Another area deserving futher investigation is the question of how the processed nuclei are mixed into the interstellar medium, and then into the objects we observe. Most work has assumed that the mixing of supernova energy and newly synthesized matter is the same. However, explosions tend toward Hubble flow (homologous expansion), so that the inner, more processed matter would move the slowest. Instabilities in the explosion complicate this picture, and the question remains open. With SN1987A we will be able to observe the process in action, but understanding will require we understand also how the symmetry was broken to make the spectacular rings.

4. Testing the Mix

Hydrogen burning at high temperature is the most favorable source for producing ^{26}Al (Arnett, 1996), which has been observed in the ISM by detection of its gamma-ray line (Oberlack *et al.*, 1996), and in presolar grains by anomalies in the abundance of its decay product ^{26}Mg; see (Bernatowicz and Zinner, 1997) and reference therein. Gamma-ray telescopes will eventually provide interesting constraints upon how much mixing can have occurred in massive stars.

Before using multidimensional hydrodynamic simulations to interpret astronomical observations, it is desirable to independently test the simulation methods in the laboratory. Such testing (Kane, Arnett and Remington, 1997) has begun; the astrophysics code PROMETHEUS and the inertial confinement fusion code CALE do equally well in reproducing experiments on the NOVA laser at Livermore National Laboratory. These data concern the Rayleigh-Taylor and Richtmyer-Meshkov instabilities well into the nonlinear regime.

We are fortunate that the laser target conditions are scalable to the conditions in the He/O interface in progenitor models of SN1987A! Thus, as a bonus, we are modelling this part of the supernova with the laser experiment. Recent work (Kane, in preparation) investigates the differences in 2D and 3D simulations for the development of these instabilities. The 3D simulations give greater penetration of processed matter into the surrounding mantle, which tends to remove the difference between simulation and observation of the ^{56}Ni velocities in SN1987A.

5. Conclusions

With such spectacular images as the rings of SN1987a and the jets in eta Carina, it is clear that the day of the spherically symmetric model is ending.

Further theoretical progress will be difficult. While it is possible to compute 2D simulations on a laptop with relative ease, well resolved 3D calculations are rare. They represent at least an increase in computer power by a factor of 100 to 1000. This in turn requires massively parallel programs and machines at present.

A second, less often mentioned difficulty is that we do not have a multidimensional equivalent of the traditional stellar evolutionary code. In the good old days we damped hydrodynamics by using implicit algorithms. This removed the causality constraint on the time step $\delta t < \ell/v$, where ℓ is the size of a region and v a characteristic velocity. If the hydrodynamics is an important part of the physics, as in the examples above, we are restricted to short evolutionary time. The problem is equivalent to weather prediction versus climate prediction. To deal with longer time intervals, more knowledge of the short time behavior is necessary.

Given these challenges, what can the theorist say to help the observer? Here are a few thoughts:

In the picture sketched above, core collapse supernovae must precede type Ia events. Observations at high redshift will test this idea, if we can distinguish type Ia events from type Ib and Ic events, which are thought to be due to core collapse of massive stars which have already shed their hydrogen.

There is a shot-noise effect in abundances, due to the finite number n of supernovae contributing to the mixing pool from which the next generation of stars gets its gas. In the extremely metal-poor stars, this effect is not diluted by metals synthesized in still earlier generations. Such a shot-noise effect gives information concerning the star forming regions, which might be tested by direct observation of such regions at high redshift.

Much information should be contained in ratios of odd-Z to even-Z elements. For metal-poor stars, mixing has a larger effect on determining the neutron excess, and hence this ratio. This effect must be untangled from the operation of the neon-magnesium cycle, which is enchanced in metal-poor stars.

Massive stars do some s-processing as well as r-processing; their s-process yield helps define the presupernova evolution instead of the explosion event itself. In particular, the neutron exposure per seed nucleus could be modified by mixing, giving a non-standard s-process.

Improvements in the physics of core silicon burning and oxygen burning will help define the precollapse core mass, entropy, and neutron excess. The abundances of iron group nuclei are sensitive to these conditions as well as to the explosion model.

Supernova searches are beginning to produce a substantive data set of events. This is of fundamental importance to us; it allows us to distiguish between a distribution of bland individuals of average properties and one having extreme individuals who are not like the mean. Theorists have focused on the 'typical supernova' so long that it affects what we expect to see. A subpopulation might be rare but have relatively large effects on particular abundances. Limits restricting such rare but fecund events are needed.

The nitrogen abundance is especially interesting. In extreme metal-poor stars the introduction of catalysts by a minor mixing can stimulate still more mixing. In other words, these nuclei are active contaminants in that they modify the flows which transport them. Because the CNO cycle is weak, any such effect should be enhanced.

We may take pleasure in the thought that our science is developing so rapidly, and that there is no dearth of fascinating problems to do!

References

Arnett, D.: 1971, *Astrophys. J.* **166**, 153.

Arnett, D.: 1978, *Astrophys. J.* **219**, 1008.

Arnett, D.: 1996, *Supernovae and Nucleosynthesis*, Princeton: Princeton University Press.

Arnett, D.: 1999, in: K. Nomoto and J.W. Truran (eds.), *Cosmic Chemical Evolution*, Kluwer Academic Publishers, Dordrecht.

Bazan, G. and Arnett, D.: 1998, *Astrophys. J.* **496**, 316.

Bernatowicz, R.J. and Zinner, E. (eds.): 1997, Astrophysical Implications of the Laboratory Study of Presolar Materials, *AIP Conference Proceedings* **402**.

Chieffi, A., Limongi, M. and Straniero, O.: 1998, *Astrophys. J.* **502**, 737.

Clayton, D.D.: 1968, *Principles of Stellar Evolution and Nucleosynthesis*, McGraw-Hill, New York.

Endal, A.S. and Sofia, S.: 1976, *Astrophys. J.* **210**, 184.

Greenstein, J.L.: 1970, Comments on astrophys. *Space Sci.* **2**, 85.

Kane, J., Arnett, D., Remington, B., *et al.*: 1997, *Astrophys. J. Letters*, in press.

Kippenhahn, R., Meyer-Hoffmeister, E. and Thomas, H-C.: 1970, *Astron. Astrophys.* **5**, 155.

Kudritski, R., 1999, in preparation.

Langer, N., Heger, A., Woosley, S.E. and Herwig, F.: 1998, in: N. Prantzos (ed.), *Nuclei in the Cosmos V*, Frontieres.

Oberlack, U., *et al.*: 1996, *Astron. Astrophys.* **120**, 311.

Parker, E.N.: 1974, *Astrophys. Space Sci.* **31**, 261.

Parker, E.N.: 1979, *Cosmical Magnetic Fields*, Oxford University Press.

Remington, B.A., Weber, S.V., Marinak, M.M., *et al.*: 1995, *Phys. Plasmas* **2**, 241.

Tassoul, J-L.: 1978, *Theory of Rotating Stars*, Princeton University Press.

Thielemann, F.K., Hashimoto, M. and Nomoto, K.: 1990, *Astrophys. J.* **349**, 222.

Thielemann, F.K., Nomoto, K. and Hashimoto, M.: 1996, *Astrophys. J.* **460**, 408.

Truran, J.W.: 1981, *Astron. Astrophys.* **97**, 391.

Wallerstein, G.: 1968, *Science* **162**, 625.

Woosley, S.E. and Weaver, T.A.: 1995, *Astrophys. J. Suppl.* **101**, 181.

SUPERNOVA NUCLEOSYNTHESIS, CHEMICAL EVOLUTION, AND COSMIC SUPERNOVA RATE

K. NOMOTO, T. NAKAMURA and C. KOBAYASHI
Department of Astronomy & Research Center for the Early Universe, School of Science, University of Tokyo, Bunkyo-ku, Tokyo 113-0033, Japan

Abstract. In the chemical evolution of the Galaxy, Type II supernovae (SNe II) have contributed to the early metal enrichment and later Type Ia supernovae (SNe Ia) have contributed to the delayed enrichment of Fe. In principle, hypothetical pre-galactic population III objects could cause the earliest heavy element enrichment. Here we present our two new findings. 1) The peculiar abundance pattern among iron peak elements (Cr, Mn, Co, and Fe) in the very metal poor can be reproduced with SN II nucleosynthesis yields without invoking the contribution from Pop III objects. 2) The observed chemical evolution in the solar neighborhood is well reproduced with the metallicity dependent occurrence of SNe Ia, where SNe Ia do not occur if the iron abundance of the progenitors is as low as $[Fe/H] \lesssim -1$. We make the prediction that the cosmic SN Ia rate drops at $z \sim 1-2$ because of the low-iron abundance, which can be observed with the Next Generation Space Telescope.

1. Introduction

There exist two distinct types of supernova explosions: One is Type II supernovae (SNe II), which are the core collapse-induced explosions of short-lived massive stars ($\gtrsim 8M_\odot$) and produce more O and Mg relative to Fe (i.e., [O/Fe] > 0), and the other is Type Ia supernovae (SNe Ia), which are the thermonuclear explosions of accreting white dwarfs (WDs) in close binaries and produce mostly Fe and little O. Note also that hypothetical pre-galactic population III objects, such as very massive stars or pair creation supernovae, could be responsible for the earliest enrichment of heavy elements.

In the present paper, we first show that SN II nucleosynthesis yields can explain the interesting trends in the abundance ratios among iron peak elements (Cr, Mn, Co, and Fe) in the very metal poor stars without invoking the contribution from Pop III objects (§2). Secondly, we introduce a metallicity dependent occurrence of SNe Ia in modeling the galactic chemical evolution and cosmic supernova rates. This SN Ia model provides a new interpretation of the evolutionary change in [O/Fe] in the solar neighborhood and the SN II-like abundance patterns of the Galactic halo and the damped Lyα systems (§3).

Astrophysics and Space Science is the original source of publication of this article. It is recommended that this article is cited as: *Astrophysics and Space Science* **265**: 37–47, 1999.
© 1999 *Kluwer Academic Publishers.*

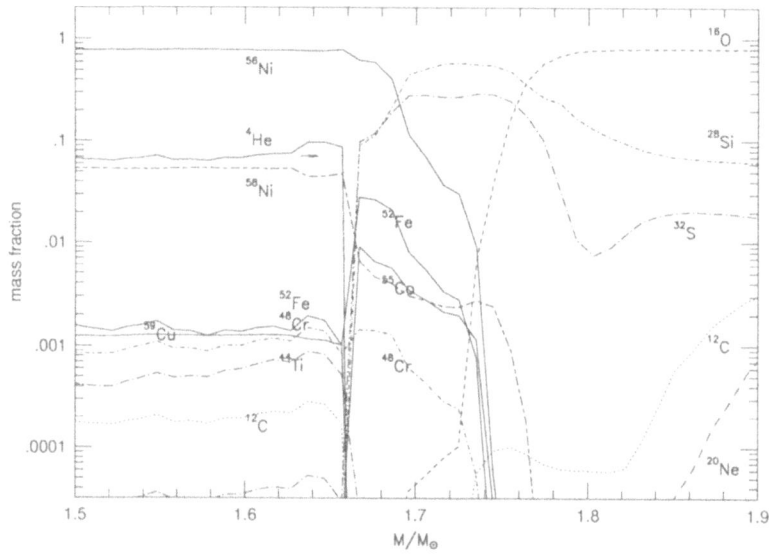

Figure 1. The isotopic composition of the ejecta from the 20 $M_\odot$ star (6 $M_\odot$ He core).

2. Nucleosynthesis in Type II Supernovae and the Abundances in Very Metal-Poor Stars

Very metal poor stars provide important clues to investigate early Galactic chemical and dynamical evolution because they contain valuable information about this time. We can learn how the Galaxy evolved chemically (as well as dynamically) in its early phase from the observation of these stars. Therefore, a large number of observations and abundance analyses have been performed. Recent high-resolution abundance surveys have discovered interesting trends of [Cr/Fe], [Mn/Fe] and [Co/Fe] with respect to [Fe/H]. Both [Cr/Fe] and [Mn/Fe] decrease with decreasing metallicity from [Fe/H] = −2.4 to −4.0, while [Co/Fe] increases (Figure 2; McWilliam, 1997; see also Ryan *et al.*, 1996 and reference therein).

We explore the effects on nucleosynthesis in SNe II of various parameters (especially, the mass cut between the ejecta and the neutron star) in order to explain the observed trends of the iron-peak element abundance ratios (Nakamura *et al.*, 1999). We will show that such a behavior can be explained by a variation of mass cuts in SNe II as a function of progenitor mass, which provides a changing mix of nucleosynthesis from an alpha-rich freeze-out of Si-burning and incomplete Si-burning.

2.1. DEPENDENCE ON MASS CUTS

In order to investigate the ratios [Cr/Fe], [Mn/Fe] and [Co/Fe], let us look into the regions where these elements are produced (Figure 1). First of all, ^{56}Ni, which decays into the most abundant Fe isotope ^{56}Fe, is produced not only in the complete Si-burning region ($M_r \lesssim 1.65\ M_\odot$) but also in the incomplete Si-burning region ($1.65 - 1.74\ M_\odot$). ^{52}Fe and ^{55}Co, which decay into ^{52}Cr and ^{55}Mn, respectively, are mostly synthesized in the incomplete Si-burning region. ^{52}Fe is also synthesized in the complete Si-burning region, but not a large amount. On the contrary, ^{59}Cu, which decays into ^{59}Co, is produced in the complete Si-burning region. Note that ^{55}Mn and ^{59}Co are the only stable isotopes of these elements, and ^{52}Cr dominates the Cr element abundance by 84% (in solar composition). Thus, it is sufficient to take these isotopes into consideration when discussing the abundances of [Cr, Mn, Co/Fe].

The above discussion suggests that the choice of the mass cut can affect the ratios [Cr/Fe], [Mn/Fe], and [Co/Fe]. For a deeper mass cut (i.e. smaller M_{cut}), the ejected mass of the complete Si-burning region is larger (i.e., the masses of Fe and Co are larger), while the ejected mass of the incomplete Si-burning region remains the same (i.e., the masses of Cr, Mn, and Fe are the same). Accordingly the ratios of [Cr/Fe] and [Mn/Fe] are smaller and [Co/Fe] is larger. For a mass cut at larger radii (larger M_{cut}) these ratios show the opposite tendency.

2.2. ABUNDANCE RATIOS IN METAL POOR STARS

More massive stars evolve faster. Thus, we expect the ejecta of the most massive stars to dominate the earliest phase of Galactic evolution, i.e. the period corresponding to the lowest [Fe/H]. If the mass cut M_{cut} between the ejecta and the neutron star tends to be smaller and thus Fe yield tends to be larger for the larger mass progenitor, one could expect to reproduce the observed trend in [Cr/Fe], [Mn/Fe] and [Co/Fe].

Figure 2 shows that the above trend of Fe yields agrees quantitatively with the observations. Here the mass cuts are chosen for the 25 $M_\odot$, 20 $M_\odot$, and 13 $M_\odot$ stars to produce Fe of 0.26 $M_\odot$, 0.09 $M_\odot$, and 0.07 $M_\odot$, respectively (see Nakamura *et al.*, 1999 for details and other elements). We consider two models to relate [Fe/H] and the progenitor masses. One is the 'well-mixed' model, where the mixing in the galactic scale is assumed to be so efficient that the chemical uniformity is achieved in the early Galaxy. The other is the 'unmixed' model which assumes that the mixing is not effective and the composition of a metal-poor star is the same as produced by a single SN II.

2.3. MIXING AND METALLICITY IN THE EARLY GALAXY

In the 'unmixed' model, the [Fe/H] of a star is determined by the amount of iron ejected from the relevant supernova and the mass of hydrogen in the mixing region. Our calculations show that in order to explain the large variations of the observed abundance ratios (e.g. -0.5 dex in [Cr/Fe]), the iron mass from SN II varies within a relatively narrow range of 0.05 $M_\odot$ to 0.26 $M_\odot$. Thus in the very low metallicity

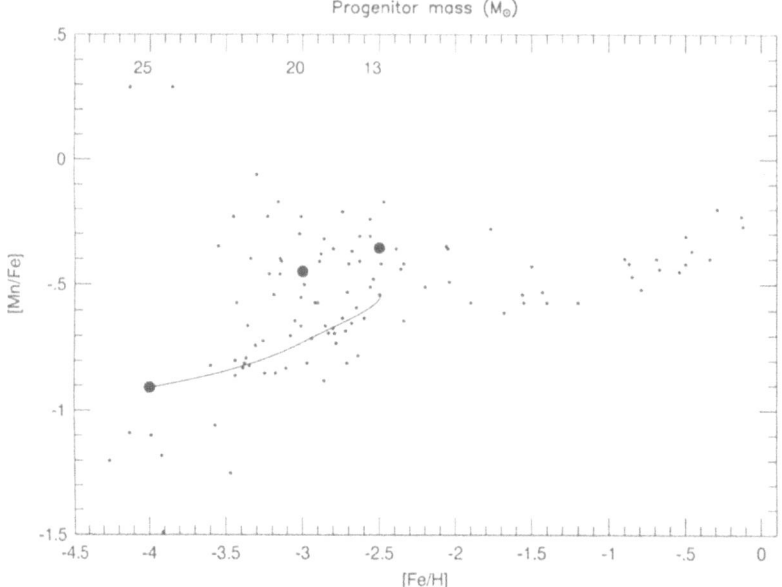

Figure 2. Evolution of the Mn/Fe ratio. The solid line shows a 'well-mixed' model, while an 'unmixed' model is shown by the filled circles.

regions, there must be an order of magnitude variation in the hydrogen mass to produce the variation of [Fe/H] in the range of -4 to -2.5.

The mass of the interstellar matter (ISM), m_{ISM}, mixed with the ejecta depends strongly on the explosion energy E of the supernova as $m_{\mathrm{ISM}} \propto E^{0.95}$ (Ryan *et al.*, 1996). In this connection, we should note the recent discovery of a hypernova SN1998bw with such a large explosion energy as $E \sim 3 \times 10^{52}$ erg from a massive progenitor (Iwamoto *et al.*, 1998). If more massive supernovae tend to produce larger explosion energy and this M-dependence is large enough, the larger M – smaller [Fe/H] relation can be obtained.

We propose another possibility. The above discussion is derived under the assumption that the ISM is uniform. However, there is a distinctive non-uniformity characterized by the 'Strömgren sphere', within which matter is ionized. The progenitors of SNe II are so hot and luminous during their main-sequence phase that their Strömgren spheres are as large as $10 \sim 100$ pc. The Strömgren sphere is likely to determine the amount of hydrogen into which the ejecta of a supernova are mixed. The shock advances easily within the Strömgren sphere, but is strongly decelerated outside the sphere because it has to ionize the matter there. Thus the shock radius is likely to coincide with the Strömgren radius, which depends on both the progenitor mass and the number density of hydrogen in the ambient ISM per cubic centimeter, n_0. We estimate that the hydrogen mass swept up by the shock is $4 \times 10^4 \cdot (n_0/1.0)^{-1} M_\odot$ for $M = 25 M_\odot$ and $4 \times 10^3 \cdot (n_0/1.0)^{-1} M_\odot$ for $M = 15 M_\odot$, using low metal progenitor models. These estimates show that

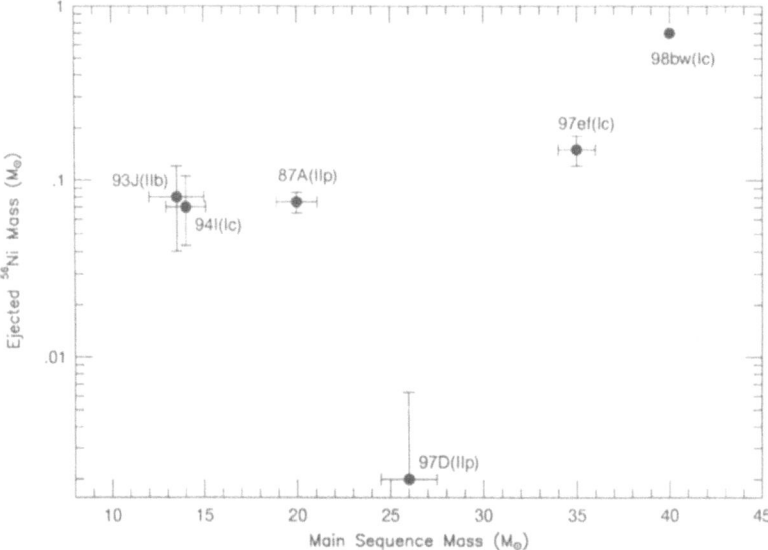

Figure 3. Ejected ^{56}Ni mass as a function of main-sequence mass, as estimated for SN II 1987A, SN IIb 1993J, SN Ic 1994I, SN II 1997D, and SN Ic 1997ef.

the hydrogen masses differ by a factor of 10, which could explain the larger M – smaller [Fe/H] relation as well as the observed range in [Fe/H].

2.4. MASS OF ^{56}NI IN TYPE II SUPERNOVA EJECTA

The above results suggest that a supernova from more massive progenitor has a deeper mass cut, thus ejecting a larger amount of ^{56}Ni mass. The mass of ^{56}Ni should be determined by the competition between the amount of neutrino absorbing matter and the depth of the gravitational potential. Accordingly, the intermediate massive stars eject a relatively large amount of ^{56}Ni because of a large neutrino absorbing region, while in a more massive star the deeper gravitational potential wins and ^{56}Ni is scarcely ejected due to fallback.

This is consistent with the recent ^{56}Ni mass estimates from the light curve modeling for SNe II and Type Ib/Ic supernovae (SNe Ib/Ic) which are the explosions of bare cores of massive stars (Figure 3). From SN IIb 1993J, SN Ic 1994I, and SN1987A, the $13 - 20\ M_{\odot}$ stars eject $\sim 0.07\ M_{\odot}\ ^{56}$Ni. For $M > 20\ M_{\odot}$, there has been little information, but recent SNe Ic and hypernovae suggest a large amount of Fe production from massive stars. The light curves of SN Ic 1997ef and 1998bw indicate the production of $\sim 0.15\ M_{\odot}$ and $\sim 0.7\ M_{\odot}$ of ^{56}Ni, respectively (Iwamoto *et al.*, 1998). On the other hand, SN II 1997D shows the synthesis of only $\sim 0.002\ M_{\odot}$ of ^{56}Ni from the $25 - 30\ M_{\odot}$ star (Turatto *et al.*, 1998). Such a variation of the observed ^{56}Ni mass is consistent with the theoretical expectation.

3. Low-Metallicity Inhibition of Type Ia Supernovae

The progenitors of the majority of SNe Ia are most likely the Chandrasekhar (Ch) mass WDs, although the sub-Ch mass models might correspond to some peculiar subluminous SNe Ia (e.g., Nomoto *et al.*, 1997 for a review). For the evolution of accreting WDs toward the Ch mass, two scenarios have been proposed: One is a double-degenerate (DD) scenario, i.e., merging of double C+O WDs with a combined mass surpassing the Ch mass limit, and the other is a single-degenerate (SD) scenario, i.e., accretion of hydrogen-rich matter via mass transfer from a binary companion. The issue of DD versus SD is still debated, although theoretical modeling has indicated that the merging of WDs does not make typical SNe Ia.

Kobayashi *et al.* (1998) have introduced the following SN Ia progenitor scenario into the cosmic and galactic chemical evolution models. In their SD scenario, if the accretion rate exceeds a certain limit, the WD blows a strong wind and burns hydrogen steadily to increase the WD mass to reach the Chandrasekhar mass limit. There are two cases of the companion star: One is a red-giant (RG) with an initial mass of $M_{RG,0} \sim 1\ M_\odot$ and an orbital period of tens to hundreds days (Hachisu, Kato and Nomoto, 1996, 1999), the other is a near main-sequence (MS) with an initial mass of $M_{MS,0} \sim 2 - 3\ M_\odot$ and a period of several tenths of a day to several days (Li and van den Heuvel, 1997; Hachisu *et al.*, 1999). This scenario involves the metallicity effect on the occurrence of SNe Ia: if the iron abundance of the accreted matter is as low as [Fe/H] $\lesssim -1$, the WD wind is too weak to increase the WD mass through the Ch mass so that few SNe Ia occur. We apply such a low-metallicity inhibition of SN Ia in modeling the chemical evolution of galaxies and cosmic supernova rates.

3.1. THE CHEMICAL EVOLUTION IN THE SOLAR NEIGHBORHOOD

The role of SNe II and SNe Ia in the chemical evolution of galaxies can be seen in the [O/Fe]-[Fe/H] relation (Figure 4: Metal-poor stars with [Fe/H] $\lesssim -1$ have [O/Fe] ~ 0.45 on the average, while disk stars with [Fe/H] $\gtrsim -1$ show a decrease in [O/Fe] with increasing metallicity. To explain such an evolutionary change in [O/Fe] against [Fe/H], we use the chemical evolution model that allows the infall of material from outside the disk region. The infall rate, the SFR, and the initial mass function (IMF) are given by Kobayashi *et al.* (1998), and the nucleosynthesis yields of SNe Ia and II are taken from Tsujimoto *et al.* (1995).

For the SD scenario, the lifetime of SNe Ia is determined from the main-sequence lifetime of the companion star. The initial mass ranges of the companions are $0.9\ M_\odot \lesssim M_{RG,0} \lesssim 1.5\ M_\odot$ for the WD+RG system and $1.8\ M_\odot \lesssim M_{MS,0} \lesssim 2.6\ M_\odot$ for the WD+MS system. The low-metallicity inhibition of SNe Ia is introduced for the metallicity of the progenitor systems lower than [Fe/H] $= -1.1$, which is determined from the metallicity dependence of the optically thick wind. For the DD

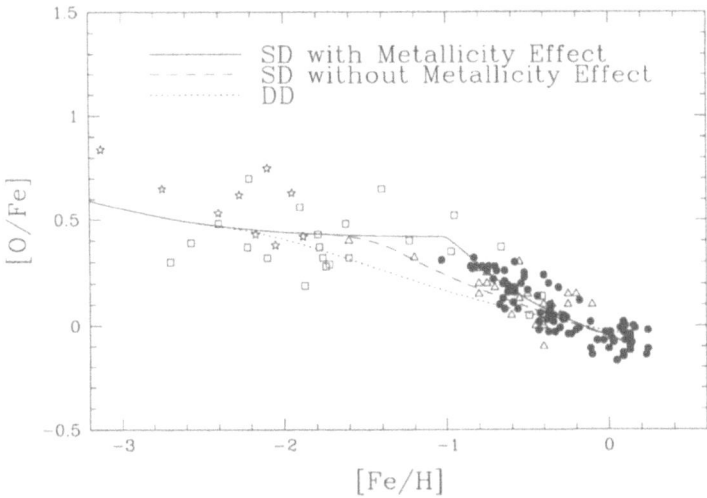

Figure 4. The evolutionary change in [O/Fe] against [Fe/H] for three SN Ia models. The dotted line is for the DD scenario, and the other lines are for our SD scenario with (solid line) and without (dashed line) the metallicity effect on SNe Ia. See Kobayashi *et al.* (1998) for observational data sources.

scenario, we adopt the distribution function of the lifetime of SNe Ia by Tutukov and Yungelson (1994), majority of which is $\sim 0.1 - 0.3$ Gyr.

Figure 4 shows the evolutionary change in [O/Fe] for three SN Ia models. The dotted line is for the DD scenario. The other lines are for our SD scenario with (solid line) and without (dashed line) the metallicity effect on SNe Ia. i) In the DD scenario the lifetime of the majority of SNe Ia is shorter than 0.3 Gyr. Then the decrease in [O/Fe] starts at [Fe/H] ~ -2, which is too early compared with the observed decrease in [O/Fe] starting at [Fe/H] ~ -1. ii) For the SD scenario with no metallicity effect, the companion star with $M \sim 2.6\ M_\odot$ evolves off the main-sequence to give rise to SNe Ia at the age of ~ 0.6 Gyr. The resultant decrease in [O/Fe] starts too early to be compatible with the observations. iii) For the metallicity dependent SD scenario, SNe Ia occur at [Fe/H] $\gtrsim -1$, which naturally reproduce the observed break in [O/Fe] at [Fe/H] ~ -1.

3.2. GALACTIC HALO AND DAMPED Lyα SYSTEMS

The low-metallicity inhibition of SNe Ia can provide a new interpretation of the SN II-like abundance patterns:
1) The Galactic halo is a low-metallicity system with [Fe/H] $\lesssim -1$, and it has an abundance pattern of genuine SN II origin, i.e., the overabundances of α-elements relative to Fe as [α/Fe] > 0 (Wheeler, Sneden and Truran, 1989). However there exist age differences of several Gyrs among the clusters as well as field stars. Since the shortest lifetime of SNe Ia is ~ 0.6 Gyr for the MS+WD close binary systems, SN Ia contamination would be seen in [α/Fe] if there were no metallicity effect on

SNe Ia. This apparent discrepancy between the age difference and the high [α/Fe] can be resolved by the low-metallicity inhibition of SNe Ia.

2) A similar interpretation also holds for DLA systems. The DLA systems observed at $0.7 < z < 4.4$ have [Fe/H] $= -2.5$ to -1 and indicate [α/Fe] > 0 (Lu et al., 1996). The age-metallicity relation in DLA systems suggested by Lu et al. (1996) implies that DLA systems have grown through a common chemical history spanning over several Gyrs. If so, the SN II-like abundance pattern in DLA systems needs the introduction of the metallicity dependent SN Ia rate to avoid the contamination of SN Ia products.

3.3. COSMIC SUPERNOVAE RATE

SNe Ia have been discovered up to $z \sim 1.2$ by the Supernova Cosmology Project and the High-z Supernova Search Team. They have given the SN Ia rate at $z \sim 0.4$ (Pain et al., 1996) but will provide the SN Ia rate history over $0 < z < 1.3$. With the Next Generation Space Telescope, both SNe Ia and II will be observed through $z \sim 4$. In a theoretical approach, the cosmic SN Ia rate as a function of redshift has been constructed for a cosmic star formation rate (SFR). In this paper, we make a prediction of the cosmic SN Ia rate as a composite of different types of galaxies which have gone through different chemical enrichment.

First, we predict the cosmic supernova rate history corresponding to the observed cosmic SFR (e.g., Madau et al., 1996) with the correction of the dust extinction (Pettini et al., 1998). The photometric evolution is calculated with the spectral synthesis population database taken from Kodama and Arimoto (1997). We adopt $H_0 = 50$ km s^{-1} Mpc^{-1}, $\Omega_0 = 0.2$, $\lambda_0 = 0$, and the redshift at the formation epoch of galaxies $z_f = 5$. We determine the initial comoving density of gas $\Omega_{g\infty}$ to reproduce the present gas fraction $\Omega_{g0} = 5 \times 10^{-4}$.

Figure 5 shows the cosmic supernova rate per $10^{10} L_\odot$ per century (SNu). The long-dashed line is for SNe II and the other lines for SNe Ia with (solid and dotted lines) and without (dashed line) the metallicity effect. The dotted line shows the SN Ia rate without the correction of the dust extinction in the SFR (Madau et al., 1996).

If we do not include the metallicity effect, the SN Ia rate is almost flat from the present to higher redshift, and decreases toward the formation epoch of galaxies. If we include the metallicity effect, the SN Ia rate drops at $z \sim 1.4 - 1.8$, where the iron abundance of the gas in the universe is too low (i.e., [Fe/H] $\lesssim -1$) for the progenitors of SNe Ia to make explosions. The redshift where the SN Ia rate drops is determined by the speed of the chemical enrichment, which depends on the effect of dust extinction on the cosmic SFR, cosmology, galaxy formation epoch, and the initial gas density. Taking into account these uncertainties, the break in the SN Ia rate occurs at $z = 1 - 2$.

In fact, galaxies being responsible for the cosmic SFR have different heavy-element enrichment timescale, so that we calculate the cosmic chemical evolution

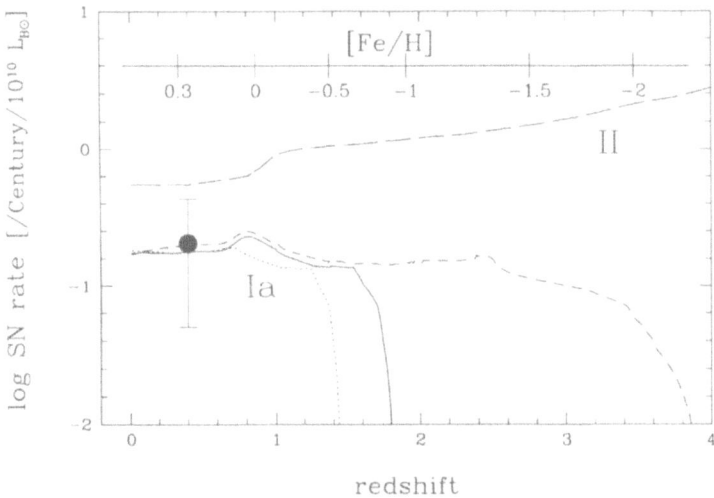

Figure 5. The cosmic supernova rate per $10^{10} L_\odot$ per century (SNu). The long-dashed line is for SNe II and the other lines for SNe Ia with (solid and dotted lines) and without (dashed line) the metallicity effect. The correction of the dust extinction is included in the SFR (Pettini *et al.*, 1998) except for the dotted line which shows the SN Ia rate without this correction (Madau *et al.*, 1996). The filled circle is the observed SN Ia rate at $z \sim 0.4$ (Pain *et al.*, 1996). The iron-abundance scale in the abscissa is calculated with the metallicity effect on SNe Ia.

as a composite of ellipticals and spirals (S0a-Sa, Sab-Sb, Sbc-Sc, and Scd-Sd). We assume that elliptical galaxies are formed by a single star burst and stop the star formation after the supernova-driven galactic wind (Kodama and Arimoto, 1997), while spiral galaxies are formed by a relatively continuous star formation. The epoch of the end of star formation in ellipticals is assumed to be 1 Gyr, which correspond to the redshift of $z \sim 3$. The infall rate and the SFR are given by Kobayashi *et al.* (1998). We then combine the contribution of ellipticals and spirals with the relative mass ratio among the types. The predicted cosmic SFR have the little smaller slope than the observed cosmic SFR from the present to the peak at $z \sim 1.4$ and the excess at $z \gtrsim 2$ corresponding to the SFR of ellipticals which may be hidden by the dust extinction.

Figure 6 shows the cosmic supernova rate (solid-line) as the composite of ellipticals (short dashed-line) and spirals (long dashed-line). The upper three lines show the SN II rates and the lower three lines show the SN Ia rates. The SN Ia rate in spirals decreases at $z \sim 2.5$ because of the low-metallicity inhibition of SNe Ia. We can find many SNe Ia at $z > 3$ in ellipticals, where the chemical enrichment takes place so early that the metallicity effect on SN Ia is not effective.

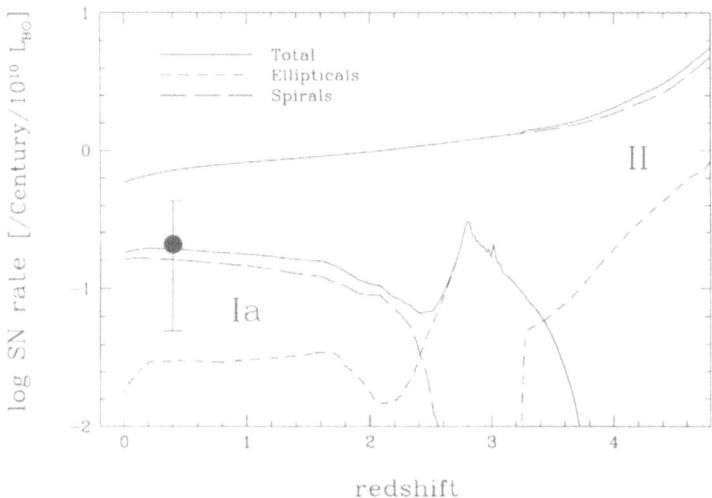

Figure 6. The cosmic supernovae rate in SNu as a composite of ellipticals and spirals. The upper three lines show SN II rates, the lower three lines show SN Ia rates. The solid-line, the short dashed-line, and the long dashed-line are for total, ellipticals, and spirals, respectively.

4. Conclusions and Discussion

1) For SNe II nucleosynthesis, we show that the observed pattern of the abundance ratios [Cr/Fe], [Mn/Fe] and [Co/Fe] can be explained by a variation of mass cuts in SNe II as a function of progenitor mass. This variation provides a changing mix of nucleosynthesis from an alpha-rich freeze-out of Si-burning and incomplete Si-burning. This explanation is consistent with the amount of ejected ^{56}Ni determined from modeling the early light curves of individual supernovae. We also suggest that the ratio [Fe/H] of halo stars is mainly determined by the mass of interstellar hydrogen mixed with the ejecta of a single supernova which is larger for larger explosion energy and the larger Strömgren radius of the progenitor.

2) For SNe Ia, we introduce a metallicity dependence of the SN Ia rate in the Galactic and cosmic chemical evolution models. In our scenario involving a strong wind from WDs, few SNe Ia occur at [Fe/H] $\lesssim$ -1. Our model successfully reproduces the observed chemical evolution in the solar neighborhood. We make the following predictions that can test this metallicity effect.

i)   SNe Ia are not found in the low iron abundance environments such as dwarf galaxies and the outskirts of spirals.

ii) The cosmic SN Ia rate drops at $z \sim 1 - 2$ because of the low-iron abundance, which can be observed with the Next Generation Space Telescope.

iii) At $z > 3$, SNe Ia can be found in ellipticals where the timescale of metal enrichment is sufficiently short, while the SN Ia rate in spirals drops at $z \sim 2.5$ due to the low-iron abundance.

Acknowledgement

This work has been supported in part by the grant-in-Aid for COE Scientific Research (07CE2002) of the Japanese Ministry of Education, Science, Sports, and Culture.

References

Hachisu, I., Kato, M. and Nomoto, K.: 1996, *Astrophys. J.* **470**, L97.

Hachisu, I., Kato, M. and Nomoto, K.: 1999, *Astrophys. J.*, submitted.

Iwamoto, K., Mazzali, P. A., Nomoto, K., Umeda, H., Nakamura, T., *et al.*: 1998, *Nature* **395**, 672.

Kobayashi, C., Tsujimoto, T., Nomoto, K., Hachisu, I. and Kato, M.: 1998, *Astrophys. J.* **503**, L155.

Kodama, T. and Arimoto, N.: 1997, *Astron. Astrophys.* **320**, 41.

Li, X.-D. and van den Heuvel, E.P.J.: 1997, *Astron. Astrophys.* **322**, L9.

Lu, L., Sargent, W.L.W., Barlow, T.A., Churchill, C.W. and Vogt, S.S.: 1996, *Astrophys. J. Suppl.* **107**, 475.

McWilliam, A.: 1997, *Annu. Rev. Astron. Astrophys.* **35**, 503.

Madau, P., Ferguson, H.C., Dickinson, M.E., Giavalisco, M., Steidel, C.C. and Fruchter, A.: 1996, *Mon. Not. R. Astron. Soc.* **283**, 1388.

Nakamura, T., Umeda, H., Nomoto, K., Thielemann, F.-K. and Burrows, A.: 1999, *Astrophys. J.* **517**, in press (astro-ph/9809307).

Nomoto, K. and Hashimoto, M.: 1988, *Phys. Rep.* **256**, 173.

Nomoto, K., Iwamoto, K. and Kishimoto, N.: 1997, *Science* **276**, 1378.

Pain, R., *et al.*: 1996, *Astrophys. J.* **473**, 356.

Pettini, M., Kellogg, M., Steidel, C., Dickinson, M., Adelberger, K.L. and Giavalisco, M.: 1998, *Astrophys. J.*, in press.

Ryan, S.G., Norris, J.E. and Beers, T.C.: 1996, *Astrophys. J.* **471**, 254.

Tsujimoto, T., Nomoto, K., Yoshii, Y., Hashimoto, M., Yanagida, S. and Thielemann, F.-K.: 1995, *Mon. Not. R. Astron. Soc.* **277**, 945.

Tutukov, A.V. and Yungelson, L.R.: 1994, *Mon. Not. R. Astron. Soc.* **268**, 871.

Wheeler, J.C., Sneden, C. and Truran, J.W.: 1989, *Annu. Rev. Astron. Astrophys.* **27**, 279.

NEW INSIGHTS INTO THE EARLY STAGE OF THE GALACTIC CHEMICAL EVOLUTION

T. TSUJIMOTO

National Astronomical Observatory, Mitaka, Tokyo 181-8588, Japan

T. SHIGEYAMA

Department of Astronomy, School of Science, University of Tokyo, Bunkyo-ku, Tokyo, 113-0033 Japan

The supernova yields of several heavy elements including α-, iron-group, and r-process elements are obtained as a function of the mass of their progenitor main-sequence stars M_{ms} from the abundance patterns of extremely metal-poor stars with a procedure recently proposed by Shigeyama and Tsujimoto (1998). The yields thus obtained are indispensable because theoretical SN II models to date can reliably predict the amounts of only a few elements. The ejected masses of α- and iron-group elements increase with M_{ms}, whereas more Eu is ejected from supernovae with lower M_{ms}. For these several heavy elements, it is shown that the average abundance ratios weighted by the Salpeter initial mass function coincide with the ratios observed in stars with $-2 < [Fe/H] < -1$ within 0.1 dex. It follows that the correlations of stellar abundance ratios with the metallicity are twofold. One is the abundance ratios for $[Fe/H] < -2.5$ imprinted by the nucleosynthesis in individual supernovae on the timescale $\sim 10^7$ yr and the other for $[Fe/H] > -2$ results from the mixing of the products from a whole site of the nucleosynthesis, taking place on the timescale longer than 10^9 yr.

References

McWilliam, A., Preston, G. W., Sneden, C., Searle, L.: 1995, *Astron. J.*, **109**, 2757
Shigeyama, T., Tsujimoto, T.: 1998, *Astrophys. J. Lett.* **507**, L135
Tsujimoto, T., Shigeyama, T.: 1998, *Astrophys. J. Lett.* **508**, L151
Yoshii, Y., Tsujimoto, T., Kawara, K.: 1998, *Astrophys. J. Lett.* **507**, L113

Astrophysics and Space Science is the original source of publication of this article. It is recommended that this article is cited as: *Astrophysics and Space Science* **265**: 49–50, 1999.
© 1999 *Kluwer Academic Publishers.*

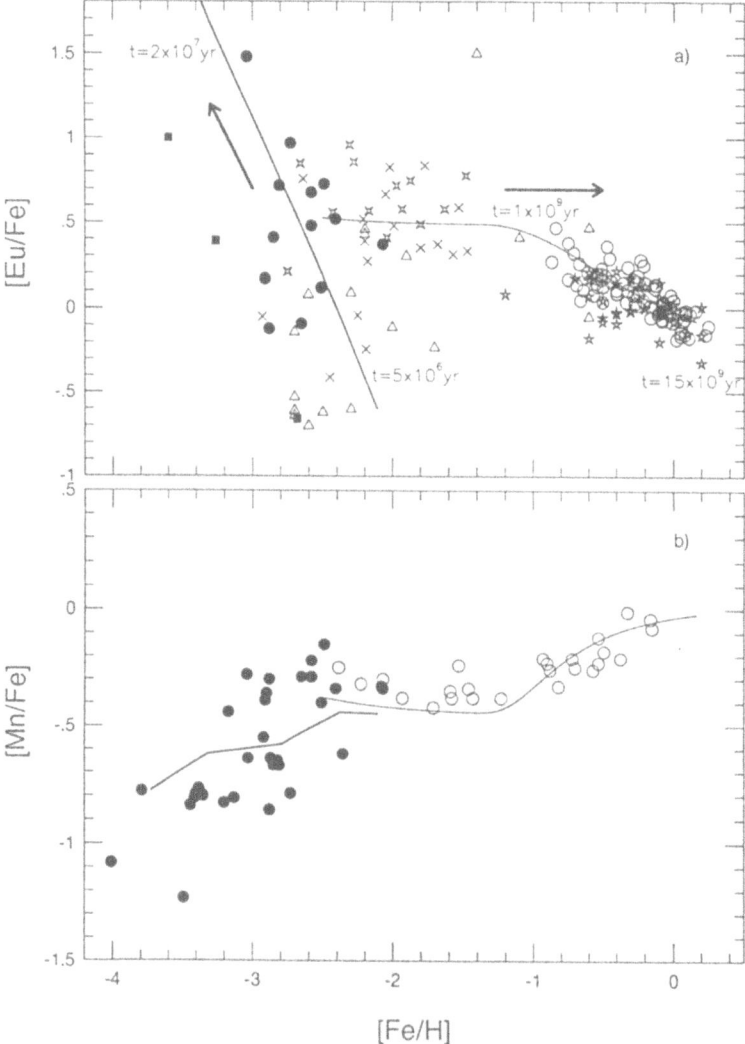

Figure 1. (a) Correlation of [Eu/Fe] with [Fe/H]. The thick line indicates the predicted nucle-osynthesis [Eu/Fe] ratios in $18 - 50$ $M_\odot$ stars. The lower mass limit corresponds to the most metal-deficient star observed by McWilliam *et al.* (1995). As a more massive star's explosion cor-responds to a higher [Fe/H], the time elapses toward lower metallicity as indicated by the arrow for [Fe/H] < -2.5. Using the nucleosynthesis yields of Fe and Eu inferred from extremely metal-poor stars, we calculate the time evolution of [Eu/Fe] with the one-zone model of Galactic chemical evolution (Yoshii *et al.*, 1998) and the result is shown in the [Eu/Fe]-[Fe/H] diagram for [Fe/H] > -2.5 (the thin line). (b) the same as (a) but for [Mn/Fe].

STELLAR EVOLUTION AND THE COSMOLOGIAL SUPERNOVAE RATES

I.E. PANCHENKO, V.M. LIPUNOV and K.A. POSTNOV
Moscow State University, Department of Physics, Sternberg Astronomical Institute, Moscow, Russia

M.E. PROKHOROV
Sternberg Astronomical Institute, Moscow, Russia

Abstract. We present the results of the population synthesis of the population of the supernovae progenitors. Both single and double degenerate progenitors of SN Ia are considered. We compute the cosmic rate histories for SN I, SN II and both classes of SN Ia, and present them in the form of redshift and magnitude distributions. These results can be compared with observational data, allowing to estimate the star formation rate history and the cosmological parameters including $\Omega_{baryons}$ which cannot be estimated from analysing the Hubble diagrams of supernovae.

We find that single degenerate (SD) SN Ia are younger than double degenerate (DD) ones, and so the SN Ia in elliptical galaxies should be mostly DD.

We propose to use the redshift dependence of relative supernovae rates in different types of galaxies, or of different supernovae types for interpretation of observations. These relative rates should be less influenced by the selection effects than the absolute ones.

1. Introduction

Supernovae, mainly SN Ia, has proved to be an important tool (a standard candle) widely used in extragalactic astronomy as a distance indicator. After the first detection of 7 'cosmological' (i.e. of $z \sim 1$) SN by Perlmutter *et al.*(1994, 1998) this role extended to the cosmological scales, allowing to measure the distance to the $z = 0.4$ galaxies and to restrict the cosmological parameters Ω_m, Ω_Λ.

The above results were obtained only from the magnitude-redshift relation (the Hubble diagram). However, a classical cosmological test $\log N - \log S$ (Weinberg, 1971; Jørgensen *et al.*, 1997) can be applied to the SN counts. As this would require knowledge of the sources number density and luminosity evolution and accounting for various selection effects, this appears more difficult than the interpretation of the Hubble diagrams.

The SN Cosmology Group (Pain *et al.*, 1996) has provided the observational estimate of the SN rate: $1.5 \cdot 10^6$ yr^{-1} for $21.3 < R < 22.3$ from the whole sky. Under the simplest cosmological assumptions, this allowed Jørgensen*et al.*(1997) to estimate the present density of the stellar matter in the Universe as $\Omega_\star = 0.004 \pm 0.001$, which agrees with other estimates (Fukugita, Hogan and Peebles, 1998).

 Astrophysics and Space Science is the original source of publication of this article. It is recommended that this article is cited as: *Astrophysics and Space Science* **265**: 51–54, 1999.
© 1999 *Kluwer Academic Publishers*.

Due to the difficulty of the detection efficiency determination, no more observational estimates of the cosmic SN rate were made since 1994.

2. Computing the SN Explosion Rate

The cosmic SNE rate density can be copmputed as

$$\mathcal{R}(z) \propto \rho(z) \int_z^\infty SFR(z) R(t(z) - t(z')) \left(\frac{dt(z')}{dz'} \right) dz', \tag{1}$$

where $\rho(z)$ is the density of stars, $SFR(z)$ is the cosmic SFR history, and $R(t)$ is
the SNE rate in the stellar population of the age t, or, in other words, the event age
distribution (AD), computed by the population synthesis.

We have used the binary star population synthesis code known as the 'Scenario Machine' (Lipunov et al., 1996) with the previously tested model parameters
(Lipunov et al., 1996), including the Sapleter IMF ($\alpha = 2.35$, $M_{min} = 0.1\ M_\odot$),
the initial mass ratio distribution $f(q) \propto q^2$, the orbital separations ($f(a) \propto a$)
and the common envelope efficiency $\alpha_{CE} = 0.5$.

The simulation included the following SN progenitors:

SN Ia. Two possible mechanisms were studied – the double degerate (DD) model
(a merger of binary CO or O-Ne-Mg white dwarves with the superchandrasek-
har total mass (Iben and Tutukov, 1984; Webbink, 1984)), and the single
degenerate (SD) model (a white dwarf reaches the Chandrasekhar limit during
the accretion (Hachisu, Kato and Nomoto, 1996; Yungelson and Livio, 1998)).

SN Ib. The collapse of the massive star core in the case if (most of) the hydrogen
envelope is lost during the mass transfer.

SN II. The collapse of the core of the massive star that did not loose its hydrogen
envelope.

In our previous work (Jørgensen et al., 1997) we approximated the SFR by
a sum of an instantaneous burst at $z_\star \sim 5$ during which ϵ of all stars has been
formed and the constant SFR lasting untill present time. However, after the works
of Madau et al. (1996) and others the SFR history became much more clear. In this
work we take the SFR history from Madau et al. (1996), still combined with an
instantaneous burst of relative weight (ϵ) at $z_\star = 5$ which is describes the formation
of the stars in spheroidals.

We used the set of cosmological parameters $\Omega_m = 0.6$, $\Omega_\Lambda = 0.4$ which agrees
with the SN Ia Hubble diagrams (Perlmutter et al., 1998), and $H_0 = 75$ km s^{-1}
Mpc^{-1} which also agrees with the SN data (Kim et al., 1997).

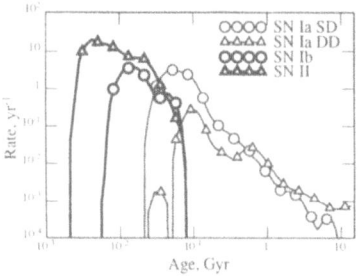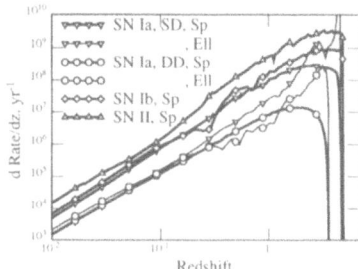

Figure 1. Left: The age distribution (AD) of the SNE. In other words, the SN rate as a funciton of the stellar population age. Right: The SN redshift distribution (for spiral and elliptical galaxies).

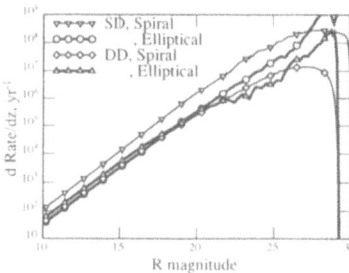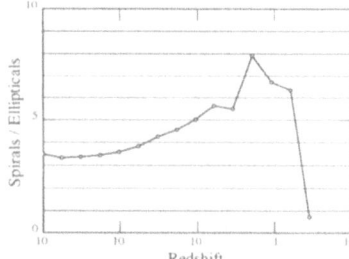

Figure 2. Left: The SN Ia R magnitude distributions. Right: Ratio of the SN Ia rates in spiral and elliptical galaxies.

3. Discussion

Let us delineate the questions that can be solved by the number counts of the cosmological SN.

The most obvious implication of the SN rate counts is the number of their progenitors, which in the case of the core collapse supernovae means the measurement of the star formation rate and in the case of SN Ia helps to choose the progenitor model. Reversly, supposing some progenitor model allows to reconstruct the cosmic SFR history from the counts of SN Ia.

The absolute number counts corrected for the detection efficiency give the absolute SFR and Ω_* (proportional to the integral of the SFR). However, the *relative* rate counts are more robust to the selection effects than the *absolute* counts, while providing almost the same information.

The host galaxy extinction blankets some of the core collapse SN which usually take part in the star formation regions with high gas content. Therefore the counts of SN II and SN Ib need a correction for this screening and are intrinsically less accurate than those of SN Ia. Nevertheless, if we suppose that the fraction of the unseen SN does not change with redshifts, or assume any other realistic model of the blanketing fraction evolution, we can use the relative number counts including the core collapse SN with the same confidence as SN Ia.

As the SN Ia and SN II/Ib originate from different ranges of inital star masses, the ratio of the rates of these types contain also the information on the initail mass function evolution.

Event age distributions produced by the population synthesis are a key point in interpreting the cosmological SN counts. Advances in the stellar evolution physics make possible to interpret the absolute and relative SN rate estimates which are expected soon with the progress in the observational techniques and instrumentation.

Acknowledgements

The authors acknowledge the French Ministery of Education and the Observatory of Paris which made possible the participation in the conference. The work was partially supported by the Russian Fund for Basic Research through Grant No 95-02-06053-a.

References

Fukugita, M., Hogan, S.J. and Peebles, P.J.E.: 1998, *Astrophys. J.* **503**, 518.
Hachisu, I., Kato, M. and Nomoto, K.: 1996, *Astrophys. J.* **470**, L97.
Iben, I. and Tutukov, A.V.: 1984, *Astrophys. J. Suppl.* **54**, 335.
Jørgensen, H.E., *et al.*: 1997, *Astrophys. J.* **486**, 110.
Kim, A.G., *et al.*: 1997, *Astrophys. J.* **467**, 63.
Lipunov, V.M., Postnov, K.A. and Prokhorov, M.E.: 1996, *The Scenario Machine: Binary Star Population Synthesis*, Harwood Academic Publishers, Amsterdam.
Lipunov, V.M., Postnov, K.A. and Prokhorov, M.E.: 1996, *Astron. Astrophys.* **310**, 489.
Madau, P., *et al.*: 1986, *Mon. Not. R. Astron. Soc.* **283**, 1388.
Pain, R., *et al.*: 1996, *Astrophys. J.* **473**, 356.
Perlmutter, S., *et al.*: 1995, *Astrophys. J. Lett.* **440**, L41.
Perlmutter, S., *et al.*: 1998, *Nature* **391**, 51.
Webbink, R.F.: 1984, *Astrophys. J.* **277**, 355.
Weinberg, S.: 1972, *Gravitation and Cosmology*, J. Wiley and Sons, New York.
Yungelson, R. and Livio, M.: 1998, *Astrophys. J.* **497**, 168.

NEW OBSERVATIONS OF GALACTIC DEUTERIUM

G. SONNEBORN

NASA Goddard Space Flight Center, Code 681, Greenbelt, MD 20771, USA

E.B. JENKINS, T. TRIPP and P. WOZNIAK

Princeton University Observatory, Princeton, NJ 08544, USA

A. VIDAL-MADJAR and R. FERLET

Institut d'Astrophysique de Paris, 98bis, Blvd. Arago, 75014 Paris, France

U.J. SOFIA

Dept. of Astronomy, Whitman College, Walla Walla, WA 99362, USA

1. Introduction

Observations made with the Interstellar Medium Absorption Profile Spectrograph (IMAPS) on the US-German ORFEUS-SPAS II mission in late 1996 provide the first new measurements of Galactic interstellar atomic deuterium beyond the local ISM since the *Copernicus* mission in the 1970s. IMAPS is an objective grating echelle spectrograph designed for high spectral resolution in the far UV (see Jenkins *et al.*, 1996). IMAPS observed δ Orionis A (O9.5 II), γ^2 Velorum (WC8+O9I), and ζ Puppis (O4 Iaf) at a spectral resolution of 4 km s^{-1} over the range 930–1160 Å, including H I Ly-δ (949.485 Å) and Ly-ϵ (937.548 Å).

2. D/H Analysis

The total D I column density (N_{DI}) toward each star was determined by simultaneously modelling the Ly-δ and Ly-ϵ D I profiles (defined by N_{DI} and temperature T), the continuum fit near the D I features, and the background (zero intensity) level near each line. The velocity profile of the gas toward each star was defined by absorption features from several multiplets of N I, covering a range of 1.80 in log(gf), recorded at good S/N in the IMAPS spectra. The N I velocity profiles were used as a template for modelling the D I profiles. This approach tightens the error limits for N_{DI}. The optimum solution for N_{DI} is determined by minimizing χ^2 for the multi-parameter fit. The details of this analysis technique is given in Jenkins *et al.* (1999).

Values of the H I column density (N_{HI}) in the literature for these stars were deemed sufficiently imprecise that new evaluations were made. N_{HI} was measured

Astrophysics and Space Science is the original source of publication of this article. It is recommended that this article is cited as: *Astrophysics and Space Science* **265**: 55–56, 1999.
© 1999 *Kluwer Academic Publishers*.

TABLE I

D/H abundance ratios

Star	l^{II}	b^{II}	d (pc)	$N_{DI}(10^{15}\,cm^{-2})$	$N_{HI}(10^{19}\,cm^{-2})$	D/H (10^{-5})
δ Ori A	203.9	-17.7	372	$1.21^{+0.29}_{-0.20}\,cm^{-2}$	$15.6 \pm 0.10\,cm^{-2}$	$0.74^{+0.19}_{-0.13}$
ζ Pup	256.0	-4.7	429	$1.38^{+0.23}_{-0.18}\,cm^{-2}$	$9.10 \pm 0.04\,cm^{-2}$	$1.48^{+0.25}_{-0.21}$
γ^2 Vel	262.8	-7.7	258	$1.08^{+0.17}_{-0.13}\,cm^{-2}$	$5.12 \pm 0.07\,cm^{-2}$	$2.11^{+0.32}_{-0.26}$

by performing a similar χ^2 analysis for Ly-α using dozens of IUE high-dispersion spectra for each star. The large number of IUE spectra significantly reduces the error in N_{HI} so that the error in the D/H measurements is dominated by the error in N_{DI}.

The D/H abundance ratios given in Table I (errors are 90% confidence limits) conclusively demonstrate that D/H in the ISM of our spiral arm can vary substantially from that measured in the local ISM (D/H$\sim 1.5 \times 10^{-5}$, see Piskunov *et al.*, 1997 and references therein). We find a difference of 3X in D/H between δ Ori and γ^2 Vel. Jenkins *et al.* (1999) also find that N/H and O/H toward δ Ori are not enhanced, indicating that the low D/H value is not the result of injection of CNO-processed (i.e. D-depleted) material into the ISM. Measuring D/H on many sightlines in the Milky Way, understanding the distribution function for D/H ratios, and its correlations with Galactic and stellar environments is a major objective of the FUSE mission.

References

Jenkins, E.B., *et al.*: 1996, *Astrophys. Space. Sci.* **239**, 315.
Jenkins, E.B., *et al.*: 1999, *Astrophys. J.*, in press.
Piskunov, N., *et al.*: 1997, *Astrophys. J.* **474**, 315.
Sonneborn, G., *et al.*: 1999, in preparation.

LIGHT ELEMENT ABUNDANCES IN FIELD STARS

SUCHITRA C. BALACHANDRAN

Department of Astronomy, University of Maryland, College Park, MD 20742, USA

Abstract. Recent results from the measurements of Li, Be and B in metal-poor and solar metallicity stars are discussed in the context of angular momentum transport models.

1. The Primordial Lithium Abundance

Of considerable interest to cosmologists and stellar spectroscopists alike is the primordial Li abundance. While cosmologists are interested in its value as a means of constraining the standard model of the Big Bang, stellar spectroscopists are intrigued by the relatively flat Li plateau seen over a range of 800 K in temperature and 2 dex in metallicity. Spectroscopists have sought to test whether the plateau is flat with effective temperature and metallicity, as any indication of a slope may yield clues to Li production and destruction proceses and thus reveal whether the plateau value is primordial. Debate has also centered around whether the small dispersion observed in the plateau is equal to or larger than the errors in the equivalent width and temperature measurements that go into the abundance determinations. Any real scatter may imply that Li has been destroyed in at least some stars and provide the basis for interpreting the plateau value as smaller than primordial.

The early work of Deliyannis *et al.* (1993) attempted to use the observational plane of Li equivalent widths versus b-y color to quantify the dispersion, but failed to account for the dependence of b-y on metallicity. Thorburn (1994) and Ryan *et al.* (1996) used the b-y color to get effective temperatures, and concluded that the dispersion in the plateau was indeed larger than the observational error, but their error estimate was disputed by Spite *et al.* (1996). The latter derived temperaures using 3 different techniques to show that extreme care was needed to fully constrain the source of the error in the abundance analysis. They concluded that the dozen or so stars in their sample all had Li abundances consistent with each other. However, as their sample size was small, statistical studies on the characteristics of the plateau could not be carried out. The most comprehensive study to date is the work by Bonifacio and Molaro (1997). With temperatures derived by the infrared flux method, they conclude there is no slope or scatter in the Li plateau with temperature or metallicity. The reader is also referred to Spite *et al.* (1998) for a review on the subject.

Astrophysics and Space Science is the original source of publication of this article. It is recommended that this article is cited as: *Astrophysics and Space Science* **265:** 57–65, 1999.
© 1999 *Kluwer Academic Publishers.*

The debate over whether the plateau represents the primordial value is far from resolved; there is a severe theoretical problem in concluding that the plateau value is primordial. In the absence of any mixing process, gravitational settling must be at work. Basic diffusion calculations indicate that settling is largest in the warmest plateau stars. As these have the thinnest surface convective zones, calculations predict that the Li plateau must curve downwards at the warmest temperatures as a result of the gravitational settling of Li (Deliyannis *et al.*, 1990; Vauclair and Charbonnel, 1998). This is not observed. The curvature may be removed by introducing some amount of mixing, but it is reasonable to infer that this mixing must then extend to cooler stars as well. Modeling the structure of the Population II stars, Vauclair and Charbonnel conclude that the primordial Li abundance is somewhat larger than the current plateau value, roughly $\log \epsilon(Li) = 2.35$. Of course, as this is the minimum initial Li required to render the plateau flat, depletion may have been considerably larger. The curious Li depletion process which leaves the plateau so uniformly flat remains to be understood.

There are some observational suggestions that the primordial value may be larger than the current plateau value. King *et al.* (1996) found the Li abundance in the halo dwarf BD+23 3912 to be about a factor of 2–3 above the plateau value and argued that its abundance was consistent with Li depletion models based on rotational braking, which begin with a larger-than-plateau primordial value. Note that these models must assume that the range in initial rotational velocities of the halo stars is small in contrast to the large spread observed in young Population I stars. Boesgaard *et al.* (1998b) have found a subgiant with a similarly high Li in the globular cluster M92. More interestingly, they point to differences in the Li abundances of several subgiants at the same T_{eff}, and presumably the same evolutionary phase. Arguing that these stars have not evolved far enough along the subgiant branch to begin dilution, the authors suggest that the dwarf progenitors of these stars did not have the same Li abundance. This is in contrast to the insignificant dispersion seen in the field Li plateau stars. However, we note that these stars have evolved from much higher main sequence temperatures than the hot end of the Li plateau and we have no information about the behavior of the Li abundances in this regime. Effects of gravitational settling, for instance, may be expected to be more severe in these stars than at the warm end of the plateau. Is the dredge-up of gravitationally-settled Li responsible for the spread in Li seen in these 3 stars? In which case, is the largest Li abundance seen in these stars representiative of the primordial value? We also note that these stars are much fainter than the field halo dwarfs and therefore the S/N ratios of the spectra are not high. Given the controversy over the errors assigned to the abundances in the far brighter halo field stars, one would expect that the errors assigned to the globular cluster stars will be challenged as well as more data are gathered with the emerging generation of 8m class telescopes. Clearly there are unanswered theoretical and observational questions that will keep us busy in this area for some more time to come.

A few other recent results are summarized. There was a suggestion by Kurucz (1995) that severe atmospheric inhomogeneities may result in large non-LTE corrections in Li abundances derived from the 6708 Å resonance line in Pop II stars, perhaps increasing the plateau value to 3.5. This has been put to rest by two studies. Bonifacio and Molaro (1998) measured the LTE Li abundance in HD 140283 from the weaker 6104 Å transition and found it to be equal to the LTE abundance measured from the resonance feature. Uitenbroek (1998) carried out a detailed study of granulation in solar type stars and found non-LTE effects to be negligible.

In stars cooler than the Li plateau, Li depletion appears to be smaller than predicted by the standard models (Ryan and Deliyannis, 1998), with the scatter at a given metallicity being reminiscent of Population I stars. In the subgiants, Li abundances are found to be consistent with simple dilution as pointed out earlier by Pilachowski *et al.* (1993).

The drop-out stars, with temperatures corresponding to the Li plateau but Li abundances far below the plateau, were examined in some detail by Norris *et al.* (1997). No trends were discovered in other abundances that may lend clues to their anomalous Li behavior.

Studies of the lighter and more fragile isotope of Li, ^{6}Li, are dealt with in a review article by R. Cayrel in these proceedings.

2. The Population I Stars

In part, the answer to the puzzle of the Li plateau must come from our improved understanding of Li depletion processes which are seen to produce a variety of patterns in Population I stars.

2.1. THE SUN

Of particular interest in this area have been recent re-determinations of the solar beryllium (Balachandran and Bell, 1998) and boron (Cunha and Smith, 1998) abundances. Li in the Sun has long been known to be depleted by a factor of about 140 relative to the meteoritic value. Previous studies, most recently King *et al.* (1997) and Primas *et al.* (1997), have suggested that Be in the Sun is depleted by a factor of 2 relative to the meteoritic value. Although this is much smaller than the Li depletion factor, the depletion of Be is important because it implies deeper mixing below the surface convective zone than is required to destroy Li, and it distinguishes critically between various depletion processes that may be at work in the solar interior. For example, it may argue for a slow mixing process (e.g., one driven by rotation) which takes surface material down efficiently to the depth required to deplete Li and less efficiently to the depth required to deplete Be. The simultaneous depletion of Li and Be in the Sun has been repeatedly cited by the Yale group as evidence for the validity of their rotational braking models (Pinsonneault *et al.*, 1989).

In Balachandran and Bell (1998), we argue that by not fully accounting for the continuous opacity in the solar UV spectrum (an age-old problem, see Gustafsson and Bell, 1979), previous studies have underestimated the solar Be abundance. We estimated the continuous opacity by requiring that infrared and ultraviolet OH features yield the same solar oxygen abundance. The robustness of our result is borne out by two independent tests. First, the same continuous opacity is estimated for 3 solar models (Kurucz, 1993; OSMARCS – Edvardsson *et al.*, 1993; Holweger and Müller, 1974) provided the appropriate oxygen abundance obtained from the infrared OH lines for each model is adopted for the test. Second, the calculated solar flux is much closer to the observed UV solar flux if the corrected continuous opacity is adopted, $(O-C)/O = -0.18$ compared to $(O-C)/O = -0.54$ for the standard opacity case. When the continuous opacity is modified by the amount demanded by the UV–OH features, the solar Be abundance is found to be essentially identical to the meteoritic value, i.e., mixing in the Sun has not extended deep enough to deplete any Be. Additional information about the source of the continuous opacity and confirmation of the validity of our technique is obtained from limb darkening observations: the depths of the Be II line at disk center ($\mu = 1.0$) and near the limb ($\mu = 0.2$) are best fit if the additional opacity is in the form of a metal bound-free source. A simple augmentation of the hydrogen opacities does not fit the line depths, neither does the standard opacity.

We chose to augment the Fe I b–f opacity, the most uncertain of the metal opacities. This had to be increased by a factor of 32 from the Dragon and Mutschlechner (1980) value to fit the solar center and limb Be II line depths and simultaneously match the solar flux OH line depths. We have since found this to be in reasonable agreement with the Opacity Project Fe I bound-free values (Bautista, 1997).

The implications of the revised solar Be abundance are important. First, it indicates that mixing below the solar surface convective zone (CZ) is superficial, sufficient to destroy Li but not deep enough to destroy Be. This is consistent with current helioseismological data which indicate the presence of a thin region of shear below the surface convective layer inferrred from the lack of accumulation of diffused helium below the CZ (Basu, 1997). The meteoritic Be abundance reveals more than current helioseismological data because it shows that mixing did not extend deep enough to destroy Be over the entire solar history, not even early in the life of the Sun, when considerable rotational spin-down generating turbulence and mixing is predicted by models.

The solar B abundance has not been subject to the same critical examination as has the Be abundance. Until the recent Cunha and Smith (1998) study, the only other detailed determination of B in the solar photosphere was by Kohl *et al.* (1977) based on the Kohl *et al.* (1978) solar spectrum acquired during an Aerobee rocket flight. The Kohl *et al.* (1977) study yielded a solar B abundance of log $\epsilon(B) = 2.6 \pm 0.3$. Within the errors, this is consistent with the meteoritic value of

log ϵ(B) = 2.78 ± 0.05 (Zhai and Shaw, 1994) although it has sometimes been interpreted as indicating a small depletion. Cunha and Smith (1998) essentially used
the same data, the Kohl *et al.* (1978) spectral atlas. Their re-determination utilized
improved solar model atmospheres and most importantly, improved opacities. In
particular, the Mg I b–f opacity is a significant contributor at the wavelength of B
I line (2497 Å) and this value has been revised by the Opacity Project calculations
(Butler *et al.*, 1993).

Cunha and Smith's (1998) revised analysis yields log ϵ(B) = $2.70^{+0.21}_{-0.12}$, which
is fully consistent with the meteoritic value and therefore indicates that the solar
photosphere has undergone no B depletion, consistent with the Be result of Balachandran and Bell (1998).

The importance of using the full set of metal opacities is highlighted in the
B analysis in Boesgaard *et al.* (1998a). In that study, the version of the progam
MOOG used to calculate abundances does not contain any metal bound-free opacity contributions. With only hydrogen and helium opacities in their analysis, Boesgaard *et al.* (1998a) obtained a B depletion of 0.7 dex relative to the meteoritic value
in a HR 3775 and HR 8340 both of which have near-meteoritic Li and are therefore
expected to have undergone no B depletion. Although Boesgaard *et al.* used a
differential analysis to eliminate the effect of this discrepancy in their analysis, the
magnitude of the discrepancy is indicative of the importance of the metal opacity
contributions at the B I wavelength and in determining the absolute B abundance.

2.2. BERYLLIUM IN OTHER STARS

Only a few measurements of Be have been carried out in stars cooler than the
Sun. Two studies, King *et al.* (1997) and Primas *et al.* (1997), measured Be in the
solar analog star α Cen A and its K1 V companion α Cen B and there is some
disagreement in their results. While King *et al.* (1997) found log ϵ(Be) in α Cen A
to be nearly 0.2 dex larger than solar, Primas *et al.* (1997) found it to be essentially
solar. Furthermore, Primas *et al.* (1997) measured the difference between the Be
abundances of α Cen A and B to be nearly 0.5 dex while according to King *et
al.* (1997) it is only 0.15 dex. Therefore there is some ambiguity about whether α
Cen B has undergone significant Be depletion or not. In a more extensive study,
García López *et al.* (1995) examined 5 Hyades G and K dwarfs between 5200 and
6000 K and concluded that Be is essentially constant over this entire temperature
range. They found the Be abundance in these stars to be consistent with the solar
Be abundance.

Note that their solar Be abundance is sub-meteoritic because they did not include all of the additional opacity we have found to be essential in the UV solar
analysis. However, in the light of our findings, the constancy of Be with temperature may be interpreted as indicating that none of the cooler G dwarfs has undergone
any Be depletion. A more thorough analysis would be most useful with estimates
of the UV opacity in the cooler stars. The lack of Be depletion is in stark contrast

with the large Li depletion seen in these stars; Li is essentially meteoritic around 6300 K and plummets down to upper limits below log ϵ(Li) = 1.0 at the cool end of their temperature range.

In contrast to the G dwarfs, mid-F stars near the Li 'dip' show very different Li and Be depletion patterns. Stephens *et al.* (1997) and Deliyannis *et al.* (1998) have shown Be depletion of nearly 1 dex in stars which have Li depletion of roughly 2 dex (of the same magnitude as in the Sun) in the mid-F stars (see Figure 2 in the latter paper). The magnitude of the Be depletion is far larger than may be accounted for by opacity errors, and is clearly real. In some of these stars where Li and Be depletion is observed, the Li I and Be II features are both detectable, providing the authors with evidence that both Li and Be are depleted simultaneously. The contrast between the F and G stars appears to indicate different depletion processes may be at work in these stars. Given the very different structures of these stars, this is not altogether surprising. According to the standard models, Li depletion is difficult to explain in the mid-F stars which have very thin convective zones. This depletion clearly requires mixing to considerable depths below the CZ. The process by which this vigorous mixing occurs may well churn the material to even deeper layers and thus deplete Be. In contrast, only shallow mixing is required below the CZ to deplete Li in the G dwarfs, and the lack of Be depletion in these stars may reflect the more superficial and milder process by which it occurs.

2.3. HELIOSEISMOLOGY, SOLAR COMPOSITION AND ANGULAR MOMENTUM TRANSPORT: COMPARISON WITH RECENT MODELS

In recent years, the mixing required for Li depletion has been tied directly to the transport of angular momentum in stars with either turbulence (Pinsonneault *et al.*, 1989) or meridional circulation (Zahn, 1992) linking the two processes. By suitably adjusting the efficiency of material transport relative to angular momentum transport, modelers found it possible to both extract sufficient angular momentum from the surface of the stars to match open cluster observations at different ages, and to reproduce some of the dispersion in Li abundances which is manifest as the clusters age. It has been suggested that the Li depletion spreads seen in stars of the same age and mass result from differences in the initial angular momentum content of these stars.

Observational results have recently revealed the limitations of these models. First, the internal structure of the Sun as revealed by helioseismology is not matched by the models. While the models predict the solar interior should be rotating rapidly with velocity increasing as a function of depth, observations indicate a flat rotational profile with relatively slow solid body rotation in regions below the surface CZ (Brown *et al.*, 1989; Thompson *et al.*, 1996). The transport of angular momentum in the Sun has therefore been far more efficient than predicted by the rotation-driven models and must have proceeded by some other means with turbulent or meridional transport providing at best a small contribution.

A second strong constraint on mixing models comes from measurements of the solar composition. Our determination that Be has not been destroyed in the solar photosphere (Balachandran and Bell, 1998) provides one limit on the depth to which mixing may be allowed in the models. A second compositional constraint comes from the recent work by Geiss and Gloecker (1998) who find from solar wind measurements that the ^{3}He/^{4}He ratio on the solar surface has not changed over the last 3 Gyr. Mixing models must therefore not extend deep enough to reach the ^{3}He peak in the solar interior. Using Zahn's formuation for angular momentum and material mixing, Richard *et al.* (1996) produced excessive mixing in the solar interior resulting in a violation of both the Be and ^{3}He constraints. Acknowledging this, Vauclair and Richard (1998) revisited the problem and determined that a much lower critical mean molecular weight gradient is required to limit the extent of the mixing in the model. Pinsonneault *et al.* (1989) also acknowldge an increase in the surface ^{3}He abundance in their rotational braking solar model. It is clear that neither the helioseismic nor the compositional constraints are met by the turbulence- and circulation-related transport models alone.

The alternatives to the transport of angular momentum via turbulence which have been modeled in some detail are transport via internal gravity waves (Kumar and Quataert, 1977; Zahn *et al.*, 1997; see also García López and Spruit, 1991) and transport via magnetic fields (Charbonneau and MacGregor, 1993; Barnes *et al.*, 1998). Models invoking the former establish uniform rotation in the solar interior within $\approx 10^7$ years. More complicated models in which gravity waves slow down the outer portion of the star while meridional circulation dominates the transfer of momentum in the interior were recently described by Talon and Zahn (1998) with the caveat that their crude treatment requires much improvement. These achieved solid body rotation on longer timescales of a few giga years. Whether or nor internal gravity waves are capable of transporting material remains to be studied. For a more extensive review, see Zahn (1998).

Barnes *et al.* (1988) compared a hydrodynamic model in which both angular momentum redistribution and the transport of light elements are treated as diffusion processes with a magneto hydrodynamic model in which angular momentum transport occurs via the interaction between non-uniform internal rotation and a large-scale poloidal field, as described by Charbonneau and MacGregor (1993), and particle transport is by turbulent diffusion (see also MacGregor and Barnes contribution in Balachandran *et al.*, 1998). They followed a 1 M$_\odot$ star beginning at the ZAMS and ending at the age of the present-day Sun. While the HD model had a rapidly rotating core at the end, the MHD model produced a state of near-unifrom rotation. More interestingly, when the two models were calibrated to reproduce observed solar Li depletion, they yielded very different Be abundances, with Be depletion occuring in the HD model but no depletion occuring in the MHD model, which only had a significant radial shear at the base of the CZ. Detailed models for different masses and initial angular momenta are awaited with much interest.

Acknowledgement

Support from NASA Grant NAG-54092 is gratefully acknowledged.

References

Balachandran, S.C. and Bell, R.A.: 1998, *Nature* **392**, 791.
Balachandran, S.C., *et al.*: 1998, Tenth Cambridge Workshop on Cool Stars, Stellar Systems and the Sun, *ASP Conf. Ser.* **154**, 111.
Barnes, G. and MacGregor, K.B.: 1998, Tenth Cambridge Workshop on Cool Stars, Stellar Systems and the Sun, *ASP Conf. Ser.* **154**, 886.
Basu, S.: 1997, *Mon. Not. R. Astron. Soc.* **288**, 572.
Bautista, M.: 1997, *Astron. Astrophys. Suppl.* **122**, 167.
Boesgaard, A.M., *et al.*: 1998a, *Astrophys. J.* **492**, 727.
Boesgaard, A.M.: 1998b, *Astrophys. J.* **493**, 206.
Bonifacio, P. and Molaro, P.: 1997, *Mon. Not. R. Astron. Soc.* **285**, 847.
Bonifacio, P. and Molaro, P.: 1998, *Astrophys. J.* **500**, L175.
Brown, T.M., *et al.*: 1989, *Astrophys. J.* **343**, 526.
Butler, K., Mendoza, C. and Zeippen, C.J.: 1993, *J. Phys B* **26** 4409.
Charbonneau, P. and MacGregor, K.B.: 1993, *Astrophys. J.* **417**, 762.
Cunha, K. and Smith, V.V.: 1999, *Astrophys. J.*, in press.
Deliyannis, C.P., Demarque, P. and Kawaler, S.D.: 1990, *Astrophys. J. Suppl.* **73**, 21.
Deliyannis, C.P., Pinsonneault, M.H. and Duncan, D.K.: 1993, *Astrophys. J.* **414**, 740.
Deliyannis, C.P., *et al.*: 1998, *Astrophys. J.* **498**, L147.
Dragon, J.N. and Mutschlecner, J.P.: 1980, *Astrophys. J.* **239**, 1045.
Edvardsson, B., *et al.*: 1993, *Astron. Astrophys.* **275**, 101.
García López, R.J. and Spruit, H.C.: 1991, *Astrophys. J.* **377**, 268.
García López, R.J., Rebolo, R. and Perez de Taoro, M.R.: 1995, *Astron. Astrophys.* **302**, 184.
Geiss, J. and Gloecker, G.: 1998, *Space Sci. Rev.* **84**, 275.
Gustafsson, B. and Bell, R.A.: 1979, *Astron. Astrophys.* **74**, 313.
Holweger, H. and Müller, E.A.: 1974, *Sol. Phys.* **39**, 19.
King, J.R., Deliyannis, C.P. and Boesgaard, A.M.: 1996, *Astron. J.* **112**, 2839.
King, J.R., Deliyannis, C.P. and Boesgaard, A.M.: 1997, *Astrophys. J.* **478**, 778.
Kohl, J.L., Parkinson, W.H. and Withbroe, G.L.: 1977, *Astrophys. J.* **212**, L101.
Kohl, J.L., Parkinson, W.H. and Kurucz, R.L.: 1978, *Center to Limb Solar Spectrum in High Resolution 225.2 nm to 319.6 nm*, Harvard University Printing Office, Cambridge.
Kumar, P. and Quataert, E.J.: 1997, *Astrophys. J.* **475**, L143.
Kurucz, R.L.: 1993, *CD-ROM No. 13: Smithsonian Astrophysical Observatory*.
Kurucz, R.L.: 1995, *Astrophys. J.* **452**, 102.
Norris, J.E., *et al.*: 1997, *Astrophys. J.* **485**, 370.
Pilachowski, C.A., Sneden, C. and Booth, J.: 1993, *Astrophys. J.* **407**, 699.
Pinsonneault, M.H., *et al.*: 1989, *Astrophys. J.*
Primas, F., *et al.*: 1997, *Astrophys. J.* **480**, 784.
Richard, O., *et al.*: 1996, *Astron. Astrophys.* **312**, 1000.
Ryan, S.G. and Deliyannis, C.P.: 1998, *Astrophys. J.* **500**, 398.
Ryan, S.G., *et al.*: 1996, *Astrophys. J.* **458**, 543.
Spite, M., *et al.*: 1996, *Astron. Astrophys.* **307**, 172.
Spite, F., Spite, M. and Hill, V.: 1998, *Space Sci. Rev.* **84**, 155.
Stephens, A., *et al.*: 1997, *Astrophys. J.* **491**, 339.

Talon, S. and Zahn, J.-P.: 1998, *Astron. Astrophys.* **329**, 315.

Thompson, M.J., *et al.*: 1996, *Science* **272**, 1300.

Thorburn, J.A.: 1994, *Astrophys. J.* **421**, 318.

Uitenbroek, H.: 1998, *Astrophys. J.* **498**, 427.

Vauclair, S. and Charbonnel, C.: 1998, *Astrophys. J.* **502**, 372.

Vauclair, S. and Richard, O.: 1998, *in Structure and Dynamics of the Interior of the Sun and Sun-like Stars, SOHO 6/GONG 98 Workshop*, in press.

Zahn, J.-P.: 1992, *Astron. Astrophys.* **265**, 115.

Zahn, J.-P.: 1998, in: J. Provost and F.-X. Schmider (eds.), *Sounding Solar and Stellar Interiors, IAU Symp. 181*, Kluwer, Dordrecht, p. 175.

Zahn, J.-P., Talon, S. and Matias, J.: 1997, *Astron. Astrophys.* **322**, 320.

Zhai, M. and Shaw, D.M.: 1994, *Meteoritics* **29**, 607.

OBSERVING BORON IN METAL-POOR STARS

F. PRIMAS

European Southern Observatory, Karl Schwarzschild str. 1, D-85748 Garching, München, Germany

Abstract. A new sample of 7 stars ranging in metallicity from [Fe/H] = −2.0 to [Fe/H] = −0.75 has been analyzed in the boron spectral region. The targets were selected according to the availability (in the literature) of their lithium and beryllium abundances, because the simultaneous knowledge of LiBeB in the same targets is a powerful diagnostic for testing depletion and internal mixing predicted by different stellar structure models. Two stars (HD 94028 and HD 194598), characterized by similar Li contents, are found to have also similar B abundances, despite a 0.3 dex difference in their Be abundances claimed by Thorburn and Hobbs (1996). Four stars out of 7 are characterized by strongly depleted Li and Be abundances: 2 of them (HD 2665 and HD 3795) are also significantly B-depleted, while two others (HD 106516 and HD 221377) have near normal B abundances despite being depleted by a factor ≥ 10 in both Li and Be abundances. These stars place strong constraints on the nature and depth of the mixing processes responsible for their light element abundances. The 7th star (HD 160617) shows the remarkable aspect of deficient B, probably deficient Be, and completely normal Li. No stellar destruction mechanism can explain this. Rather, chemical inhomogeneities in the halo could be the cause.

1. Introduction

The importance of studying the light elements lithium, beryllium, and boron is at least twofold: they are useful probes of stellar interiors and useful tools in better constraining the chemical evolution of the Galaxy. The selection of the sample of stars here presented was aimed at investigating stellar depletion and internal mixing predicted by stellar structure models. Since Li, Be, and B cannot survive in deep stellar interiors because they burn at progressively higher temperatures at densities found in F and G stars near the base of the surface convection zone, they are a powerful diagnostic for testing stellar structure models. These new 7 data points enlarge by more than 50% our previous B sample (cf. Duncan *et al.*, 1997).

2. Observations and Analysis

This new set of boron spectra was observed during HST Cycle 6, with the Large Science Aperture (LSA) combined with the G270M grating on the Goddard High Resolution Spectrograph (GHRS). Such an instrumental set-up gives a resolution of ≈ 25 000 at the wavelengths of interest. The spectral range covers a window

 Astrophysics and Space Science is the original source of publication of this article. It is recommended that this article is cited as: *Astrophysics and Space Science* **265**: 67–70, 1999.
© 1999 *Kluwer Academic Publishers*.

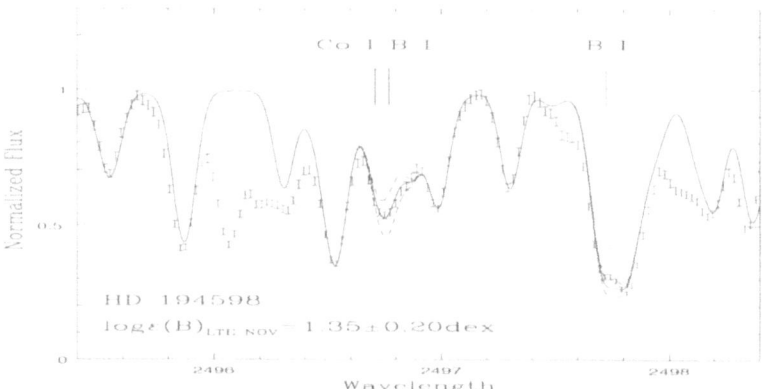

Figure 1. Best-fit for HD 194598 (T_{eff} = 5950 K, log g = 4.3, [Fe/H] = −1.15). The observed data are represented by photon statistics error bars. Overplotted are also spectrum syntheses computed with no boron (dotted line) and with ± the net final error (dot-dashed line).

of approximately 40 Å centered around the 2 main resonant atomic transitions of neutral boron at λ 2496.772 and λ 2497.725.

Data reduction was performed following the standard HST procedure, using the IRAF stsdas package, combining the quarter-stepped, FP-SPLIT data within each visit with the tasks poffsets and specalign.

The data were analyzed via spectrum synthesis technique, as required by the severe crowding of spectral lines that characterizes the near-UV region and the presence of blends. The B abundances were determined by making use of the Kurucz model atmospheres and the ATLAS and SYNTHE codes (Kurucz, 1993; cf. Figure 1 for a sample spectrum).

The computation of a synthetic spectrum also requires a list of the atomic and molecular lines present in the spectral region of interest, and the determination of the main stellar parameters of the objects under investigation, *i.e.* effective temperature, gravity, and metallicity, in order to select the closest model atmosphere in the available grid. Each of these introduces different uncertainties, which have been carefully taken into account in our error analysis.

We analyzed all the stars with Kurucz model atmospheres computed with (OV) and without (NOV) overshooting, in order to estimate the influence of this new treatment of convection on the final B abundances. The final list of atomic and molecular lines adopted during this analysis derives from the list compiled by Duncan *et al.* (1998), that had been originally taken from the official list of atomic and hydrides lines of Kurucz (1993). Slight adjustments to the oscillator strengths of some of the lines around the boron doublet were allowed in order to improve the overall fit. Accurate and very consistent values of the main stellar parameters were found in the literature for all the stars of the sample (cf. Primas *et al.*, 1998, 1999 for more details). Finally, once LTE B abundances were determined, NLTE

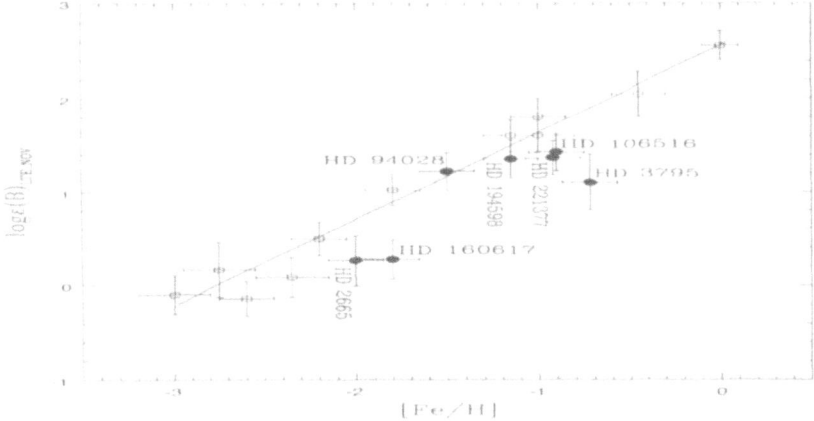

Figure 2. [B] vs. [Fe] for all the stars analyzed in this work (filled symbols) and in Duncan *et al.* (1997, open symbols). The linear chi-squares fit determined by Duncan *et al.* (1997) has been superimposed.

B abundances were computed by interpolating in T_{eff}, log g, [Fe/H], and LTE B abundances the NLTE corrections computed by Kiselman and Carlsson (1995). Our discussion relies on the LTE NOV B abundances only (shown in Figure 2), but similar conclusions hold in the NLTE case (cf. Primas *et al.*, 1999).

3. Results

Our main results can be summarized as follows:

- Among the 4 stars known to be Li- and Be-poor (compared to stars of similar metallicities), 2 of them were found to be substantially B-poor. HD 2665 is a metal-poor giant star for which dilution is able to account for the observed depletion factor of all 3 light elements. For the subgiant HD 3795 additional non-standard mixing is required to explain the low Li, Be, and B abundances.
- Another Li- and Be-poor star of the sample (HD 221377) shows a probable B depletion, which would also require non-standard mixing for its explanation. This probable B-depletion strongly constrains the extra-mixing at disk metallicities. The observed pattern of its LiBeB abundances qualitatively confirms what one should expect according to their burning temperatures: Be is substantially depleted (a factor of ≥ 30) before mixing starts to penetrate the layer where B is preserved.
- A similar scenario to HD 221377 was found for HD 106516, but in this case the interpretation of its light element abundances might be different in the light of the recent finding that the star is a single-lined spectroscopic binary, that has likely undergone mass transfer (Carney, 1998, *private communication*).

– HD 94028 and HD 194598 have been found to have comparable amounts of boron, despite a 0.3 dex difference in Be claimed by Thorburn and Hobbs (1996). Because these 2 stars have also very similar Li abundances, we stress the need for re-observing the Be spectral region of this pair of objects (or at least HD 194598).

– HD 160617 turned out to be the most critical object of the entire sample, being characterized by a lower B content with respect to other stars with similar stellar parameters, *but* normal Li. Such a scenario is not predicted by any destruction and/or mixing mechanism. It suggests that stars very similar in their stellar characteristics can have different amounts of boron (and beryllium), which becomes a strong indication of a spatially different efficiency in producing beryllium and boron in the Galaxy.

References

Duncan, D.K., Primas, F., Rebull, L.M., *et al.*: 1997, *Astrophys. J.* **488**, 338.
Duncan, D.K., Peterson, R.C., Thorburn, J.A., *et al.*: 1998, *Astrophys. J.* **499**, 871.
Kiselman, D. and Carlsson, M.: 1996, *Astron. Astrophys.* **311**, 680.
Kurucz, R.L.: 1993, CD-ROM # 1,13,18.
Primas, F., Duncan, D.K. and Thorburn, J.A.: 1998, *Astrophys. J.*, **506**, L51.
Primas, F., Duncan, D.K., Peterson, R.C. and Thorburn, J.A.: 1999, *Astron. Astrophys.*, **343**, 545.
Thorburn, J.A. and Hobbs, L.M.: 1996, *Astron. J.* **111**, 2106.

Li Be B ENERGETICS AND COSMIC RAY ORIGIN

REUVEN RAMATY
NASA/GSFC, Greenbelt, MD 29771, USA

RICHARD E. LINGENFELTER
UCSD, LaJolla, CA 92093, USA

Abstract. Three different models have been proposed for LiBeB production by cosmic rays: the CRI model in which the cosmic rays are accelerated out of an ISM of solar composition scaled with metallicity; the CRS model in which cosmic rays with composition similar to that of the current epoch cosmic rays are accelerated out of fresh supernova ejecta; and the LECR model in which a distinct low energy component coexists with the postulated cosmic rays of the CRI model. These models are usually distinguished by their predictions concerning the evolution of the Be and B abundances. Here we emphasize the energetics which favor the CRS model. This model is also favored by observations showing that the bulk (80 to 90%) of all supernovae occur in hot, low density superbubbles, where supernova shocks can accelerate the cosmic rays from supernova ejecta enriched matter.

1. Introduction and Overview

In a series of papers (Ramaty *et al.*, 1997; Ramaty, Kozlovsky and Lingenfelter, 1998a; Lingenfelter, Ramaty and Kozlovsky, 1998; Higdon, Lingenfelter and Ramaty, 1998; Ramaty, Lingenfelter and Kozlovsky, 1998b) we developed a LiBeB (Li, Be, B) and cosmic-ray origin paradigm (CRS) in which at all epochs of Galactic evolution the cosmic rays are accelerated out of fresh supernova ejecta, and all of the Be and part of the B are produced by interactions of such cosmic rays with the ambient interstellar medium (ISM). This model differs from the CRI paradigm which posits that the current epoch cosmic rays are accelerated out an ambient medium of solar composition (suggested to be the ISM, Meyer, Drury and Ellison, 1997), and that at all past epochs the composition of the source particles of the cosmic rays was that of the average ISM at that epoch. Hybrid models of LiBeB origin were also suggested (Cassé, Lehoucq and Vangioni-Flam, 1995; Vangioni-Flam *et al.*, 1996; Ramaty, Kozlovsky and Lingenfelter, 1996). In these models (LECR) a Galaxy-wide separate low energy cosmic-ray component, also accelerated out of fresh nucleosynthetic matter, coexists with the CRI cosmic rays and dominates the Be and B production, particularly in the early Galaxy.

The excess of the observed Be abundances in low metallicity stars over the CRI predictions was discussed by Pagel (1991), and as additional Be data accumulated (see Vangioni-Flam *et al.*, 1998), it became clear that the dependence of log(Be/H)

 Astrophysics and Space Science is the original source of publication of this article. It is recommended that this article is cited as: *Astrophysics and Space Science* **265**: 71–76, 1999.
© 1999 *Kluwer Academic Publishers.*

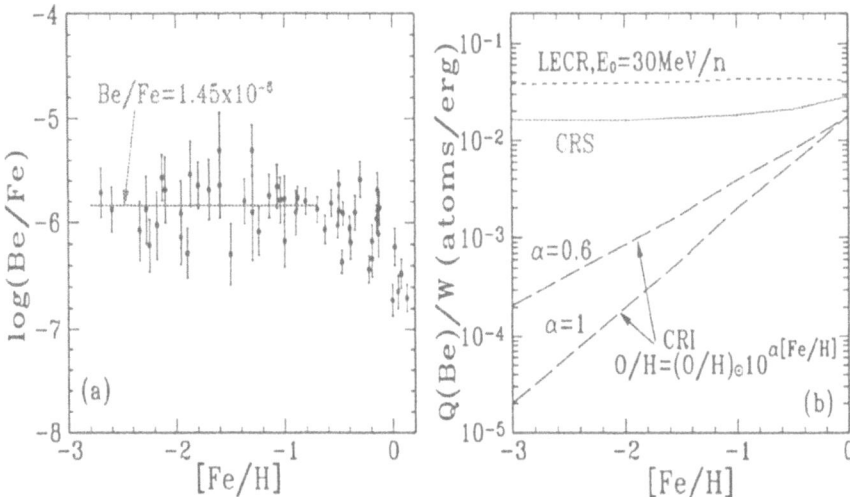

Figure 1. Panel (a): observed Be-Fe abundance ratio as a function of [Fe/H]; data compilation by Vangioni-Flam *et al.* (1998). Panel (b): number of Be atoms produced per erg of cosmic-ray source kinetic energy; the ambient medium is neutral and the cosmic-ray escape length from the Galaxy is 10 g cm^{-2}. If no cosmic-ray escape is allowed (closed Galaxy), Q(Be)/W would increase by a factor of $\simeq 2$, except for the LECR model for which most of the accelerated particles stop in 10 g cm^{-2}. The CRI curve for $\alpha = 0.6$ takes into account the less rapid decrease of O/H at low [Fe/H] (Israelian *et al.*, 1998).

on [Fe/H] is essentially linear, not quadratic as predicted by this model. On the other hand, both the CRS and LECR models predict a linear evolution, consistent with the observations, although Fields and Olive (1998) recently suggested that the CRI model should still be considered as viable, based on their re-analysis of the data including O in low metallicity stars (Israelian, Garcia Lopez and Rebolo, 1998).

We showed previously (see references above) that the energy W_{SN} in cosmic rays per supernova required to produce the observed Be abundance is a powerful diagnostic of the models. This can be seen by considering log(Be/Fe) as a function of [Fe/H] in Figure 1a (Ramaty *et al.*, 1998a; Vangioni-Flam *et al.*, 1998), where for [Fe/H] < -1 the data are consistent with a constant, log(Be/Fe) $= -5.84 \pm 0.05$. Since Fe production in this epoch is dominated by core collapse supernovae (SNII), the constancy of Be/Fe strongly suggests that Be production is also due to SNIIs, which is eminently reasonable since supernova shocks are the most likely accelerators of the cosmic rays (e.g. Axford, 1981). The decrease of Be/Fe for [Fe/H] $\gtrsim -1$ probably results from the additional Fe production in Type Ia supernovae (e.g. Matteucci and Greggio, 1986). The essentially constant Be/Fe, together with information on the average Fe yield per SNII, allows the determination of the Be yield per SNII which, coupled with calculations of LiBeB production by cosmic rays (Ramaty *et al.*, 1997), leads to the energy in cosmic rays per SNII for the various models. We have shown that for the CRS model $W_{SN} \simeq 10^{50}$ erg, a value which is quite consistent with that required to produce the current epoch cosmic rays, based on direct cosmic ray measurements and supernova statistics. We have also shown that the LECR model is energetically less favored, and that the CRI model faces very severe problems of energetics (Ramaty et al., 1998b).

The three models (CRI, CRS, LECR) imply different current epoch cosmic-ray origin scenarios. While the CRI scenario posits acceleration from the ambient ISM (Ellison, Drury and Meyer, 1997), the CRS model implies that the cosmic rays are accelerated out of fresh supernova ejecta. We have shown (Lingenfelter et al., 1998) that the standard arguments against such a cosmic-ray origin (Webber, 1997; Meyer et al., 1998) can be answered, and that the most likely scenario involves the collective acceleration by successive supernova shocks of ejecta-enriched matter in the interiors of superbubbles (Higdon et al., 1998). This scenario is based on observations (summarized by Higdon et al., 1998) showing that most of the Type II and Ibc supernova progenitors (O and B stars) are produced in giant OB associations, that the subsequent supernova explosions produce giant superbubbles that make up the hot, low density phase filling roughly half of the ISM, and that the bulk (80 to 90%) of all supernovae occur in these superbubbles enabling their shocks to mostly accelerate fresh ejecta matter. These results, by themselves and quite apart from the LiBeB origin arguments, favor the CRS over the CRI scenario for cosmic ray origin. Independent arguments that the cosmic rays are accelerated from supernova ejecta were given by Erlykin and Wolfendale (1997).

The LECR scenario was motivated by the reported (Bloemen et al., 1994) detection with COMPTEL/CGRO of C and O nuclear gamma-ray lines from Orion. These gamma rays were attributed to a low energy cosmic-ray component highly enriched in C and O relative to protons and α particles (see Ramaty, 1996 for review and Ramaty et al., 1996 for extensive calculations of LiBeB production by LECRs). It was suggested (Bykov, 1995; Ramaty et al., 1996; Parizot, Cassé and Vangioni-Flam, 1997) that such enriched LECRs might be accelerated out of metal-rich winds of massive stars and the ejecta of supernovae from massive star progenitors which explode within the bubble around the star formation region due to their very short lifetimes. But since the validity of these Orion observations has been questioned by the COMPTEL team (private communication, V. Schönfelder, 1998), the determination of the role of LECRs in LiBeB origin must await future nuclear gamma-ray line observations.

2. The Energetics of Be Production

We calculate the energy in cosmic rays, W, needed to produce a given number of Be atoms, $Q(\text{Be})$ (Ramaty et al., 1997). The calculation assumes a cosmic-ray source generating accelerated particles with given composition and energy spectrum, which then propagate and interact in an ambient medium of given composition. The transport of the particles is characterized by a target thickness, X, measured in g cm^{-2}. Results are shown in Figure 1b for a neutral ambient medium and $X = 10$ g cm^{-2}, the approximate Galactic target thickness for the current epoch cosmic rays. The accelerated particle source energy spectra are taken proportional to $(p^{-2.2}/\beta)e^{-E/E_0}$, where p, $c\beta$ and E are particle momentum, velocity and energy/nucleon, respectively; except for the LECR case, E_0 is ultrarelativistic. For both the CRS and LECR cases, the accelerated particle composition is independent of [Fe/H] and the same as that of the current epoch cosmic rays, except that there are no protons and α-particles for the latter. The ambient medium composition is solar scaled with Fe/H, except that for the CRI case we also consider a slower decrease of the O abundance, O/H=(O/H)$_\odot 10^{0.6[Fe/H]}$, which fits the recent Israelian *et al.* (1998) data. Such a modification of the ambient medium abundances has only a negligible effect on the calculations for the CRS and LECR models. The accelerated particle composition for the CRI case varies with [Fe/H], being equal to the ambient medium abundances increased by factors consistent with ISM shock acceleration theory (Ellison *et al.*, 1997). For both the CRS and LECR cases Q(Be)/W is essentially constant. On the other hand, for the CRI case Q(Be)/W decreases with decreasing [Fe/H], becoming very low at low [Fe/H] in spite of the increase by as much as an order of magnitude due to the incorporation of the enhanced ISM O abundance.

The required energy per SNII is given by

$$W_{SN} = Q_{SN}(Be)/(Q(Be)/W) \; , \tag{1}$$

independent of the details of the employed Galactic chemical evolution model. Thus, for any given cosmic-ray scenario, the main uncertainty is due to Q_{SN}(Be), the Be yield per SNII. As Figure 1a indicates that the Be and Fe yields should be well correlated, we take Q_{SN}(Be) = (Be/Fe)Q_{SN}(Fe), where Q_{SN}(Fe) is the number of Fe nuclei ejected per SNII. The problem then is the determination of this number.

Using the calculations of Woosley and Weaver (1995, WW95), we calculate the ejected Fe mass (mostly from ^{56}Ni) per SNII averaged over the Salpeter IMF for progenitor masses $M_{low} < M < 40$ M$_\odot$. The results are shown by curves A, B and C in Figure 2. These correspond to the WW95 cases A, B and C which give different ^{56}Ni yields for progenitor masses above 30 M$_\odot$, due to different assumed final ejecta kinetic energies, typically 1.2, 2 and 2.5 $\times 10^{51}$ ergs for cases A, B and C, respectively. Also shown in Figure 2 (the TS98 curve) is a similar average based on the results of Tsujimoto and Shigeyama (1998). We see that the ejected mass averaged over the entire 10 to 40 M$_\odot$ range is about 0.1 M$_\odot$ for all four cases. Taking into account the main sequence lifetimes of the SNII progenitors in this mass range, such an average would be appropriate for evolutionary scenarios in which [Fe/H] reaches 10^{-3} in 10 Myrs or more. Since this is quite reasonable (e.g. Ramaty *et al.*, 1998b), we shall use 0.1 M$_\odot$ in our subsequent estimates. However we note that if [Fe/H] = 10^{-3} is reached in just a few Myrs, only SNIIs

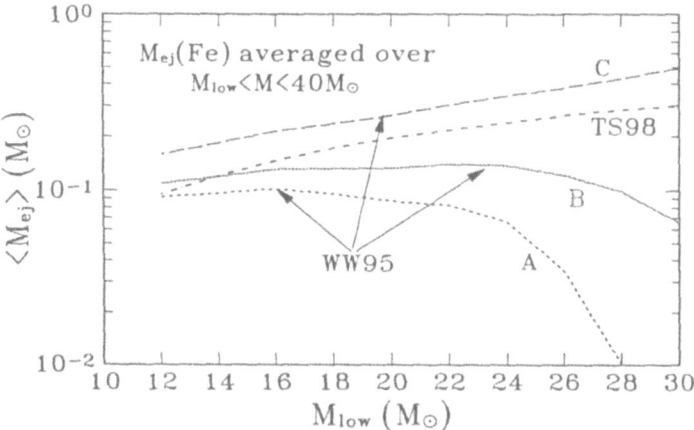

Figure 2. IMF averaged Fe mass ejected mostly as ^{56}Ni per SNII for progenitor masses in the range M_{low} to 40 $M_{\odot}$. Curves A, B and C, corresponding to different final ejecta kinetic energies, are based on the calculations of Woosley and Weaver (1995) for metallicity 10^{-4}. The TS98 curve employs the results of Tsujimoto and Shigeyama (1998).

from progenitors more massive than about 25 $M_{\odot}$ can contribute, allowing ejected Fe masses lower than 0.1 $M_{\odot}$, but only for case A.

Combining the average ejected Fe mass of 0.1 $M_{\odot}$ with the constant Be/Fe (Figure 1), we obtain the required Be yield per SNII, 3×10^{48} atoms. As the recent analysis of Fields and Olive (1998) indicates somewhat lower Be/Fe values at the lowest [Fe/H], we assign a downward uncertainty to this value of about a factor of 3. Using $Q_{SN}(Be) \simeq 3 \times 10^{48}$ in Equation (1), we obtain $W_{SN}(CRS) \simeq 1.5 \times 10^{50}$ erg, which as already mentioned is in excellent agreement with the current epoch value. On the other hand, using the $\alpha = 0.6$ curve in Figure 1b at [Fe/H] $= 10^{-3}$, we obtain $W_{SN}(CRI) \simeq 1.5 \times 10^{52}$ erg, a highly excessive value, even if it were possible to reduce it by the above mentioned factor of 3. For the LECR model of Vangioni-Flam *et al.* (1996), in which only the $> 60\ M_{\odot}$ progenitors contribute, we obtain $W_{SN}(LECR) \simeq 8 \times 10^{50}$ erg. This energy, however, is just that residing in the metals. If protons and α particles accompany the metals with abundances equal to those of the current epoch cosmic rays, $W_{SN}(LECR) \simeq 5 \times 10^{51}$ erg for this model. But this energetic efficiency can be improved by relaxing the $> 60\ M_{\odot}$ progenitor constraint, and by allowing E_0 (defined above) to exceed 30 MeV/nucleon but still be nonrelativistic. Observations of Galaxy-wide nuclear gamma-ray lines are needed to determine the contribution of the LECR component to LiBeB production.

RR wishes to acknowledge Sean Scully for useful discussions.

References

Axford, W.I.: 1981, *Annals N.Y. Acad. Sci.* **375**, 297.

Bloemen, H., *et al.*: 1994, *Astron. Astrophys.* **281**, L5.

Bykov, A.M.: 1995, *Space Sci. Rev.* **74**, 397.

Cassé, M., Lehoucq, R. and Vangioni-Flam, E.: 1995, *Nature* **373**, 318.

Ellison, D.C., Drury, L.O'C. and Meyer, J-P.: 1997, *Astrophys. J.* **487**, 197.

Erlykin, A.D. and Wolfendale, A.W.: 1997, *Astropart. Phys.* **7**, 203.

Fields, B.D. and Olive, K.A.: 1998, *Astrophys. J.*, **516**, 797.

Higdon, H.C., Lingenfelter, R.E. and Ramaty, R.: 1998, *Astrophys. J. Lett.* **509**, L33.

Israelian, G., Garcia Lopez, R.G. and Rebolo, R.: 1998, *Astrophys. J.* **507**, 805.

Lingenfelter, R.E., Ramaty, R. and Kozlovsky, B.: 1998, *Astrophys. J.* **500**, L153.

Matteucci, F. and Greggio, L.: 1986, *Astron. Astrophys.* **154**, 279.

Meyer, J-P., Drury, L. O'C. and Ellison, D.C.: 1997, *Astrophys. J.* **487**, 182.

Pagel, B.E.J.: 1991, *Nature* **354**, 267.

Parizot, E.M.G., Cassé, M. and Vangioni-Flam, E.: 1997, *Astron. Astrophys.* **328**, 107.

Ramaty, R.: 1996, *Astron. Astrophys. Suppl.* **120**, C373.

Ramaty, R., Kozlovsky, B. and Lingenfelter, R.E.: 1996, *Astrophys. J.* **456**, 525.

Ramaty, R., Kozlovsky, B. and Lingenfelter, R.E.: 1998a, *Phys. Today* **51**, No. 4, 30.

Ramaty, R., Kozlovsky, B., Lingenfelter, R.E. and Reeves, H.: 1997, *Astrophys. J.* **488**, 730.

Ramaty, R., Lingenfelter, R.E. and Kozlovsky, B.: 1998b, in N. Prantzos and S. Harrisopulos (eds.), Frontières, Paris, p. 52. *Nuclei In the Cosmos V*.

Tsujimoto, T. and Shigeyama, T.: 1998, *Astrophys. J.* **508**, L151.

Vangioni-Flam, E., Cassé, M., Fields, B.D. and Olive, K.A.: 1996, *Astron. Astrophys.* **468**, 199.

Vangioni-Flam, E., Ramaty, R., Olive, K. and Cassé, M.: 1998, *Astron. Astrophys.* **337**, 714.

Webber, W.R.: 1997, *Space Sci. Rev.* **81**, 107.

Woosley, S.E. and Weaver, T.A.: 1995, *Astrophys. J. Suppl.* **101**, 181.

COSMIC LITHIUM-BERYLLIUM-BORON STORY

ELISABETH VANGIONI-FLAM

Institut d'Astrophysique de Paris, CNRS, 98 bis bd Arago, Paris, France

MICHEL CASSÉ

Service d'Astrophysique, CEA, Orme des Merisiers, 91191 Gif/Yvette France and Institut d'Astrophysique de Paris, Paris, France

Abstract. Light element nucleosynthesis is an important chapter of nuclear astrophysics. Specifically, the rare and fragile light nuclei Lithium, Beryllium and Boron (LiBeB) are not generated in the normal course of stellar nucleosynthesis (except ^{7}Li) and are, in fact, destroyed in stellar interiors. This characteristic is reflected in the low abundance of these simple species. Up to recently, the most plausible interpretation was that Galactic Cosmic Rays (GCR) interact with interstellar CNO to form LiBeB. Other origins have been also identified: primordial and stellar (^{7}Li) and supernova neutrino spallation (^{7}Li and ^{11}B). In contrast, ^{9}Be, ^{10}B and ^{6}Li are pure spallative products. This last isotope presents a special interest since the ^{6}Li/^{7}Li ratio has been measured recently in a few halo stars offering a new constraint on the early galactic evolution of light elements. Optical measurements of the beryllium and boron abundances in halo stars have been achieved by the 10 meter KECK telescope and the Hubble Space Telescope. These observations indicate a quasi linear correlation between Be and B vs Fe, at least at low metallicity, which, at first sight, is contradictory to a dominating GCR origin of the light elements which predicts a quadratic relationship. As a consequence, the theory of the origin and evolution of LiBeB nuclei has to be refined. Aside GCRs, which are accelerated in the general interstellar medium (ISM) and create LiBeB through the break up of CNO by fast protons and alphas, Wolf-Rayet stars (WR) and core collapse supernovae (SNII) grouped in superbubbles could produce copious amounts of light elements via the fragmentation in flight of rapid carbon and oxygen nuclei colliding with H and He in the ISM. In this case, LiBeB would be produced independently of the interstellar medium chemical composition and thus a primary origin is expected. These different processes are discussed in the framework of a galactic evolutionary model. More spectroscopic observations (specifically of O, Fe, Li, Be, B) in halo stars are required for a better understanding of the relative contribution of the various mechanisms. Future tests on the injection and acceleration of nuclei by supernovae and Wolf Rayet relying on gamma-ray line astronomy will be invoked in the perspective of the European INTEGRAL satellite.

1. Introduction

A general trend in nature is that complex nuclei are not proliferating: the abundance of the elements versus the mass number draws a globally decreasing curve. In the whole nuclear realm, LiBeB are exceptional since they are both simple and rare. Typically, in the Solar System, Li/H = 2. 10^{-9}, B/H = 7. 10^{-10}, Be/H = 2.5 10^{-11} (Anders and Grevesse, 1989). Indeed, they are rare because they are fragile and apparently a selection principle at the nuclear level has operated in nature. Due to

Astrophysics and Space Science is the original source of publication of this article. It is recommended that this article is cited as: *Astrophysics and Space Science* **265**: 77–86, 1999.
© 1999 *Kluwer Academic Publishers.*

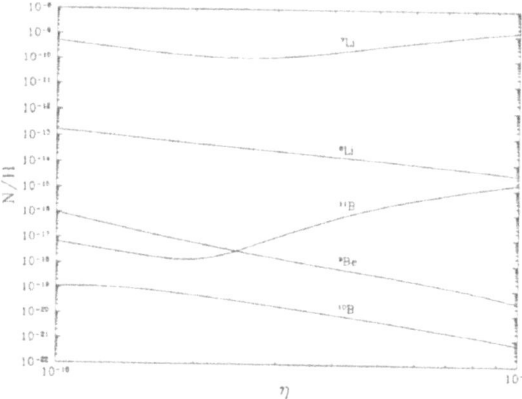

Figure 1. Big Bang nucleosynthesis of Lithium, Beryllium and Boron vs photon over baryon ratio.

the fact that nuclei with mass 5 and 8 are unstable, the Big-Bang nucleosynthesis (BBN) has stopped at A = 7, and primordial thermonuclear fusion has been unable to proceed beyond lithium. The standard BBN is hopelessly ineffective in generating ^{6}Li, ^{9}Be, ^{10}B, ^{11}B (Figure 1). Thus, stars were necessary to pursue the nuclear evolution bridging the gap between ^{4}He and ^{12}C much later, through nuclear fusion.

Stellar nucleosynthesis, quiescent or explosive, forge the whole variety of nuclei from C to U but they destroy LiBeB in the interior of stars, except ^{7}Li which is produced in AGB and novae. The destruction temperature are 2, 2.5, 3.5 and 5.3 millions of degrees for ^{6}Li, ^{7}Li, ^{9}Be, and ^{10}B respectively. Finally, ^{7}Li and ^{11}B could be produced by neutrino spallation in carbon shells of core collapse supernovae (Woosley *et al.*, 1990; Vangioni-Flam *et al.*, 1996); however, this mechanism is particularly uncertain depending strongly on the neutrino energy distribution.

It is clear that another source is necessary to generate at least ^{6}Li, ^{9}Be, ^{10}B and this non thermal mechanism is the break up of heavier species (CNO, mainly) by energetic collisions, also called spallation.

The LiBeB story has been rich and moving. The genesis of LiBeB was so obscure to Burbidge *et al.*(1957) that they called X the process leading to their production. Then came Hubert Reeves and his students. In a seminal work, Meneguzzi, Audouze and Reeves (1971) identified the production process, i.e. the Galactic Cosmic Rays – Interstellar Medium interaction. Exploiting the fast p,α in the GCRs interacting with CNO in the ISM, they were able to make quantitative estimates of the LiBeB production on the basis of cross section measurements notably made in Orsay (Raisbeck and Yiou, 1971, 1975). However, this estimate, based on the local and present observations (LiBeB and CNO abundances, cosmic ray flux and spectrum) was based on an extrapolation over the whole galactic lifetime assuming that all the parameters are constant. This result accounted fairly well for the cumulated light element abundances but obviously not for their evolution which, at that time, was unknown. The pertinence of their idea is illuminated by the simple and

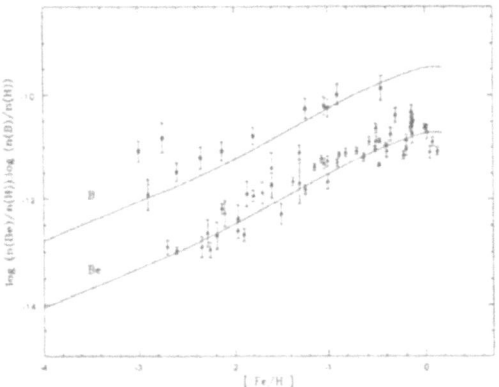

Figure 2. Beryllium and Boron evolution vs [Fe/H].

beautiful fact that the hierarchy of the abundances $^{11}B > {}^{10}B > {}^{9}Be$ is reflected in the cross sections (Read and Viola, 1984). This is another proof that nature follows the rules of nuclear physics. ^{6}Li, ^{9}Be and ^{10}B were nicely explained but problems were encountered with ^{7}Li and ^{11}B. The calculated $^{7}Li/^{6}Li$ ratio was 1.2 against 12.5 in meteorites. Stellar sources of ^{7}Li appeared necessary. The estimated $^{11}B/^{10}B$ ratio was 2.5 instead of 4 in meteorites. An ad-hoc hypothesis drawing on unobservable low energy proton operating through the $^{14}N(p,x)^{11}B$ reaction was advocated (see Reeves, 1994 for a review).

New measurements of Be/H and B/H from KECK and HST, together with [Fe/H] (Rebolo *et al.*, 1988; Gilmore *et al.*, 1992; Duncan *et al.*, 1992; Boesgaard and King, 1993; Ryan *et al.*, 1994; Duncan *et al.*, 1997; García-López *et al.*, 1998) in very low metallicity halo stars came to set strong constraints on the origin and evolution of light isotopes.

The evolution of BeB was suddenly known over about 10 Gyr, taking [Fe/H] as an evolutionary index. A compilation of Be and B data is presented in Figure 2. The most striking point is that log(Be/H) and log(B/H) are both quasi proportional to [Fe/H], at least up to [Fe/H] = −1 and that the B/Be ratio lies in the range 10–30 (Duncan *et al.*, 1997).

This linearity came as a surprise since a quadratic relation was expected from the GCR mechanism. It was a strong indication that the standard GCRs are not the main producers of LiBeB in the early Galaxy. A new mechanism of primary nature was required to reproduce these observations: low energy fast CO nuclei produced and accelerated by massive stars (WR and SN II) fragment on H and He at rest in the ISM. This low energy component (LEC) has the advantage of coproducing Be and B in good agreement with the ratio observed in Pop II stars, (Figure 2 and Vangioni-Flam *et al.*, 1998).

A primary origin, in this language, means a production rate independent of the interstellar metallicity. In this case, the cumulated abundance of a given light isotope L is approximately proportional to Z. At variance, standard GCRs offer a secondary mechanism because it should depend both on the CNO abundance of the ISM at a given time and on the intensity of cosmic ray flux, itself assumed to be proportional to the SN II rate.

Note however, the two discrepant points in the boron diagram at the lowest [Fe/H]. This is mainly due to the huge NLTE correction on the data (Kiselman and Carlsson, 1996) that increases the departure from a straight line. It is important to take a carefull look to this delicate correction.

The Be-Fe and B-Fe correlations taken at face value show a contradiction between theory and observation. But, since oxygen is the main progenitor of BeB, the apparent linear relation between BeB and Fe could be misleading if O were not strictly proportional to Fe (Israelian et al., 1998 and Boesgaard et al., 1999). Thus the pure primary origin of BeB could be questionned (Fields and Olive, 1999). However, the oxygen measurements themselves are confronted to many difficulties (Mac Williams, 1997; Cayrel, Spite and Spite, private communication). On the theoretical side, the situation is not better. The $[\alpha/Fe]$ vs [Fe/H] where $\alpha = Mg$, Si, Ca, S, Ti (Cayrel, 1996; Ryan et al., 1996) show a plateau from about [Fe/H] = -4 to -1. On nucleosynthetic grounds, it would be surprising that oxygen does not follow the Si and Ca trends. Moreover, using the published nucleosynthetic yields (Woosley and Weaver, 1995; Thielemann, Nomoto and Hashimoto, 1996) it is impossible to fit the log(O/H) vs [Fe/H] relation of Israelian et al.(1998) and Boesgaard et al.(1999) since the required oxygen yields are unrealistic. Thus the subject is controversial.

Concerning lithium, a compilation of the data is shown in Lemoine et al.(1997) and Molaro et al.(1997). The Spite plateau extends up to [Fe/H] = -1.3. Beyond, Li/H is strongly increasing until its solar value of 2.10^{-9}.

A stringent constraint to any theory of Li evolution is avoiding to cross the Spite's plateau below [Fe/H] = -1. Accordingly, the Li/Be production ratio should be less than about 100.

Recent measurements of ^6Li have been made successfully in two halo stars, HD84937 and BD+26 3578 at about [Fe/H] = -2.3 (Hobbs and Thorburn, 1997; Cayrel et al., 1999; Smith et al., 1993, 1998), yielding $^6Li/^7Li$ about 0.05. The great interest of ^6Li, besides of being an indicator of stellar destruction (Pinsonneault et al., 1998; Chaboyer, 1998; Cayrel et al., 1999; Vauclair and Charbonnel, 1998) is to represent a pure spallation product as ^9Be.

Lithium-6 has different sources, a secondary one and two primary ones: i) fast (p,α) on CNO at rest (this secondary process related to GCRs should not be efficient in the early galaxy), ii) fast CO on H, He (primary) and iii) specifically $\alpha + \alpha$ at low energy (primary). The non thermal reaction $\alpha + \alpha$ produces almost equal amounts of ^6Li and ^7Li at low energy and the cross section above 100 MeV is specially low (Read and Viola, 1984). The second and third processes are associated to LEC. Consequently, this LEC is specially fertile in lithium isotopes.

TABLE I

Source composition

Element	SS	CRS	W40	GR	SN40(low Z)	SN35(Zo)
H	1200	220	80	2	37	27
He	120	22	25	0	8.8	7.6
C	0.47	0.87	1.6	0.3	0.09	0.08
N	0.13	0.04	–	0.03	–	–
O	1	1	1	1	1	1

Preliminary estimates of the $^6Li/^9Be$ ratio have been performed for few Pop II stars but with a large uncertainty (20–80). This range of values is much higher than the $^6Li/^9Be$ ratio generated by the present GCR (about 6). This could indicate that this ratio is varying all along the galactic evolution (Vangioni-Flam et al., 1999; Fields and Olive, 1999).

To summarize, we can give six observational constraints on LiBeB evolution:

1. Be and B proportional to Fe
2. Li/Be < 100 up to [Fe/H] = −1
3. B/Be = 10–30
4. $^{11}B/^{10}B = 4$ at solar birth
5. $^7Li/^6Li = 12.5$ at solar birth
6. $^6Li/^7Li = 0.05$ and $^6Li/^9Be = 20$ to 80 (to be confirmed) at [Fe/H] about −2.3.

We recall that the observational O–Fe relation is central to the interpretation since specifically the production of Be is related to O. Most of the observers find oxygen proportional to Fe.

2. Basic Physical Parameters of Non Thermal Nucleosynthesis of LiBeB

Four parameters are influential to the spallative production of light elements: the reaction cross sections, the energy spectrum of fast nuclei, the composition of the beam and the composition of the target.

Cross sections are well measured (Read and Viola, 1984) and have been updated recently by Ramaty et al.(1997).

The adopted spectra are of two kinds:

1. GCR: $N(E)dE = kE^{-2.7}$ above a few GeV/n with a flatenning below (e.g. Lemoine et al., 1998).

2. LEC: Shock wave acceleration with a cut at Eo of the form $N(E) dE = kE^{-1.5} exp(-E/Eo) dE$ (Ramaty et al., 1996), propagated in the ISM.

The source composition of GCR is well determined (e.g. Du Vernois, 1996). It is p and α rich (H/O = 200, He/O = 20) contrary to the possible source composition of the LEC. The most obvious contributors to LEC are supernovae, Wolf-Rayet and mass loosing stars (Cassé *et al.*, 1995; Ramaty *et al.*, 1996, 1997, 1998). It is worth noting that in the early galaxy, supernovae play a leading role since at very low metallicity the stellar winds are unsignificant. Table I shows a sample of compositions used by different authors: solar system (SS) for comparison (Ramaty *et al.*, 1996 from Anders and Grevesse, 1989), cosmic ray source (CRS) (Ramaty *et al.*, 1996 from Mewaldt, 1983), wind of massive stars (W40) (Parizot *et al.*, 1997 from Meynet *et al.*, 1994), composition of grain products (GR) (Ramaty *et al.*, 1996 and Lingenfelter *et al.*, 1998), 40 Mo supernova at $Z = 10^{-4}$ Zo (from Woosley and Weaver, 1995), 35 Mo supernova of solar metallicity (Ramaty *et al.*, 1996 from Weaver and Woosley, 1993). The two supernovae, though at different metallicities, (SN40 and SN35), give similar yields due to the fact that metallicity dependent mass loss has not been taken into account. Resulting elemental and isotopic ratios (B/Be, ^{11}B/^{10}B, ^{6}Li/^{9}Be) for different compositions and Eo can be found in Ramaty *et al.*(1996) and Vangioni-Flam *et al.*(1997).

Note that N is unsignificant and that Type II supernovae are O-rich whereas the winds of massive stars are C and He-rich. These abundance differences are important since the highest ^{11}B/^{10}B ratios are produced by C-rich beams through ^{12}C(p,x)^{11}B and the highest ^{6}Li/^{9}Be ratios are produced by He and O rich compositions.

The fourth parameter, i.e., the composition of the target (ISM) varies from the birth of the galaxy up to now. The extensive study of Ramaty *et al.*(1996) and Vangioni-Flam *et al.*(1997) shows that there are only slight differences in the results when the ISM metallicity is varied between 0 (early galaxy) and Zo (now), except perhaps concerning the ^{11}B/^{10}B ratio.

3. LiBeB production mechanisms and galactic evolution

Analyzing all the physical parameters discussed above, two main LiBeB producers emerge, the first one is the standard GCR (spectrum 1) in which fast p,α nuclei interact with CNO in the ISM. This process seems unable to produce sufficient amounts of LiBeB at the level observed in the halo stars (Vangioni-Flam *et al.*, 1996). A recent study (Fields and Olive, 1999) based on the O–Fe relation derived by Israelian *et al.* (1998) and Boesgaard *et al.*(1999) at low metallicity (still controversial) try to fit the observational constraints with a pure standard GCR (secondary production) component, but has problems with the B/Be ratio among other things.

The second one, reverse to the GCR mechanism, invokes fragmentation of CO nuclei in flight by collision with H and He in the ISM. Massive stars are able in principle to furnish freshly synthesized C and O and accelerate them via the shock waves they induce. This mechanism is related to superbubbles (S) through a scenario proposed by Bykov (1995), Parizot (1998) and Vangioni-Flam et al.(1998). Here only the most massive stars (greater than about 60 Mo) contribute due their short lifetime. Originally proposed in relation with the observation of gamma ray line from Orion it remains as a distinctive possibility after the withdrawal of the COMPTEL results on this molecular complex (Bloemen et al., 1999) and the announcement of a possible excess of gamma rays in the 3–7 MeV range from the Vela region. Possible observations in X-rays seem to substantiate this scenario (Tatischeff et al., 1998). The observation or non observation of C, O lines at 4.4 and 6.1 MeV and of the Li-Be feature close to 500 keV by the INTEGRAL satellite (Winkler, 1997) will be the strongest test of the superbubble hypothesis (Parizot et al., 1997).

The acceleration of grain debris in supernovae (scenario G) is another version of a primary mechanism (Lingenfelter et al., 1998; Ramaty, these proceedings). The G model leads to quite different predictions concerning the evolution of Be, B and ^{6}Li only at very low metallicity (Vangioni-Flam et al., 1998, 1999) because all SN II (10–100 Mo) are implied rather than the most massive ones in the S scenario.

Finally, neutrino spallation (Woosley et al., 1990; Woosley and Weaver, 1995; Vangioni-Flam et al., 1996) is helpful to increase the ^{11}B/^{10}B ratio up to the value observed in meteorites. It is also a primary process since it implies the break up of ^{12}C within supernovae and not in the ISM. However it cannot be the unique mechanism to produce light elements since it does not produce ^{9}Be. Moreover, it has been shown that its contribution is only marginal (Vangioni-Flam et al., 1996).

These different mechanisms are included in a galactic evolutionary model (Van-gioni-Flam et al., 1996, 1998, 1999) to follow the whole evolution of each isotope.

Concerning beryllium and boron, in this context, the main results are the following: the quasi-linearity (Be-B vs Fe) is easily reproduced (Figure 2). Standard GCR contribute no more than about 30 per cent to Solar System values. The B/Be ratio is in the range 10–30 as observed. The value 30 leaves enough room for neutrino spallation to reach ^{11}B/^{10}B = 4 at solar birth.

The ^{6}Li/H ratio can be explained in the framework of the same superbubble model (Vangioni-Flam et al., 1999; Cayrel et al., 1999) this without piercing the Spite plateau (Figure 3). In this figure, showing the evolution of ^{6}Li/H vs [Fe/H], it can be seen that GCR is overwhelmed by LEC. The decrease of the ^{6}Li/^{9}Be ratio could be explained in terms of the variation of the composition of superbubbles in the course of the galactic evolution, being O rich at start due to SNII and becoming more and more C rich due to the increasing contribution of mass loosing stars (Vangioni Flam et al., 1999). Moreover, the evolutionary curve of ^{6}Li crosses the halo observations (Figure 3) meaning that ^{6}Li is almost essentially intact in the envelope of stars in which it is measured. ^{7}Li in turn, more tightly bound than ^{6}Li, is even less destroyed, thus the mean value of the Spite plateau reflects nicely the Big Bang ^{7}Li abundance. This reinforce the use of ^{7}Li as a cosmological baryometer.

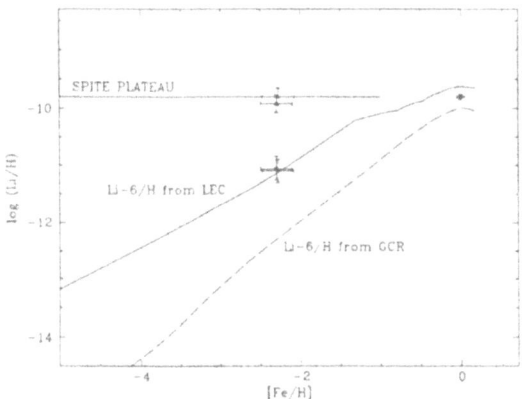

Figure 3. Evolution of Lithium vs [Fe/H].

4. Conclusion

Recent LiBeB observations indicate that a primary component is probably at work in the early Galaxy, presumably related to core collapse supernovae. Promising scenarios have been presented implying respectively the acceleration of freshly synthesized C and O in superbubbles and/or grains debris around supernovae (see also Ramaty in these proceedings).

New ^{6}Li observations put strong constraints on the composition and spectrum of an early population of fast particles. The superbubble model is also able to reproduce the ^{6}Li observation, until the local meteoritic value. A low energy component originally O rich and becoming progressively C rich due to the strengthening of stellar winds at increasing metallicities is required to explain the high ^{6}Li/^{9}Be observed in a few halo stars with respect to the one measured in meteorites. But a definitive conclusion should wait confirmation.

Essentially no destruction of ^{6}Li and ^{7}Li is implied by the evolutionary curve of ^{6}Li. As a consequence, ^{7}Li is a good baryonic density indicator for cosmology. The needs for the future are the following:

On the theoretical side it would be necessary:

i) to check NLTE corrections on B abundances since two bothering points remain at very low Z.

ii) to develop and refine SN II models, specially at very low Z and high mass, M > 60 Mo.

On the observational side, it would be desirable to get measurements of ^{6}Li, ^{7}Li, ^{9}Be, B, O, Fe in the same halo stars and to get ^{11}B/^{10}B ratios in various stars.

Finally, the observation of C,O and Li-Be gamma-ray lines are important objectives of the INTEGRAL satellite which will open up an European era in gamma-ray astronomy.

Acknowledgements

We have been very much honored to participate to the celebration of the scientific work of Giusa and Roger Cayrel, who are an example for every researcher in Astronomy. Moreover, we thank warmly Monique and François Spite for the excellent organization of the meeting. This work was supported in part by the PICS 319, CNRS.

References

Anders, E. and Grevesse, N.: 1989, *Geochim. Acta* **35**, 197.

Bloemen, H.: 1999, in: *The Extreme Universe*, Third INTEGRAL Worshorp, Taormina ESA proceedings, to appear.

Boesgaard, A.M. and King, J.R.: 1993, *Astron. J.* **106**, 2309.

Boesgaard, A., *et al.*: 1999, *Astrophys. J.S.* **17**, 492.

Burbidge, E.M., Burbidge, G.R., Fowler, W.A. and Hoyle, F.: 1957, *Rev. Mod. Phys.* **29**, 547.

Bykov, A.M.: 1995, *Space Sci. Rev.* **74**, 397.

Cassé, M., Lehoucq, R. and Vangioni-Flam, E.: 1995, *Nature* **373**, 318.

Cayrel, R.: 1996, *Astron. Astrophys. Rev.* **7**, 217.

Cayrel, R., Spite, M., Spite, F., Vangioni-Flam, E., Cassé, M. and Audouze, J.: 1999, *Astron. Astrophys.* **343**, 923.

Cayrel, R., Lebreton, Y. and Morel, P.: 1999, this proceedings.

Chaboyer, B.: 1999, astroph/9803106.

Duncan, D.K., Lambert, D. and Lemke, M.: 1992, *Astrophys. J.* **401**, 584.

Duncan, D.K., *et al.*: 1997, *Astrophys. J.* **488**, 333.

Du Vernois, M.A.: 1996, in: *Cosmic Abundances*, Edts ASP Conferences series, Vol. 99, 385.

Fields, B. and Olive, K.: 1999, *Astrophys. J.*, **516**, 801.

García-López, R.J., *et al.*: 1998, *Astrophys. J.*, **500**, 241.

Gilmore, G., Gustafsson, B., Edvardsson, B. and Nissen, P.E.: 1992, *Nature* **357**, 379.

Hobbs, L.M. and Thorburn, J.A.: 1997, *Astrophys. J.* **491**, 772.

Israelian, G., García-López, R.J. and Rebolo, R.: 1998, *Astrophys. J.*, **507**, 805.

Kiselman, D. and Carlsson, M.: 1996, *Astron. Astrophys.* **311**, 681.

Lemoine, M., Schramm, D.N., Truran, J.W. and Copi, C.J.: 1997, *Astrophys. J.* **478**, 554.

Lemoine, M., Vangioni-Flam, E. and Cassé, M.: 1998, *Astrophys. J.* **499**, 735.

Lingenfelter, R.E., Ramaty, R. and Kozlovsky, B.: 1998, *Astrophys. J. Lett.* **500**, 153.

Mac Williams, A.: 1997, *Annu. Rev. Astron. Astrophys.* **55**, 503.

Meneguzzi, M., Audouze, J. and Reeves, H.: 1971, *Astron. Astrophys.* **15**, 337.

Mewaldt, R.: 1983, *Rev. Geophys. Space Phys.* **21**, 295.

Meynet, G., *et al.*: 1994, *Astron. Astrophys. Suppl.* **103**, 97.

Molaro, P., Bonifacio, P., Castelli, F. and Pasquini, L.: 1997, *Astron. Astrophys.* **319**, 593.

Parizot, E., Cassé, M. and Vangioni-Flam, E.: 1997, *Astron. Astrophys.* **328**, 107.

Parizot, E.: 1998, *Astron. Astrophys.*, **331**, 726.

Pinsonneault, M.H., *et al.*: 1998, *Astrophys. J.*, astroph/9803073.

Raisbeck, G. and Yiou, F.: 1971, *Phys. Rev. A* **4**, 1848.

Raisbeck, G. and Yiou, F.: 1975, in: B.S.P. Shenand and M. Merker (eds.), *Spallation Nuclear Reactions and their Applications*, Dordrecht, Reidel, p. 83.

Ramaty, R., Kozlovsky, B. and Lingenfelter, R.E.: 1996, *Astrophys. J.* **456**, 525.

Ramaty, R. Kozlovsky, B, Lingenfelter, R.E. and Reeves, H.: 1997, *Astrophys. J.* **488**, 730.

Ramaty, R., Kozlovsky, B. and Lingenfelter, R.E.: 1998, *Physics Today* **51**, No. 4, 30.

Read, S.M. and Viola, V.E.: 1984, *Atomic Data and Nuclear Data Tables* **31**, 359.

Reeves, H.: 1994, *Rev. Mod. Phys.* **66**, No. 1, 193.

Rebolo, R., Molaro, P. and Beckman, J.E.: 1988, *Astron. Astrophys.* **192**, 192.

Ryan, S., Norris, I., Bessel, M. and Deliyannis, C.: 1994, *Astrophys. J.* **388**, 184.

Ryan, S.G., Norris, J.E. and Beers, T.C.: 1996, *Astrophys. J.* **471**, 254.

Smith, V.V., Lambert, D.L. and Nissen, P.E.: 1993, *Astrophys. J.* **408**, 262.

Smith, V.V., Lambert, D.L. and Nissen, P.E.: 1998, *Astrophys. J.* **506**, 405.

Tatischeff, V., Ramaty, R. and Kozlovsky, B.: 1998, *Astrophys. J.*, **504**, 874.

Thielemann, F.K., Nomoto, K. and Hashimoto, M.: 1996, *Astrophys. J.* **460**, 108.

Vangioni-Flam, E., Cassé, M., Olive, K. and Fields, B.D.: 1996, *Astrophys. J.* **468**, 199.

Vangioni-Flam, E., Cassé, M. and Ramaty, R.: 1997, in: *Proceedings of the 2nd INTEGRAL Workshop, 'The Transparent Universe'*, Saint Malo, France, ESA, SP.382 pp. 123.

Vangioni-Flam, E., Ramaty, R., Olive, K. and Cassé, M.: 1998, *Astron. Astrophys.* **337**, 714.

Vangioni-Flam, E., Cassé, M., Cayrel, R., Audouze, J., Spite, M., Spite, F.: 1999, *New Astronomy* **123**.

Vauclair, S. and Charbonnel, C.: 1998, *Astron. Astrophys., Astrophys. J.* **502**, 372.

Weaver, T.A. and Woosley, S.E.: 1993, *Phys. Repts* **227**, 65.

Winkler, C., 1997, in: *The Transparent Universe*, 2nd INTEGRAL Workshop, Saint Malo, Edts, ESA, SP-382, pp. 573.

Woosley, S.E., Hartmann, D., Hoffman, R. and Haxton, W.: 1990, *Astrophys. J.* **356**, 272.

Woosley, S.E. and Weaver, T.A.: 1995, *Astrophys. J. Suppl.* **101**, 181.

SURVIVAL OF ^{6}Li, AND ^{7}Li, IN METAL-POOR STARS

R. CAYREL

Observatoire de Paris, 61 av. de l'Observatoire F-75014 Paris, France

Y. LEBRETON

Observatoire de Paris, Section de Meudon, F-92195 Meudon Cedex, France

P. MOREL

Observatoire de la Côte d'Azur, BP 229, F-06304 Nice Cedex 4, France

Abstract. The relationship between the depletions of ^{6}Li and ^{7}Li is studied for two models of lithium burning, below the convective zone. The parameters of the depletion models are submitted to the constraint that the slope of the ^{7}Li theoretical depletion curve agrees with the slope of the observed depletion curve, for cool subdwarfs. Other less restrictive models are also considered.

In all cases, a ^{6}Li depletion less than 0.5 dex implies a ^{7}Li depletion less than 0.1 dex. With the constraint on the slope of the ^{7}Li curve, the depletion of ^{7}Li for the same depletion of ^{6}Li is below 0.05 dex.

The still unsolved problem for the true ^{7}Li abundance in subdwarfs is the possible influence of temperature inhomogeneities, raised by Kurucz, subsequently shown to be small in the solar case, but not yet computed with the inclusion of departure from LTE for metal-poor stars.

1. Introduction

Lithium is one of the great fossil records available in our local stellar environment. It was discovered by the Spites (1982) in subdwarfs.

Apparently, the ^{7}Li/H ratio, has survived all the way from 100 seconds after the Big Bang, to the birth of the first stars, and, more or less, subsequently in the atmospheres of these stars, in which we observe it, now, 13 Gyr later. The value of this ratio, based on the analysis of 40 metal-poor stars (Bonifacio and Molaro, 1997), is about 1.6×10^{-10} (by number, or 1.1×10^{-9} by mass), close to the Standard Big Bang Nucleosynthesis prediction (Schramm and Turner, 1998), for the η ratio (baryons/photons) set by the other products of the SBBN.

The debated question is that of the survival of ^{7}Li in the atmospheres of the stars, during the 12 or 13 Gyr they have spent on the main sequence. Two processes can alter the initial lithium content: gravitional settlling at the bottom of the convective zone (CZ hereafter) (Vauclair and Charbonnel, 1998), and poster at this meeting), and mixing of the matter in the CZ with deeper layers, in which the temperature allows nuclear burning of ^{7}Li, (Schatzman, 1977; Pinsonneault *et al.*, 1992; Charbonnel *et al.*, 1992; Montalban and Schatzman, 1996). Furthermore, Kurucz

 Astrophysics and Space Science is the original source of publication of this article. It is recommended that this article is cited as: *Astrophysics and Space Science* **265:** 87–93, 1999.
© 1999 *Kluwer Academic Publishers.*

TABLE I

Depletion of ^{6}Li and ^{7}Li, in dex, during the PMS and MS phases (values between parentheses are extrapolated and less reliable)

| M | [Fe/H] | ^{6}Li (PMS) | ^{6}Li (MS) | ^{6}Li (tot) | ^{7}Li (PMS) | ^{7}Li (MS) | ^{7}Li (tot) |
$M_\odot$	dex	dex	dex	dex	dex	dex	dex
0.85	−1.5	0.20	0.48	(.74)	0.0	0.06	0.06
0.85	−2.0	0.11	0.01	0.12	0.0	0.00	0.00
0.80	−1.5	0.411	1.06	1.47	0.01	0.37	0.38
0.80	−2.0	0.32	.52	0.84	0.00	0.06	0.06
0.75	−1.5	0.92	(2.3)	(3.22)	0.01	0.87	0.88
0.75	−2.0	0.81	1.23	2.04	0.01	0.56	0.57

(1995) has suggested that the abundance of lithium determined with plane-parallel atmospheres might be strongly in error, due to effects of temperature inhomogeneities in the convective zone, closer to the surface in metal-poor stars than in solar composition stars.

We shall discuss here only the main process, the burning process by mixing of the CZ with deeper layers, clearly acting on metal-poor stars of effective temperature below 5700 K (see Figure 1), and responsible of a depletion of ^{7}Li by a factor of 200 in the Sun.

As ^{6}Li has been also observed in two turnoff metal-poor stars, HD 84937 and HD 338529, and burns at a lower temperature than ^{7}Li, it is interesting to see what constraints the observation of ^{6}Li brings on the maximum depletion of ^{7}Li.

2. Burning of ^{6}Li and ^{7}Li During Pre-Main-Sequence

The burning of these elements during pre-main-sequence has been computed using the CESAM code (Morel, 1997). Opacities are from Rogers and Iglesias (1992), Rogers et al. (1996), Alexander and Fergusson (1994), or Kurucz (1995). The EFF equation of state is used (Eggelton et al., 1973). The results are given in columns 3 and 6 of Table I.

It must be noted that there is in all cases some depletion of ^{6}Li during the PMS, and that this depletion reaches 0.4 dex at metallicity −1.5 for a mass of 0.80 $M_\odot$. PMS depletion of ^{7}Li is negligible in the range of masses considered here. Depletions are counted as positive for a reduction in lithium abundance (e.g. a depletion of 1.0 dex means that the present abundance is one tenth of the initial abundance).

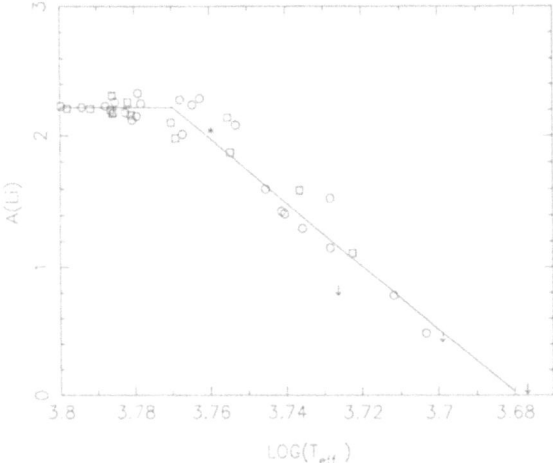

Figure 1. Abundance of Li as a function of log T_{eff}. The sources are Bonifacio and Molaro (1997) (squares) and Ryan and Deliyannis (1998) (circles). The abundances are given in the usual logarithmic scale, in which the abundance of hydrogen is 12.0. These abundances are the total abundance of Li, in practice, they are abundances of ^{7}Li, ^{6}Li representing at the most 5 per cent of the abundance of ^{7}Li, and most of the time much less.

3. Burning of ^{6}Li and ^{7}Li During MS Lifetime and After

3.1. PREDICTIONS BASED ON A SIMPLE CIRCULATION MODEL

We shall use first a very simple model of destruction based on the fact that the rate of destruction of ^{6}Li and ^{7}Li varies as a function of depth almost as a step function. In the convective zone itself, and just below, the destruction rate is so small that no depletion of these elements occur in the time available (12 to 14 Gyr). On a short length, of about one hundredth of the radius of the star, the rate increases to such a large value that the element is burnt everywhere below, in a time much smaller than the age of the object. The depletion in the well-mixed convective zone depends only upon its exchange of matter with the zone in which the element is totally burnt. The corresponding level is of course deeper for ^{7}Li than for ^{6}Li. For example, for a star of mass 0.80 $M_\odot$ and [Fe/H] = −2.0, the transition occurs for ^{7}Li at a depth of 0.155 below the bottom of the CZ, in units of the stellar radius, but only at a depth of 0.10 for ^{6}Li. Under the assumption that the exchange of matter is driven by convective plumes penetrating the radiative zone, as suggested by hydrodynamical simulations (Stein and Nordlund, 1989; Rieutord and Zahn, 1995), the critical parameter for the exchange of matter between the CZ and the burning zone is of course the separation between the two levels.

On Figure 1 we have plotted depletions of ^{7}Li according to two recents papers (Bonifacio and Molaro, 1997; Ryan and Deliyannis, 1998), as a function of log(T_{eff}). We have also computed the distance d of the top of the burning zone

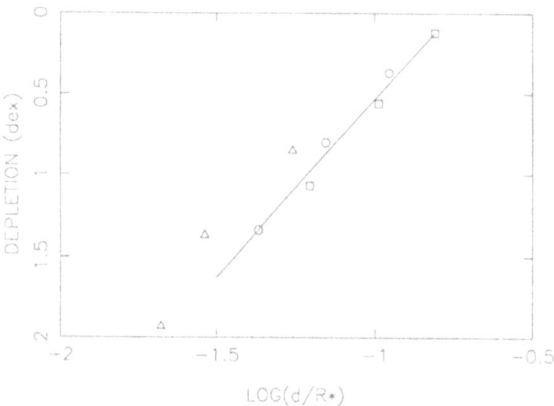

Figure 2. Depletion of ^{7}Li as a function of the distance d between the bottom of the convective zone and the top of the burning region. The unit of length is the radius of the star. The symbols are: squares: metallicity [Fe/H] = −2.0; circles: [Fe/H] = −1.5; triangles: [Fe/H] = −1.0. The straight line is a global fit for the two lowest metallicities.

to the bottom of the convective zone, on the main sequence, with the CESAM code, for a grid of models, with masses 0.70[0.05]0.85 $M_\odot$ (the value between brackets is the step), metallicities [Fe/H] = −2.0[0.5]−1.0 an helium abundance near $Y_p = 0.23$, and an enhancement of α-elements by 0.4 dex. As T_{eff} is a result of the computation, this allows to redraw the depletion of ^{7}Li as a function of d. In this transformation we have allowed a shift of 0.01 between the theoretical $\log(T_{eff})$ and the observed $\log(T_{eff})$, as shown necessary in Cayrel *et al.* (1997).

The result, shown in Figure 2, confirms the idea that d is a key parameter for the control of depletion by burning. The two straight lines for the two metallicities [Fe/H] = −1.5 and −2.0 can be approximated by the single relation:

$$D(^7\text{Li}) = 0., \text{ for } d > 0.16R_* \tag{1}$$

$$D(^7\text{Li}) = -2.7(\log(d/R_*) + 0.83) \text{ for } d < 0.16R_* \tag{2}$$

The physical interpretation of this is that if the rate of exchange of matter between the convective zone and the burning zone is $a_{7\text{Li}}$, in fraction of mass of the CZ per Gyr, the depletion factor of ^{7}Li after a time t will be:

$$R = \exp(-a_{7\text{Li}}t)$$

as the matter returning from the burning zone has a zero-content in ^{7}Li. In decimal exponent the depletion becomes $D = 0.4343 * a_{7\text{Li}}t$ or, for an age of 13 Gyr: $D = 5.65 \times a_{7\text{Li}}$. For a depletion of 0.5 dex this gives $a_{7\text{Li}} = 0.088$, or a circulation rate a little below one tenth of the mass of the CZ per Gyr. Because the circulation rate as a function of depth is independent of the element considered, formulae (1) and (2) are applicable as well to ^{6}Li, at the only condition that d refers to the

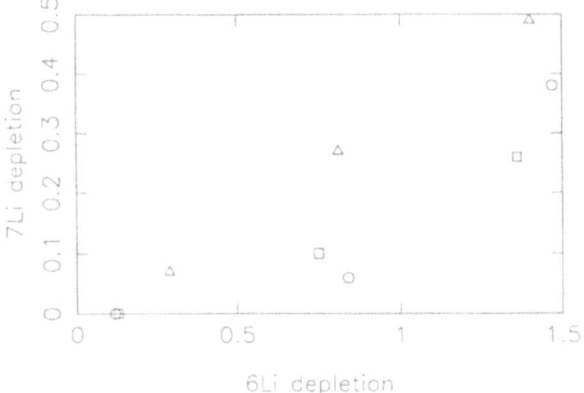

Figure 3. Relationship between the respective depletions of ^{6}Li and ^{7}Li. The open circles are the results of the simple circulation model of Section 3.1. The squares are the results obtained with a mixing by random-walk, with a diffusion coefficient adjusted to reproduce the depletion curve of ^{7}Li in Figure 1. The triangles are for a model with a diffusion coefficient independent of depth.

depth corresponding to the top of the burning zone of ^{6}Li and not to the top of the burning zone of ^{7}Li. This is how we have computed the predicted values for ^{6}Li, actually derived from the observed depletions of ^{7}Li for the same distance between the bottom of the CZ and the top of the corresponding burning zone. The depletions of ^{6}Li and ^{7}Li obtained are plotted in Figure 3, as open circles. Up to a depletion of 1.0 for ^{6}Li the depletion of ^{7}Li remains below 0.1 dex. The main cause of this is the strong depletion of ^{6}Li in the PMS phase.

3.2. PREDICTIONS BASED ON A RANDOM-MOTION MIXING MODEL

The first model of mixing was proposed by E. Schatzman (1977), and is generally described as mixing by 'turbulence', or 'diffusion'. We shall adopt here such a model, phenomenogically described by random-motions characterized locally by the value of a diffusion coefficient, function of depth. If the source of the stirring is convection, one expects the diffusion coefficient to decrease strongly with depth below the bottom of the CZ (see Montalban and Schatzman, 1996). Actually we can adjust the value at some arbitrary depth and the scale-height of an assumed exponential decline of the diffusion coefficient in such a way that the abundance of ^{7}Li follows the behaviour seen in Figure 1. In Figure 3 the squares show the relationship between the two depletions for this model. The conclusions are the same as for the first model. An extremely conservative bound for the depletion of ^{7}Li at a given depletion of ^{6}Li is given by a model for which the diffusion coefficient is constant with depth (triangles in Figure 3). Even in that case the depletion of ^{7}Li is les than 0.1 up to depletions of ^{6}Li of about 0.4. For this last model the constraint of reproducing the slope of the depletion curve of Figure 1 is not fullfilled.

4. Conclusions

Models of destruction by burning, reproducing the slope of the depletion curve of ^{7}Li, in unevolved metal-poor stars cooler than 5800 K, show a fairly robust ratio of the depletions of ^{6}Li and ^{7}Li in the range of interest. The depletion of ^{7}Li remains below 0.1 dex, as soon as the depletion of ^{6}Li is below the range 0.5–1.0.

This result is valid both for a random motion type of mixing or for a stationary circulation type of mixing. A large fraction of ^{6}Li burning occurs during the pre-main sequence.

The burning of ^{6}Li leads to an important depletion as soon as one of these two conditions is met: either a mass below 0.80 $M_\odot$, or [Fe/H] > -1.5. This helps to understand why ^{6}Li has been found so far only in TO or subgiants more metal-poor than -1.5.

Although there are some indications that other effects, as gravitational settling, and temperature inhomogeneities in convective elements are not very important for ^{7}Li, a global and consistent treatment including microscopic diffusion, mixing, mass loss and temperature inhomogeneities is a must for the coming years.

Acknowledgements

Part of this work has been performed using the computing facilities provided by the OCA program 'Simulations Interactives et Visualisation en Astronomie et Méca-nique (SIVAM)'.

References

Alexander, D.R. and Ferguson, J.W.: 1994, *Astrophys. J.* **437**, 879.

Bonifacio, P. and Molaro, P.: 1997, *Mon. Not. R. Astron. Soc.* **285**, 847.

Cayrel, R., Lebreton, Y., Perrin, M.-N. and Turon, C.: 1997 in: *Hipparcos: 'Venice '97'*, ESA-SP 402, pp. 219.

Charbonnel, C., Vauclair, S. and Zahn, P.: 1992, *Astron. Astrophys.* **255**, 191.

Eggleton, P.P., Faulkner, J. and Flannery, B.P.: 1973, *Astron. Astrophys.* **23**, 325.

Kurucz, R.L.: 1991, in: L. Crivellari, I. Hubeny and D.G. Hummer (eds.), *Stellar Atmospheres: Beyond Classical Models*, NATO ASI Series, Kluwer, Dordrecht, p. 440.

Kurucz, R.L.: 1995, *Astrophys. J.* **452**, 102.

Montalban, J. and Schatzman, E.: 1996, *Astron. Astrophys.* **305**, 513.

Morel, P.: 1997, *Astron. Astrophys. Suppl.* **124**, 597.

Pinsonneault, M.H., Delyannis, C.P. and Demarque, P.: 1992, *Astrophys. J. Suppl.* **78**, 179.

Rieutord, M. and Zahn, J.P.: 1995, *Astron. Astrophys.* **296**, 127.

Rogers, F.J. and Iglesiais, C.A.: 1992, *Astrophys. J. Suppl.* **79**, 507.

Rogers, F.J., Swenson, F.J. and Iglesias, C.A.: 1996, *Astrophys. J.* **456**, 902.

Ryan, S.G. and Deliyannis, C.P.: 1998, *Astrophys. J.* **500**, 398.

Schatzman, E.: 1997, *Astron. Astrophys.* **56**, 211.

Schramm, D.M. and Turner, M.S.: 1998, *Rev. Mod. Phys.* **70**, 303.
Spite, F. and Spite, M.: 1982, *Astron. Astrophys.* **115**, 357.
Stein, R.F. and Nordlund, A.: 1989, *Astrophys. J.* **342**, L95.
Vauclair, S. and Charbonnel, C.: 1998, *Astrophys. J.* **502**, 372.

THE LITHIUM PLATEAU ENIGMA

C. CHARBONNEL and S. VAUCLAIR

Laboratoire d'Astrophysique de Toulouse – OMP, 14, av. E. Belin, 31400 Toulouse, France

1. The Lithium Plateau Enigma May Be Summarized As Follows

Main-sequence Pop II stars stars with effective temperatures between 5500 K and 6500 K show a remarquably constant value for the lithium abundance ($\Longleftrightarrow$ the lithium plateau). Furthermore the dispersion around this value is very small (Bonifacio and Molaro, 1997). However, from the observations of Pop I stars, there is strong evidence that the lithium abundance highly varies from star to star. The variation with T_{eff} and age clearly appears in the galactic cluster data.

From the theory and modelisation of stellar internal structure, lithium is expected to vary from star to star due to both nuclear destruction and/or element settling. These effects account well for the observations of Pop I stars, although more quantitative comparisons between observational and theoretical results are still needed. Helioseimology now provides a spectacular confirmation of the precision we have attained in the modelisation of the solar internal structure, including element settling (Richard *et al.*, 1996).

So why is the lithium abundance constant in the so-called lithium plateau while all predictions suggest that it should vary from star to star? Is there an 'abundance attractor' which would work in Pop II stars but not in Pop I stars?

2. Hints For a Solution

Several models have been proposed to account for the lithium plateau. The 'old standard model' in which no settling was introduced is excluded as unphysical. In the 'mass loss model' (Vauclair and Charbonnel, 1995), a stellar wind is supposed to prevent element settling during the stellar lifetime. In the 'rotation model' (Pinsonneault *et al.*, 1990; Charbonnel and Vauclair, 1999), the Pop II stars are supposed to have been mildly mixed below the convection zone due to rotation-induced shears. In any case the solution seems somewhat 'ad hoc' as it assumes that some parameters are fixed in all stars (the mass loss rate or the rotation rate) for the lithium value to remain constant along the plateau. It would be much more

Astrophysics and Space Science is the original source of publication of this article. It is recommended that this article is cited as: *Astrophysics and Space Science* **265**: 95–96, 1999.
© 1999 *Kluwer Academic Publishers.*

satisfying to find a 'lithium abundance attractor' which would remain stable in halo stars while fundamental parameters (M_*, T_{eff}, [Fe/H]) vary.

3. Lithium Abundance Attractor

Such an attractor may exist (Vauclair and Charbonnel, 1998): Indeed, the lithium profiles inside the Pop II standard stellar models including element segregation present a maximum value, Li_{max}, which remains constant all over the range in T_{eff} and metallicity of the plateau while the surface value is expected to change.

This result leads to the idea that the observed lithium abundances may be related to Li_{max}. Since the observations of the lithium in the plateau reveal a very small dispersion around a stable value, this value must indeed lie close to Li_{max}. In this case the derived primordial value is 2.35. When compared to BBN computations (Copi *et al.*, 1995) this result leads to a baryonic number between 1.2 and 5 10^{-10}. For H = 50, this value corresponds to $0.018 < \Omega_b < 0.075$. The macroscopic process which would act in the way of moving this lithium up to the surface is still to be found.

References

Bonifacio, P. and Molaro, P.: 1997, *Mon. Not. R. Astron. Soc.* **285**, 847.

Charbonnel, C. and Vauclair, S.: 1999, in preparation.

Copi, C.J., Schramm, D.N. and Turner, M.S.: 1995, *Science* **267**, 192.

Pinsonneault, M.H., Kawaler, S.D. and Demarque, P.: 1990, *Astrophys. J. Suppl.* **74**, 501.

Richard, O., Vauclair, S., Charbonnel, C. and Dziembowski, W.A.: 1996, *Astron. Astrophys.* **312**, 1000.

Vauclair, S. and Charbonnel, C.: 1995, *Astron. Astrophys.* **295**, 715.

Vauclair, S. and Charbonnel, C.: 1998, *Astrophys. J.* **502**, 372.

EFFECT OF GRAVITY WAVES ON TRANSPORT PROCESS IN STELLAR RADIATIVE REGION

E. SCHATZMAN
Observatoire de Paris-Meudon, F92195 Meudon Cedex, France

Abstract. The aim of this work is to present a transport process which is likely to have a great importance for the internal constitution of the stars.

In order to set the problem, we first give a short presentation of the physical properties of the Sun and stars, described usually under the names of *'Standard Solar Model'* or *'Standard Stellar Models'* (SSM). Next we show that an important question about SSM is that they do not explain the age dependance of lithium deficiency of stars of known age: stars of galactic clusters and the Sun. It has been suggested a long time ago to assume the presence of a macrosocpic diffusion process in the radiative zone, below the surface convective zone of solar like stars. It is then possible for the lithium present in the convective zone to be carried to the thermonuclear burning level below the convective zone. The first assumption was that differential rotation generates turbulence and therefore that a turbulent diffusion process takes place. However, this model predicts a lithium abundance which is strongly rotation dependant, contrary to the observations. Furthermore, the diffusion coefficient being large all over the radiative zone, it prevents the possibility of gravitational separation by diffusion and consequently leads to an impossibility of explaining the difference of helium abundance between the surface and the center of the Sun. The consequence is obviously that we need to take into account another physical process.

Stars having a mass $M < 1.3 \, M_\odot$ have a convective zone which begins close to the stellar surface and extends down to a depth which is an appreciable fraction of stellar radius. In the convective zone, strong stochastic motions take care, at least partially, of heat transfer. These motions do not vanish at the lower boundary and generate internal waves into the radiative zone. These random internal waves are at the origin of a diffusion process which can be considered as responsible of the diffusive transport of lithium down to the lithium burning level. This is certainly not the only physical process responsible of lithium deficiency in main sequence stars, but its properties open the way to a completely consistent analysis of lithium deficiency.

The model of generation of gravity waves is based on a model of heat transport in the convective zone by diving plumes. The horizontal component of the turbulent motion at the boundary of the convective zone is supposed to generate the horizontal motion of internal waves. The result is a large horizontal component of the diffusion coefficient, which produces in a short time an horizontal uniform chemical composition. It is known that gravity waves, in the absence of any dissipative process, cannot generate vertical mixing. Therefore, the vertical component of the diffusion coefficient is entirely dependant of radiative damping. It decreases quickly in the radiative zone, but is large enough to be responsible of lithium burning.

Due to the radial dependance of velocity amplitude, the diffusion coeficient increases when approaching the stellar center. However, very close to the center, non-linear dissipative and radiative damping of internal waves become large and the diffusion coefficient vanishes at the very center. The development of this abstract can be found in E. Schatzman (1996, *J. Fluid Mech.* **322**, 355).

 Astrophysics and Space Science is the original source of publication of this article. It is recommended that this article is cited as: *Astrophysics and Space Science* **265**: 97, 1999.
© 1999 *Kluwer Academic Publishers.*

PHOSPHORUS OVERABUNDANCE IN QUASARS RE-EXAMINED

E. MARGARET BURBIDGE

University of California, San Diego, Center for Astrophysics & Space Sciences, La Jolla, CA 92093-0424, USA

Abstract. Absorption features in the UV spectra in about 10–15% of quasars show very broad profiles, indicating outflow velocities from the quasars of one or even two tenths the speed of light. Determinations of the abundances of some elements, particularly phosphorus, have yielded apparent large overabundances relative to solar. However, it has recently been shown that the absorbing gas which does not completely cover the continuum source of the emission-line clouds may have large optical depth so that the absorption lines are heavily saturated, leading to spurious overabundances of some elements, including phosphorus.

1. Introduction

The QSO PG 0946+301 ($V = 16$, $z_e = 1.223$) is one of the 10–15% of QSOs in whose spectra very broad absorption troughs displaced shortward of the broad emission lines are seen. Spectra of this object, obtained with the Faint Object Spectrograph on the Hubble Space Telescope, show the characteristic broad absorptions of CIV, SiIV and OVI, and a firm identification of the PV$\lambda\lambda$1118, 1128 doublet and PIV λ951 in absorption has been made by Junkkarinen *et al.* (1997). The strength of these features is surprising.

Abundances in the solar photosphere tabulated by Anders and Grevesse (1989) give log N for ^6C, ^7N, and ^{15}P, relative to log N for hydrogen = 12.00, as 8.56 (^6C), 8.05 (^7N), and 5.45 (^{15}P), i.e. the abundance ratio $P/C \simeq 0.001$. The presence of the PV and PIV features, combined with a calculation of the ionization equilibrium of P, yielded an apparent ratio in PG 0946+301 of $P/C \sim 60(P/C)_\odot$. This is almost certainly too large, and improved estimates, taking into account the effects of saturation and incomplete covering of the source of radiation by the outflowing BAL gas, are underway (Junkkarinen, 1999).

2. Where is Phosphorus Synthesized?

Little attention has until recently been paid to nucleosynthetic processes that can produce phosphorus in late stages of stellar evolution. Its synthesis was virtually ignored by Burbidge *et al.* (1957), but enormous advances in the power of present-day computers and the sophistication of programming for networks of nuclear

 Astrophysics and Space Science is the original source of publication of this article. It is recommended that this article is cited as: *Astrophysics and Space Science* **265**: 99–101, 1999.
© 1999 *Kluwer Academic Publishers*.

reactions have made it possible to re-examine some elements to which little attention has hitherto been paid. The case of phosphorus is particularly interesting, because of the observations of PG 0946+301.

Shields (1996) compared the abundances of C, N, O, and Si relative to H, measured in some BALQSOs, with over-abundances found in the ejected shells of galactic novae. Using data published by Politano *et al.* (1995), he suggested that the apparent great overabundance of P in PG 0946+301 might be due to enormous bursts of nova activity in stellar clusters embedded in the active QSO center producing the ejecta.

Novae, however, represent late evolutionary stages of low-mass stars, and the amount of material ejected in nova explosions is not large. Massive stars, with short lifetimes ending in Type II supernovae would seem a more likely source. Nucleosynthesis in massive stars in advanced evolutionary stages involves carbon, neon, oxygen, and silicon burning, and has been discussed by C.A. Barnes in Chapter IX of Wallerstein *et al.* (1997). The O-burning reactions at $T \sim 2 \times 10^9$ K that produce ^{31}P and yield most energy are:

$$^{16}O + {^{16}O} \longrightarrow {^{28}Si} + {^4He} + 9.594 \text{ MeV}$$
$$^{16}O + {^{16}O} \longrightarrow {^{31}P} + p + 7.678 \text{ MeV}.$$

It is my impression that such reactions in evolved stars with masses $M \geq 13M_\odot$ shortly before they become Type II supernovae may be responsible for phosphorus production in the Galaxy.

Also, calculations made for Ne-burning, which follows C-burning and O-burning, sets in at $T = 2.3 \times 10^9$ K and density $\rho = 1.521 \times 10^5$ g cm^{-3} (Arnett, 1996). Arnett's Figure 9.4 presents the element abundances produced as 'explosive Ne ashes'. The abundance of ^{31}P in this calculation is, as Arnett comments, larger than the solar system abundance, thus could account for the presence of P in an object at $z = 1.2$. The largest production of P in SNeII probably occurs in explosions of stars with masses $M \geq 25M_\odot$.

3. Conclusion

The PV and PIV transitions from the ground level to the first excited levels lie shortward of Lyα emission, and thus are best observed in QSOs without a strong Lyα forest. PG 0946+301 is thus an ideal object for observation. Higher-resolution spectra with the Space Telescope Imaging Spectrograph on the Hubble Space Telescope will enable more accurate observations of the continuum level surrounding these transitions to be made.

An accurate determination of the flux at the bottom of the CIV and SiIV BAL profiles is important. Since variability in BAL absorptions is known to occur, it will be important to monitor PG 0946 + 301 for variations in total flux and in the depth of the absorption troughs. To determine the degree of saturation at the bottoms of the absorption troughs requires simulation of the effect of adding incremental amounts of phosphorus to the profiles (Junkkarinen, 1999).

References

Anders, E.H. and Grevesse, N.: 1989, *Geochim. Cosmochim. Acta.* **53**, 197.

Arnett, D.: 1996, *Supernovae and Nucleosynthesis*, Princeton Univ. Press, Princeton, N.J.

Burbidge, E.M., Burbidge, G.R., Fowler, W.A. and Hoyle, F.: 1957, *Rev. Mod. Phys.* **29**, 547.

Junkkarinen, V.T.: 1999, in preparation.

Junkkarinen, V.T., Beaver, E.A., Burbidge, E.M., Cohen, R.D., Hamann, F. and Lyons, R.W.: 1997, Mass Ejection from Active Galactic Nuclei, in: N. Arav, I. Shlosman and R.J. Weymann (eds.), *ASP Conference Series* **128**, 220.

Politano, M., Starrfield, S., Truran, J.W., Weiss, A. and Sparks, W.M.: 1995, *Astrophys. J.* **448**, 807.

Shields, G.A.: 1996, *Astrophys. J.* **461**, L9.

Wallerstein, G. and 14 co-authors: 1997, *Rev. Mod. Phys.* **69**, 995.

III – THE OLD HALO POPULATIONS

THE METALLICITY DISTRIBUTION FUNCTION OF EXTREMELY LOW-METALLICITY STARS

T.C. BEERS

Department of Physics & Astronomy, Michigan State University, East Lansing, MI 48824, USA

Abstract. We discuss new results based on the many thousands of extremely metal-poor stars discovered in the ongoing HK survey of Beers and collaborators. The present status of the photometric and spectroscopy follow-up efforts are summarized, and the nature of the halo metallicity distribution function is considered. We point out the existence of apparent complexities in the kinematics of the lowest abundance stars in the Galaxy, and discuss the presence of a large fraction of carbon-enhanced stars among the HK survey stars with $[\mathrm{Fe/H}] \leq -2.0$.

1. Introduction

The intermediate- and low-metallicity stars of the Galactic halo and thick disk contain precious chemical and kinematic information from which we can learn much about the formation and evolution of the Milky Way, and galaxies like it. In this review we discuss a number of interesting results based on recent spectroscopic and photometric follow-up of stars selected in the HK survey of Beers and collaborators.

Space precludes a full discussion of topics which might be explored with the presently-available sample of metal-poor (MP) stars, but a listing would include:

1. The nature of the halo metallicity distribution function (MDF) – how low can we go?
2. Tests of the primordial lithium hypothesis via measures of the dispersion in lithium abundance on the Spite Plateau
3. Relative abundances of heavy metal species in the first generations of stars – unravelling the mass spectrum of early supernovae
4. Identification of the astrophysical site of the r-process
5. Identification of the astrophysical site(s) for the main and weak s-process
6. Efficiency of mixing processes in the early Galaxy
7. Comparison with elemental species (such as Zn and Cr) used to explore gas abundances along lines of site to various QSOs
8. Estimation of lower limits on the age of Galaxy and the Universe via stellar chronometers such as thorium
9. Kinematic evidence for the existence (or not) of the metal-weak tail of the thick disk

 Astrophysics and Space Science is the original source of publication of this article. It is recommended that this article is cited as: *Astrophysics and Space Science* **265**: 105–113, 1999.
© 1999 *Kluwer Academic Publishers.*

10. Tests of the stellar accretion hypothesis by well-selected subsamples of low-metallicity dwarfs
11. Measures of the local halo (+ thick disk) velocity ellipsoid.

2. A Synopsis of the HK Survey

The HK interference-filter/objective-prism survey was initiated in 1978 by George Preston and Stephen Shectman of the Carnegie Observatories. I joined the project in 1983, and was able to expand the number of plates obtained in both the southern and northern hemispheres. The final set of plates was obtained in 1992. There are 275 (unique) plates in the HK survey, obtained with the Curtis and Burrell Schmidt telescopes at CTIO and KPNO, covering a total of some 7000 square degrees of sky.

The plates are deep (90 minute) widened exposures through an interference filter which passes roughly 150 Å of spectrum (at a dispersion of 180 Å /mm) centered on the CaII H and K lines at $\lambda \sim 3950$ Å. The plates reach a limiting magnitude of around $B \sim 15.5$, several magnitudes fainter than was obtained by previous searches for metal-weak stars using the objective-prism technique. Visual scans of the HK survey plates with a 10X microscope (originally by Preston, later by myself) was used to identify a variety of interesting candidates for later investigation, based on the appearance of the calcium lines and/or stellar continuum in this narrow slice of spectrum. Objects identified on the plates include:
1. Metal-poor F and G stars
2. Field Horizontal Branch (FHB) and other A-type stars
3. Subdwarf O- and B-type stars
4. White Dwarfs
5. Emission-line objects.

When the nature of the MP stars discovered in the HK survey is considered, it is important to keep in mind that the visual identification procedure (in the absence of color information) leads to a severe selection bias with declining temperature. The strength of the CaII H and K lines for stars with $B - V > 0.6$ is large enough, even for stars with quite low abundances, that numerous MP candidates will be mistaken for hotter stars of higher abundance, and hence not tagged for later study.

3. Photometry, Spectroscopy, and Abundance Determinations

There are some 30 000 visually-identified stars selected from the HK survey, roughly 10 000 of which were classified as MP candidates. Photometric and spectroscopic follow-up of such long lists of stars requires collaborative efforts, and this project has been fortunate to benefit from the efforts of a number of groups worldwide.

3.1. THE PHOTOMETRY

Broadband UBV photometry has been reported for a total of some 5000 HK survey stars (MP, FHB/A) by Preston, Shectman and Beers (1990), Doinidis and Beers (1990, 1991), and most recently, by Norris, Ryan and Beers (1998). Additional collaborations are underway to obtain more such measurements.

UBV photometry plays a vital role in the estimation of metal abundance, but it is of even greater importance in estimating the distances to the MP stars. In the color regime $0.3 < B - V < 0.5$ we can use UBV photometry to separate out stars at the main-sequence turnoff from the lower-gravity (and far more distant) cooler halo FHB stars which occur with high frequency among the MP candidates.

A limited amount of photometry (some 750 stars) has been obtained for MP candidates from the HK survey using the Strömgren system, which provides a sharper tool for the assignment of stellar surface gravity estimates and effective temperatures. Such information is of particular value when carrying out fine analysis of high-resolution spectroscopy. Schuster et al. (1996, 1999) report $uvby$ photometry for several hundred MP stars with abundances [Fe/H] ≤ -2. Anthony-Twarog et al. (1999) report photometry on the $Ca - by$ system for another 500 MP candidates.

3.2. THE SPECTROSCOPY

Medium-resolution spectroscopy has been obtained to date for some 5000 MP candidates, at resolution of typically 2 Å, over the wavelength region $3700 < \lambda < 4500$ Å. The first large set of spectroscopic results was reported in Beers, Preston and Shectman (1992, BPS). In the past five years, collaborations using facilities at Siding Spring Observatories (Norris et al., 1999), ESO (Cayrel et al., 1999), the INT telescope on La Palma (Rebolo et al., 1999), and KPNO (Beers et al., 1999b) have resulted in a much larger sample of MP stars with available spectroscopy.

3.3. ABUNDANCE DETERMINATION

Abundance estimates (on the [Fe/H] scale) are made with two techniques, used in combination. The first technique relies on the assumption that the strength of the CaII K line tracks the overall stellar [Fe/H], an assumption which is particularly good for stars with [Fe/H] ≤ -1.5. The second is based on an Auto-Correlation Function (ACF, originally described by Ratnatunga and Freeman, 1989) of a stellar spectrum. The ACF method is particularly good for stars with [Fe/H] > -1.5, where the CaII K line begins to saturate with increasing metal abundance.

Beers et al. (1999a) discuss this new calibration, and demonstrate, based on comparisons with some 550 stars with external high-resolution abundance estimates, that this approach yields abundance determinations with small scatter (on the order of 0.15–0.20 dex) over the entire range of stellar abundances we expect to find in the Galaxy ($-4.0 \leq$ [Fe/H] ≤ 0.0).

TABLE I

Preliminary abundances from the HK survey

	BPS	INT	ESO	ANU	KPN	UNIQUE
TOT	1296	811	880	992	1103	4754
0.0	1287	797	852	968	1079	4660
−0.5	1075	547	655	911	844	3835
−1.0	846	242	472	713	443	2546
−1.5	680	115	352	485	244	1724
−2.0	419	62	241	313	149	1065
−2.5	131	22	97	127	60	373
−3.0	32	7	30	42	13	98
−3.5	5	...	4	13	2	18
−4.0	...	...	...	...	...	...

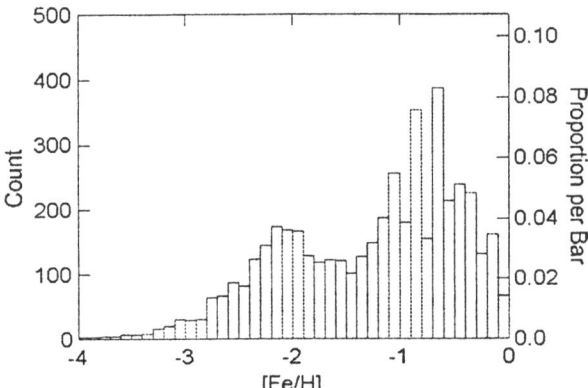

Figure 1. Abundance distribution of candidate MP stars from the HK survey. Bins are 0.1 dex in width.

4. Abundance Distribution of the Metal-Poor Stars

In Table I we summarize preliminary numbers for abundances determined for the MP stars in the HK survey. The column headers indicate the source of the spectroscopy, while the table entries indicate the numbers of stars in the sample LESS than the listed [Fe/H] values on the left side of the table. The last column of the table lists the numbers of stars from all the various collaborations, after elimination of non-unique targets.

Figure 1 shows a histogram of this distribution. Two things are clear – (1) The biases mentioned above are affecting the nature of the MDF for [Fe/H] > −2.0, and (2) there are NO stars identified in our extensive survey with [Fe/H] ≤ −4.0.

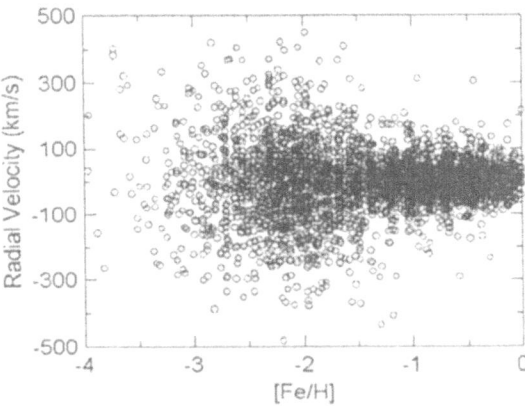

Figure 2. Line of sight velocity distribution as a function of estimated [Fe/H] for MP stars in the HK survey.

We believe that *if* the Galaxy had stars with this low abundance within a few kpc of the sun, the HK survey should have detected them. We conclude that we may have identified the low-metallicity end of the MDF. The observed MDF is *no longer consistent* with Simple Models of Galactic chemical evolution in which star formation proceeds from zero metal-abundance gas in the early Galaxy (e.g., Hartwick, 1976). The very first generations of stars which contributed to the proto-Galactic cloud were apparently able to pollute the remaining gas to a metallicity on the order of [Fe/H] ~ -4.0.

5. Kinematics of the Metal-Poor Stars

Radial velocities are obtained from our medium-resolution spectroscopy with an accuracy on the order of 10 km s^{-1}. Figure 2 is a plot of the radial velocity as a function of estimated [Fe/H] for the sample of 4660 stars below [Fe/H] = 0.0. The expected change of the dispersions of Galactic populations as we move from stars normally associated with the disk (thin and thick) to the halo is evident. A number of quite high velocity stars are noticeable, even for [Fe/H] > -0.5, which is of some interest.

Roughly 25% of the MP stars with abundance estimates lie in directions within 20° of the NGP or SGP. Analysis of this large sample of stars indicates that there are not-yet-understood complexities in the kinematics of even quite metal-poor stars. For example, Figure 3 shows a histogram of the 300 stars in this polar subsample with [Fe/H] < -2.0, and a superposed best-fit single Gaussian distribution. A mixture-model analysis reveals that this distribution is much more likely to be described in terms of a low-dispersion (~ 60 km s^{-1}) population and a high-dispersion (~ 130 km s^{-1}) population of roughly equal portions. The implications of the existence of two (or more) metal-weak populations with very different kin-

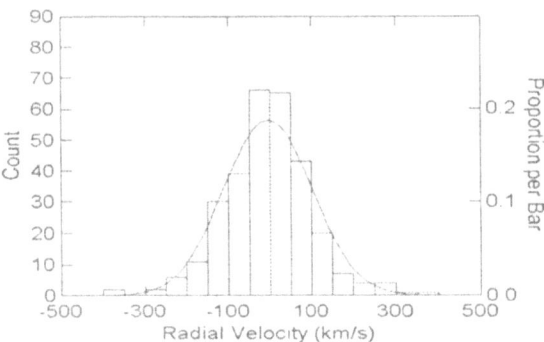

Figure 3. Distribution of radial velocities for 300 HK survey stars near the SGP and NGP with abundance [Fe/H] ≤ -2.0. Bins are 50 km s^{-1} in width. The solid line is a best-fit single Gaussian model, which *does not* fit the data well.

ematics is still being pursued, but it could be tied to several possible explanations – the metal-weak tail of thick disk (e.g., Beers and Sommer-Larsen, 1995), a 'dual halo' (flattened + accreted) configuration (e.g., Norris, 1994), or some other peculiarity in the nature of extremely metal-poor stars in the nearby volume of the Galaxy, which dominate the present sample.

6. The Carbon-Enhanced Stars

The fraction of MP stars from the HK survey which are enhanced in their carbon abundances (based on an estimate obtained from the strength of the CH feature at $\lambda \sim 4300$ Å) appears to be larger at lower metal abundance. As many as 10% of the HK survey stars with [Fe/H] ≤ -2.5 exhibit carbon abundances in excess of 1 dex above the solar value, and this fraction rises to 25% of the sample with [Fe/H] ≤ -3.0.

Rossi, Beers and Sneden (1999) show that, for the *most* carbon-rich MP stars, the envelope of carbon-enhancement increases rapidly with decreasing metal abundance. Fujimoto, Aikawa and Iben (1999) report on new stellar evolution models for extremely-metal-poor stars which suggest that enhanced mixing at the time of helium-core flash may result in the conversion of some quite metal-deficient stars into carbon-enhanced stars, a mechanism which might explain some of the peculiar stars identified in the HK survey. In any case, *some* additional mechanism for carbon enhancement, other than pollution from an evolved binary companion, needs to be found, as several of the carbon-enhanced stars we have studied at high resolution do *not* appear to be binaries (Norris *et al.*, 1997a), nor do they all exhibit neutron-capture elements which might have been expected from thermal pulsing on the AGB – the best counter-example being CS 22957-027 (Norris *et al.*, 1997b; Bonifacio *et al.*, 1998).

7. Future Work

A great deal of work remains to be done.

Once the data from the various collaborations has been fully analyzed and combined, we can take a close look at the functional form of the tail of the halo MDF. Although there may exist sufficient number of MP stars in the sample (below [Fe/H] $= -2.0$) to perform a detailed analysis, if evidence continues to indicate that stars of low abundance may NOT form a single population (either kinematically or chemically), we may have to subdivide the MDF analysis into two (or more) subsamples. If forced to do this, the statistical reliability of an MDF analysis will once again be stretched to the limit. To put it clearly – *we STILL need additional metal-poor stars!*

The kinematics of the full sample of MP stars (and FHB stars) will be analysed shortly (along the lines of Beers and Sommer-Larsen, 1995). For the results of such an analysis to be meaningful, reliable separation of the MP main-sequence turnoff stars and the cooler FHB stars must be obtained, so that reasonable distance estimates can be assigned. We are working on collecting the required UBV photometry for the hotter MP stars which are presently lacking this information. Proper motions are becoming available, at least for the brighter HK survey stars, and with precise distance estimates, we should be able to obtain a large sample of MP stars with full space motions to analyse.

Carbon abundances will be obtained for the several thousand MP stars in the present database with detectable CH G bands, following the line-synthesis approach described by Rossi *et al.* (1999). A more detailed understanding of the trends of [C/Fe] for stars with [Fe/H] ≤ -2.0 will surely emerge.

To supplement the numbers of candidates for later spectroscopic and photometric follow-up, we are presently obtaining digital scans of the HK survey plates using the APM facility in Cambridge. Color calibration of direct sky survey plates in the northern and southern hemisphere is almost completed. With this data in hand, we should be able to recover a large fraction of the cooler MP stars on the HK survey plates which were missed during the visual selection procedure. Preliminary results suggest that we should be able to identify on the order of 5000–10000 *new* MP candidates from the HK survey plates.

The large numbers of newly-discovered MP and FHB stars from the HK survey, as well as from other surveys such as the Hamburg-ESO survey (Christlieb *et al.*, 1999), demand that we utilize wide-field multiple-fiber instrumentation to obtain the required medium-resolution spectroscopy. The planned upgrade of the existing FLAIR II instrument on the UK Schmidt Telescope (Watson, 1998) should provide an excellent resource for work of this sort.

Ultimately, of course, deeper understanding of the chemical history of the Galaxy will only come from high-resolution analysis of MP stars, from which one can recover a much more complete knowledge of the distribution of elemental abundances which were produced in the early Galaxy. The next-generation of 6.5 m to 10 m telescopes presently coming online will be able to obtain the required high S/N data with essentially complete wavelength coverage from the near-UV to the near-IR. With a collaborative effort similar to that which has resulted in the discovery of the large number of MP stars in the HK survey, astronomers will be rewarded with a vastly more detailed picture of chemical evolution than we have at present.

Acknowledgements

TCB acknowledges support for this work received from grants AST 92-22326 and AST 95-29454 from the National Science Foundation.

TCB would also like to express his gratitude to Roger Cayrel, whose enthusiastic support of the HK survey follow-up effort over the years has significantly increased the scientific payoff of this endeavor. Roger loves the low metallicity stars of the Galactic halo, and they love him as well.

References

Anthony-Twarog, B., Twarog, B., Sarajedini, A., Beers, T.C. and Hawley, S.: 1999, in preparation.
Beers, T.C. and Sommer-Larsen, J.: 1995, *Astrophys. J. Suppl.* **96**, 175.
Beers, T.C., Preston, G.W. and Shectman, S.A.: 1992, *Astron. J.* **103**, 1987.
Beers, T.C., Rossi, S., Norris, J.E., Ryan, S.G. and Shefler, T.: 1999a, *Astron. J.*, in press.
Beers, T.C., Rossi, S., Anthony-Twarog, B., Twarog, B., Hawley, S., Rhee, J., Tourtelot, J., Wilhelm, R. and Sarajedini, A.: 1999b, in preparation.
Bonifacio, P., Beers, T.C., Molaro, P. and Vladilo, G.: 1998, *Astron. Astrophys.* **332**, 672.
Cayrel, R., Beers, T.C., Nissen, P.E., Andersen, J., Nordstrom, B., Rossi, S., Spite, M., Spite, F. and Barbuy, B.: 1999, in preparation.
Christlieb, N., Wisotzki, L., Reimers, D., Gehren, T., Reetz, J., Gehren, T., Grasshoff, G. and Beers, T.C.: 1999, this volume.
Doinidis, S.P. and Beers, T.C.: 1990, *Publ. Astron. Soc. Pacific* **102**, 1392.
Doinidis, S.P. and Beers, T.C.: 1991, *Publ. Astron. Soc. Pacific* **103**, 973.
Fujimoto, M., Aikawa, M. and Iben, I.: 1999, The Third Stromlo Symposium: The Galactic Halo, in: B. Gibson, T. Axelrod and M. Putman (eds.), *ASP Conference Series*, Astronomical Society of the Pacific, San Francisco, in press.
Hartwick, F.D.A.: 1976, *Astrophys. J.* **209**, 418.
Norris, J.E.: 1994, *Astrophys. J.* **431**, 645.
Norris, J.E., Ryan, S.G. and Beers, T.C.: 1997a, *Astrophys. J.* **488**, 350.
Norris, J.E., Ryan, S.G. and Beers, T.C.: 1997b, *Astrophys. J.* **489**, L169.
Norris, J.E., Ryan, S.G. and Beers, T.C.: 1998, *Astrophys. J. Suppl.*, submitted.
Norris, J.E., Beers, T.C., Ryan, S.G. and Rossi, S.: 1999, in preparation.
Preston, G.W., Beers, T.C. and Shectman, S.A.: 1994, *Astron. J.* **108**, 538.
Preston, G.W., Shectman, S.A. and Beers, T.C.: 1991, *Astrophys. J. Suppl.* **76**, 1001.
Ratnatunga, K.U. and Freeman, K.C.: 1989, *Astrophys. J.* **339**, 126.
Rebolo, R., Beers, T.C., Allende Prieto, C., Molaro, P., Rossi, S., Garcia-Lopez, R. and Bonifacio, P.: 1999, in preparation.

Rossi, S., Beers, T.C. and Sneden 1999: The Third Stromlo Symposium: The Galactic Halo, in: B. Gibson, T. Axelrod and M. Putman (eds.), *ASP Conference Series*, Astronomical Society of the Pacific, San Francisco, in press.

Schuster, W.J., Nissen, P.E., Parrao, L., Beers, T.C. and Overgaard, L.P.: 1996, *Astron. Astrophys. Suppl.* **117**, 317.

Schuster, W.J., Beers, T.C., Nissen, P.E., Parrao, L. and Franco, A.: 1999, in preparation.

Watson, F.: 1998, *AAO Newsletter* **85**, 11.

THE (SUB)STRUCTURE OF THE HALO

S.R. MAJEWSKI

*Department of Astronomy, University of Virginia, P.O. Box 3818, Charlottesville, VA 22903-0818,
USA*

Abstract. The fossil record of the Milky Way indicates an evolution including periodic accretions of
smaller galaxies and clusters, consistent with hierarchical models of galaxy formation. I discuss three
observational programs that demonstrate that the phase space distribution of stars, clusters and dwarf
galaxies in the Galactic halo contains degrees of substructure left by the débris of tidally disrupted
stellar systems.

1. Introduction

This symposium has featured the great advances being made in the area of chemical
abundance studies of Galactic stellar populations. Among these advances (represented by numerous contributions to this proceedings) are the discovery of halo
stars with metal abundances below $10^{-3} Z_\odot$ and the exploitation of fossil evidence
in elemental abundance patterns capable of fingerprinting the birth conditions of
stars, particularly low abundance stars, in unprecedented detail. Such enterprises
are revolutionizing our ability to understand the early history of the Galaxy and
represent a great step forward from the traditional populations studies' *modus
operandi* via interpretation of connections between *mean* abundances and global
spatial and kinematical distributions of stars (à la Eggen *et al.*, 1962).

Just as delving into the detailed chemical patterns in the atmospheres of some
stars reveals new layers of insight into the star's origins, so too can analyzing
the spatio-kinematical distributions of stellar populations at new degrees of detail
reveal complexities that are hallmarks of past dynamical events. Here I concentrate on fossilized dynamical evidence for accretion in the Galactic halo. Unlike
chemical fingerprints, which, in general, remain indelibly imprinted in stellar spectra, dynamical fingerprints are subject to processes that alter stochastically the
kinematical attributes of stars and make detailed information on their birth sites
hopelessly irrecoverable. In the least, when, by Liouville's theorem, the débris of
stellar systems remain forever dynamically linked and, in principle, recoverable,
the contortion of the distribution of débris in phase space may be extremely complicated and unobvious by conventional means (Aguilar, 1997; Helmi and White,
1999, 'HW99'). For example, one may need to abandon looking for fossil substructure by spatial and velocity clumpings, and instead search the parameter space

Astrophysics and Space Science is the original source of publication of this article. It is recommended that this article is cited as: *Astrophysics and Space Science* **265:** 115–122, 1999.
© 1999 *Kluwer Academic Publishers.*

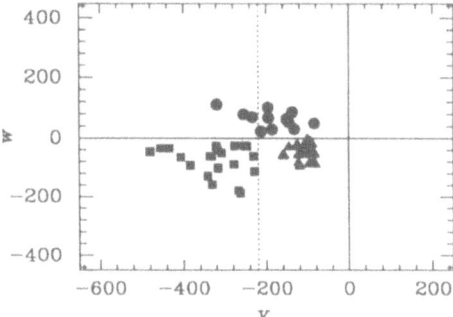

Figure 1. Halo clumpiness as found by Majewski *et al.,* (1996). v is the rotational velocity with respect to the Local Standard of Rest, and w is the velocity perpendicular to the plane. To remove most disk stars, we show only stars with $V < -80$ km s^{-1}.

of conserved attributes, like the integrals of motion (Poveda *et al.*, 1992, HW99). Unfortunately, such a strategy requires full 6-D space-velocity information, and is presently limited to nearby samples of stars (previous references; Section 2). However, dynamical timescales in a large volume of the outer Galactic halo are long enough that little phase mixing has occurred and at least some fossil evidence remains unerased in simple phase space representations (Johnston, 1998, 'J98'). We discuss possible manifestations of this preserved information in Sections 3 and 4.

2. Substructure in the Nearby Halo

Nearby, proper motion selected samples containing full spatial and velocity information have already been used to argue for the presence of accreted débris near the sun (Norris, 1994; Carney *et al.*, 1996; Nissen and Schuster, 1997). Majewski (1992), upon finding a significant retrograde mean rotational velocity for a sample of more distant halo stars in a pencil-beam survey of proper motions towards the North Galactic Pole (NGP), appealed to accretion as an explanation. A potential retrograde moving group was identified in this halo sample, and shown to have coherence in all three dimensions of motion in Majewski *et al.* (1994). Further work to obtain radial velocities for stars in this region of the sky (Majewski *et al.*, 1996) reveals what looks to be a clumpy halo distribution (Figure 1), rather than one dominated by what would be called "randomly mixed, field halo" stars. The metal-poor, retrograde moving group (*squares* in Figure 1) is apparently falling down upon the Galactic plane as well as inward toward the Galactic center. In spite of the general uniformity of the bulk motion of the "moving group" stars, their relatively large velocity spread is beyond traditional expectations for long-lived coherence of "moving groups".

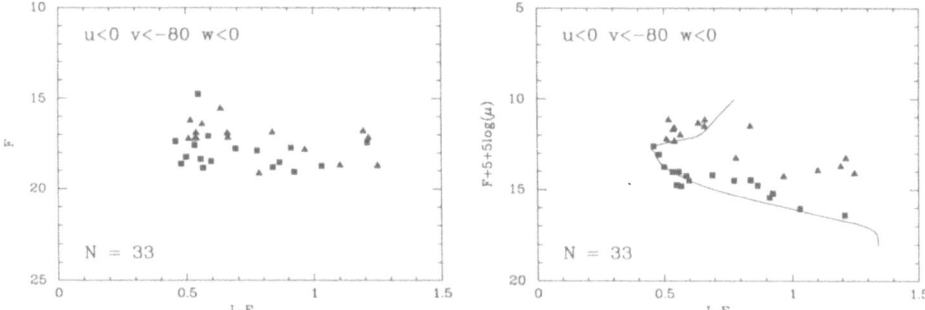

Figure 2. Photographic JF color magnitude diagram (left) and reduced proper motion diagram for a subsample of the Majewski *et al.* (1996) data in Figure 1. The *solid line* is the M92 locus adapted from Stetson & Harris (1988) at $V_{trans} = 245$ km s^{-1}.

A shortcoming of the analysis leading to Figure 1 is that without data to gauge the intrinsic luminosities of the stars, Majewski *et al.*, (1996) derived photometric parallaxes (after accounting for ultraviolet excess) under the assumption that all stars were dwarfs. Thus, the distances of unresolved binaries, evolved stars, and stars with peculiar chemical abundance patterns will be systematically incorrect. Systematic errors in distances of some survey stars will artificially spread out coherent substructures in the velocity dimensions derived from proper motions (in this case, u and v). As one example, Majewski *et al.* point out that if the apparently separate clump of stars shown as *triangles* in Figure 1 were subgiants rather than dwarfs, their measured proper motions would translate to V velocities coincident with the retrograde group (shown as *squares*). One would expect just such a population of subgiants (and a fewer number of giants, and, of course, some number of binaries) along with the population of dwarfs in any coherent moving group. What one would like is an independent way to test the veracity of the putative moving group that does not resort to the uncertain distances, and, instead, is based only on observables.

If the moving group stars were at a similar distance, one might expect them to trace out a population "isochrone" in the color-magnitude diagram and also to have nearly identical proper motion vectors. Unfortunately these stars are found throughout a $\sim$ 10 kpc long pencil beam, and, consequently, their isochrone is smeared out in the CMD, while *their proper motions are scaled to different values based on their distances.* The latter point may be exploited, however, since in a tidal stream the stars should have very similar transverse velocities. Thus, the proper motions depend *solely* on the stellar distances, and one should be able to recover a population isochrone through use of the reduced proper motion diagram (RPMD). Applied to the stars in Figure 1, this technique does indeed reveal a rather nice distribution similar to the M92 isochrone. Moreover, a number of the stars suspected as moving group subgiants with underestimated distances (*triangles*) do indeed move to the proper subgiant location in the RPMD (Figure 2).

The RPMD analysis suggests at least some of the velocity spread may be contributed by systematic distance errors; however, this should *not* affect the w velocities (obtained from radial velocities). A dynamical explanation for the velocity spread may come from HW99. While they demonstrate how the disruption of a satellite in the inner halo leads to increasing *spatial* incoherence but velocity structures with *decreasingly cold* dispersions, HW99 are able to account for the NGP moving group as the débris from a single object if member stars come from multiply wrapped tidal streams. They derive initial characteristics for the parent object of 2×10^8 M$_\odot$, an initial radius of ~ 1 kpc and a velocity distribution of 48 km s^{-1}.

3. Substructure in the Distant Halo

The outer halo field star population has been probed primarily via the use of bright, evolved stars. Surveys of horizontal branch stars have led to the suggestion of several putative halo moving groups manifest as radial velocity and distance clumpings (cf. summary in Majewski *et al.,* 1996). Of course, the tidal streamers of the Sagittarius dwarf galaxy provide us with the most obvious example of substructure in the halo that is still spatially coherent. Both RR Lyrae (e.g., Alcock *et al.,* 1997) and red clump stars (e.g., Ibata *et al.,* 1995), have been used to map the Sgr tidal stream, though Sgr débris is near enough that it is also practical to take advantage of the more populous main sequence turn off stars to follow the tidal débris to rather low surface brightnesses (Mateo *et al.,* 1998, Siegel *et al.,* 1997) some 25° away from the core. According to models by J98, it would take only 3 Gyr for Sgr débris to circle all of the way around the Galaxy, and in 10 Gyr Sgr stellar débris may cover a substantial fraction of the sky (5–15%).

But this is several times smaller than J98's model results for potential sky coverage of débris from the break-up of the Large Magellanic Cloud (LMC) after 10 Gyr. Although there have been several failed attempts to find stellar Magellanic débris in the past (Mathewson *et al,.* 1979; Recillas-Cruz, 1982; Brück and Hawkins, 1983), Totten and Irwin (1998) now report a possible planar alignment of carbon stars in the outer halo that may be associated with the Magellanic Stream. The verification of Magellanic tidal débris is clearly important, as it bears on a potentially important contributor of Galactic halo stars. However, because models suggest Magellanic débris may be highly dispersed (Moore and Davis, 1994, J98), its contrast against the Galactic background may be very low. Indeed, if a major contributor to the halo, Magellanic débris may *be* the "Galactic background" in some directions of the sky! These speculations and model results argue for a new search covering much larger areas than surveyed before.

We (Majewski *et al.,* 1999) have been conducting such a large area survey with the Las Campanas 1-m telescope. Giants are identified photometrically using Washington filters with the $DDO51$ filter centered on the gravity-sensitive

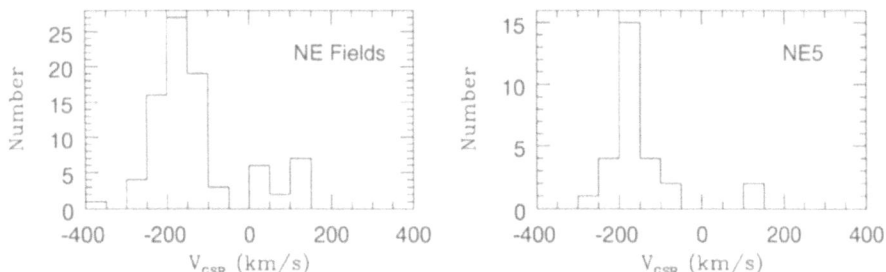

Figure 3. Evidence that the outer halo is very clumpy from a survey of outer halo giants in progress by Majewski *et al.* (1999). (*Left*) Distribution of radial velocities (corrected to the Galactic Standard of Rest) for selected giant star candidates in six fields (approximately magnitude limited to $M \sim 18$) spread out across almost $20°$ in the general direction to the northeast of the LMC. (*Right*) Radial velocity distribution for giants to $M \sim 18$ in one of the fields, located more than $15°$ to the northeast of the center of the LMC. The velocity dispersion of the left hand group is only 42 km s^{-1} in the NE5 field alone and 44 km s^{-1} in the data for all fields (left panel).

MgH+Mgb triplet (see Geisler, 1984). Radial velocities are obtained with 1-2Å resolution spectroscopy on the du Pont 2.5-m. Presently 29 fields of ~ 1 deg^{-2} each have been photometered as part of a $20°$ radius survey ring around the Clouds. Most evident in this survey of the outer Milky Way is the paucity of any significant contribution by a "random halo field" population with expected mean (Galactic standard of rest) velocity near 0 km s^{-1} and $\sigma \sim 110$ km s^{-1}. Instead, we find rather skewed velocity distributions and "moving group" clumps with relatively tight dispersions (Figure 3). Similar findings are reported in other parts of the sky by Harding *et al.* (1998). In our survey, the radial velocity-position distributions are found to be consistent with expectations for débris in models of the disruption of the LMC into multiply-wrapped streamers around the Milky Way over the past 10 Gyr. Our peculiar velocity distributions, whether or not ultimately associated with the tidal disruption of the Magellanic Clouds, are the reflection of an outer halo containing a high degree of substructure in the form of correlated streams of tidal débris.

4. Substructure in the Galactic Satellites and Clusters

Alignments among Milky Way satellite galaxies and globular clusters have long been posited as evidence for possible associations via the tidal break-up of larger Milky Way companions (Kunkel, 1979; Lynden-Bell, 1982). Correlations specific to Galactic globular clusters showing a second parameter effect in their horizontal branch morphology have also been suggested among the Galactic alignments (Majewski, 1994; Lynden-Bell and Lynden-Bell, 1995; Fusi Pecci *et al.,* 1995). Association of these clusters with dwarf satellite galaxies as joint tidal débris could provide a natural explanation for the second parameter effect, es-

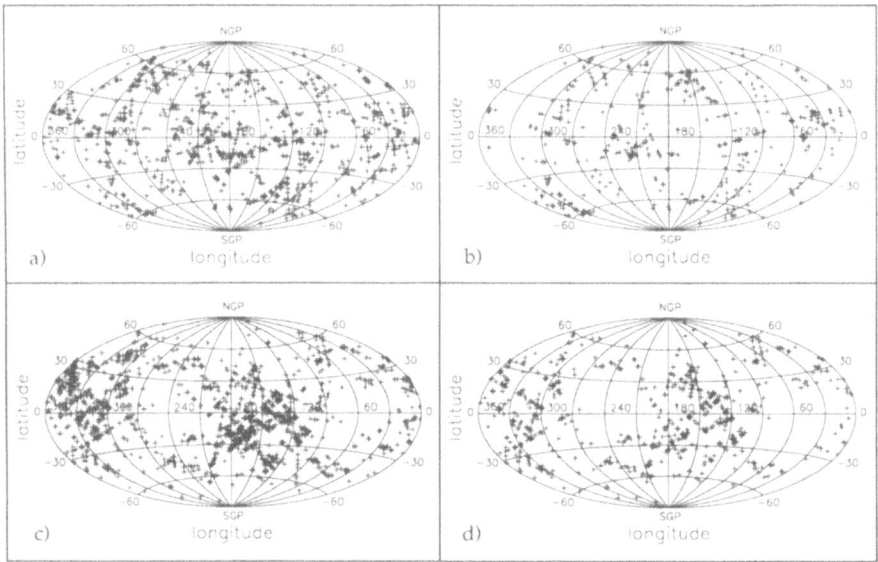

Figure 4. Crossing points for (a) non-second parameter globular clusters and satellite galaxies, (b) non-second parameter globulars alone, (c) second parameter globulars and satellite galaxies, (d) second parameter globulars alone. All samples for $R_{GC} > 8$ kpc. The diagrams are symmetric about 180° in the Galactocentric coordinates shown.

pecially (though not necessarily solely) if the direct source of that effect is age. Among those globular clusters specifically cited as dynamically connected to dwarf galaxies – Ter 7, Arp 2, Ter 8 and M54 with Sagittarius (Ibata *et al.,* 1995) and Pal 12 and Rup 106 with the Magellanic Clouds (Lin and Richer, 1992) – as well as those associated with Fornax (see Zinn, 1993), all but one (Ter 8) shows a significant second parameter effect and several are widely accepted as examples of younger clusters. Rup 106 and Pal 12 also show low [α/Fe] ratios characteristic of enrichment by Type Ia supernovae in an environment with multiple star formation events like the Magellanic Clouds (Brown *et al.,* 1997). King (1997) has found low [α/Fe] ratios, comparable to those found in Pal 12 and Rup 106, in the common proper motion pair HD134439/134440, stars that also have kinematics that suggest they were accreted and a photometric age similar to these two young halo clusters.

Recently, Palma *et al.* (1999) have analyzed the statistical likelihood for Galactic cluster/satellite associations via the orbital pole methodology of Lynden-Bell and Lynden-Bell (1995). In most cases, proper motions, and hence orbits, for these objects are unknown. However, even without knowledge of kinematics we may confine the family of possible orbital poles for each Galactic satellite to a great circle in the Galactocentric system. Conjunctions of intersections of these great circles indicate possible dynamical families. Figure 4 shows the crossing points of great circle pole families for outer halo ($R_{GC} > 8$ kpc) globular clusters divided into Zinn's (1993) groupings by horizontal branch morphology. The difference in distributions is clear: the second parameter halo clusters show a strong nexus of crossing points near $(l, b) = (150, 0)°$ – a similar area of concentration for the crossing points of Galactic satellite galaxies alone – while the non-second parameter clusters show a more homogeneous distribution. A comparison of the angular two-point correlation functions for the crossing point distributions shows the discrepancy to be highly significant. The orbital pole analysis reinforces suspicions of associations between second parameter clusters (which predominate in the outer halo) and the Galactic satellites.

Clearly our understanding of potential connections among Galactic satellites will increase as proper motions are obtained and orbits thereby derived. Unfortunately, the number of distant clusters (and especially second parameter clusters) with proper motions is very small. For some clusters and satellites, we may have to await the abilities of GAIA and the Space Interferometry Mission to get a handle on their orbits. For others, orbits may be estimated if tidal tails, as are now being found (Grillmair, 1998), are present. Even rudimentary proper motions can help tremendously. For example, proper motions have been measured to various precisions for about half of the Galactic satellites, and these support the notion of at least one compelling dynamical family of satellite galaxies consisting of the LMC, SMC, Ursa Minor and Draco (Palma et al., 1999). The true orbital poles for members of this possible family lie near the nexus of crossing points for second parameter clusters in Figure 4, and, again, suggest a connection of satellite galaxies to some Galactic clusters. Such a family of satellites and clusters might have derived from the break-up of a formerly larger Magellanic Galaxy. Of course, such a process would naturally have led to associated débris streams of Magellanic *stars* (Section 3).

No doubt some surprises await our having complete velocity information for the entire family of Milky Way globulars and satellites. An example of the new insights to expect may be the newly derived (Dinescu et al., 1999) orbit of ω Centauri – strongly retrograde but in the plane of the Galactic disk. When considered in light of its other characteristics (large mass, extended metallicity distribution, flattened shape), this orbit supports the notion that ω Cen may itself be the nucleus of a disrupting satellite galaxy on a tidally decaying orbit (à la Quinn et al., 1992), much as M54 has been proposed to be the nucleus of Sagittarius (and perhaps similar to the description of nucleated dwarf galaxies by Freeman, 1993).

Acknowledgements

I would like to thank Giusa and Roger Cayrel and Monique and Francois Spite for their invitation to this Symposium and for their wonderful hospitality and assistance during this enjoyable conference.

References

Aguilar, L.A.: 1997, in: *IAUS 186: Galaxy Interactions at Low and High Redshift*, in press.

Alcock, C., *et al.*: 1997, ApJ, **474**, 217.

Brown, J.A., Wallerstein, G. and Zucker, D.: 1997, *AJ* **114**, 180.

Brück, M.T. and Hawkins, M.R.S.: 1983, *A&A* **124**, 216.

Carney, B.W., Laird, J.B., Latham, D.W. and Aguilar, L.A.: 1996, *AJ* **112**, 668.

Dinescu, D.I., Girard, T.M. and van Altena, W.F.: 1999, *AJ* **117**, 1792.

Eggen, O.J., Lynden-Bell, D. and Sandage, A.R.: 1962, *ApJ* **136**, 748.

Freeman, K.: 1993, in: G.H. Smith and J.P. Brodie (eds.), *The Globular Cluster-Galaxy Connection*, ASP Conf. Ser. Vol. **48**, p. 608.

Fusi Pecci, F., Bellazzini, M., Cacciari, C. and Ferraro, F.R.: 1995, *AJ* **110**, 1664.

Geisler, D.: 1984, PASP, 96, 723.

Grillmair, C.J.: 1998, in: D. Zartisky (ed.), *Galactic Halos*, ASP Conf. Ser. Vol. **136**, p. 45.

Harding, P., Freeman, K.C., Mateo, M., Morrison, H.L., Olszewski, E. and Norris, J.: 1998, *BAAS* **30**, 1408.

Helmi, A. and White, S.D.M.: 1999, MNRAS, **307**, 495 (HW99).

Ibata, R., Gilmore, G. and Irwin, M.: 1994, *Nature* **370**, 194.

Ibata, R.A., Gilmore, G. and Irwin, M.J.: 1995, *MNRAS* **277**, 781.

Johnston, K.V.: 1998, *ApJ* **495**, 297 (J98).

King, J.: 1997, *AJ* **113**, 2302.

Kunkel, W.E.: 1979, *AJ* **228**, 718.

Lin, D.N. and Richer, H.B.: 1992, *ApJ* **388**, L57.

Lynden-Bell, D.: 1982, *Observatory* **102**, 202.

Lynden-Bell, D. and Lynden-Bell, R.M.: 1995, *MNRAS* **275**, 429.

Majewski, S.R.: 1992, *ApJSuppl* **78**, 87.

Majewski, S.R.: 1994, *ApJ* **431**, L17.

Majewski, S., Kunkel, W., Ostheimer, J., Johnston, K. and Patterson, R.: 1999, in preparation.

Majewski, S.R., Munn, J.A. and Hawley, S.L.: 1994, *ApJ* **427**, L37.

Majewski, S.R., Munn, J.A. and Hawley, S.L.: 1996, *ApJ* **459**, L73.

Mateo, M., Olszewski, E. and Morrison, H.: 1998, *ApJL* **508**, L55.

Mathewson, D.S., Ford, V.L., Schwarz, M.P. and Murray, J.D.: 1979, in: W.B. Burton (ed.), *The Large-Scale Characteristics of the Galaxy*, Reidel, Dordrecht, p. 547.

Nissen, P.E. and Schuster, W.J.: 1997, *A&A* **326**, 751.

Norris, J.: 1994, *ApJ* **431**, 645.

Palma, C., Majewski, S.R. and Johnston, K.V.: 1999, *ApJ*, submitted.

Poveda, A., Allen, C. and Schuster, W.: 1992, in: B. Barbuy and A. Renzini (eds.), *IAUS 149: The Stellar Populations of Galaxies*, Kluwer, Dordrecht, p. 471.

Quinn, P.J., Hernquist, L. and Fullager, D.P.: 1992, *ApJ* **403**, 74.

Recillas-Cruz, E.: 1982, *A&A* **124**, 216.

Siegel, M.H., Majewski, S.R., Reid, I.N., Thompson, I., Landolt, A.U. and Kunkel, W.E.: 1997, *BAAS* **29**, 1341.

Stetson, P.B. and Harris, W.E.: 1988, *AJ* **96**, 909.

Totten, E.J. and Irwin, M.J.: 1998, *MNRAS* **294**, 1.

Zinn, R.: 1993, in: G.H. Smith and J.P. Brodie (eds.), *The Globular Cluster-Galaxy Connection*, ASP Conf. Ser. Vol. **48**, p. 38.

KINEMATICS AND DYNAMICS OF THE GALACTIC STELLAR HALO

JESPER SOMMER-LARSEN

Theoretical Astrophysics Center, Juliane Maries Vej 30, DK-2100, Copenhagen Ø, Denmark

Abstract. The structure, kinematics and dynamics of the Galactic stellar halo are reviewed including evidence of substructure in the spatial distribution and kinematics of halo stars. Implications for galaxy formation theory are subsequently discussed; in particular it is argued that the observed kinematics of stars in the outer Galactic halo can be used as an important constraint on viable galaxy formation scenarios.

1. Introduction

Although the stellar halo accounts for at most a few percent of the luminous mass of the Galaxy, it plays a crucial role in studies of the Galaxy's formation, evolution, and present-day structure. The halo has long been considered the Galaxy's oldest component. Age estimates can be obtained for its most conspicuous constituent, the metal-weak globular clusters, as well as for individual metal-weak stars (with less certainty). Thus the dynamical and chemical state of the luminous halo population provides information on the formation of large disk galaxies such as the Milky Way.

2. Is There Substructure in the Halo?

In the currently favoured hierarchical galaxy formation scenarios big disk galaxies like the Milky Way are built up through merging of smaller subsystems. In particular it is thought that the Galactic halo was formed and, to some extent, still is being formed by accretion of many small satellites.

Likely examples of ongoing *stellar* accretion are the Sagittarius dwarf (Ibata, Gilmore and Irwin 1994) and the Magellanic Stream (Majewski *et al.*, 1998). Furthermore one would generally expect strong substructure in the distribution of halo stars in phase-space due to past accretion events (e.g., Johnston, Spergel and Hernquist, 1995). This substructure is difficult to detect due to the very low number density of halo stars ('observational shot-noise') and the physical effect of phase-mixing, but accreted systems with extreme kinematics should be detectable – one very likely example of this is the moving group of Majewski, Munn and Hawley (1994). The blue metal-poor stars (BMPs) of Preston, Beers and Shectman (1994)

 Astrophysics and Space Science is the original source of publication of this article. It is recommended that this article is cited as: *Astrophysics and Space Science* **265**: 123–131, 1999.
© 1999 *Kluwer Academic Publishers.*

may well be another example of a population of halo stars accreted in one (major) event – see also Carney *et al.* (1996).

3. The Local Halo Velocity Ellipsoid and the Structure of the Inner Halo

The inner halo, defined here as the part of the halo inside of or about at the solar distance from the center of the Galaxy ($r \lesssim r_\odot \simeq 8$ kpc) consists of at least two components: the flat and the round halo. The ratio of the local density of flat to round halo is not well determined ranging from 0.5–1 (Sommer-Larsen and Zhen, 1990) to $\sim 4-8$ (Kinman, Suntzeff and Kraft, 1994; Hartwick, 1987). The velocity ellipsoid of local halo stars has been determined by Beers and Sommer-Larsen (1995). They used radial velocities of stars selected without kinematic bias, making the result very robust, also to distance errors. For almost 900 stars with [Fe/H] < -1.5 they find a velocity ellipsoid for local halo stars of $(\sigma_r, \sigma_\phi, \sigma_\theta) = (153 \pm 10, 93 \pm 18, 107 \pm 7)$ km s^{-1} in spherical polars. This velocity ellipsoid is fairly radially elongated, but is still characterized by a quite large vertical (and horizontal) tangential velocity dispersion of ~ 100 km s^{-1}. It follows from the tensor virial theorem that the flat component can not be both very flat *and* locally dominant – see Sommer-Larsen and Christensen (1989).

Carney *et al.* (1996) analyzed two classes of halo stars from their local sample: The 'low' halo stars have $\langle |z_{max}| \rangle \leq 2$ kpc and the 'high' halo stars have $\langle |z_{max}| \rangle \geq 5$ kpc, where $\langle |z_{max}| \rangle$ is the typical distance a star reaches from the plane of the disk – as the Carney *et al.* stars are all local z_{max} was calculated from the observed space motions by orbit integration. These two classes of halo stars can be seen as representative of the flat and round halo respectively. The 'low' halo is characterized by a slight prograde average rotation $\langle v_\phi \rangle \simeq 20 \pm 15$ km s^{-1} with respect to the Galactocentric restframe assuming a circular speed at the solar distance from the Galactic center of $v_\odot = 220$ km s^{-1}. The 'high' halo, on the other hand, is characterized by a net retrograde rotation $\langle v_\phi \rangle \simeq -50 \pm 15$ km s^{-1}. The latter result was first found by Majewski (1992) from proper motions of halo stars situated at the north galactic pole (NGP) and at least 5 kpc from the plane of the disk. The results above are further indications of kinematic substructure in the halo. Surprisingly a similar retrograde net rotation has not been detected so far for halo stars *in situ* at $|z| > 5$ kpc at the SGP (Beers, private communication). As the retrograde halo stars are found both at high z at the NGP and locally these stars should be well mixed, so from stellar dynamics considerations it follows that such stars *should* be found at the SGP in the future.

Martin and Morrison (1998) find in an interesting recent study of the kinematics of local ($d \lesssim 1$ kpc) RR-Lyrae stars with accurate 3-D velocities that $\sigma_r = 193 \pm 15$ km s^{-1} for the halo stars in their sample. As noted by the authors this differs from the result of Beers and Sommer-Larsen (1995) for halo stars in general by more than 2 σ (see above). A possible explanation of this difference is that the spatial distribution of the RR-Lyrae halo stars differs from that of the halo stars in general: Approximating the spatial distributions by power-laws $\rho \propto r^{-\alpha}$, then $\alpha_{RR} = 3.2 \pm 0.1$, whereas for the halo stars in general $\alpha_{HALO} \simeq 3.5$ – see, e.g., Preston, Shectman and Beers (1991). Pushing α_{RR} to 3.1 and assuming the Beers and Sommer-Larsen value for σ_r of halo stars in general it follows from the Jeans equation that one would expect $\sigma_r \simeq 179 \pm 12$ km s^{-1} for the RR-Lyrae halo stars, in good agreement with the results of Martin and Morrison.

Martin and Morrison used the 'old' distance scale with $M_V(RR) = 0.73$ at [Fe/H] $= -1.9$. A 'new' distance scale, based on *Hipparcos* trigonometrical parallaxes of Cepheid variables and subdwarfs, has recently been advocated by Feast and Catchpole (1997) and Chaboyer *et al.* (1998) resulting in $M_V(RR) \simeq 0.32$ at [Fe/H] $= -1.9$. With this distance scale Martin and Morrison obtain $\sigma_r \simeq 225 \pm 18$ km s^{-1} for their RR-Lyrae halo stars. This differs from the value predicted above for $\alpha_{RR} = 3.1$ by more than 2 σ. Hence, as also noted by Martin and Morrison, one should perhaps be somewhat cautious about using the 'new' distance scale, despite its many merits, like resulting in cosmologically 'reasonable' globular cluster ages of about 12 Gyrs, a fairly low value of the Hubble constant etc.

4. The Outer Halo

The outer stellar halo ($r \gtrsim r_\odot$) is approximately spherical – see references in Sommer-Larsen *et al.* (1997), but also Sluis and Arnold (1998). Several types of stars have been used as tracers of the outer halo: K-giants (e.g., Ratnatunga and Freeman, 1989), RR-Lyrae stars (e.g., Hawkins, 1984) and blue horizontal branch stars (e.g., Sommer-Larsen *et al.*, 1997, and references therein). The blue horizontal branch field (BHBF) stars have proven to be particularly useful tracers of the the outer halo for two main reasons: (a) They are easy to identify in the halo because of their blue colours and (b) Using a medium-sized telescope, spectra of sufficient quality for accurate line-of-sight velocity and Balmer line-width determination can be obtained fairly easily, even for quite distant stars ($d \sim 30 - 60$ kpc), because of their intrinsic brightness. Furthermore they seem representative of the stellar halo in general since the density fall-off of the BHBF stars in the outer halo can be well approximated by the power-law relation $\rho(r) \propto r^{-\alpha}$, $\alpha = 3.4 \pm 0.3$ – see Sommer-Larsen, Flynn and Christensen (1994).

Sommer-Larsen *et al.* (1997) analyzed a sample of almost 700 BHBF stars with good line-of-sight velocity determinations and situated at Galactocentric distances $r \sim 7 - 70$ kpc. At distances $d \gtrsim 20 - 30$ kpc the line-of-sight velocity is close to being the radial component of the velocity in Galactocentric coordinates. Hence it is possible to determine the radial velocity dispersion at large Galactocentric distances from line-of-sight velocities only. Whereas the radial velocity dispersion of local halo stars is 140–150 km s^{-1} σ_r is found to drop to about 100 km s^{-1}

at large r. This kinematic behaviour is modelled in the following simple way by Sommer-Larsen et al.:

It is assumed that both the *outer* Galactic halo and gravitational potential are spherically symmetric (moderate departures from this does not affect the outcome of the analysis in any significant way). The radial velocity dispersion as a function of r is modelled as

$$\sigma_r = \left(\sigma_0^2 + \frac{\sigma_+^2}{\pi} \left(\frac{\pi}{2} - \text{Arctan}(\frac{r - r_0}{l}) \right) \right)^{1/2} . \tag{1}$$

Adopting this form gives good flexibility in modelling the decrease in $\sigma_r(r)$ with increasing r. It follows from eq. [1] that σ_0 is the asymptotic value of the radial velocity dispersion for $r \gg (r_0 + l)$ and that $\sqrt{\sigma_+^2 + \sigma_0^2}$ approximately is the radial velocity dispersion in the inner halo ($r \lesssim r_0$). The physical meaning of the two scale parameters r_0 and l is given in Sommer-Larsen, Flynn and Christensen (1994) and is fairly straightforward.

The rotation curve of the Galaxy is approximately flat to at least $R \simeq 20$ kpc (Fich and Tremaine, 1991) and most likely to much larger distances as shown by Kochanek (1996). Consequently the gravitational potential of the outer halo is approximated by

$$\Phi(r) = V_c^2 \ln(r) , \tag{2}$$

corresponding to a flat rotation curve with $v_c(r) \equiv V_c = 220$ km s^{-1}. Substituting this and eq. [1] into the Jeans equation

$$\frac{1}{\rho} \frac{d(\rho \sigma_r^2)}{dr} + \frac{2(\sigma_r^2 - \sigma_t^2)}{r} = -\frac{d\Phi}{dr} , \tag{3}$$

where σ_t is the (1-D) tangential velocity dispersion, yields the following expression for $\sigma_t(r)$

$$\sigma_t = \left(\frac{V_c^2}{2} - \sigma_r^2 \frac{(\alpha - 2)}{2} - \frac{1}{2\pi} \frac{r}{l} \frac{\sigma_+^2}{(1 + [(r - r_0)/l]^2)} \right)^{1/2} , \tag{4}$$

where α is the BHBF star density power-law index defined above. The line-of-sight velocity dispersion of a set of stars located at a distance d in a field at Galactic coordinates (l, b) where the velocity ellipsoid has components σ_r and σ_t is

$$\sigma_{\text{los}} = \sqrt{\gamma^2 \sigma_r^2 + (1 - \gamma^2) \sigma_t^2} , \tag{5}$$

where γ is a simple geometric projection factor

$$\gamma \equiv (d - r_\odot \cos l \cos b)/r . \tag{6}$$

The parameters $(\sigma_0, \sigma_+, r_0, l)$ are determined by maximum likelihood fitting of expressions [1] and [4] to the data using Equation [5]. A surprisingly good fit is

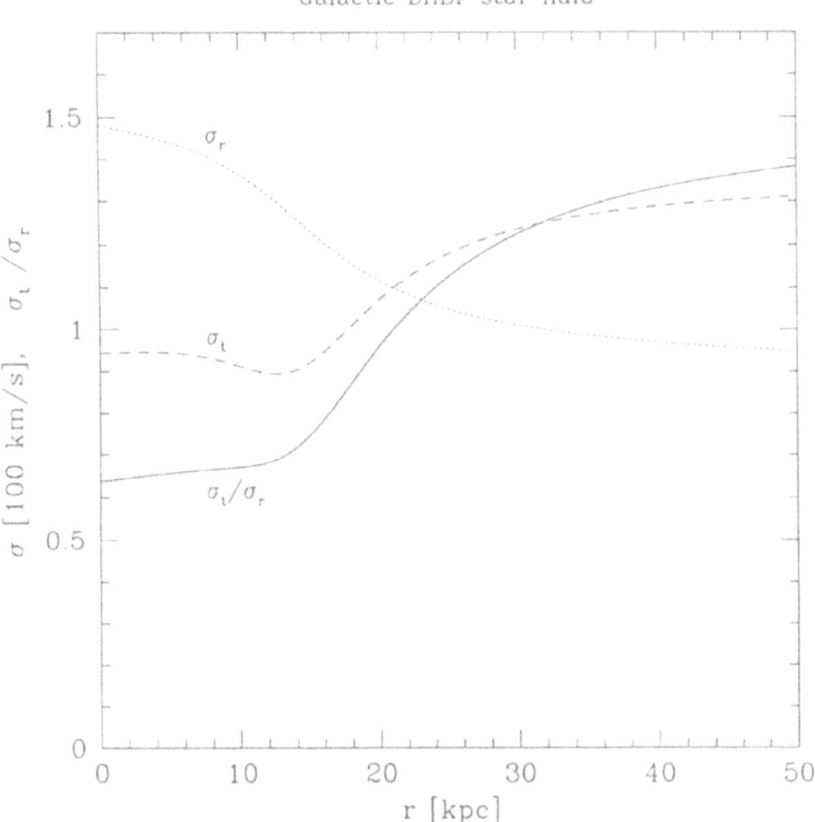

Figure 1. Best-fit model. The dotted line shows the radial velocity dispersion $\sigma_r(r)$ and the dashed line shows the tangential velocity dispersion $\sigma_t(r)$. Also shown is σ_t/σ_r as a function of r (solid line).

obtained for this simple model – see Sommer-Larsen *et al.* The resulting $\sigma_r(r)$, $\sigma_t(r)$ and $\eta(r) \equiv \sigma_r(r)/\sigma_t(r)$ are shown in Figure 1. The asymptotic value of $\sigma_r(r)$ at large r is found to be $\sigma_0 = 89 \pm 19$ km s^{-1}.

The main result of the analysis is that σ_r decreases from 140–150 km s^{-1} at $r = r_\odot$ to 89 ± 19 km s^{-1} at large r – matched by a corresponding increase in σ_t. Hence, the most important kinematic feature of the model is that the velocity ellipsoid changes from *radial* anisotropy in the solar vicinity ($\eta \simeq 0.65$) to *tangential* anisotropy in the outer halo ($\eta \sim 1.4$).

5. Outer Stellar Halo Kinematics: Clues About the Formation of the Milky Way

The results concerning the dynamics and kinematics of the outer stellar halo are of considerable interest in relation to theories of the formation of the Milky Way, in particular, and galaxies in general:

If the Galaxy formed from a single collapsing over-dense region in the early universe, then one might expect the outer halo to be characterized by radially anisotropic kinematics (see, e.g., van Albada, 1982), whereas the data show that quite the opposite is the case. If, on the other hand, at least the outer parts of the proto-Galaxy were assembled by accretion of small subsystems, then a large tangential velocity dispersion in the outer parts of the Galaxy is possible, depending on the nature of the accretion (Norris, 1994; Freeman, 1996). So the results indicate that the outer stellar halo formed by some sort of accretion and merging processes. The kinematics of stars in the inner halo are, at least locally, radially anisotropic, possibly indicating that the inner parts of the halo formed during a more dissipative and coherent collapse.

Chemical evolution arguments lead to a similar conclusion on the basis of the finding that there is a significant abundance gradient in the inner halo, but essentially none in the outer halo. This was discussed in the pioneering work by Searle and Zinn (1978) and in much subsequent work – see, e.g., Norris (1996) and references therein.

In the following I will attempt to quantify the connection to galaxy formation theory, more specifically the Cold Dark Matter (CDM), hierarchical galaxy formation scenario:

Sommer-Larsen, Gelato and Vedel (1999) carried out cosmological (CDM), gravitational/hydrodynamical, Tree-SPH simulations of the formation and evolution of large (Milky Way sized) disk galaxies. For the dark matter haloes of four different model galaxies (at the present epoch) the ratio $\eta(r)$ is shown in Figure 2. As can be seen from the Figure $\eta \simeq 0.9 \pm 0.1$. This is quite different from what is found in the outer stellar halo of the Milky Way and close to isotropic. Furthermore the rotation curves of the model galaxies are approximately flat over the range $r \sim 10 - 100$ kpc. Hence it should be a reasonable approximation for the present purpose to represent a dark matter halo by an isothermal sphere with phase-space distribution function

$$f_{DM}(E) \propto \exp\left(-\frac{\vec{v}^2 + 2\Phi(r)}{V_c^2}\right) \ , \tag{7}$$

where $\Phi(r)$ is the gravitational potential given by Equation (2). The dark matter density falls of like $\rho_{DM} \propto r^{-2}$, whereas the stellar halo density profile is considerably steeper, $\alpha \simeq 3.4$, so the halo star formation efficiency $\epsilon_\star$ must have had a dependence on some radial property r_{orb} of the dark matter orbits such that

$$\epsilon_\star \propto r_{orb}^{2-\alpha} \ . \tag{8}$$

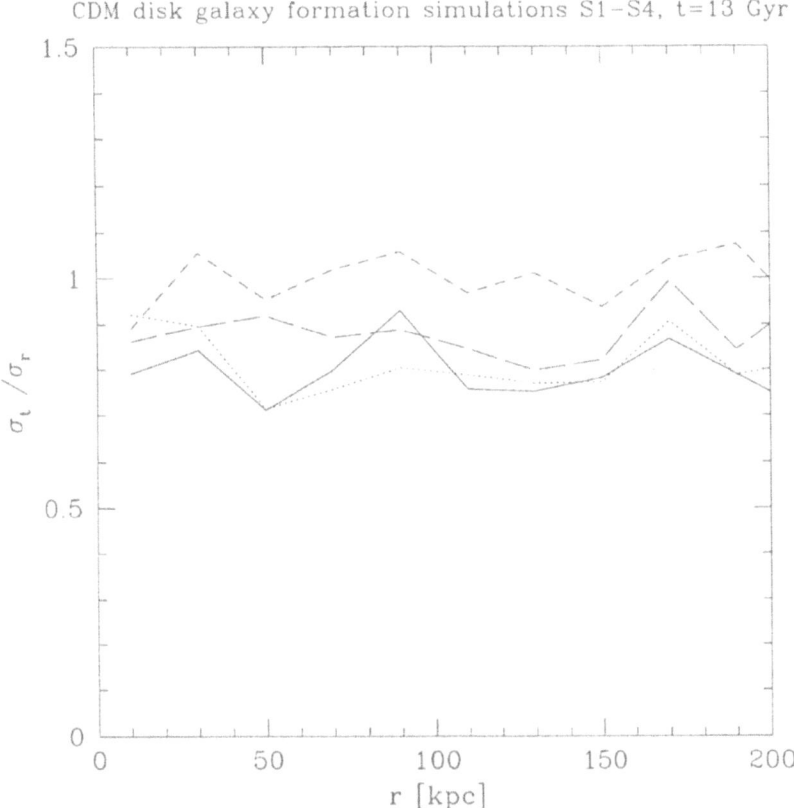

Figure 2. σ_t/σ_r as a function of r for the dark matter haloes of four Milky Way sized model galaxies.

One can then 'build' the stellar halo using dark matter orbits weighted by $\epsilon_\star$. If the halo stars were formed in small, proto-galactic subsystems before or as they were accreted onto the main dark matter halo it would be reasonable to take $r_{orb} = r_a$, where r_a is the apocenter distance of the orbit in the main dark matter halo. If, on the other hand, the halo star-formation was triggered by disk and halo shocking of the gas in the subsystems when these were near the inner turning points of their orbits one would take $r_{orb} = r_p$, where r_p is the pericenter distance of the orbit. Finally one might also study an intermediate case like $r_{orb} = (r_a + r_p)/2$. Figure 3 shows the resulting values of η as a function of the power-law index α of the halo stars for $r_{orb} = r_a$ (thick solid line), $r_{orb} = r_p$ (thick long-dashed line) and $r_{orb} = (r_a + r_p)/2$ (thick short-dashed line) – η does not depend on r for the models I consider here since these are scale-free. Also shown is the asymptotic value (at large r) of η from the observations of the outer halo (solid line) and 1 σ and 2 σ deviations (dotted lines). Finally the local value of $\eta \simeq 0.65$ is indicated by a short solid line. As can be seen from the Figure the model predictions do not agree well with the observations: for $r_{orb} = r_a$ the model predictions can be rejected with 82%

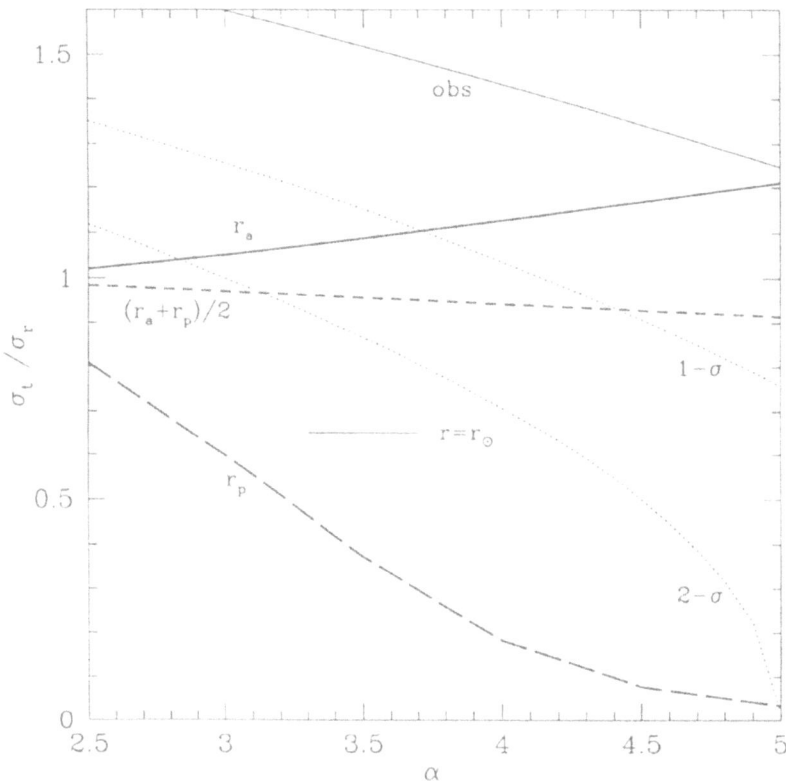

Figure 3. σ_t/σ_r as a function of power-law index α for models and observations – see text for details.

confidence for $\alpha = 3.4$ and the $r_{orb} = r_p$ case can almost completely be excluded. Hence it appears unlikely that the formation of the halo stars was controlled by disk and halo shocking, at least in the framework of the models presented here. More observations of BHBF stars in the outer part of the Galactic halo ($r \gtrsim 30 - 40$ kpc) are needed in order to reduce the statistical uncertainties, but outer stellar halo kinematics clearly has the potential of becoming an important constraint on viable galaxy formation scenarios.

Acknowledgements

I have benefited from discussions with Tim Beers, Per Rex Christensen, Chris Flynn, Ken Freeman, Sergio Gelato, Steve Majewski, John Norris and Bernard Pagel.

References

Beers, T.C. and Sommer-Larsen, J.: 1995, *Astrophys. J. Suppl.* **96**, 175.

Carney, B.W., Laird, J.B., Latham, D.W. and Aguilar, L.A.: 1996, *Astron. J.* **112**, 668.

Chaboyer, B., Demarque, P., Kernan, P.J. and Krauss, L.M.: 1998, *Astrophys. J.* **494**, 96.

Feast, M.W. and Catchpole, R.M.: 1997, *Mon. Not. R. Astron. Soc.* **286**, L1.

Fich, M. and Tremaine, S.: 1991, *Annu. Rev. Astron. Astrophys.* **29**, 409.

Freeman, K.C.: 1996, in: H.L. Morrison and A. Sarajedini (eds.), *Formation of the Galactic Halo – Inside and Out*, ASP, San Francisco, p. 3.

Hawkins, M.R.S.: 1984, *Mon. Not. R. Astron. Soc.* **206**, 433.

Ibata, R.A., Gilmore, G. and Irwin, M.J.: 1994, *Nature* **370**, 194.

Johnston, K.V., Spergel, D.N. and Hernquist, L.: 1995, *Astrophys. J.* **451**, 598.

Kinman, T.D., Suntzeff, N.B. and Kraft, R.P.: 1994, *Astron. J.* **108**, 1722.

Kochanek, C.S.: 1996, *Astrophys. J.* **457**, 228.

Majewski, S.R.: 1992, *Astrophys. J. Suppl.* **78**, 87.

Majewski, S.R., Munn, J.A. and Hawley, S.L.: 1994, *Astrophys. J.* **427**, L37.

Majewski, S.R., Kunkel, W.E., Ostheimer, J.C., Johnston, K.V. and Patterson, R.J.: 1998, in preparation.

Martin, J.C. and Morrison, H.L.: 1998, *Astron. J.* **116**, 1724.

Norris, J.E.: 1994, *Astrophys. J.* **431**, 645.

Norris, J.E.: 1996, in: H.L. Morrison and A. Sarajedini (eds.), *Formation of the Galactic Halo – Inside and Out*, ASP, San Francisco, p. 14.

Preston, G.W., Shectman, S.A. and Beers, T.C.: 1991, *Astrophys. J.* **375**, 121.

Preston, G.W., Beers, T.C. and Shectman, S.A.: 1994, *Astron. J.* **108**, 538.

Ratnatunga, K. and Freeman, K.C.: 1989, *Astrophys. J.* **339**, 126.

Searle, L. and Zinn, R.: 1978, *Astrophys. J.* **225**, 357.

Sluis, A.P.N. and Arnold, R.A.: 1998, *Mon. Not. R. Astron. Soc.* **297**, 732.

Sommer-Larsen, J. and Christensen, P.R.: 1989, *Mon. Not. R. Astron. Soc.* **239**, 441.

Sommer-Larsen, J. and Zhen, C.: 1990, *Mon. Not. R. Astron. Soc.* **242**, 10.

Sommer-Larsen, J., Flynn, C. and Christensen, P.R.: 1994, *Mon. Not. R. Astron. Soc.* **271**, 94.

Sommer-Larsen, J., Beers, T.C., Flynn, C., Wilhelm, R. and Christensen, P.R.: 1997, *Astrophys. J.* **481**, 775.

Sommer-Larsen, J., Gelato, S. and Vedel, H.: 1999, *Astrophys. J.*, **519**, 501.

van Albada, T.S.: 1982, *Mon. Not. R. Astron. Soc.* **201**, 939.

ABUNDANCE RATIOS IN EXTREME METAL-POOR STARS

ANDREW MCWILLIAM and LEONARD SEARLE

Carnegie Observatories, 813 Santa Barbara Street, Pasadena, CA 91101, USA

Abstract. In extremely metal-poor stars ([Fe/H]≤−2.5) the neutron capture elements are character-ized by a 300-fold dispersion in M/Fe ratios which decreases with increasing metallicity, the median M/Fe ratio increases with increasing [Fe/H], but the average M/Fe number ratio is approximately constant. These observations are consistent with a highly dispersed intrinsic yield of neutron-capture elements in supernova (SN) events, and a progression to increasing metallicity by stochastic chemical evolution.

The abundance trends indicate that the synthesis of elements heavier than barium was dominated by the r-process. The Sr/Ba ratio shows a dispersion which suggests a stochastic source of Sr in excess of the r-process value; possibly due to the alpha-rich freeze out.

The iron-peak elements Cr, Mn, and Co show non-solar abundance ratios for extreme metal-poor stars, and no measurable intrinsic dispersion relative to iron. We discuss chemical evolution models which explain these observations.

1. Neutron-Capture Elements

Two well established characteristics of neutron-capture heavy element ($Z > 30$) abundances in extreme metal-poor stars are: *(i)* the trend of the median to lower [M/Fe] with decreasing [Fe/H] and *(ii)* a dispersion in [M/Fe] which increases with decreasing [Fe/H].

Spite and Spite (1978) were the first to find evidence of a trend of declining [Ba/Fe] and [Y/Fe] below [Fe/H]∼−2. The earliest evidence for a dispersion at low metallicity comes from Griffin *et al.* (1982), who found very strong Eu lines in the halo star HD 115444 ([Fe/H]∼−3). Studies by Gilroy *et al.* (1988) and Ryan *et al.* (1991) reported a dispersion in neutron-capture abundances. McWilliam *et al.* (1995, hereafter MPSS95) and Ryan *et al.* (1996) found a dispersion greater than observational uncertainties in large samples of extreme metal-poor stars. Figure 1 shows the results for [Sr/Fe] from MPSS95; the open squares provide a comparison with more metal-rich stars from Gratton and Sneden (1988, 1994). In the interval $-4 \le$ [Fe/H] ≤ -2 there is a 2.5 dex (300-fold) range in [Sr/Fe] at a given [Fe/H], much larger than the measurement uncertainty of ∼ 0.2 dex. The dispersion in-creases with decreasing [Fe/H] and is especially obvious near [Fe/H] ∼ −3. The data for [Ba/Fe] also shows a large dispersion, but with fewer points at super-solar ratios (McWilliam, 1998).

Astrophysics and Space Science is the original source of publication of this article. It is recom-mended that this article is cited as: *Astrophysics and Space Science* **265**: 133–140, 1999.
© 1999 *Kluwer Academic Publishers.*

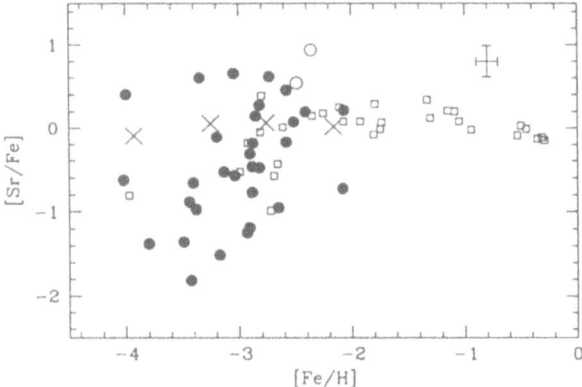

Figure 1. A plot of [Sr/Fe] versus [Fe/H] in extreme metal-poor stars. Filled circles from MPSS95; boxes from Gratton and Sneden, 1988, 1994; open circles are CH stars; crosses indicate average Sr/Fe in 0.5 dex bins.

Figure 1 shows that at lower [Fe/H] most stars have sub-solar [Sr/Fe] values. Thus we may understand why Spite and Spite (1978) found an apparent trend of decreasing neutron-capture abundances with declining [Fe/H]; their small sample was not large enough to find the rarer stars with super-solar neutron-capture abundances, or to detect the intrinsic dispersion. The large crosses in Figure 1 indicate the average [Sr/Fe] number ratio for stars taken in 0.5 dex wide bins; the average is remarkably constant, and similar to the asymptotic value for metal-poor stars. The data for [Ba/Fe] are consistent with a constant average value, but this is not well established.

A fundamental question is whether the observed neutron-capture abundance dispersion reflects the original composition from which the stars were formed, or is due to late stages of stellar evolution, such as the barium star phenomenon seen in pop. I stars. Three lines of evidence suggest that the dispersions are primordial in origin: the abundance pattern for elements with $Z \geq 56$ is consistent with pure r-process nucleosynthesis, not the s-process pattern expected from red giant nucleosynthesis (e.g. MPSS95, Cowan *et al.*, 1995, 1998; McWilliam, 1998); large [Sr/Fe] dispersions are also seen in extreme metal-poor dwarfs (Ryan *et al.*, 1996); to explain the observed spread requires that almost half of the known extreme metal-poor stars are contaminated by stellar evolution effects, which is much larger than the 1–2% frequency of barium stars among pop. I GK giants (MacConnell and Frye, 1972).

Truran (1981) was the first to suggest that heavy elements in the halo were produced by r-process nucleosynthesis; Truran assumed that s-process elements are only produced on time scales longer than the halo formation time (by red giant stars in the AGB phase), and that r-process elements can be made on short time scales by type II SN.

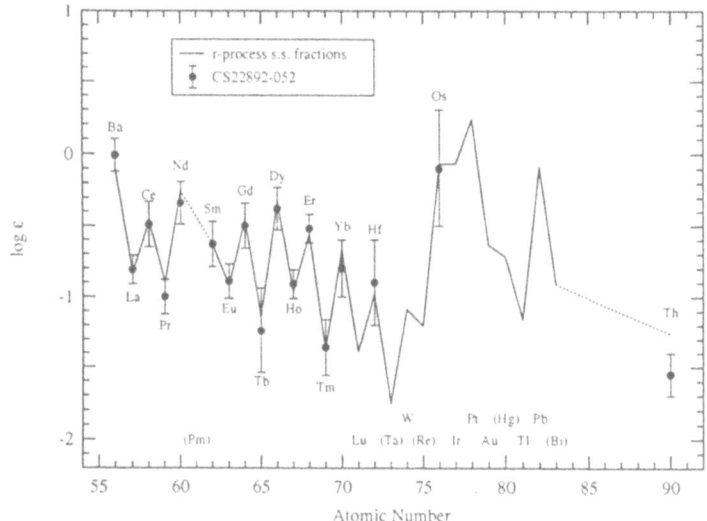

Figure 2. Observed neutron-capture element abundances in CS 22892–052, from Sneden *et al.* (1996); the line represents solar system r-process ratios, normalized to Ba.

Early observational evidence in favor of r-process abundance ratios in halo stars came from Sneden and Parthasarathy (1983) and Gilroy *et al.* (1988). Accurate abundance results for many elements (with $Z \geq 56$) in individual extreme metal-poor stars include MPSS95, Cowan *et al.* (1995), Sneden *et al.* (1996) and Cowan *et al.* (1998); although presently only 4 stars have been studied in this way (CS22892–052, HD115444, HD122563, HD126238). Figure 2 shows the abundance pattern for CS 22892–052 from Sneden *et al.* (1996), compared the the solar system r-process abundances (normalized to Ba); the match to the solar system r-process is remarkable, and only one element, Th, shows any significant deviation.

MPSS95 and McWilliam (1998) used [Ba/Eu] abundance ratio to investigate the r-/s-process ratio in 13 extreme metal-poor stars, as indicated in Figure 3. The observed [Ba/Eu] ratios are consistent with pure r-process nucleosynthesis for stars with [Fe/H] ≤ -2.4, with the exception of two stars thought to be CH giants, self-polluted with s-process material.

Cowan *et al.* (1998) suggested that the r-process in SN is universal and always produces the same abundance pattern. If correct this assumption simplifies the effort required to estimate the amount of thorium produced in the r-process by extrapolation from the stable nuclei. Comparison of the extrapolated original Th abundance with the observed Th abundance then gives the age. The estimated minimum age of the Galaxy, determined by Cowan *et al.* (1997, 1998), from the Th chronometer is 13 ± 5 Gyr.

It is not universally accepted that the neutron capture elements in extreme metal-poor stars follow the r-process pattern. Magain (1995) used high S/N, high resolution spectra, to measure the ratio of odd to even barium isotopes in HD140283 with

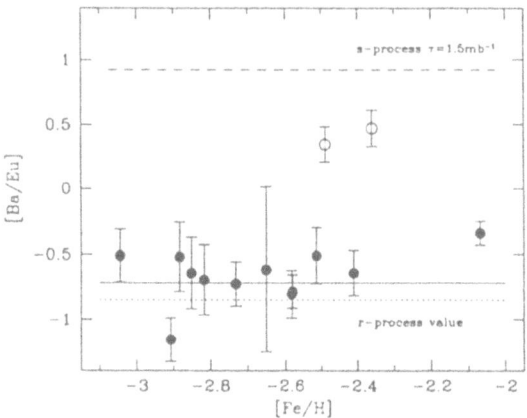

Figure 3. [Ba/Eu] in extreme metal-poor stars, from McWilliam (1998). solid line: solar system r-process value of Wisshak *et al.* (1996); dotted line: solar system r-process value from Käppeler *et al.* (1989); dashed line: extreme s-process value from Malaney (1987); open circles are CH stars.

line profile fits. An r-process isotopic mixture would have been dominated by odd Ba isotopes; however Magain found an odd/even isotope ratio consistent with the solar system composition. A re-investigation by Gacquer (1997) did not reproduce Magain's result, but rather favored a pure r-process mix. Thus, at the present time the conclusion regarding the Ba isotopes is not settled. This clever idea by Magain is clearly a very useful tool and should be investigated further.

1.1. TWO SOURCES FOR SR?

If the light neutron-capture element strontium ($Z = 38$) is also produced by the r-process, then its abundance should scale with Ba, plus scatter due to measurement error. Figure 4 (from McWilliam, 1998) shows that Sr does scale roughly with Ba, but with an intrinsic dispersion. In particular the star CS 22897–008 exhibits a massive Sr overabundance relative to Ba ([Sr/Ba] $=\sim$ +2.0 dex); the Sr lines in this star are very strong, but the Ba lines are only just above the detection threshold.

The r-process rich star CS 22892–052, which has strong Sr and Ba lines and for which excellent spectra have been acquired (Sneden *et al.*, 1996), lies on the lower envelope of the distribution in Figure 4 (−3.04, −0.28). This value of [Sr/Ba] is consistent with the solar-system r-process (actually the solar r-process value is probably not as well determined as the ratio for CS 22892–052). The bulk of the stars in Figure 4 lie at +0.33 dex [Sr/Ba], completely inconsistent with the value for CS 22892–052. It seems difficult to imagine that the abundance results for CS 22892–052 suffered from the measurement errors required to place it on the lower envelope, because this star has been analyzed many times (MPSS95, Sneden *et al.*, 1996; Cowan *et al.*, 1998) with the highest quality data. It is also difficult to understand how the bulk of the stars could have suffered a systematic error of 0.6 dex.

The observations are perfectly well understood if there is a second source for Sr in addition to the normal r-process. The large Sr abundance in CS 22897–008 is dominated by the second source, whereas the large Sr abundance in CS 22892–052 is dominated by the r-process. There is always at least the Sr abundance level produced by the r-process; although there is a large r-process Sr dispersion it scales with the Ba abundance (which is only made by the r-process). The star CS 22892–052 contains such a large r-process excess that any second Sr component is drowned-out, and the [Sr/Ba] ratio is equal to the r-process value; CS 22897–008 has a small r-process Sr abundance, negligible in comparison to the second Sr source. Most stars in Figure 4 contain Sr from both the r-process and the second source, with most Sr produced by the second source. Some stars, for example CS 22891–200, are particularly low in total r-process abundance, but still show the r-process [Sr/Ba] ratio; this can be explained if the yield from the second Sr source is variable.

An important question is weather the second source Sr is primordial or produced in the stellar interiors and then dredged-up to the surface, or transferred from a companion. A secure answer must await studies of the [Sr/Ba] ratios in extreme metal-poor dwarf stars. Transfer from a companion is unlikely, as it would require nearly all stars in Figure 4 to be affected. Self-pollution might have occurred during the He core flash, but only about half of the sample in Figure 4 have experienced the He core flash, even though many more show enhanced [Sr/Ba] ratios. The likely self-pollution mechanism is s-process nucleosynthesis, which tends to produce more Ba than Sr (as witnessed by the two CH stars in the sample, see Busso *et al.*, 1995).

Therefore, it seems likely that the additional Sr has a primordial origin. Two potential mechanisms are: the weak s-process in the cores of massive stars (see Raiteri *et al.*, 1993), and the alpha-rich freeze-out in type II SN (e.g. Woosley and Hoffman, 1992). Because s-process enrichment is strongly metallicity dependent, with low production at low [Fe/H], the weak s-process is a less favored mechanism than the alpha-rich freeze-out.

1.2. STOCHASTIC CHEMICAL EVOLUTION

It is clear from Figure 1 that the Galaxy was far from homogeneous at low metallicity; the intrinsic [Sr/Fe] dispersion signals the break down of the instantaneous mixing approximation of the Simple model; it also provides proof that the [Sr/Fe] yield ratios from SN events are highly variable.

Obviously, a successful chemical evolution model must explain the features of Figure 1, including: the trend of increasing dispersion with decreasing [Fe/H], the decline of the median [Sr/Fe] ratio with decreasing [Fe/H], and the approximately constant average Sr/Fe number ratio with metal content. The model must

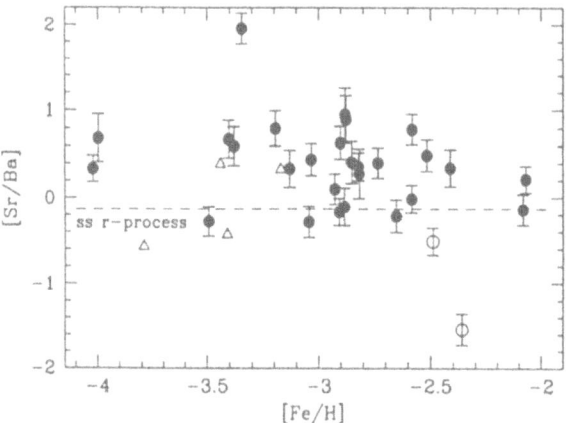

Figure 4. [Sr/Ba] for metal-poor stars (from McWilliam, 1998). Open triangles: lower limits; dashed line: solar system r-process value from Käppeler *et al.* (1989); Open circles are CH stars.

also explain the asymmetry in the [Sr/Fe] dispersion, although this is a redundant constraint.

Searle and McWilliam (1998) have developed a model of stochastic chemical evolution which fits the observed abundances. Suppose a star is composed of ejecta from N SN events, then the Sr/Fe number ratio is:

$$(Sr/Fe)_N = \frac{\sum_j q_j (Sr/Fe)_j}{\sum_j q_j}$$

Where q_j is the weight of the jth event, equal to its metal increment. It is assumed that the Sr/Fe number ratio, r, is distributed according to a probability density function, or yield function, $g(r)$. An initial estimate of $g(r)$ may be obtained from the most metal-poor stars in the sample; Searle and McWilliam (1998) chose to use those stars with [Fe/H] ≤ -3. Because the number of stars with [Fe/H] ≤ -3 is small the data are too coarse to use a straight-forward histogram plot for $g(r)$, so a Gaussian kernel smoothing was applied to the data, with $\sigma = 0.90$ dex; the precise value of σ does not greatly affect the predictions.

Figure 5 shows the result of a Monte-Carlo type simulation; the dotted box indicates the stars used to generate the yield function, $g(r)$. It is remarkable that the model, based on the distribution of points for [Fe/H] ≤ -3, predicts the features of the [Sr/Fe] versus [Fe/H] diagram at higher metallicity so well.

There is a decrease in dispersion towards higher [Fe/H], the median [Sr/Fe] increases with increasing [Fe/H], the cumulative probability levels are approximately consistent with the observations, and the dispersion is asymmetric. Also, since the calculations utilized a single yield function, $g(r)$, (and constant q) the average Sr/Fe number ratio is independent of [Fe/H] by design, as observed.

The model assumed that the yield function is approximately given by stars with [Fe/H]~ -3.4. However, it could easily be the case that the yield function is much

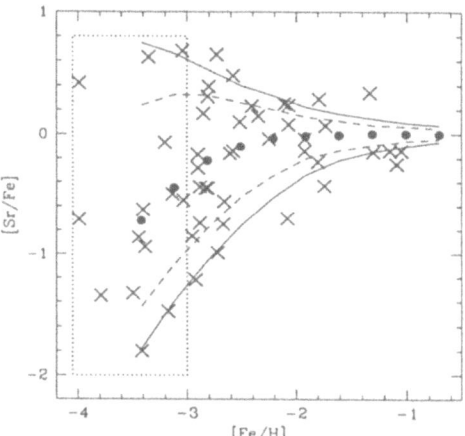

Figure 5. Comparison of observed [Sr/Fe] values (crosses) with the results of a stochastic chemical evolution model. The dotted box indicates the sub-sample used to determine $g(r)$. Solid line: 5 and 95th percentiles; dashed lines: 15 and 85th percentiles; filled circles are medians. The value of q is assumed constant at [Fe/H] $= -3.42$.

more dispersed than the observations, and that the stars with [Fe/H]~ -3.4 already reflect homogenization; the composition being drawn from many SN events. To test this idea we employed an analytical yield function, with $g(r)\alpha r^{-1.5}$; the exponent of approximately -1.5 is required to fit the data. The results showed that, indeed, for a given observed dispersion it is possible to select an arbitrarily large intrinsic [Sr/Fe] dispersion with the appropriate choice of the number of SN events; although it is possible that the enrichment parameter, q, may break the degeneracy.

2. Iron-Peak Elements

The deviation of elements in the iron-peak from the solar composition, specifically Cr, Mn and Co, is well known (MPSS95, Ryan *et al.*, 1996); [Mn/Fe] and [Cr/Fe] decrease below [Fe/H] ~ -2.5 and [Co/Fe] increases towards lower metallicity. The remarkable observation is that [Co/Cr] is tightly correlated with [Fe/H], despite the large range in neutron-capture element abundances. The challenge for chemical evolution models is to explain both the large dispersion of neutron-capture elements, and the tight correlation of the iron-peak. Three proposals have been put forward:

- A metallicity-dependent [Co/Cr] yield ratio, with few enough generations in the interval $-4 \leq$ [Fe/H] ≤ -2.5 that the neutron-capture element dispersion was not fully homogenized (MPSS95).
- Stars in the interval $-4 \leq$ [Fe/H] ≤ -2.5 produced by individual SN events, with the lowest mass SN producing the lowest [Fe/H] and the highest [Co/Cr]

ratios by an unpredicted correlation of [Co/Cr] with SN progenitor mass (Ryan *et al.*, 1996; Shigeyama and Tsujimoto, 1998).

• Dilution of primordial material (pop. III) characterized by high [Co/Cr] ratios, with SN ejecta of normal [Co/Cr] yield ratio (Searle and McWilliam, 1998).

References

Cowan, J.J., Burris, D.L., Sneden, C., McWilliam, A. and Preston, G.W.: 1995, *Astrophys. J.* **439**, L51.

Busso, M., Lambert, D.L., Beglio, L., Gallino, R., Raiteri, C.M. and Smith, V.V.: 1995, *Astrophys. J.* **446**, 775.

Cowan, J.J., McWilliam, A., Sneden, C. and Burris, D.L.: 1997, *Astrophys. J.* **480**, 246.

Cowan, J.J., Pfeiffer, B., Kratz, K.-L., Thielemann, F.-K., Sneden, C., Burles, S., Tytler, D. and Beers, T.C.: 1998, *Astrophys. J.* (submitted).

Gacquer, W.: 1997, *IAUS* **187**, 102.

Gilroy, K.K., Sneden, C., Pilachowski, C.A. and Cowan, J.J.: 1988, *Astrophys. J.* **327**, 298.

Gratton, R.G. and Sneden, C.: 1988, *Astron. Astrophys.* **204**, 193.

Gratton, R.G. and Sneden, C.: 1994, *Astron. Astrophys.* **287**, 927.

Griffin, R., Griffin, R., Gustafsson, B. and Vieira, T.: 1982, *Mon. Not. R. Astron. Soc.* **198**, 637.

MacConnell, D.J. and Frye, R.L.: 1972, *Astron. J.* **77**, 384.

Magain, P.: 1995, *Astron. Astrophys.* **297**, 686.

McWilliam, A., Preston, G.W., Sneden, C. and Searle, L.: 1995, *Astron. J.* **109**, 2736 (MPSS95).

McWilliam, A.: 1998, *Astron. J.* **115**, 1640.

Raiteri, C.M., Gallino, R., Busso, M., Neuberger, D. and Käppeler, F.: 1993, *Astrophys. J.* **419**, 207.

Ryan, S.G. and Norris, J.E.: 1991, *Astron. J.* **101**, 1865.

Ryan, S.G., Norris, J.E. and Beers, T.C.: 1996, *Astrophys. J.* **471**, 254.

Searle, L. and McWilliam, A.: 1998, in progress.

Shigeyama, T. and Tsujimoto, T.: 1998, *Astrophys. J.* **507**, L135.

Sneden, C. and Parthasarathy, M.: 1983, *Astrophys. J.* **267**, 757.

Sneden, C., McWilliam, A., Preston, G.W., Cowan, J.J., Burris, D.L. and Armosky, B.J.: 1996, *Astrophys. J.* **467**, 819.

Truran, J.W.: 1981, *Astron. Astrophys.* **97**, 391.

Woosley, S.E. and Hoffman, R.D.: 1992, *Astrophys. J.* **395**, 202.

ABUNDANCES IN VERY METAL-POOR STARS

M. SPITE, F. SPITE, R. CAYREL and V. HILL
DASGAL, Observatoire de Paris, Paris, France

B. NORDSTRÖM
Astronomical Observatory, Copenhagen University, Copenhagen, Denmark

B. BARBUY
Univ. of São Paulo-IAG, Dep. de Astronomia Brazil, São Paulo, Brazil

T. BEERS
Michigan State University, Dep. of Astronomy, East Lansing, MI, USA

P.E. NISSEN
Inst. of Phys. and Astron., Aarhus Univ., Aarhus, Denmark

1. Introduction

Stars with a metallicity lower than [Fe/H] = −3 are of particular interest since they were formed early in the history of the Galaxy. The chemical compositions of their atmospheres represent the yields of the first massive supernovae (Ryan *et al.*, 1991, 1996).

In the BPS survey (Beers *et al.*, 1992) many very metal poor stars have been found. We decided to obtain high resolution spectra of a sample of these stars in order make a homogeneous study with emphasis on:
- Spread in abundance ratios
- Relations between abundances of different elements
- Occurrence of binarity
- Lithium abundance

In fact two different kinds of stars have been observed: CH CN rich stars which are evolved stars, and turnoff stars.

2. CH CN Rich Stars

These stars although they are very metal deficient ([Fe/H] < −3) present strong molecular bands in their spectra (Barbuy *et al.*, 1997; Hill *et al.*, 1999). An example is given in Figure 1 where we compare the spectrum of a normal metal poor giant with one of these CH CN rich stars.

 Astrophysics and Space Science is the original source of publication of this article. It is recommended that this article is cited as: *Astrophysics and Space Science* **265**: 141–144, 1999.
© 1999 *Kluwer Academic Publishers.*

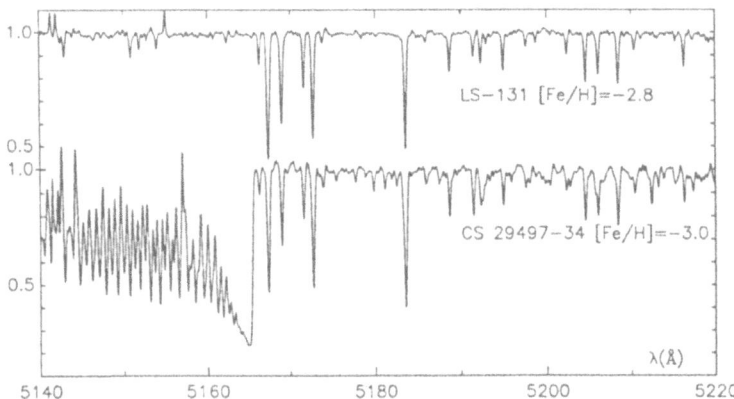

Figure 1. Comparaison of a normal metal-poor giant (top) and of a CH CN rich giant (lower spectrum) in the region of the Mg triplet and of a C2 band.

In these stars the temperatures are difficult to determine. The stars are cool and thus the wings of the hydrogen lines cannot be used. The relation color/temperature depends on the model used, the molecular bands influence so strongly on the continuum, that it is necessary to compute special models. We found for CS 22948-27 and CS 29497-34 a metallicity of -3 and -3.3 respectively, but the the ratios [C/Fe] and [N/Fe] amount to +2.

The $^{12}C/^{13}C$ ratio in these stars is about 13 which shows that they are strongly mixed.

From the blue spectra recently obtained, we could determine the abundance patterns of the neutron-rich heavy elements like Sr, Y, Ba, La, Ce Pr, Nd, Sm, Eu and Dy. These abundance patterns appear to be intermediate between s-process and r-process enrichment.

3. Turnoff Stars

Eleven turnoff stars have been studied. Two of them have been found to be double lined binaries: CS 22171-16 and CS 22873-139.

MARCS models (Edvardsson *et al.*, 1993) have been used and the temperatures of the stars have been deduced from the wings of the H_α line.

In the following table we give preliminary results for the lithium abundance. We note that in seven stars the lithium abundance was found to be constant: $N_{Li} = 2.15 \pm 0.05$.

In the double lined binaries the lithium abundance is low: $N_{Li} = 1.90$ in CS 22171-16, and $N_{Li} = 1.75$ in CS 22873-139. However these values do not take into account the veiling of the lithium lines by the light of the companion. It is probable that the abundance of lithium will become normal after correction at least in CS 22171-16.

TABLE I

Lithium abundance in the turnoff stars. (Preliminary results). The iron abundance [Fe/H] is from Beers *et al.* (1992)

Star	T_{eff}	[Fe/H]	W_{Li}	N_{Li}
Normal stars (as far as we know)				
CS 22175-13	6100	−3.	30	2.18
CS 22186-17	6200	−3.14	20	2.06
CS 22876-42	6300	−2.7	23	2.20
CS 22943-95	6200	−2.55	27	2.20
CS 22953-37	6300	−	21	2.15
CS 29518-43	6150	−4.	23	2.09
CS 29520-89	6200	−3.7	24	2.14
CS 22952-11	6000	−3.16	15	1.86
CS 29527-15	6050	−4.	17	1.86
Double lined binaries				
CS 22171-16	≈ 6100	−3.2	17	1.90
CS 22873-139	≈ 6200	−3.14	10	1.75

Two other stars appear to have a low abundance of lithium, (N_{Li} = 1.9), their temperature has to be carefully checked (an error of 200K on the temperature correspond to an error of 0.2 dex on N_{Li} but they could also be binaries.

The detailed study of these turnoff stars is underway.

Acknowledgements

This work was partially supported by CNPq and FAPESP as well as the Carlsberg Foundation and the Danish Science Research Council.
We also acknowledge allocation of telescope time at ESO, La Silla.

References

Barbuy, B., Cayrel, R., Spite, M., Beers, T., Spite, F., Nordström, B. and Nissen, P.E.: 1997, *Astron. Astrophys.* **317**, L63.
Beers, T.C., Preston, G.W. and Schectman, S.A.: 1992, *Astron. J.* **103**, 1987 (BPS).

Edvardsson, B., Andersen, J., Gustafsson, B., Lambert, D., Nissen, P.E. and Tomkin, J.: 1993, *Astron. Astrophys.* **275**, 101.

Hill, V., Barbuy, B., Spite, M., Spite F., Cayrel R., Plez, B., Beers, T., Nordström, B. and Nissen, P.E.: 1999, *Astron. Astrophys.*, submitted.

Ryan, S., Norris, J. and Bessell, M.: 1991, *Astron. J.* **102**, 303.

Ryan, S., Norris, J. and Beers, T.: 1996, *Astrophys. J.* **471**, 254.

ABUNDANCES IN GLOBULAR CLUSTERS

C. SNEDEN

Department of Astronomy & McDonald Observatory, University of Texas, Austin, TX 78712, USA

Abstract. Large inter- and intra-cluster abundance variations of several light elements are observed among Galactic globular cluster stars. Correlations among these abundances suggest that many of the chemical inhomogeneities have been caused by high temperature advanced proton-capture fusion reactions. The evidence for this proton-capture reshuffling is addressed, along with comments on possible sites for this nucleosynthesis.

1. Introduction

Knowledge of globular cluster chemistry has increased rapidly over the last four decades, tracking significant enhancements in instruments, detectors, and telescope apertures. Improved broad-band photometry, proper motions, and radial velocities have yielded good target samples for many globular clusters over a large M_V range. Spectroscopists have responded by deriving abundances for tens to hundreds of stars in the brighter globular clusters. These new data have answered some of the early questions about the chemical histories of clusters, and in so doing have raised even more complex questions.

Abundance inhomogeneities probably exist in all clusters. Their abundance patterns share some of the characteristics of halo field stars, but there are enough differences to warrant serious questions on whether field and cluster stars are drawn from the same stellar populations. Important topics (e.g., the cluster metallicity scale) must of necessity be neglected in this short review. The massive multi-metallicity globular cluster ω Cen has unique abundance features (Norris and Da Costa, 1995) and deserves its own review. Early work on clusters often dealt with low-resolution spectroscopy of CH, CN and NH molecular bands. There were pioneering high resolution spectroscopic research on cluster giants by Cohen, Peterson, Pilachowski and Wallerstein. These studies have received several excellent reviews (Kraft, 1979; Freeman and Norris, 1981; Smith, 1987) and so will receive little additional attention here. Rather, we discuss the rapidly changing views of cluster chemical compositions from spectroscopy of the last decade.

Astrophysics and Space Science is the original source of publication of this article. It is recommended that this article is cited as: *Astrophysics and Space Science* **265**: 145–152, 1999.
© 1999 *Kluwer Academic Publishers.*

2. Oxygen Depletion in Cluster Giants

Oxygen is overabundant in metal-poor field stars. The only question is the magnitude of the overabundance and whether it is constant with metallicity. The [O I] lines in cool giant stars consistently yield [O/Fe] $\simeq$ +0.4 over the whole metallicity range $-1 \geq$ [Fe/H] ≥ -3, while the high excitation O I triplet lines in less evolved warmer stars suggest much larger overabundances: [O/Fe] $\simeq$ +0.8 (e.g., Tomkin *et al.*, 1992), and recent work on these lines and on OH near-UV transitions (e.g., Israelian *et al.*, 1998) argues for increasing [O/Fe] with decreasing [Fe/H]. This issue is far from settled; here we concentrate on abundances from the [O I] lines, because those are the usual oxygen abundance indicators in cluster giants.

In the top panel of Figure 1 we have combined data for halo field giants from several literature sources and from a new survey by Carretta *et al.*(this conference; their sample includes stars from a variety of evolutionary states). No attempt has been made to normalize the different studies to a common abundance scale, and duplicate points have not been eliminated. But the $\sim$ +0.4 dex oxygen overabundance for field giants with [Fe/H] ≤ -1 is seen in all investigations, and much of the $\sim \pm0.15$ dex scatter at a given [Fe/H] is due to observational/analytical uncertainties. There are rare cases of halo stars with truly lower [O/Fe]; one such star in Figure 1 is HD 134439 ([Fe/H] = −1.39, [O/Fe] $\simeq$ 0.0), which is known to be deficient in all light α-capture elements (King, 1997). These low-α stars, probably members of a separate halo sub-population, are discussed by Nissen and Schuster (1997).

By contrast, oxygen abundances are quite varied among globular cluster giants. For clusters with sufficiently large sample size ($\geq$ 10 stars), [O/Fe] scatters of 0.5 dex are common, and in M13 (Kraft *et al.*, 1997) the range in [O/Fe] is from +0.4 to –0.9. Stars with low [O/Fe] may be found all over the giant branch, from the tip down to at least the luminosity of the HB. We assume that the oxygen contents of O-poor stars have at some past time been depleted from initial ratios [O/Fe] $\sim$ +0.4. This is because the maximum [O/Fe] values of cluster giants agree well with those of field-stars, and because of correlations with other abundance ratios (see below).

3. CNO Abundance Sums

The simplest way to deplete oxygen in stellar interiors is through the ON-cycle branch of CNO-cycle proton-capture fusion. This can be tested observationally, by checking for constancy of the abundance sum C+N+O. Sufficient data are now available to make this test, and in Figure 2 we plot the results of several studies. Note that in reality [(C+N+O)/Fe] $\simeq$ [(N+O)/Fe] in most globular cluster giants, since carbon will be severely depleted in the CN-cycle prior to the onset of ON-cycle burning. With that qualifier, the data of Figure 2 exhibit constancy of the

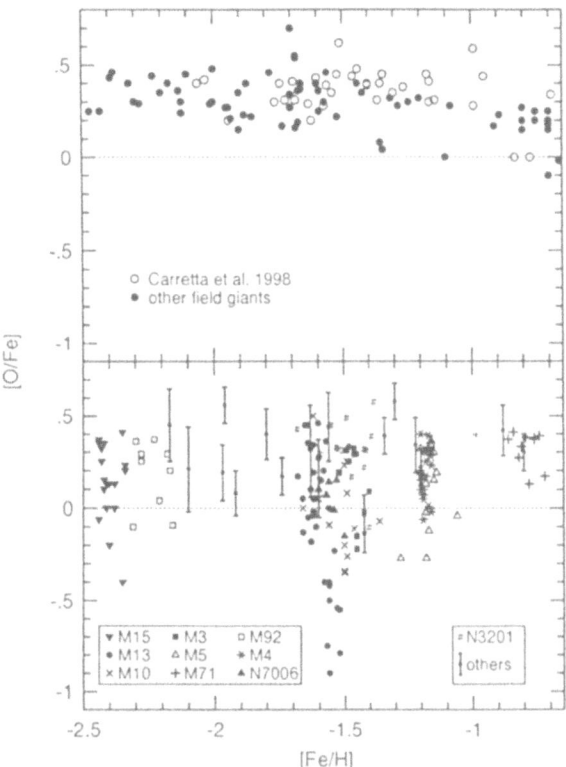

Figure 1. Relative [O/Fe] abundance ratios in field giants (upper panel) and globular cluster giants (lower panel). The data for field stars are taken from Carretta *et al.*(this conference) and from Gratton and Ortolani (1986), Barbuy (1988), and Barbuy and Erdelyi-Mendes (1989). Data for the members of the clusters named in the left-hand legend of the lower panel are from the Lick/Texas group (Kraft *et al.*, 1998; Sneden *et al.*, 1997; and references therein). The points for NGC3201 are from Gonzalez and Wallerstein (1998). For clusters with abundance analyses of only a few giants, we simply plot mean [O/Fe] values and sample deviations about the means; sources for these points (labeled 'others' in the legend) are Gratton (1987), Gratton and Ortolani (1989), Brown *et al.*(1990), and Minniti *et al.*(1996). ω Cen stars (not plotted here) also exhibit a large [O/Fe] scatter throughout its [Fe/H] range (see Norris and Da Costa, 1995).

C+N+O abundance sum, except for one anomalous star. That star is M15 K969, which has several abundance peculiarities: it is the only star in that very metal-poor cluster to have detectable a CN λ4215 Å bandhead, leading to [N/Fe] = +1.6 (Trefzger *et al.*, 1983), and it has by far the largest Ba abundance of the 18 giants studied by Sneden *et al.*(1997). Clearly K969's material has a more complex nucleosynthesis history, and can be safely set aside from the CNO-cycle arguments, pending discovery of further examples of this type. For now, it appears that cluster giants conserve the C+N+O abundance sum, and this conservation holds in the face of often large scatter in [O/Fe] ratios among stars of each cluster.

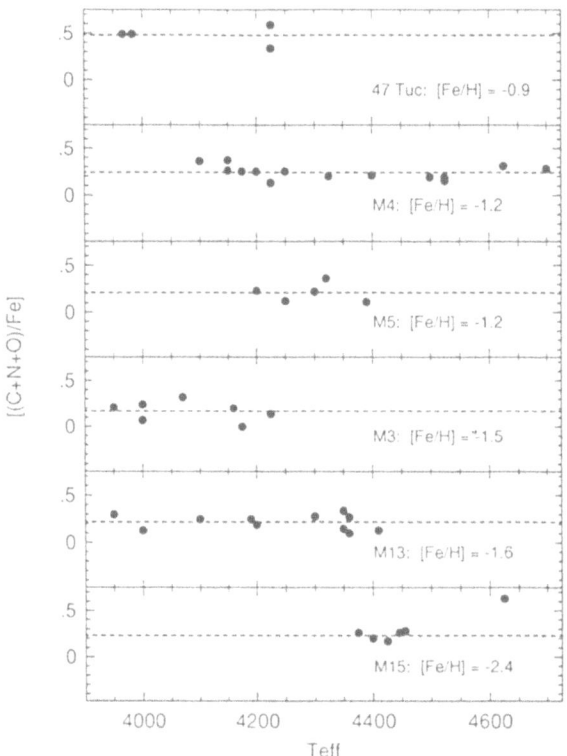

Figure 2. C+N+O abundance sums plotted as a function of T_{eff}. The dashed line in each panel represents the mean abundance sum for a cluster. Data are taken from Brown *et al.*(1990), Ivans *et al.*(this conference), Smith *et al.*(1996, 1997), and Kraft *et al.*(1997).

4. Light Element Correlations and Anticorrelations

Variations in the abundances of sodium and aluminum have been known for some time, as well as their positive correlations with CN band strengths, and Cottrell and DaCosta (1981) suggested that ^{23}Na could be synthesized by neutron captures by ^{22}Ne during helium fusion episodes. More recent work attributes Na production to high-temperature NeNa proton-capture fusion cycles. This explanation is indicated by the mass of observational data demonstrating that the abundances of O and Na are anticorrelated. In Figure 3 we summarize these data. Field giants occupy only the high-O, low-Na area of this figure. But the cluster giants with low [O/Fe] ratios (especially those of M13) regularly accompany them with large values of [Na/Fe]. Moreover, those cluster stars with high [O/Fe] have low [Na/Fe] values in good agreement with the field giants. From this, it is easy to argue that the *ab initio* [Na/Fe] values in clusters were low (like the field giants) and that more recent high-temperature NeNa cycle processing has enhanced Na at the same time that the ON cycle has depleted O. But the inclusion of abundances from ω Cen alerts

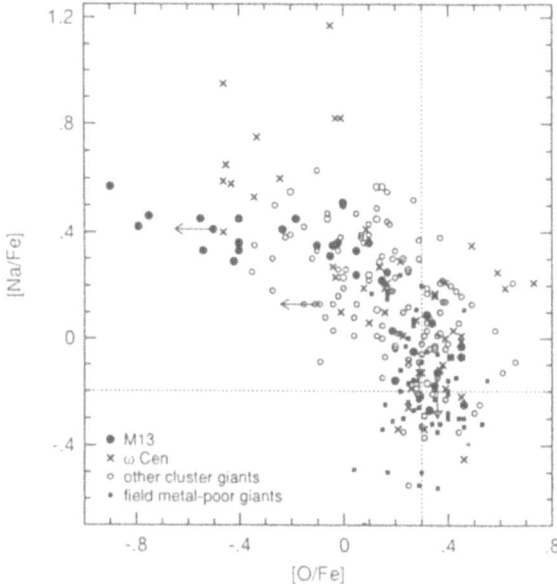

Figure 3. Abundances of oxygen and sodium in field and cluster giant stars. The Na abundances for field giants are from Pilachowski *et al.*(1996) and the O abundances for these stars are from the literature sources cited for Figure 1. The vertical and horizontal dotted lines intersect at ([O/Fe],[Na/Fe]) = (+0.3, −0.2), the approximate mean for the field giants. The globular cluster data are from the same same studies cited in Figure 1, and attention is drawn to M13 by the use of filled circles for its data. In this figure only, abundances for ω Cen from Norris and DaCosta (1995) are displayed also.

one that the connection among various light element abundances is more complex in some clusters. In ω Cen, O and Na are anticorrelated but some stars have much larger Na abundances than might be imagined from the simple action of the NeNa cycle. In fact, it is remarkable that ω Cen abundances show this anticorrelation at all, given the extreme variations in overall Fe metallicity.

Aluminum and magnesium abundances can be quite variable in some clusters, and there is a rough anticorrelation between these two elements. However, the relationship is not nearly as clean as that of O vs. Na. The clusters M13 and ω Cen have the most obvious Al vs. Mg anticorrelations discovered to date (the very large samples of stars analyzed in these clusters may be partly the reason). This anticorrelation also suggests a simple scenario in which at interior temperatures even more extreme than those needed to operate the ON and NeNa cycles, the MgAl cycle drives production of fresh Al at the expense of Mg. Part of the messy nature of the Al vs. Mg relationship in other clusters may be purely an observational headache: the few Mg I lines upon which the Mg abundances are based tend to be very strong even in metal-poor giant stars, and the abundance derivations have substantial uncertainties. The rest of the confusion in this anticorrelation may simply be trying to signal a multiplicity in the origin of Al in globular clusters.

In M4 for example, Al abundances vary a lot from star to star but they all are large: $+0.4 \leq$ [Al/Fe] $\leq +0.8$ (Brown and Wallerstein, 1992; Ivans *et al.*, in this conference), whereas in M13 the range is larger and begins from underabundance: $-0.2 \leq$ [Al/Fe] $\leq +1.2$.

5. Supporting Data and Comments on the Nucleosynthesis

Isotopic abundance ratios provide additional clues. Carbon isotopic ratios of upper red giant branch stars in clusters with [Fe/H] < -1 are nearly always low: $^{12}C/^{13}C \simeq 4$ (e.g., Suntzeff and Smith, 1991; Shetrone, 1996; Brown and Wallerstein, 1992; Ivans *et al.*, in this conference). This ratio is about the equilibrium value expected in the CN-cycle hydrogen fusion layers. Also, for M13 Shetrone (1996) has determined Mg isotopic ratios, finding unexpectedly high values of $^{25}Mg/^{24}Mg$ and $^{26}Mg/^{24}Mg$; the overabundances of Al from the MgAl cycle are primarily draining very abundant ^{24}Mg.

The isotopic abundance ratios, the O-Na, and Al-Mg anticorrelations, and large depletions of C and enhancements of N in cluster giants all point to high-temperature proton-capture nucleosynthesis. This seems to be firmly established. Remaining questions revolve around where and when it occurred. An opening assumption is that the cluster stars we observe have altered their surface abundances through internal nucleosynthesis and envelope mixing events as they have ascended the giant branch. In M15 and M92, for example, Carbon *et al.* (1992) and Trefzger *et al.*(1993) showed that the abundance of C declines and N rises with decreasing M_V. Kraft *et al.*(1997 and references therein) demonstrated the same effect in O abundances for M13. But theoretically, the ON and NeNa cycles require temperatures of $\sim 40 \times 10^6$ K; the MgAl cycle needs $\sim 70 \times 10^6$ K (e.g., Langer *et al.*, 1997). Such temperatures are not easily attained in hydrogen-burning shell regions of low mass, low metallicity model stars. Cavallo *et al.*(1998) have recently explored the parameters involved in such models. They show that some observed abundances can be matched, such as the low carbon isotopic ratios, while others like the magnesium abundances and isotopic ratios are more difficult to understand, and may depend on assumed nuclear reaction rates. A continuing frustration of these kinds of studies is the lack of ability to completely model the hydrogen shell fusion; it is usually assumed to proceed at a steady rate for computational simplicity, but probably is time-variable, with accompanying temperature variations.

The interior temperature problem may be mitigated by blaming the proton-capture nucleosynthesis on an earlier generation of higher-mass stars: the giants observed today simply were born with the abundances we now see. The most compelling observational evidence favoring this idea comes from the data gathered on the mildly metal-poor cluster 47 Tuc. Repeated investigations (Briley, 1997; Briley *et al.*, 1994 and references therein) have shown that CH and CN band strengths vary all the way from the main sequence to the giant branch tip; the fractions of

CN strong to CN weak stars do not change with M_V. CH and CN are anticorrelated, implicating the CN cycle, but it is difficult to imagine mixing episodes that would alter the surface C and N abundances in main sequence stars. So at least some of the variations appear to have been created by earlier cluster generations of stars. This idea has its own unanswered questions. For example, how do chemical inhomogeneities created by nucleosynthesis in individual prior stars manage to not be homogenized in the intra-cluster gas that forms subsequent stars?

More extensive data on the relationship of the abundance anomalies to other cluster properties (overall metallicity, c-m diagram morphology, . . .) are needed. A complete scenario should invoke multiple generations of stellar element donors, including the presently observed stars. With that said, we call attention to recent data on neutron-capture elements in two clusters. In many ω Cen stars, there is a very surprising deficiency in europium, but barium is plentiful (Norris and Da Costa, 1995; Smith *et al.*, 1995). But in M15, europium varies from star-to-star and is always more overabundant than barium. Europium is a nearly pure 'r-process' neutron-capture element, probably synthesized only in supernovae explosions. This element is not altered in quiescent stellar evolution, and variations in its abundance must be due to an earlier generation of cluster stars. Clearly, a complex past and present must be considered in the chemistry of globular clusters.

Acknowledgements

The support of my colleagues in the Lick/Texas cluster abundance collaboration is gratefully acknowledged, as is the hospitality and encouragement of R. and G. Cayrel to me during an extended visit to the Paris Observatory in 1998. This work has been sponsored in part by NSF grant AST-9618364. I am also grateful for a grant from the University of Texas Rom Rhome International Travel Fund.

References

Barbuy, B.: 1988, *Astron. Astrophys.* **191**, 121.
Barbuy, B. and Erdelyi-Mendez, M.: 1989, *Astron. Astrophys.* **214**, 239.
Briley, M.M.: 1997, *Astron. J.* **114**, 1051.
Briley, M.M., Hesser, J.E., Bell, R.A., Bolte, M. and Smith, G.H.: 1994, *Astron. J.* **108**, 2183.
Brown, J.A. and Wallerstein, G.: 1990, *Astron. J.* **98**, 1643.
Brown, J.A., Wallerstein, G. and Oke, J.B.: 1990 *Astron. J.* **101**, 1693.
Carbon, D.F., Romanishin, W., Langer, G.E., Butler, D., Kemper, E., Trefzger, C.F., Kraft, R.P. and Suntzeff, N.B.: 1982, *Astrophys. J. Suppl.* **49**, 207.
Cavallo, R.M., Sweigart, A.V. and Bell, R.A.: 1998, *Astrophys. J.* **492**, 575.
Cottrell, P.L. and Da Costa, G.S.: *Astrophys. J.* **245**, L79.
Freeman, K.C. and Norris, J.: 1981, *Annu. Rev. Astron. Astrophys.* **19**, 319.
Gonzalez, G. and Wallerstein, G.: 1998, *Astron. J.* **116**, 765.

Gratton, R.G.: 1987, *Astron. Astrophys.* **179**, 181.

Gratton, R.G. and Ortolani, S.: 1986, *Astron. Astrophys.* **169**, 201.

Gratton, R.G. and Ortolani, S.: 1989, *Astron. Astrophys.* **208**, 171.

Israelian, G., García López, R.J. and Rebolo, R.: 1998, *Astrophys. J.* **507**, 805.

King, J.R.: 1997, *Astron. J.* **113**, 2302.

Kraft, R.P.: 1979, *Annu. Rev. Astron. Astrophys.* **17**, 309.

Kraft, R.P., Sneden, C., Smith, G.H., Shetrone, M.D., Langer, G.E. and Pilachowski, C.A.: 1997, *Astron. J.* **113**, 279.

Kraft, R.P., Sneden, C., Smith, G.H., Shetrone, M.D. and Fulbright, J.: 1998, *Astron. J.* **115**, 1500.

Langer, G.E., Hoffman, R.E. and Zaidins, C.S.: 1997, *Publ. Astron. Soc. Pacific* **109**, 244.

Minniti, D., Peterson, R.C., Geisler, D. and Claria, J.J.: 1996, *Astrophys. J.* **470**, 953.

Nissen, P. and Schuster, W.J.: 1997, *Astron. Astrophys.* **326**, 751.

Norris, J.E. and Da Costa, G.S.: 1995, *Astrophys. J.* **447**, 680.

Shetrone, M.D.: 1996, *Astron. J.* **112**, 2639.

Smith, G.H., Shetrone, M.D., Bell, R.A., Churchill, C.W. and Briley, M.M.: 1996, *Astron. J.* **112**, 1511.

Smith, G.H., Shetrone, M.D., Briley, M.M., Churchill, C.W. and Bell, R.A.: 1997, *Publ. Astron. Soc. Pacific* **109**, 236.

Smith, V.V., Cunha, K. and Lambert, D.L.: 1995, *Astron. J.* **110**, 2827.

Sneden, C., Kraft, R.P., Shetrone, M.D., Smith, G.H., Langer, G.E. and Prosser, C.F.: 1997, *Astron. J.* **114**, 1964.

Suntzeff, N.B. and Smith, V.V.: 1991, *Astrophys. J.* **381**, 160.

Tomkin, J., Lemke, M., Lambert, D.L. and Sneden, C.: 1992, *Astron. J.* **104**, 1568.

Trefzger, C.F., Carbon, D.F., Langer, G.E., Suntzeff, N.B. and Kraft, R.P.: 1983, *Astrophys. J.* **266**, 144.

COMPOSITION DIFFERENCES BETWEEN GLOBULAR CLUSTER AND HALO FIELD GIANTS

R.P. KRAFT

UCO/Lick Observatory, University of California at Santa Cruz, Santa Cruz, CA 95064, USA

Abstract. Low metallicity $(-3 <= [Fe/H] <= -1)$ halo field giants exhibit the expected correlation of Na and Mg abundances, based on the assumption that Na is produced in the same nucleosynthetic sites as are the alpha elements, confirming a result noted by Sneden (1998). On the other hand, giants in at least some globular clusters (especially M13, but also M15 and NGC 6752) do not exhibit the Mg vs Na correlation found among halo field giants (Hanson *et al.*, 1998). The very large [Na/Fe]-ratios and widely scattered [Mg/Fe]-ratios found among M13 giants depend, on the average, on evolutionary state and are probably induced by deep mixing of stellar envelopes through the CNO hydrogen-burning shell. Why M13 (and M15 and NGC 6752) giants should experience deep mixing whereas field halo giants in the same evolutionary state mix not at all is an anomaly unexplained by current theories of stellar evolution. By contrast, giants in the outer halo cluster NGC 7006 show little evidence of deep mixing (Kraft *et al.*, 1998). These differences in the degree of deep mixing among stars in related, but different, stellar populations may be connected to the so-called 'second parameter effect'.

On the basis of a medium spectral resolution survey of Bond's (1980) sample (Pilachowski *et al.,* 1996a), Sneden (1998) discovered that the Na and Mg abundances of halo field giants were roughly correlated (Figure 1a, filled circles) (cf. Hanson *et al.*, 1998). The scatter about the mean line in Figure 1a is rather large, but is believed to result largely from observational errors in the derived [Mg/Fe]-ratios. The correlation probably reflects the fact that Na and Mg were produced in the same nucleosynthetic sites, and that on the average, the neutron excess varies over a fairly small range. The giants in question lie in the metallicity interval $-3 \leq$ [Fe/H] ≤ -1; a similar correlation had been found earlier among stars having [Fe/H] > -1.2 (Nissen and Schuster, 1997).

The correlation between [Na/Fe] and [Mg/Fe] shown by these halo field giants is not exhibited by giants in at least some globular clusters having stars of comparable metallicity, e.g., M13 (Pilachowski *et al.*, 1996b; Kraft *et al.*, 1997) and M15 (Sneden *et al.*, 1997). However, in making the comparison between halo field and cluster giants, one must ensure that the samples refer to stars in comparable evolutionary states. Because the luminosity function of giants increases rapidly with increasing M_v, Bond's (1980) objective prism survey is heavily weighted toward giants of relatively low luminosity (i.e., those having $\log g > +1$ or $M_v > -1.7$). Thus some 75 percent of the halo field giants exhibited in Figure 1a have $\log g > +1$. On the other hand, flux limitations generally confine abundance surveys among

Astrophysics and Space Science is the original source of publication of this article. It is recommended that this article is cited as: *Astrophysics and Space Science* **265**: 153–156, 1999.
© 1999 *Kluwer Academic Publishers*.

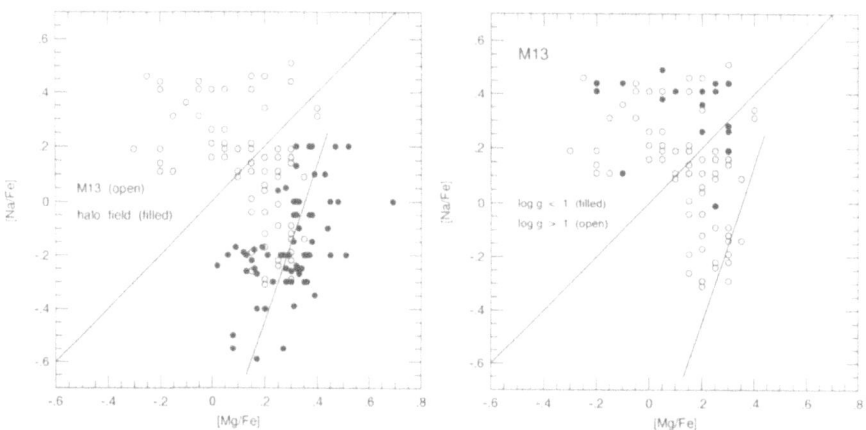

Figure 1. (a) M13 and field giants and (b) M13 giants alone.

globular cluster stars to bright giants (i.e., those having $\log g < +1$ or $M_v < -1.7$), which occupy a more advanced evolutionary state.

Fortunately the deep survey of Na and Mg abundances among M13 giants (Pilachowski et al 1996b) contains 80 giants having $\log g > +1$; these are plotted as open circles in Figure 1a. Except for a few stars, the distribution of open and filled circles is remarkably dissimilar. A clue to the disagreement is found from inspection of Figure 1b, in which we plot [Mg/Fe] vs [Na/Fe] for M13 giants alone, but now we add the 18 giants of M13 (a complete sample) which have $\log g < +1$. The M13 giants in the more advanced evolutionary state generally lie in the regime of high Na abundances; the less advanced stars can have low Na abundances similar to those found among halo field giants.

Inspection of Figure 1b leads to the conclusion (Pilachowski *et al.*, 1996b; Kraft *et al.*, 1997) that the change in the Na distribution of M13 giants is a function of evolutionary state, and therefore reflects an activity taking place within the stars themselves. 'Deep mixing' is the most plausible mechanism, the abundance alterations having resulted from the processing of the stellar envelope through the CNO hydrogen-burning shell (Langer *et al.*, 1997; Sweigart, 1997a,b). In this picture, the final abundance pattern depends upon the depth to which the mixing proceeds, and this may vary randomly from star to star in the same evolutionary state. Relatively shallow mixing drives the $^{12}C/^{13}C$-ratio toward equilibrium and converts C into N. Deeper mixing converts O into N and Ne into Na; still deeper mixing converts Mg into Al (see reviews by Kraft, 1994; Briley *et al.*, 1994). The 'hook-like' distribution of points in Figure 1b may be understood if the mixing is deep enough to process Ne into Na in all M13 giants, but the still deeper mixing required to convert Mg into Al takes place in some, but certainly not all, of these stars.

It has long been supposed that halo field and globular cluster stars of comparable metallicity are simply surrogates of one another, but this appears not to be the case.

It may be that M13 stars leave the main sequence satisfying the same Mg vs Na correlation as do halo field stars. But by the time they reach the mid-giant branch deep mixing sets in, which drastically alters the surface abundances of C, N, O, Ne, and Na, and in some instances Mg and Al as well. Similar deep-mixing effects are found among giants in other globular clusters, notably M15 (Sneden *et al.*, 1997), NGC 6752 (Shetrone, 1997) and NGC 3201 (Gonzalez and Wallerstein, 1998). But halo field giants do not experience such dramatic deep mixing effects, the alterations being mostly confined to the conversion of ^{12}C to ^{13}C and C to N (cf. Gratton, 1998).

The driving mechanism underlying the deep mixing phenomenon is not firmly identified, but is suspected to be meridional circulation driven by stellar angular momentum (Sweigart and Mengel, 1979). Support for this view is found in the unusually high rotational velocities of M13 horizontal branch (HB) stars (Peterson, 1983; Peterson *et al.*, 1995). The rotational velocities of HB stars in M3, a cluster of comparable metallicity, are smaller than those of M13, and it is notable that M3 giants appear to have less evidence for deep mixing than is the case for M13 giants (Kraft *et al.*, 1993). An obvious corollary then is the expectation that halo low-metallicity HB stars should have rotational velocities as low, or lower, than HB stars in M3. A rotational survey of a pure sample of halo HB stars (Kinman *et al.*, 1994) is conceivable, but high spectral resolution is required and the stars are faint ($V > 14$).

References

Bond, H.: 1980, *Astrophys. J. Suppl.* **44**, 517

Briley, M.M., Bell, R.A., Hesser, J.E. and Smith, G.H.: 1994, *Can. J. Phys.* **72**, 772.

Gonzalez, G. and Wallerstein, G.: 1998, *Astron. J.* **116**, in press.

Gratton, R.G.: 1998, *Poster Paper*, this Conference, in press.

Hanson, R.B., Sneden, C., Kraft, R.P. and Fulbright, J.P.: 1998, *Astron. J.* **116**, 1286.

Kinman, T.D., Suntzeff, N.B. and Kraft, R.P.: 1994, *Astron. J.* **108**, 1722.

Kraft, R.P.: 1994, *Publ. Astron. Soc. Pacific* **106**, 553.

Kraft, R.P., Sneden, C., Langer, G.E. Shetrone, M.D.: 1993, *Astron. J.* **106**, 1490.

Kraft, R.P., Sneden, C., Smith, G.H., Shetrone, M.D., Langer, G.E. and Pilachowski, C.A.: 1997, *Astron. J.* **113**, 279.

Langer, G.E., Hoffman, R.D. and Zaidins, C.S.: 1997, *Publ. Astron. Soc. Pacific* **109**, 244.

Nissen, P.E. and Schuster, W.J.: 1997, *Astron. Astrophys.* **326**, 751.

Peterson, R.C.: 1983, *Astrophys. J.* **275**, 737.

Peterson, R.C., Rood, R.T. and Crocker, D.A.: 1995, *Astrophys. J.* **453**, 214.

Pilachowski, C.A., Sneden, C. and Kraft, R.P.: 1996a, *Astron. J.* **111**, 1689.

Pilachowski, C.A., Sneden, C., Kraft, R.P. and Langer, G.E.: 1996b, *Astron. J.* **112**, 545.

Shetrone, M.D.: 1997, in: *IAU Symp.* **189**, Sydney, in press.

Sneden, C.: 1998, in: *IAU Symp.* **187**, Kyoto, in press.

Sneden, C., Kraft, R.P., Shetrone, M.D., Smith, G.H., Langer, G.E. and and Prosser, C.F.: 1997, *Astron. J.* **114**, 1964.

Sweigart, A.V.: 1997a, *Astrophys. J.* **474**, L23.

Sweigart, A.V.: 1997b, in: A.G.D. Philip, J. Liebert and R. Saffer (eds.), *Third Conference on Faint Blue Stars*, L. Davis Press, Schenectady, N.Y., p. 3.
Sweigart, A.V. and Mengel, J.G.: 1979, *Astrophys. J.* **229**, 624.

O AND α−ELEMENT ABUNDANCES IN METAL-POOR STARS

R.G. GRATTON

Osservatorio Astronomico di Padova, Vicolo dell'Osservatorio 5, 35122 Padova, Italy

Abstract. Since about thirty years it is known that Oxygen and other α−elements are overabundant in metal-poor stars. In this talk I briefly review the historical and theoretical background, discuss reliability of present abundance determinations for O, and finally comment about the implications relevant to galactic chemical evolution.

1. Introduction

The first explicit reference to an overabundance of O in (moderately) metal-poor star is in Conti *et al.* (1967); the first modern determination (in very metal-poor stars) is the work by Lambert *et al.* (1974) on HD122563. Their value (based on the forbidden [OI] line at 6300 Å) is still valid today. Shortly after, Sneden *et al.* (1979) presented the first extensive set of O abundances in metal-poor stars, now based on the near-IR triplet. These lines are more sensitive to the adopted temperatures and require corrections for departures from LTE which were not included in the Sneden et al. analysis. For this reason, more reliable results were obtained in the next few years using the forbidden lines (Gratton and Ortolani, 1986; Barbuy, 1988), which showed the existence of a plateau in the [O/Fe] ratio at low metal abundances. This result was confirmed by analogous findings for the other α−elements (Mg, Si, Ca, Ti: François, 1986; Gratton and Sneden, 1987, 1988; Magain, 1987).

In the early '90s, additional important features were discovered: (i) a number of papers from the Lick-Texas group (Sneden *et al.*, 1991, and several more) revealed the presence of an O-Na anticorrelation amongst globular cluster giants; (ii) Gratton *et al.* (1996) noticed the presence of a change in [Fe/O] at constant [O/H] at the thick-thin disk transition; and finally (iii) very recently several papers have shown the existence of a small but significant intrinsic scatter in [Fe/O] values amongst field halo stars (King, 1997; Carney *et al.*, 1997; Jehin *et al.*, 1998).

The first interpretation of the overabundances of O and α−elements in metal-poor stars was based on variations of the initial mass function (IMF) (see e.g. Arnett, 1978). The idea is that yields of O and Fe from massive stars exploding as type II SNe might be a strong function of metal abundance. However, Fe yields from these models are largely uncertain due to poor knowledge of the explosion mechanism, and hence of the mass cut. Furthermore, models with flat IMF required to explain the large [O/Fe] ratio in metal-poor stars overproduce O in the disk. To

Astrophysics and Space Science is the original source of publication of this article. It is recommended that this article is cited as: *Astrophysics and Space Science* **265**: 157–164, 1999.
© 1999 *Kluwer Academic Publishers.*

R.G. GRATTON

TABLE I

Abundances with Holweger and Muller (1974) model atmosphere

Lines	Abundance	Source
Forbidden	8.92	Lambert (1978)
	8.86	Revised gf's (Baluja and Zeippen, 1988)
Permitted	8.86 ± 0.04	Biemont et al. (1991)
OH	8.91 ± 0.01	Sauval et al. (1984)

TABLE II

Role of the model atmosphere

	HM	K95 no	K95 over	OSMARCS
Forbidden	8.84	8.78	8.81	8.77
Permitted	8.81	8.73	8.82	8.89

solve this problem, Tinsley (1979) proposed that a rather large fraction of Fe is produced by type Ia SNe, whose progenitors are intermediate mass stars in binary systems (Whelan and Iben, 1973); a quantitative recipe was given by Matteucci and Greggio (1986): type Ia SNe produce about half of solar Fe but nearly no O and α−elements. Within this scenario, production of a significant part of Fe is delayed; stars forming before explosion of a significant number of type Ia SNe are Fe-poor. This provides a connection between timescales for stellar and galactic evolution: while O production is strictly related to star formation (SF), Fe results from the integral of SF over some former period.

Theoretical uncertainties are still large: they mainly concern problems in reproducing type II SN explosions; and the nature of the precursors of type Ia SNe, affecting their timescale (see discussion in e.g. Ruiz-Lapuente et al., 1995). In spite of this, element-to-element abundance ratios provide strong constraints on SF history during early phases of our own Galaxy.

2. Reliability of O Abundance Determinations

The solar O abundance can be obtained only using the photospheric spectrum (Anders and Grevesse, 1989). Table I lists values provided by different indicators, using the Holweger and Müller (1974, HM) model atmosphere. Agreement is quite good; however, when comparing these results with those obtained for stars, we should remind that derived abundances depend on the adopted model atmosphere.

TABLE III

Sensitivity of O abundances to adopted atmospheric parameters

Lines	$\Delta T = +100$ K		$\Delta \log g = +0.3$ dex		$\Delta[A/H] = +0.2$ dex	
	[O/H]	[O/Fe]	[O/H]	[O/Fe]	[O/H]	[O/Fe]
Permitted	−0.09	−0.07	−0.11	+0.06	+0.02	+0.11
Forbidden	+0.01	+0.03	−0.17	−0.01	0.00	+0.09
OH	+0.20	+0.10	−0.10	−0.08	−0.03	+0.01

In Table II abundances obtained with the HM model atmosphere and my own code are compared with those obtained using Kurucz model atmospheres with and without overshooting, and with OSMARCS model atmosphere. We remind that currently the best theoretical model atmospheres for solar type stars are those by Kurucz (1995) with no overshooting (Castelli *et al.*, 1997)). Abundances provided by permitted lines are very sensitive to the adopted model, and are affected by uncertainties in the adopted collisional damping parameter and possibly by some departures from LTE.

Meyer *et al.* (1998) have recently determined the abundance of O in the ISM from HST observations. They found a value of $\log n(O) = 8.70$, significantly lower than the value obtained with the HM solar model. They propose different solutions for this apparent solar peculiar O abundance: a type II SN event just before birth of the Sun, infall of oxygen-poor material, or effects of the galactic gradient (the galactic gradient has been measured as [O/H] = -0.07 dex kpc^{-1}: Smartt and Rolleston, 1997). However, given sensitivity of solar O abundance value to the adopted model atmospheres, we think more studies are required before assessing a real solar anomaly.

Table III gives typical sensitivity of O abundances on the atmospheric parameters adopted in the stellar analysis. When preparing this table, abundances from permitted and forbidden lines were compared to Fe abundances from ionized lines (this reduces sensitivity to gravity, since both O I and Fe II are dominant species); OH abundances with Fe abundances from neutral lines. There are large and opposite dependences of abundances from permitted lines and OH on adopted temperatures.

It has been suggested in the past that there are systematic differences between O abundances provided by permitted and forbidden lines. However, since dependences on temperature are different (see Table III) this is likely an artefact of a too low temperature scale (Gratton, 1991; see also Spite and Spite, 1991; Spiesman and Wallerstein, 1991). When modern temperature scales are adopted, and the appropriate, moderate non-LTE corrections are included in the analysis of permitted lines, a good agreement is achieved between abundances from permitted and

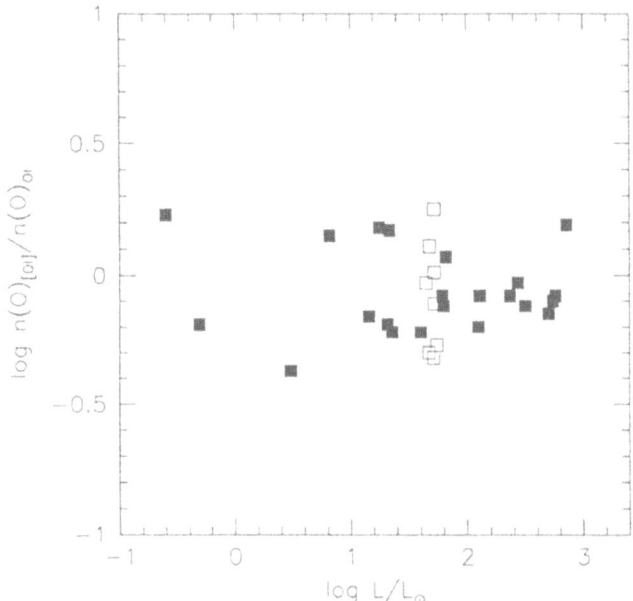

Figure 1. Run of the difference between O abundances determined from permitted and forbidden lines as a function of luminosity.

forbidden lines (see Figure 1): both these indices yield a moderate Fe deficiency of [Fe/O] ~ −0.4 with no clear trend with metal abundance (or luminosity).

Larger Fe deficiencies (Fe/O] ~ −0.7) in metal-poor stars have been obtained by Nissen *et al.* (1994) and Israelian *et al.* (1998) from the OH band in the ground-based UV. A comparison with abundances from permitted and forbidden lines (Figure 2) shows that the [O/H] values obtained by different studies are similar; however there are quite large differences in the [Fe/H] values, due to the adopted atmospheric parameters and model atmospheres. Once these differences are con-sidered, the agreement appears reasonable (in view of the large uncertainties in the analysis of this difficult band). Note however that OH lines are observable also in extremely metal-poor stars where other indicators are vanishingly weak: for this reason, analysis of OH bands is very promising.

3. Implications for Galactic Evolution

An issue often discussed in the literature is the constancy of the [Fe/O] ratio in metal-poor stars. It is generally assumed that a constant ratio implies a fast SF rate, while a non-zero slope signals a slow SF. However, galactic evolution models predict some trend even if SF is very fast, due to the different yields from type II SNe of different mass. In this last case some scatter is also expected (although of course not in one-zone models), similar to that observed for the neutron capture

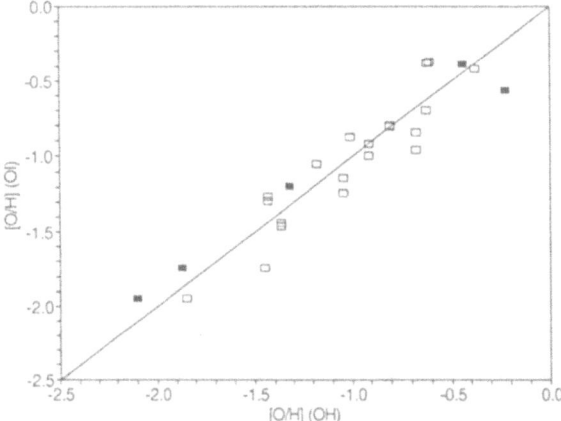

Figure 2. Comparison between O abundances from OH and OI triplet as a function of metal abundance: data for OH are from Nissen *et al.* (1994) (filled squares), and Israelian *et al.* (1998) (open squares); data for OI are from Carretta *et al.* (1998).

elements (McWilliam mechanism). The issue is then related to that of scatter in the Fe/O values: in case some intrinsic scatter exist, it would signal an independent chemical evolution, and might even be considered as a signature of accretion. Indeed, some stars (King, 1997; Carney *et al.*, 1997) and globular clusters (Brown *et al.*, 1996) have [Fe/O] values higher than or nearly equal to solar, indicative of a significant contribution by ejecta of type Ia SNe. Carney *et al.* suggested a possible correlation with apogalactic distance. However, as shown by Gratton (1997) Fe-rich halo stars are rare, and the correlation with apogalactic distance not clear. In spite of this, accretion remains a valid hypothesis to explain these objects.

On a smaller scale, how much is the intrinsic scatter in [Fe/O] ratios? The answer to this question has been generally searched for by looking at correlations between overabundances for different species. Laird and Xiao (1997) found a considerable scatter amongst the most metal-rich halo stars; more recently, the very careful study by Jehin *et al.* (1998) allowed them to divide halo stars into two groups according to overabundances of α− and s−process elements.

More insight into this matter is gained by looking at correlations between element-to-element abundance ratios and kinematics. An interesting contribution has been given by Nissen and Schuster (1997), who found that some metal-rich halo stars have lower α−elements to iron ratios than disk stars of similar metal abundance. In the discussion, they considered this result as possibly due to the accretion of individual dwarf galaxies. This has been criticized by Gilmore and Wyse (1998), who noticed that the low pericentric distances of objects poor in α−elements require that the parent galaxy disrupted very close to the galactic center, and then be very dense, differently from typical dwarf galaxies. They then suggested that Nissen and Schuster result is due to self-enrichment scenarios and survival of star forming complexes for ∼ 1 Gyr.

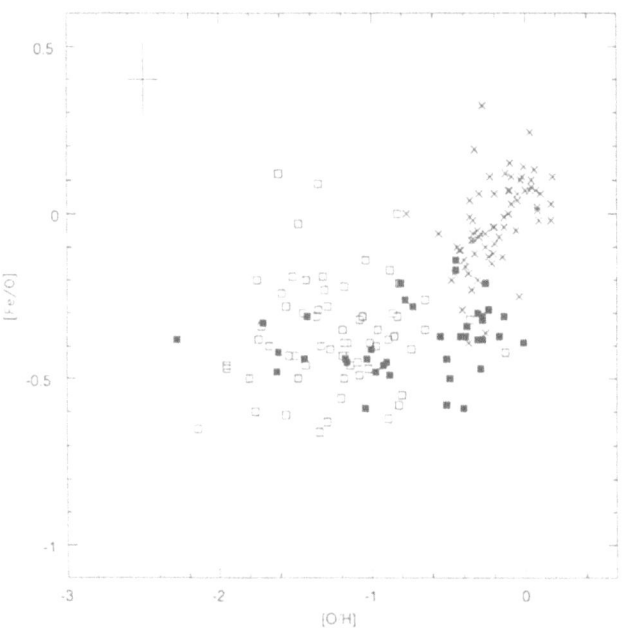

Figure 3. Run of the [Fe/O] ratio as a function of [O/H] for stars in three different kinematic populations: halo (open squares), thick disk (filled squares), and thin disk (crosses).

Another related issue concerns the transition between thick and thin disk. Gratton *et al.* (1996) noticed that the raise of the [Fe/O] ratio at constant [O/H] implies a hiatus in SF; a similar conclusion has been reached by Fuhrmann (1998). Gratton *et al.* argues that this is an evidence that the collapse of the Galaxy was not smooth; however, when combined with other evidences (not universality of thick discs), the scenario proposed included both dissipational collapse and cahotic accretion. In this scenario an original disk formed by dissipational collapse was heated (to a thick disk) by accretion of small satellite; the large induced SF rate (and consequently high SN rate) halted star formation that later started again in a newly formed (thin) disk.

In order to have some more insight into these problems, I considered a more extended sample of stars of different populations. This (biased and uncomplete) sample included data from Gratton *et al.* (1996; in part themselves a reanalysis of literature values), Carretta *et al.* (1998), Nissen *et al.* (1994), and Israelian *et al.* (1998). All data were put in a homogeneous scale applying suitable corrections when required. We then divided stars into three kynematics group: (a) Thin Disk ($V_{rot} > 170$ km s^{-1}, $z_{max} < 0.5$ kpc); (b) Thick Disk ($V_{rot} > 100$ km s^{-1}, $z_{max} < 1.5$ kpc); and (c) Halo ($V_{rot} < 100$ km s^{-1}, $z_{max} > 1.5$ kpc). The [O/H] vs [Fe/O] diagram obtained for this sample is shown in Figure 3. The main results are: (i) Halo and thick Disk have a similar O overabundance; (ii) the scatter is larger (and significant) for the halo; (iii) the transition from thick to thin disk occurs at increasing Fe while constant O (evidence for a phase of low SF). The scatter of [Fe/O] values observed in the halo (but not in the thick disk) suggests that the halo was made of individual fragments which evolve individually from chemical point of view. This might have two explanations: (i) contribution by type Ia SNe (as suggested by Gilmore and Wyse); (ii) the number of SNe in each individual fragment was small enough that significant random scatter arises. The second explanation is more appealing because it explains stars with [Fe/O] values below average, and because the dynamical timescale for 10^6 $M_\odot$ structures (10^7 yr) is shorter than galactic collapse time ($\sim 10^8$ yr). In order to explain the observed scatter of [Fe/O] values for the halo (0.15 dex at [O/H] = -1), fragment masses of 3×10^5 $M_\odot$ are required. These are of the same order of magnitude as globular clusters. The smaller scatter (< 0.1 dex) observed for thick disk stars at the same metallicity implies that the ISM was well mixed over masses $> 10^6$ $M_\odot$.

References

Anders and Grevesse: 1989, *Geochim. Cosmochim. Acta* **53**, 197.

Arnett, W.D.: 1978, *Astrophys. J.* **219**, 1008.

Baluja, K.L. and Zeippen, C.J.: 1988, *J. Phys. B* **21**, 1455.

Barbuy, B.: 1988, *Astron. Astrophys.* **191**, 121.

Biemont, E., Hibbert, A., Godefroid, M., Vaeck, N. and Fawcett, B.C.: 1991, *Astrophys. J.* **375**, 818.

Brown, J.A., Wallerstein, G. and Zucker, D.: 1997, *Astron. J.* **114**, 180.

Carney, B.W., Wright, J.S., Sneden, C., Laird, J.B. and Aguilar, L.A.: 1997, *Astron. J.* **114**, 363.

Carretta, E., Gratton, R.G., Sneden, C. and Bragaglia, A.: 1998, this meeting.

Castelli, F., Gratton, R.G. and Kurucz, R.L.: 1997, *Astron. Astrophys.* **318**, 841.

Conti, P.S., Greenstein, J.L., Spinrad, H., Wallerstein, G. and Vardja, M.S.: 1967, *Astrophys. J.* **148**, 105.

François, P.: 1986, *Astron. Astrophys.* **160**, 264.

Fuhrmann, K.: 1998, *Astron. Astrophys.* **338**, 161.

Gilmore, G. and Wyse, R.: 1998, *Astron. J.* **166**, 748.

Gratton, R.G.: 1991, in: G. Michaud and A.V. Tutukov (eds.), *Evolution of Stars: The Photometric Abundance Connection, IAU Symp* **145**, Univ. Montreal, Montreal, p. 27.

Gratton, R.G.: 1997, in: F. Caputo (ed.), *Views on Distance Indicators*, in press.

Gratton, R.G. and Ortolani, S.: 1986, *Astron. Astrophys.* **169**, 201.

Gratton, R.G. and Sneden, C.: 1987, *Astron. Astrophys.* **178**, 179.

Gratton, R.G. and Sneden, C.: 1988, *Astron. Astrophys.* **193**, 218.

Gratton, R.G., Carretta, E., Matteucci, F., Sneden, C.: 1996, in: *Formation of the Galactic Halo ... Inside and Out, ASP Conf. Ser.* **92**, 307.

Holweger, H. and Muller, E.A.: 1974, *Sol. Phys.* **39**, 19.

Israelian, G., García Lopez, R.J. and Rebolo, R.: 1998, *Astrophys. J.* **507**, 805.

Jehin, E., Magain, P., Neuforge, C., Noels, A. and Thoul, A.A.: 1998, *Astron. Astrophys.* **330**, L33.

King, J.R.: 1997, *Astron. J.* **113**, 2302.

Kurucz, R.L.: 1995, *CD-ROM*, 23.

Laird, J.B. and Xiao, Z.: 1997, *Astron. Astrophys. Suppl.* **191**, 1302.

Lambert, D.L.: 1978, *Mon. Not. R. Astron. Soc.* **182**, 249.

Lambert, D.L., Sneden, C. and Ries, L.: 1974, *Astrophys. J.* **188**, 97.

Magain, P.: 1987, *Astron. Astrophys.* **179**, 176.

Matteucci, F. and Greggio, L.: 1986, *Astron. Astrophys.* **154**, 279.

Meyer, D.M., Jura, M., Cardelli, J.A.: 1998, *Astrophys. J.* **493**, 222.

Nissen, P.E. and Schuster, W.J.: 1997, *Astron. Astrophys.* **326**, 751.

Nissen, P.E., Gustafsson, B., Edvardsson, B. and Gilmore, G.: 1994, *Astron. Astrophys.* **285**, 440.

Ruiz-Lapuente, P., Burkert, A. and Canal, R.: 1995, *Astrophys. J. Lett.* **477**, 69.

Sauval, A.J., Grevesse, N., Brault, J.W., Stokes, G.M. and Zander, R.: 1984, *Astrophys. J.* **282**, 330.

Smartt, S.J. and Rolleston, W.R.J.: 1997, *Astrophys. J.* **481**, L47.

Sneden, C., Lambert, D.L. and Whitaker: 1979, *Astrophys. J.* **234**, 964.

Sneden, C., Kraft, R.P., Prosser, C.F. and Langer, G.E.: 1991, *Astron. J.* **102**, 2001.

Spiesman, W.J. and Wallerstein, G.: 1991, *Astron. J.* **102**, 1790.

Spite, M. and Spite F.: 1991, *Astron. Astrophys.* **252**, 689.

Tinsley, B.M.: 1979, *Astrophys. J.* **229**, 1046. **460**, 408.

Whelan, I. and Iben, I. Jr.: 1973, *Astrophys. J.* **186**, 1007. **101**, 181.

NEW VIEWS ON THE EARLY EVOLUTION OF OXYGEN IN THE GALAXY

R. REBOLO, G. ISRAELIAN and R.J. GARCÍA LÓPEZ

Instituto de Astrofísica de Canarias, E-38200 La Laguna, Tenerife, Spain

Abstract. We discuss results on the oxygen abundance in a sample of 23 metal-poor ($-3.0 <$ [Fe/H] < -0.3) unevolved stars and one giant. High resolution spectroscopy of OH lines in the near UV allowed us to trace the early evolution of oxygen versus metallicity. Contrary to previous expectations, we find that oxygen abundances derived from these low excitation lines agree well with those derived from the high excitation lines of the O I IR triplet and from the [O I] λ 6300 Å line. Our new oxygen abundances show a smooth extension of previously known trends of [O/Fe] versus [Fe/H] in disk stars to much lower metallicities, with a slope of -0.31 ± 0.11. The [O/Fe] ratio increases from 0.6 to 1 between [Fe/H] $= -1.5$ and -3.0. Comparison with oxygen abundances in giant stars of the same metallicity imply that the latter may have suffered a process of oxygen depletion. We briefly discuss the impact of these results on the yields of Type II SNe in the early Galaxy and on the age of globular clusters.

1. Introduction, Observations and Analysis

Type II SNe are expected to produce significant amounts of oxygen. Iron is produced in both, Type II and in Type I SNe. Since the latter come from longer lifetime progenitors, it has been argued for a long time that oxygen must be overabundant in very old stars. Evidence for high [O/Fe] ratios in many metal-poor stars has been reported during the last decades, however, up to date it is unclear whether the [O/Fe] ratio increases at very low metallicities or it shows a *plateau*. Here we present the results of abundances derived from near-UV OH lines for 24 metal-poor stars and make a comparison with abundances from the O I IR triplet and the [O I] λ 6300 Å line that we have derived from data available in the literature. The details of the work and discussion of the results can be found in Israelian, García López and Rebolo (1998).

The observations were carried out in different runs using the UES ($R = \lambda/\Delta\lambda \sim$ 50000) of the 4.2-m WHT at the Observatorio del Roque de los Muchachos (La Palma), and the UCLES ($R \sim 60000$) of the 3.9-m AAT. The final signal-to-noise ratio (S/N) varies for the different echelle orders, being in the range 30–50 for most of the stars.

The high-resolution solar flux atlas of Kurucz *et al.* (1984) and the Moore *et al.* (1966) atlas were examined to identify relatively strong unblended OH lines. Twelve OH lines in the wavelength range 3080–3300 Å were selected for de-

Astrophysics and Space Science is the original source of publication of this article. It is recommended that this article is cited as: *Astrophysics and Space Science* **265:** 165–170, 1999.
© 1999 *Kluwer Academic Publishers.*

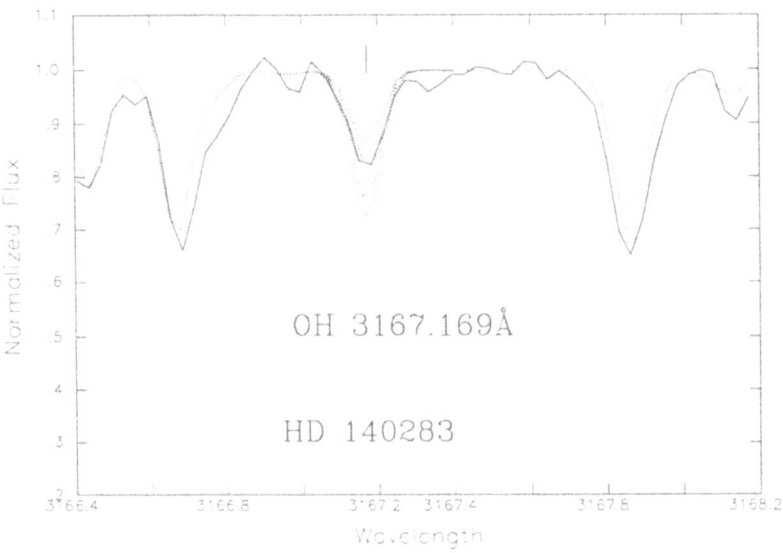

Figure 1. Comparison between observed (solid line) and synthetic spectra (dotted lines, correspond-ing to [O/H] = −2.1, −2.0, −1.9, −1.8, and −1.7) of HD 140283.

tail analysis in the stellar spectra. Laboratory wavelengths were taken from Stark *et al.* (1994) and line parameters from Goldman and Gillis (1981). Atomic data were obtained from the VALD database (Piskunov *et al.*, 1995). A grid of LTE, plane-parallel, constant flux, and blanketed model atmospheres provided by Kur-ucz (1992), computed with ATLAS9 without overshooting, and interpolated for given values of T_{eff}, log *g*, and [Fe/H]) was used to adjust the oscillator strengths (log *gf*) of the lines. These were changed until we reproduced the Kurucz *et al.* solar spectrum in the neighborhood of the selected OH lines using solar abundances from Anders and Grevesse (1989). The maximum change in log *gf* that we applied was 0.2 dex while most of the lines were modified by less than 0.1 dex. These changes had negligible effect on our differential analyses with respect the solar spectrum.

Effective temperatures (T_{eff}) for our stars were estimated using the Alonso *et al.* (1996) calibrations versus $V - K$ and $b - y$ colors, which were derived applying the infrared flux method, and cover a wide range of spectral types and metal content. Surface gravities (log *g*) were determined by comparing the observed Strömgren $b - y$ and c_1 indices with synthetic ones generated using the corresponding filter transmissions and a grid of Kurucz (1992) blanketed model atmospheres fluxes. Metallicities were adopted from literature values obtained from high resolution spectra.

Using the set of *solar gf* values we proceeded to derive the oxygen abundances of our stars by computing synthetic spectra with the adopted stellar parameters, and

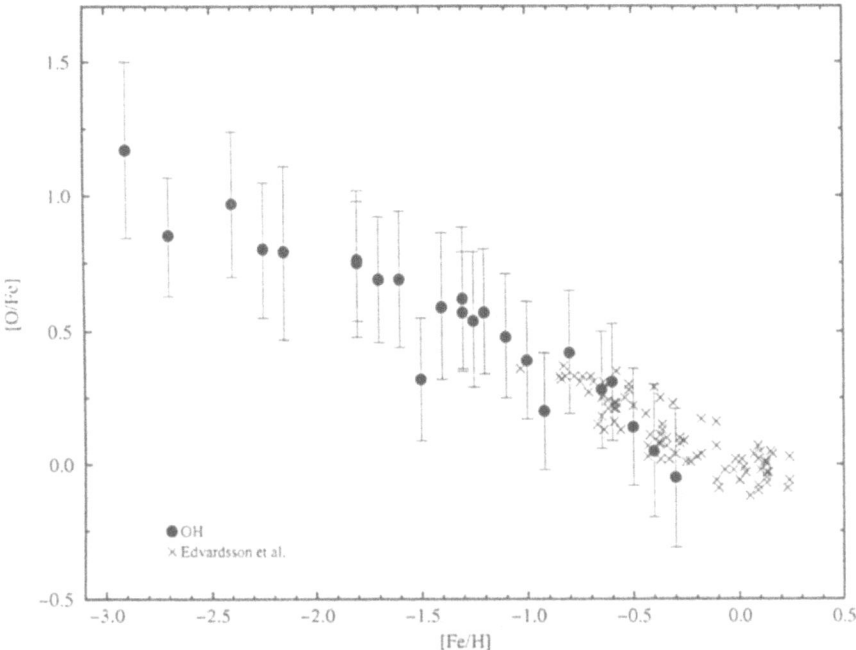

Figure 2. Oxygen overabundances (derived from OH lines) with respect to iron against metallicity for the 24 stars analised in this work. Also included in the plot are the values derived by Edvardsson *et al.* (1993) from the IR O I triplet but shifted by +0.05 dex to move them to the same abundance scale than our analysis. Note the smooth continuous increase in [O/Fe] with decreasing metallicity, which reaches a value ∼ +1 for [Fe/H] = −3.

changed the oxygen abundance until a good fit (e.g. Figure 1) to the observations was obtained (see Israelian *et al.*, 1998 for details).

For two stars (HD 140283 and HD 84937) in common with the OH analysis of Nissen *et al.* (1994) we find oxygen abundances that differ by less than 0.2 dex. The difference, which is within error bars, can be explained by the use of different model atmospheres (Kurucz versus OSMARCS models), different line lists, and slightly different stellar parameters.

2. OH Versus IR Triplet & Forbidden Lines

To compare our OH oxygen abundances with those coming from the O I IR triplet, we have computed them using the equivalent widths measured by Tomkin *et al.* (1992) and our stellar parameters. The same LTE code and model atmospheres used for the OH lines were employed, while the oscillator strengths were taken from Tomkin *et al.* (1992). The agreement between both sets of abundances is very good (with a mean difference of 0.00 ± 0.11 dex) and they delineate the same trend of [O/H] with [Fe/H].

We have found only four dwarfs (HD 22879, HD 76932, HD 103095 and HD 134169) in our sample for which oxygen abundances have been derived using [O I]. We have synthesized the forbidden oxygen line adopting the same set of stellar parameters than for the OH analysis, the gf value given by Lambert (1978) and the equivalent widths provided in the literature for the λ 6300 Å line. We found abundances in reasonable agreement with those derived from OH but still slightly lower. The sensitivity of the forbidden line to the adopted gravity is higher than for the OH lines. Gravities derived using accurate parallaxes measured by *Hipparcos* (Nissen *et al.*, 1998) are larger by 0.22 dex in average than the values adopted in our analysis, which implies a reduction of the oxygen abundance inferred from OH lines and an increase of the abundance derived from the forbidden line. The abundances found using these gravities are much closer and strongly suggests that an *Hipparcos*-based gravity scale may indeed be key to explain the discrepancies on oxygen abundances from forbidden and permitted lines in unevolved metal-poor stars.

3. Dependence of [O/H] and [O/Fe] on Metallicity

The oxygen abundances derived ([O/H]) show a clear decrease with decreasing metallicity and a smooth continuation of the trend found by Edvardsson *et al.* (1993) from their study of disk stars with $-1.0 < $ [Fe/H] < 0.3 using the O I IR triplet. Simple linear regressions performed between [O/H] and [Fe/H] on our, and Edvardsson et al.'s samples show a common slope of 0.63 ± 0.03, and a systematic difference of 0.05 ± 0.06 dex in the oxygen abundances. Our [O/Fe] ratios are plotted versus [Fe/H] in Figure 2. A shift of $+0.05$ dex has been applied to the values of Edvardsson *et al.* to move them, on average, to the same abundance scale as that of our analysis. The data do not indicate any clear flattening out of [O/Fe] with decreasing metallicity below [Fe/H] ~ -1, and can be fitted using a straight line with a slope -0.31 ± 0.11 (taking into account the error bars in both oxygen abundances and metallicities) in this metallicity range. We have investigated the presence of possible systematic dependences of [O/H] on $T_{\rm eff}$ or $\log g$ and have found no trends. The most metal-deficient star in our sample, BD $+23^o$ 3130), is a giant, and it is possible that the [O/Fe] derived for this star is not primordial, but, even removing this object from the plot, the evidence for any flattening of the distribution of points at very low metallicities is very marginal, and it seems that the [O/Fe] ratio continues increasing below [Fe/H] ~ -3. We are currently working on high S/N additional spectra obtained for this star in order to determine the oxygen abundance from the O I IR triplet and from the [O I] λ 6300 Å. The preliminary analysis of these lines gives results consistent with the high oxygen abundance obtained from the OH lines.

Boesgaard *et al.* (1999) have obtained high quality Keck spectra of metal-poor stars in the near UV and recently concluded a similar analysis of a different set of

OH lines. They find very good agreement with our results, and basically the same dependence of [O/Fe] versus metallicity confirming independently our work. The mean difference in oxygen abundance for ten stars in common with our sample is 0.00 ± 0.06 dex.

4. Implications and Conclusions

1. We have found that the [O/Fe] ratio of metal-poor stars increases from 0.6 to 1 between [Fe/H] $= -1.5$ and -3.0 with a slope of -0.31 ± 0.11. Contrary to the previously accepted picture, our oxygen abundances, derived from low-excitation OH lines, agree well with those derived from high-excitation lines of the triplet.
2. Our oxygen abundances for unevolved stars when compared with values in the literature for giants of similar metallicity imply that the latter may have suffered a process of oxygen depletion. As a result, unevolved metal-poor stars shall be considered better tracers of the early evolution of oxygen in the Galaxy.
3. The extrapolation of our results to very low metallicities indicates that the first Type II SNe in the early Galaxy provided oxygen to iron ratios [O/Fe] ~ 1. This high ratios can perfectly be explained using available yields for Type II SNe (Thielemann et al., 1996). Chemical evolution models of the early Galaxy can also explain the evolution of oxygen delineated in Figure 2. (see e.g. Chiappini et al., 1999.). The evolution of oxygen proposed in this paper also helps to understand the evolution of ^{6}Li versus [Fe/H] and the ^{6}Li/Be ratio at low metallicities in the framework of standard Galactic Cosmic Ray Nucleosyntehsis (Fields and Olive, 1999).
4. Our observations suggest a reduction of current estimates of ages of very metal-poor globular clusters by about 1–2 Gyr which may contribute significantly to remove any conflict among them and values of the age of the Universe inferred from recent measurements of the Hubble constant (Mould et al., 1995). The abundance of oxygen affects both the internal opacity and the energy generation (thermonuclear burning) of cool metal-poor stars near the turn-off loci of the color-magnitude diagram. Ages derived from comparison of theoretical evolutionary tracks with observational color-magnitude diagrams will be significantly affected. If our oxygen values are representative of the initial oxygen abundances of globular-cluster stars, it would imply a considerable reduction (1–2 Gyr) of globular cluster ages.

References

Alonso, A., Arribas, S. and Martinez-Roger, C.: 1996, *Astron. Astrophys. Suppl.* **313**, 873.
Anders, E. and Grevesse, N.: 1989, *Geochim. Cosmochim. Acta* **53**, 197.
Boesgaard, A.M., King, J.R., Deliyannis, C.P. and Vogt, S.S.: 1999, *Astron. J.*, in press.

Chiappini, C., Matteucci, F., Beers, T.C. and Nomoto, K.: 1999, *Astrophys. J.*, in press.

Edvardsson, B., Andersen, J., Gustafsson, B., Lambert, D.L., Nissen, P.E. and Tomkin, J.: 1993, *Astron. Astrophys.* **275**, 101.

Fields, B.D. and Olive, K.A.: 1999, *Astrophys. J.*, in press.

Goldman, A. and Gillis, J.R.: 1981, *JQSRT* **25**, 111.

Israelian, G., García López, R. and Rebolo, R.: 1998, *Astrophys. J.* **507**, 805.

Kurucz, R.L.: 1992, private communication.

Kurucz, R.L., Furenlid, I., Brault, J. and Testerman, L.: 1984, *Solar Flux Atlas from 296 to 1300 nm*, NOAO, Atlas No. 1.

Lambert, D.L.: 1978, *Mon. Not. R. Astron. Soc.* **183**, 249.

Moore, C.E., Minnaert, M.G.J. and Houtgast, J.: 1966, *The Solar Spectrum 2935 Å to 8770 Å*, National Bureau of Standards, Monograph 61.

Mould, J., *et al.*: 1995, *Astrophys. J.* **449**, 413.

Nissen, P.E., Gustafsson, B., Edvardsson, B. and Gilmore, G.: 1994, *Astron. Astrophys.* **285**, 440.

Nissen, P.E., Høg, E. and Schuster, W.J.: 1998, Proceedings of the "Hipparcos Venice '97 Symposium', ESA SP-402, in press.

Pavlenko, Ya.V.: 1991, *Soviet Astron.* **35**, 212.

Piskunov, N.E., Kupka, F., Ryabchikova, T.A., Weiss, W.W. and Jeffrey, C.S.: 1995, *Astron. Astrophys. Suppl.* **112**, 525.

Stark, G., Brault, J.W. and Abrams, M.C.: 1994, *J. Opt. Soc. Amer. B* **11**, 3.

Thielemann, F.K., Nomoto, K. and Hashimoto, M.: 1996, *Astrophys. J.* **460**, 408.

Tomkin, J., Lemke, M., Lambert, D.L. and Sneden, C.: 1992, *Astron. J.* **104**, 1568.

OXYGEN ABUNDANCES IN SOLAR-TYPE STARS

JOHANNES REETZ

Universitäts-Sternwarte München, München, Germany

Abstract. The *absolute* solar oxygen abundance, $\log \varepsilon_\odot = 8.80 \pm 0.06$, has been determined from various oxygen abundance indicators in different solar atlases, and a new method is proposed to test collision rate coefficients for the NLTE model of O I. Using effective temperatures derived from Balmer lines, oxygen abundances from O I 7773 triplets in 83 solar-type stars within the solar neighborhood spanning a metallicity range of [Fe/H] $= -2.3 \ldots +0.4$ have been determined. NLTE effects are not negligible, especially in warm stars ($T_{\mathrm{eff}} > 5800$) with [Fe/H] > -0.5.

1. Introduction

Element abundances in solar-type stars are often determined *differentially* with respect to the Sun. An accurate solar oxygen abundance is required e.g. to compare theoretical SN II yields with [Mg/O] or [Fe/O] determined for metal-poor stars. The determination of oxygen abundances in solar-type stars, including the Sun, has been subject of many investigations. Both, the question whether the solar oxygen abundance is representative for thin disk stars and the run of the O/Fe abundance ratio with metallicity is still a matter of debate. There are not many oxygen abundance indicators, and it is well known that each spectral window is problematic to analyze due to blends ([O I] 6300, 6363, O I 8446), NLTE effects (O I 7773, 8446 triplets), insufficient understanding of line formation (ultraviolet OH A-X band), and strong dependencies on the temperature structure (nearly all lines, in particular the high excited O I transitions and the infrared rotational-vibrational band of OH). In the following I present results which are discussed in detail elsewhere (Reetz, 1999).

2. The Solar Oxygen Abundance

Sedlmayr (1974) demonstrated that the center-to-limb variation of the solar O I 7773 triplet *can* be explained as a NLTE effect, i.e. a non-thermal population of the lower and upper level of the transitions. The center-to-limb behavior of the model profiles depends significantly on the choice of the temperature structure, thus a simultaneous representation of center and limb intensity profiles, as shown in Figure 1, may be not sufficient to estimate appropriate NLTE corrections. Some investigators

Astrophysics and Space Science is the original source of publication of this article. It is recommended that this article is cited as: *Astrophysics and Space Science* **265**: 171–174, 1999.
© 1999 *Kluwer Academic Publishers.*

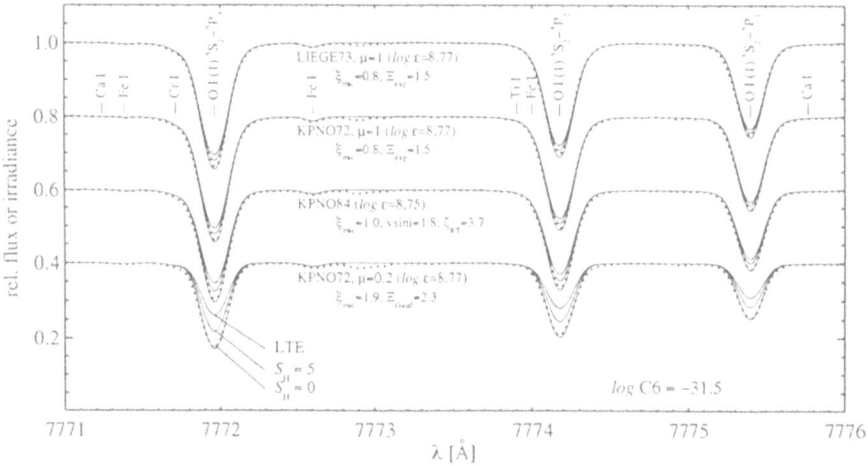

Figure 1. Theoretical LTE and NLTE profiles of O I 7773 (solid lines) calculated with the semi-empirical HM74 model (Holweger and Müller, 1974) compared with various observed solar atlases: KPNO72 (Brault and Testerman, 1972), LIEGE73 (Delbouille *et al.*, 1973) and KPNO84 (Kurucz *et al.*, 1984). μ is the cosine of the inclination angle (for intensity spectra). For convenience the spectra have been shifted. Parameters for microturbulence ξ_{mic} and macroturbulence (Ξ_{exp}: exponential profile, ζ_{RT}: radial-tangential profile, $\Xi_{Gauß}$: Gaussian) are given in km s^{-1}. $S_H = 0$ corresponds to the NLTE model without hydrogen collisions. Electron collision rates have been calculated according to van Regemorter (1962). The van der Waals constant according to the definition $(2\pi)^{-1}\Delta\omega = C_6/r^6$. Depending on C_6 – which corresponds to Unsöld's approximation in the present case – and the macroturbulence profile, $\log\varepsilon_\odot = 8.74\ldots 8.82$ is derived.

noted the contradiction between the 'high' abundance value derived by Lambert (1978, $\log\varepsilon_\odot = 8.92$) and Grevesse *et al.* (1984, $\log\varepsilon_\odot = 8.93$), respectively, and the 'low' value of $\log\varepsilon_\odot \sim 8.7\ldots 8.8$ determined from O I 7773 under consideration of NLTE without taking hydrogen collisions into account (Tomkin *et al.*, 1992; Kiselman, 1993). As a consequence of the apparent necessity to enhance the collision strengths in the NLTE model to achieve consistency with the 'high' value, NLTE corrections became negligible – even in warm stars like Procyon. The situation changed a little since Grevesse *et al.* (1996) recommended $\log\varepsilon_\odot = 8.87 \pm 0.07$, but questions remain: How strong are the 'primary' indicators [O I] 6300, 6363 blended (Lambert, 1978), and how can the adopted collision strengths be estimated more independently from the underlying temperature structure?

Lambert (1978) emphasized that the estimate of the contribution of the Ni I line at 6300.34 Å and the CN blend at 6363.8Å suffers from *systematic* uncertainties. Reinvestigating the Ni I blend inspecting multiplet 246 and lines of the CN red band in the solar spectrum, I found that the oxygen abundance derived from both forbidden transitions ($\log\varepsilon_\odot \sim 8.81$) is consistent with those determined from O I 7773, provided hydrogen collisions are neglected (cf. Figure 1). It is notable that Goldman *et al.* (1983) derived $\log\varepsilon_\odot = 8.84$ from pure rotational OH transitions in the far infrared.

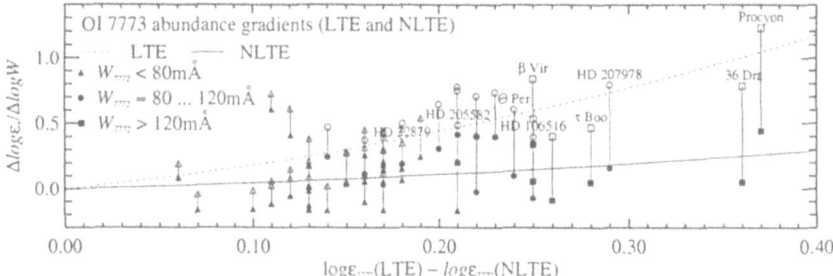

Figure 2. Correlation between abundance gradients and the NLTE effect for O I 7772 Å. The dotted line fits to the LTE abundance gradients (open symbols), and the solid line fits the NLTE abundance gradients ($S_H = 0$). Both curves are fixed at the origin, assuming that all gradients vanish if the NLTE effect vanishes. The scatter around the fitting curves can be explained by errors of equivalent widths and microturbulence parameters.

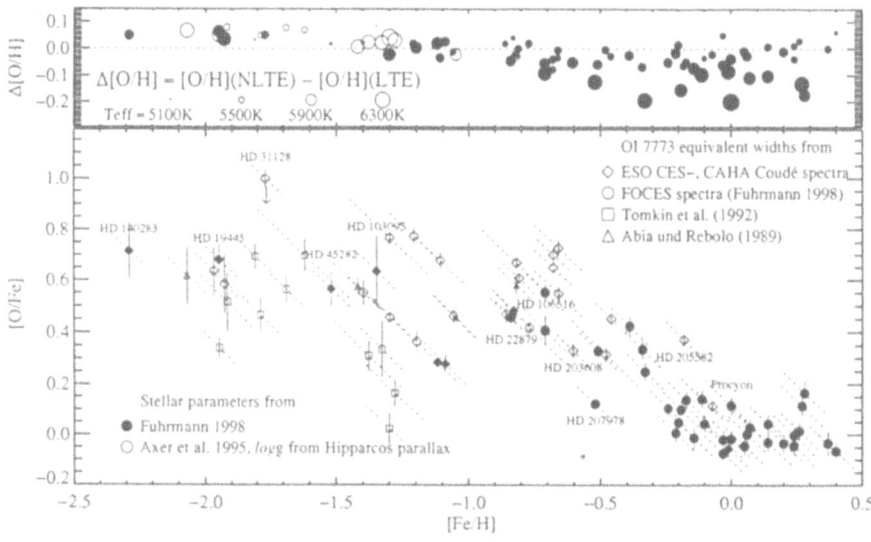

Figure 3. Distribution of differential O/Fe abundance ratios (derived from the O I 7773 triplet) over the metallicity range. *Upper panel*: differential NLTE effects (with respect to the NLTE effect in the Sun). The symbol diameters correspond to T_{eff}. *Lower panel*: symbols denote the source of spectra, equivalent widths and stellar parameters. Solid error bars correspond to uncertainties of the equivalent widths; the dotted error bars correspond to uncertainties of [Fe/H]. CES spectra of HD 140283 and HD 128279 have been kindly provided by Dr P. Magain.

The new method to test the adopted collision strengths is based on the fact that the NLTE effects depend on the mean formation depth of each line. An LTE analysis of the O I 7773 triplet therefore must lead to inconsistent abundances. Figure 2 shows the abundance trends derived from stellar O I 7773 triplets. The LTE abundance gradients correlate with the NLTE effect, and the gradients decrease if hydrogen collisions are neglected. The benefit of this examination is that the abundance gradients are not sensitive to *small* variations of the broadening constants, the adopted gravity, effective temperature and the temperature structure. Figure 2 justifies that NLTE model which leads to $\log \varepsilon_\odot = 8.74 \ldots 8.82$ (cf. Figure 1).

3. Consequences for $[O/Fe]_{7773}$

From the results of the NLTE analysis displayed in Figure 3 the following conclusions can be drawn:

- NLTE corrections are negligible for metal-poor stars but they become important for warm stars with $[Fe/H] > -0.5$.
- No significant scatter is detected above $[O/Fe] \sim 0.8$, and the slope of $[O/Fe]$ in the halo ($[Fe/H] < -1.5$) seems to be flat.
- The O/Fe ratio in metal-rich stars is solar in the average, and the distribution of $[O/Fe]$ shows no significant trend for $[Fe/H] > -0.1$.

References

Abia, C. and Rebolo, R.: 1989, *Astrophys. J.* **347**, 186.

Axer, M., Fuhrmann, K. and Gehren, T.: 1995, *Astron. Astrophys.* **300**, 751.

Brault, J. and Testerman, L.: 1972, *Preliminary Kitt Peak photoelectric atlas*, KPNO.

Delbouille, L., *et al.*: 1973, *Photometric Atlas of the Solar Spectrum*, Liège.

Fuhrmann, K.: 1998, *Astron. Astrophys.* **338**, 161.

Goldman, A., *et al.*: 1983, *Mon. Not. R. Astron. Soc.* **203**, 767.

Grevesse, N., Sauval, A. and van Dishoeck, E.: 1984, *Astron. Astrophys.* **141**, 10.

Grevesse, N., Noels, A. and Sauval, A.: 1996, *ASP Conf. Series* **99**, 117.

Holweger, H. and Müller, E.: 1974, *Sol. Phys.* **39**, 19.

Kiselman, D.: 1993, *Astron. Astrophys.* **275**, 269.

Kurucz, R., *et al.*: 1984, *Solar Flux Atlas from 296 to 1300 nm*, KPNO, Tucson.

Lambert, D.: 1978, *Mon. Not. R. Astron. Soc.* **182**, 249.

Reetz, J.: 1999, *PhD thesis*, Ludwig-Maximilians-Universität München.

Sedlmayr, E.: 1974, *Astron. Astrophys.* **31**, 23.

Tomkin, J., *et al.*: 1992, *Astron. J.* **104**, 1568.

van Regemorter, H.: 1962, *Astrophys. J.* **136**, 906.

KINEMATICS OF NEARBY SUBDWARFS

B. FUCHS, H. JAHREISS and R. WIELEN

Astronomisches Rechen-Institut, Mönchhofstraße 12-14, D-69120 Heidelberg, Germany

Abstract. A sample of subdwarfs with accurate space velocities and standarized metallicities is presented. This was constructed by combining Hipparcos parallaxes and proper motions with radial velocities and metallicities from Carney *et al.* (1994; CLLA). The accurate Hipparcos parallaxes lead to an – upward – correction factor of 11% of the photometric distance scale of CLLA. The kinematical behaviour of the subdwarfs is discussed in particular in relation to their metallicities. Most of the stars turn out to be thick disk stars, but the sample contains also many genuine halo stars. While the extreme metal poor halo does not rotate, a population of subdwarfs with metallicities in the range $-1.6 < [\text{Fe/H}] < -1.0$ dex appears to rotate around the galactic center with a mean rotation speed of about 100 km s^{-1}.

1. Data

The data material, which we have analyzed, is based on the sample of high proper motion stars by Carney *et al.* (1994; CLLA). CLLA have measured photometric parallaxes, radial velocities, and metallicities of most of the A to early G stars, many late G and some early K stars in the *Lowell Proper Motion Catalogue*. This data set of 1464 stars has been cross-identified with the Hipparcos catalogue (ESA 1997) and we found 767 stars in common. However, the sample had to be cleaned under various aspects. For instance, for some stars not all data were available or some Hipparcos parallaxes were not accurate enough. In Figure 1 the colour-magnitude diagram of the remaining stars is shown. Despite the efforts of CLLA to avoid them the sample is still contaminated by giants and subgiants, which can be detected now due to the improved accuracies of the absolute magnitudes. We have removed all stars lying above a line in the CMD defined by the zero age main sequence of stars with solar metallicity shifted upwards by $\Delta M_V = 0.8$ mag.

Furthermore we have determined an overall correction factor of the photometric distance scale of CLLA by analyzing parallax differences $\pi_{\text{Hipp}} - \pi_{\text{CLLA}}$. A linear regression analysis $\Delta \pi = f \cdot \pi_{\text{CLLA}}$ gives a coefficient $f = -0.113 \pm 0.007$, indicating an upward correction of the photometric distances by 11%. Jahreiß *et al.* (1997) have already found a similar correction on the basis of a smaller sample of subdwarfs.

Astrophysics and Space Science is the original source of publication of this article. It is recommended that this article is cited as: *Astrophysics and Space Science* **265**: 175–178, 1999.
© 1999 *Kluwer Academic Publishers.*

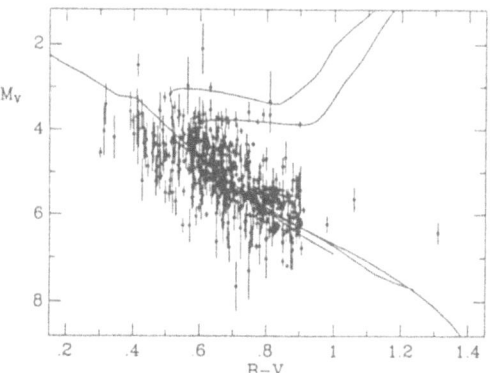

Figure 1. CM-diagram of 617 identified CLLA stars. Hipparcos parallaxes were used to determine the absolute magnitudes. The full lines indicate the main sequence and the CMDs of the old clusters M 67 and NGC 188, respectively.

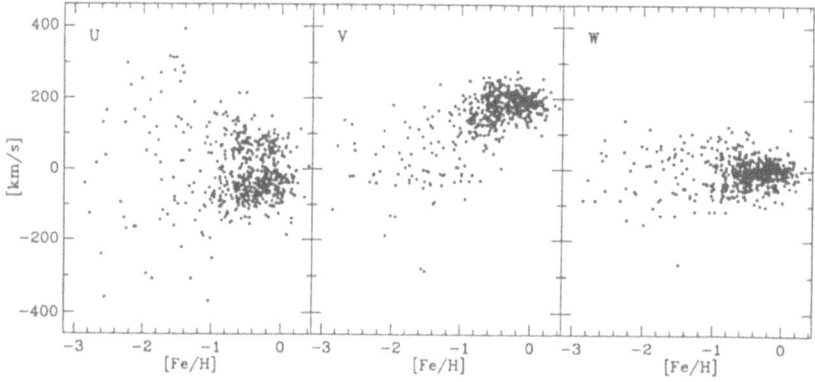

Figure 2. Space velocity components versus metallicity [Fe/H]. U points towards the galactic center, V in the direction of galactic rotation, and W towards the galactic north pole. The space velocities have been reduced to the local standard of rest and a rotation velocity of the LSR of 220 km s^{-1} has been assumed.

2. Kinematics

We have determined reliable space velocities of 560 nearby subdwarfs. These are shown in Figure 2 versus the metallicities of the stars as scatter plots. The populations of thick disk and halo stars can be clearly distinguished by the widths of the distributions and the mean rotational velocities. Since our sample is based originally on a proper motion survey, the sample is biased in the sense that low velocity stars are underrepresented. This can be seen clearly as a trough in the U-velocity distribution in Figure 2 or the asymmetric V-velocity distribution of the disk stars in Figure 3. However, since we are mainly interested in the kinematics of the halo stars, this bias is of no consequence in the present context.

TABLE I

Kinematic parameters

dex	N	$\overline{U}$	$\overline{V}$	$\overline{W}$	σ_U	σ_V	σ_W
				km s^{-1}			
[Fe/H] > −1	477	−21	−54	−10	72	46	38
−1.6 < [Fe/H] < −1	48	6	−186	−17	168	105	73
[Fe/H] < −1.6	35	4	−216	−15	181	83	74

In order to determine mean velocities and velocity dispersions we have considered three metallicity groups. Results are given in Table I. The corresponding V-velocity distributions are shown in Figure 3. The first and largest group ([Fe/H] > −1 dex) represents obviously thick disk stars. The third group ([Fe/H] < −1.6 dex) has the same kinematics as other tracers of the extreme metal poor halo (cf. the article by Norris in this volume). The second group (−1.6 < [Fe/H] < −1) shows peculiar kinematics. The V-velocity distribution is rather skewed with respect to the mean. The V-velocity distribution at velocities less than the mean velocity of the extreme metal poor halo of −220 km s^{-1} is statistically indistinguishable from that of the extreme metal poor halo, while at velocities larger than −220 km s^{-1} there is an excess of 15 stars compared to the corresponding distribution of the extreme metal poor halo. Indeed, 19 out of the sample of 48 stars lie beyond one standard deviation of the velocity distribution of the extreme metal poor halo, i.e. at velocities less than −133 km s^{-1}. We have carried out Monte-Carlo simulations, which show that such a distribution has *not* been drawn with a probability of 99.99% from a gaussian distribution resembling that of the extreme metal poor halo. This is illustrated further in the fourth panel of Figure 3, where we show the resulting distribution, if the left part of the V-velocity distribution is folded at V = −220 km s^{-1}, subtracted from the right part of the distribution and smoothed by a running mean. This gives the distinct impression of an excess population of subdwarfs in this metallicity range, which rotates with a velocity of about 100 km s^{-1} around the galactic center, while the rest of the subdwarfs in this metallicity range have the same kinematics as the extreme metal poor halo.

The nature of this excess population is not clear at present. It may well be that these stars represent a metal weak – and dynamically hot – tail of the thick disk. It is conspicuous that their asymmetric drift coefficient, $\overline{V}/\sigma_U^2 = -1.2 \cdot 10^{-2}$ (km s^{-1})$^{-1}$, is very similar to that of the thick disk $\overline{V}/\sigma_U^2 = -1 \cdot 10^{-2}$ (km s^{-1})$^{-1}$. According to this interpretation late stages of halo formation and early phases of disk formation would have overlapped. Alternatively, this population may be a relic of earlier accretion events (cf. the article by Majewski in this volume).

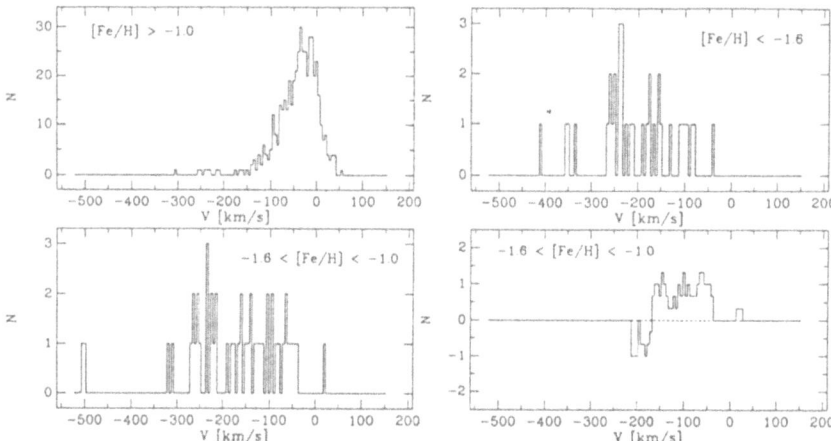

Figure 3. V-velocity distributions of the subdwarfs grouped according to their metallicities. The velocities are reduced to the local standard of rest. The panel in the lower right shows the excess population (see text for details).

References

Carney, B.W., Latham, D.W., Laird, J.B. and Aguilar, L.A.: 1994, *Astron. J.* **102**, 2240.
ESA: 1997, *The Hipparcos and Tycho Catalogues*, ESA SP-1200.
Jahreiß, H., Fuchs, B. and Wielen, R.: 1997, in: M.A.C. Perryman and P.L. Bernacca (eds.), *Hipparcos-Venice '97'*, ESA SP-402, pp. 587.

AN ESTIMATION OF THE GALACTIC ESCAPE VELOCITY FROM HIGH-VELOCITY METAL-POOR STARS

L. MEILLON

URA 335 du CNRS, DASGAL, Observatoire de Paris, F-92195 Meudon, France

Abstract. A large number of high-velocity stars was included in the Hipparcos Input Catalogue for the purpose of determining the galactic escape velocity in the solar vicinity. However, the 'fastest' stars known, listed by Carney *et al.* (1994, CLLA) were not included because they are too faint. In the intersection between the CLLA list and the Hipparcos Catalogue (770 common stars), the metal-deficient stars with the most reliable parallaxes $((\frac{\sigma_\pi}{\pi})_{\mathrm{HIP}} \leq 0.15)$ are used for recalibrating CLLA absolute magnitudes and photometric parallaxes, using metallicities and VandenBerg *et al.* (1998) isochrones. In this way, about twenty non-Hipparcos stars get improved parallaxes and are added to our primary sample of Hipparcos high-velocity stars, for a better determination of the escape velocity from the Galaxy.

1. Introduction

We have searched in the Hipparcos catalogue (ESA, 1997) the highest velocity stars. Only 10 stars with $V_{\mathrm{tot}} \geq 350$ km s^{-1} have been found. Due to the limiting magnitude of the satellite, faint high-velocity stars are absent. In order to improve the number of such stars, we have undertaken a calibration of photometric absolute magnitudes (Meillon *et al.*, 1998) for the proper motion stars from CLLA, using VandenBerg *et al.* (1998) isochrones and Hipparcos subdwarfs. The larger star number allows a new determination of the galactic escape velocity in the solar neighborhood, following the method developed by Leonard and Tremaine (1990, LT).

2. Results

Using new photometric distances for CLLA stars, we have computed their velocities* and associated errors. Adding high-velocity Hipparcos stars, we obtain 98 stars with $|V_r| \geq 250$ km s^{-1}, 33 with $V_{\mathrm{tg}} \geq 300$ km s^{-1} and 24 with $V_{\mathrm{tot}} \geq 350$ km s^{-1}. We find only four stars with $V_{\mathrm{tot}} \geq 400$ km s^{-1} (Figure 1, left) and the fastest star has $V_{\mathrm{tot}} = 458$ km s^{-1}. Following the LT method, it seems that the escape velocity cannot exceed 530 km s^{-1} (Figure 1, right). However, in the future, possible biases

---

* In a reference frame centred at the Sun, but at rest with respect to galactic center

 Astrophysics and Space Science is the original source of publication of this article. It is recommended that this article is cited as: *Astrophysics and Space Science* **265**: 179–180, 1999.
© 1999 *Kluwer Academic Publishers*.

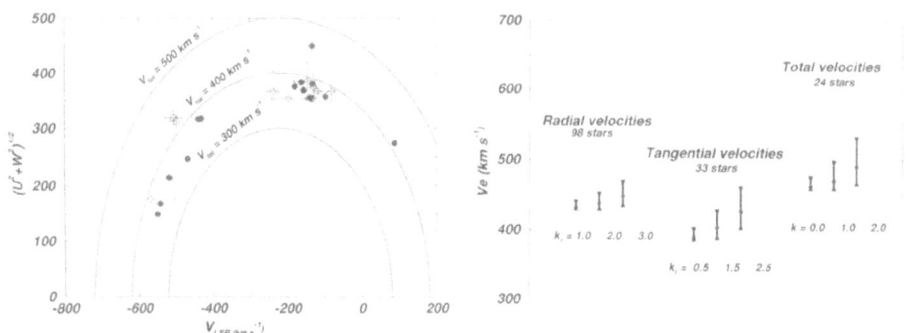

Figure 1. Left: Toomre diagram for the 24 stars with $V_{tot} \geq 350$ km s^{-1}. Filled circles are for Hipparcos stars and open circles for CLLA stars. Curves represent different constant V_{tot}. Right: estimates of the escape velocity and the corresponding 90% confidence intervals, using radial, tangential and total velocities.

will be considered and an estimation of the mass of the Galaxy will be given. New faint very high-velocity stars may change this result.

Acknowledgements

We thank Drs M. Mayor and S. Udry for kindly providing Coravel radial velocities and Drs F. Crifo, A.E. Gómez, R. Cayrel and H. Jahreiß for useful discussions.

References

Carney, B.W., Latham, D.W., Laird, J.B. and Aguilar, L.A.: 1994, *Astron. J.* **107**, 2240, CLLA.
ESA: 1997, *The Hipparcos Catalogue*, ESA SP-1200.
Leonard, P.J.T. and Tremaine, S.: 1990, *Astron. J.* **353**, 486, LT.
Meillon, L., Crifo, F., Cayrel, R., Perrin, M.-N. and Gómez, A.E.: 1998, to be published in the proceedings of the colloquium 'Harmonizing Cosmic Distance Scales in a Post-Hipparcos Era'.
VandenBerg, D.A., Swenson, F.J., Rogers, F.J., Iglesias, C.A. and Alexander, D.R.: 1998, in preparation.

SEARCHING FOR LOW MASS STARS IN THE HALO

A.C. ROBIN
CNRS-ERA 6091 – Observatoire de Besançon, 25010 Besançon Cedex, France

M. CRÉZÉ
Université de Bretagne-Sud, F-56000 Vannes, France

V. MOHAN
U.P. State Observatory, Nainital, 263 129, India

From microlensing events detected at high galactic latitude as well as from direct measurements of the halo stellar density, estimations of a potential contribution of low mass stars to the mass budget is questioned. Up to now no clear answer has been found for one main reason: the difficulty to observe them in a sufficient volume, which requires deep detection limit and/or wide field of view. While the HST fulfills the first requirement, its field of view is small therefore the problem requires to observe a very large number of WFC2 fields. Wide field CCD mosaics on large ground based telescopes offer a promising alternative, provided that image quality is sufficient to allow reliable star/galaxy separation.

Here we attempt to estimate the luminosity and mass functions in the stellar halo from deep ground based star counts on a wide field. Observations were performed in 1996 at CFHT, using the University of Hawaii CCD mosaic which covers about 29 × 29 arcminutes, in three bands, V, R and I. Data have been treated using the IRAF package, photometrically reduced using DAOPHOT-II. Star/galaxy separation was performed by the SExtractor package (Bertin and Arnouts, 1996).

Model predicted V-I histograms in various magnitude bins show that the 3 populations (disc, thick disc and halo) are suitably separated in colour allowing to independantly estimate their luminosity functions.

The number of halo stars at these magnitudes depends mostly on the luminosity function. Data have been compared with predictions of the Besançon model of population synthesis (Bienaymé *et al.*, 1986; Robin *et al.*, 1996; Haywood *et al.*, 1997), assuming various initial mass function (IMF). The IMF is transformed into luminosity functions and color magnitude diagrams using Bergbush and Vandenberg (1992) isochrones. Then colour/magnitude counts are simulated according to the observed conditions (photometric errors and area surveyed). Eventually, the simulated distributions are compared with the data using a maximum likelihood test.

The fit was limited to magnitude 23.5 since the seeing and signal to noise of our 1996 images do not allow reliable star/galaxy separation beyond this limit. The

Astrophysics and Space Science is the original source of publication of this article. It is recommended that this article is cited as: *Astrophysics and Space Science* **265**: 181–182, 1999.
© 1999 *Kluwer Academic Publishers.*

best fit model in this condition has a spheroid mass function slope of 0.5. A slope larger than 1.5 is clearly excluded.

Our result is in good agreement with Chabrier and Méra (1997) value of 0.7 ± 0.2 deduced from Dahn *et al.* local sample. We disagree with Gould *et al.* (1997) who derive an IMF slope of -0.25 ± 0.32 (in our notation) from HST-WFC2 observations. We suspect that the assumed high metallicity in their calibration of the halo may explain this discrepancy.

From our IMF determination, 8 objects (among which 7 halo stars) with $26 < V < 28$ would have been observed in the HDF, compatible with Elson *et al.* detection of 9 stellar objects at this magnitude. With an IMF slope of 1.0 and 1.5 we obtain respectively 11 and 19 objects. Clearly, while our newly determined IMF slope is in very good agreement with HDF star counts but the constraint is weak because of the small area covered and the small number of detected stars.

With such an IMF, single red and brown dwarfs make about 0.5% of the local density of the dark halo. They are therefore excluded as candidates for baryonic dark matter unless a strong discontinuity in the IMF, leaving place to a number of brown dwarfs. Moreover halo red dwarfs can account for less than 1% of the microlensing optical depth. Even with a higher IMF slope, such as 2.5, its optical depth would only be 3% of the observed one.

Since the probable masses of the microlensing candidates should range between 0.1 and 1 solar mass, the only stellar candidates for these events remain halo white dwarfs in certain conditions of star formation history in the early Galaxy.

References

Bergbush, P.A. and Vandenberg, D.A.: 1992, *Astrophys. J.* **81**, 163.

Bertin, E. and Arnouts, S.: 1996, *Astron. Astrophys. Suppl.* **117**, 393.

Bienaymé, O., Robin, A.C. and Crézé, M.: 1987, *Astron. Astrophys.* **180**, 94.

Chabrier, G. and Méra, D.: 1997, *Astron. Astrophys.* **328**, 83.

Elson, R.A., Santiago, B.X. and Gilmore, G.F.: 1996, *NewA* **1**, 1.

Gould, A., Bahcall, J. and Flynn, C.: 1997, *Astrophys. J.* **482**, 913.

Haywood, M., Robin, A.C. and Crézé, M.: 1997, *Astron. Astrophys.* **320**, 428.

Robin, A.C., *et al.*: 1996, *Astron. Astrophys.* **305**, 125.

STRÖMGREN 4-COLOR PHOTOMETRY OF VERY METAL-POOR STARS

W.J. SCHUSTER, L. PARRAO and A. FRANCO
Instituto de Astronomía, UNAM, México

T.C. BEERS
Dept. of Physics and Astronomy, Michigan State Univ., East Lansing, MI, USA

P.E. NISSEN
Institute of Physics and Astronomy, Univ. of Aarhus, Aarhus, Denmark

Abstract. Strömgren $uvby$ photometry has been observed for an additional 140 very metal-poor stars from the survey of Beers *et al.* (1992). These Galactic stars of very-low metallicity provide crucial information for the investigation of the formation and evolution of the Galaxy, as well as on the nature of the early Universe. The Strömgren $uvby - \beta$ system allows the measurement of stellar atmospheric parameters as a prelude to detailed abundance studies which will make use of high-resolution spectroscopy and the new generation of large telescopes. The photometric techniques developed by Schuster *et al.* (1996) are used not only to classify these very metal-poor stars but also to derive effective temperatures, surface gravities, and improved estimates for their interstellar reddenings. In particular, photometric diagrams such as $[c_1]$, $[m_1]$ and c_0, $(b-y)_0$ are used to classify these stars, especially those near the main-sequence turnoff, where contamination from slightly-evolved subgiants, lower surface-gravity horizontal-branch stars, and even a few supergiant or AGB candidates is found.

References

Beers, T.C., Preston, G.W. and Shectman, S.A.: 1992, *Astron. J.* **103**, 1987.
Schuster, W.J., Beers, T.C., Nissen, P.E., Parrao, L. and Franco, A.: 1998a, *Astron. Astrophys. Suppl.*, in preparation.
Schuster, W.J. and Nissen, P.E.: 1988, *Astron. Astrophys. Suppl.* **73**, 225 (SN).
Schuster, W.J. and Nissen, P.E.: 1989, *Astron. Astrophys.* **221**, 65.
Schuster, W.J., Nissen, P.E., Parrao, L., Beers, T.C. and Overgaard, L.P.: 1996, *Astron. Astrophys. Suppl.* **117**, 317 (Paper VIII).
Schuster, W.J., Parrao, L. and Contreras-Martinez, M.E.: 1993, *Astron. Astrophys. Suppl.* **97**, 951 (SPC).
Schuster, W.J., Parrao, L. and Garcia-Cole, A.: 1998b, *Astron. Astrophys. Suppl.*, in preparation.
Strömgren, B.: 1966, *Annu. Rev. Astron. Astrophys.* **4**, 433.

Astrophysics and Space Science is the original source of publication of this article. It is recommended that this article is cited as: *Astrophysics and Space Science* **265**: 183–184, 1999.
© 1999 *Kluwer Academic Publishers.*

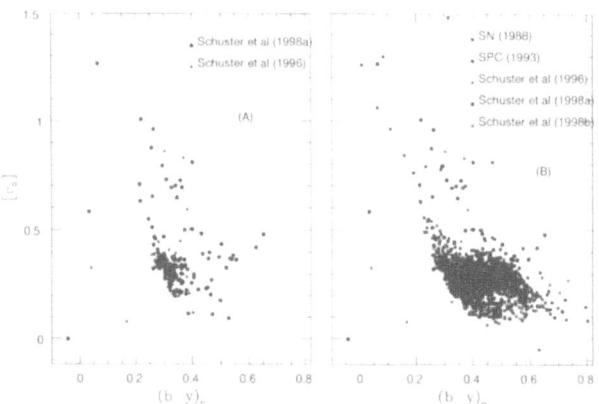

Figure 1. (A) The c_0, $(b - y)_0$ diagram for the two catalogues of very metal-poor stars, mostly stars with [Fe/H] $\leq$ −2.0 from the survey of Beers *et al.* (1992). The $uvby - \beta$ data have been taken at the San Pedro Mártir observatory using the 1.5m H.L. Johnson telescope and the same observing and reduction techniques as discussed in Schuster *et al.* (1996). Whenever possible the photometry has been dereddened using the stars' Hβ values plus the intrinsic-color calibration of Schuster and Nissen (1989). (B) For comparison, the c_0, $(b - y)_0$ diagram is plotted for five catalogues of $uvby - \beta$ data for high-velocity and metal-poor stars. The last catalogue contains approximately 450 high-velocity and metal-poor stars from the literature and from lists provided by B.W. Carney, S.G. Ryan, and T.C. Beers. It can be appreciated that the two very metal-poor catalogues provide a significant proportion of the bluest TO stars and of the greatly evolved stars.

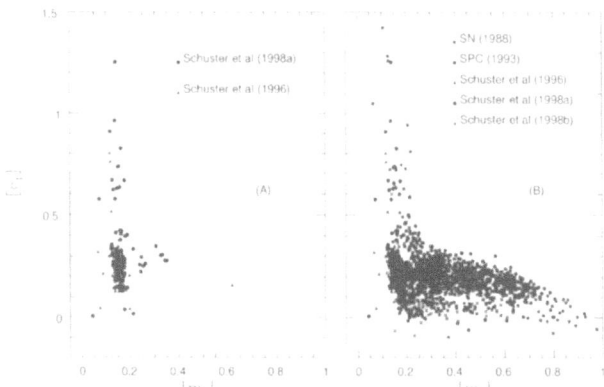

Figure 2. The $[c_1]$, $[m_1]$ diagrams are plotted for the same catalogues of $uvby - \beta$ data as in Figure 1. This is the figure suggested by Strömgren (1966; Figure 1) for the classification of B-G type stars, where $[m_1]$ and $[c_1]$ are reddening-free indices. These figures emphasize the TO, very metal-poor nature of the stars in these samples. One can also appreciate the higher percentage of greatly evolved stars in the most recent very metal-poor catalogue.

HIGH RESOLUTION SPECTROSCOPY OF SELECTED
FAINT STARS

V.G. KLOCHKOVA, S.V. ERMAKOV and V.E. PANCHUK

Special Astrophysical Observatory, Nizhnij Arkhyz, Karachaevo-Cherkesia 357147, Russia

This work is the next step of our spectroscopic research (Klochkova and Panchuk, 1996) of fundamental parameters and atmospheric chemical composition of stars at the 6-meter telescope and presents results of our investigation of 20 metal–deficient stars. For the determination of the model atmospheres parameters and for the computation of the chemical composition the models grid (Kurucz, 1993) was applied. Equivalent widths of Fe I and Fe II lines (only with W $<$ 200 mÅ) were used to determine values of T_{eff} and log g. The abundance ratios with respect to iron for 22 chemical elements are determined to a precision ranging from 0.05 to 0.25 dex.

In Table I we present model atmosphere parameters, the average abundance of the iron group elements $[M/H]_{\odot}$ and abundances derived for the another main groups of chemical elements. For 6 metal–poor stars ($[Fe/H]_{\odot} \leq -1.4$) the abundance of the iron group elements corresponds to that of iron: the average abundance of scandium, titanium, cromium, manganese and nickel is $[X/Fe]_{\odot} = +0.1$ dex. A number of elements produced by the α–process have been revealed to be slightly overabundant relative to iron for these metal–poor stars: $[Mg/Fe]_{\odot} = +0.27$ dex, $[Si/Fe]_{\odot} = +0.64$ dex, $[Ca/Fe]_{\odot} = +0.25$ dex. At the same time the sodium abundance is normal relative to that of iron $[Na/Fe]_{\odot} = +0.09$ dex. The abundance ratio of odd/even elements produced by the α-process is $[Na/Mg]_{\odot} = -0.34$ dex.

The barium abundance, most of which is synthesized in the processes of slow neutronization, is significantly lower in the investigated halo stars: $[Ba/Fe]_{\odot} = -0.44$ dex. For the halo stars with $[Fe/H]_{\odot} \approx -2$ the Ba and Eu abundance ratio is $\log(Ba/Eu) = -0.27$, which indicates the r–process to be the main mechanism of production of heavy metals at the early epoch of nucleosynthesis. Considerable overabundances of lanthanoides (La, Ce, Pr, Nd) have been found: the average value of these heavy metals is $[hmet/Fe]_{\odot} = +0.83$ dex.

As in our previous work, part of the stars, classified earlier (Carney *et al.*, 1994) as non-evolved halo objects, we rank among subgiants, members of the halo or thick disk populations.

 Astrophysics and Space Science is the original source of publication of this article. It is recommended that this article is cited as: *Astrophysics and Space Science* **265**: 185–186, 1999.
© 1999 *Kluwer Academic Publishers.*

TABLE I

The atmospheric parameters for stars studied and abundances of the main groups of chemical elements. Here [M/H]$_\odot$ means the average abundance of iron group elements. In the following columns are given the average abundances of some groups elements relative to iron [X/Fe]$_\odot$: α – for Mg, Si, Ca, s – for Y, Zr, Ba, hmet – for La, Ce, Pr, Nd. The same data we determined also for the standard halo star HD 122563

Star	T_{eff}	lg g	ξ_t	[M/H]$_\odot$	α	Na,Al	s	hmet	Eu
G 29-20	5030	2.0	1.2	−0.91	.30	.01	.31	.02	.19
G 30-52	5025	2.5	3.2	−0.60	.10	.22	−.08	.04	.76
G 90-25	5370	3.0	2.5	−1.67	.48	.06	−.15	.45	
G 122-57	5040	3.0	1.0	−0.33	.32	−.04	.46	.26	.20
G 126-62	6640	4.55	0.9	−0.93	.25	.15	.21		
G 170-47	5000	1.9	1.1	−2.47	.35		−.88		
G 182-7	5500	4.2	2.0	−0.14	.17	−.06	.07	.46	.20
G 188-22	5860	3.5	1.7	−1.43	.59	.16	−.20	1.09	.48
G 231-52	5750	4.4	2.1	−1.49	.37		−.04		
G 245-32	6725	4.65	0.3	−0.72	.19	−.13	−.24		
G 246-38	5240	3.5	3.5	−2.18	.43	−.08	−.13	1.12	.88
G 265-1	5500	3.5	1.5	−0.66	.30	.18	−.06	−.05	.17
HD 65583	5800	5.0	1.1	−0.20	.21	.03	.31	.50	
HD 88609	4325	0.0	2.3	−2.82	.54		−.68	.15	.76
HD 175305	5000	2.2	1.2	−1.35	.19	−.06	.11	.17	.49
BD 09°3223	5900	2.6	3.2	−1.53	.34	.24	−.11		
BD 11°2998	5625	2.4	1.8	−0.88	.33		.20	.19	.33
BD 17°3248	5590	2.4	1.1	−1.15	.20	−.01	.11	.53	.86
BD 23°3912	5880	3.6	1.4	−1.22	.34	.12	.15	.62	
HD 122563	4395	0.0	1.8	−2.77	.45	.47	−.98	.52	.62

References

Carney, B.W., Latham, D.W., Laird, J.B. and Aguilar, L.A.: 1994, A survey of proper motion stars. XII. An expanded sample, *Astron. J.* **107**, 2240–2289.

Klochkova, V.G. and Panchuk, V.E.: 1996, The atmospheric chemical composition of stars with large proper motions, *Astron. Rep.* **40**, 829–844.

Kurucz, R.L.: 1993, CDROMs.

G-DWARF STARS IN THE GALACTIC HALO

C.J. CORBALLY
Vatican Observatory, University of Arizona, Tucson AZ 85721, USA

R.F. GARRISON
David Dunlap Observatory

W. RUEGER
Vatican Observatory

C.R. STAGG
Mount Royal College, Calgary

1. The Project and Preliminary Results

Since G-dwarf stars are unambiguous, long-lived probes of Galactic evolution, we are extending an earlier, spectroscopic study of G-dwarf stars in the Galactic Pole regions (Corbally and Garrison, 1988a, 1988b; Garrison and Corbally, 1993, 'GC'). Currently, we are in the process of selecting candidate G-dwarf stars from *UBVRI* photometry in fields at the North Galactic Pole, while again relying on the Hertzsprung gap to filter out giant stars.

The CCD data is coming from the 1.8 m Vatican Advanced Technology Telescope (VATT). To date, we have had only a few nights suitable for all-sky photometry. In an area of 220 sq. arcmin (using a 6.4×6.4 arcmin CCD) a count of 74 stars have magnitudes. Among these there are about 10 candidates (14%) in the range $0.50 < B - V < 0.70$ or $0.45 < V - R < 0.55$.

All candidate stars will be classified in the MK System, as was done for the earlier study, to determine their metallicity and characteristics. Resulting G-dwarf stars will reach to about 5 kpc from the Galactic plane.

2. Comparisons with Other Photometry

The NGP is a well-studied region, and so other photometry is available for comparison with ours. We selected two of the most recent CCD studies, Majewski *et al.* (1994, 'Maj.') and Lu *et al.* (1998, 'Lu'). Rather few stars are in common with our own ('CGRS' in Figure 1), but we can say the following: the $U - B$ and $B - V$ comparisons show the expected agreement within the errors of the CGRS and Maj.

 Astrophysics and Space Science is the original source of publication of this article. It is recommended that this article is cited as: *Astrophysics and Space Science* **265:** 187–188, 1999.
© 1999 *Kluwer Academic Publishers.*

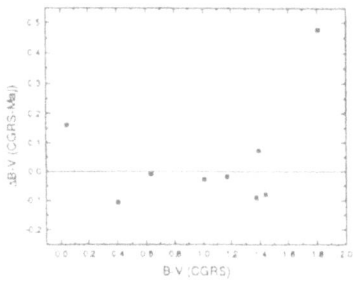 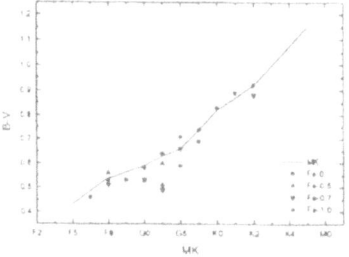

Figure 1. Comparison of CGRS and Majewski photometry for (*B-V*) colors.

Figure 2. Comparison of standard relation MK-(*B-V*) with photometry from Lu, binned according to spectral class and metallicity.

studies; the CGRS *V* magnitudes may have an offset (w.r.t. Maj.) and a limit of accuracy at 19th (Lu) or 20th (Maj.) magnitude; an offset is especially clear for the CGRS *R* − *I* colors w.r.t. Maj, showing the errors are getting large and unstable redder than the *V* filter (we are investigating how to obtain more stability).

3. Comparison of Lu's Photometry with Classifications

A comparison could also be made between the expectation from GC's MK classifications, based on standard MK-*BVRI* relations, and a second, brighter set of Lu's (*et al.*, 1998) photometry. This set contained the same 192 stars of GC's original NGP survey. The data, binned according to spectral subclass and metallicity, are presented in Figure 2 for *B* − *V*. We found for the three colors that: Lu's *B* − *V* for solar metallicity stars ('Fe 0') quite closely follows the expected MK relation; Lu's *V* − *R* and *R* − *I* are offset from the expected relation, but in the opposite sense to the offsets seen in the comparison between our data and their deep 'snapshot' KPNO data; no significant trends with metallicity are seen, even for *B* − *V*.

References

Corbally, C.J. and Garrison, R.F.: 1988a, *Astron. J.* **95**, 739.

Corbally, C.J. and Garrison, R.F.: 1988b, *Astron. J.* **95**, 745.

Garrison, R.F. and Corbally, C.J.: 1993, *Astron. J.* **106**, 2301 (GC).

Lu, P.K., *et al.*: 1998, preprint (Lu).

Majewski, S.R., Kron, R.G., Koo, D.C. and Bershady, M.A.: 1994, *Publ. Astron. Soc. Pacific* **106**, 1258 (Maj.).

FIRST DREDGE-UP AND FURTHER MIXING MECHANISMS ALONG THE RED GIANT BRANCH IN FIELD STARS

E. CARRETTA and R.G. GRATTON
Osservatorio Astronomico di Padova, Padova, Italy

C. SNEDEN
Dept. of Astronomy, Univ. of Texas at Austin, TX, USA

A. BRAGAGLIA
Osservatorio Astronomico di Bologna, Bologna, Italy

1. Mixing along the Red Giant Branch (RGB)

As a small mass star evolves up the RGB, the outer convective envelope expands inward and penetrates into the CN-cycle processed interior regions (**first dredge-up**). Some further mixing is possible in the latest phases of the RGB (see Charbonnel, 1995: C95), when the advancing H-burning shell reaches the chemical discontinuity left behind by the receding convective envelope. Thereafter, no mean molecular weight gradient is present between the convective envelope and the near vicinity of the shell, and is possible that circulation currents give rise to further mixing.

Here we test this scenario presenting abundances for Fe, Li, C, N, O for about 60 metal-poor ($-2 <$ [Fe/H] < -1) field stars in different evolutionary stages. Details of the analysis will be given in a forthcoming paper. The main results of our study are presented in Figure 1, where abundances as a function of luminosity are shown.

There are two distinct mixing/dilution episodes along the RGB evolution of small mass field stars:

1) the *first dredge-up* follows canonical predictions (see e.g. C95). This is quite clear in the pattern shown by elements and isotopic ratios in panels a,b,c. The luminosity is about $\log L/L_\odot = 1$; Li is diluted by about a factor of 20, ^{12}C abundance decreases by ~ 0.1 dex.

2) a *second mixing episode* occurs when the star becomes brighter than the RGB bump (and the molecular weight barrier is canceled) at $\log L/L_\odot \sim 2$, again in agreement with predictions by C95. Most of the remaining Li is destroyed, ^{12}C abundance further decreases by a factor ~ 2, and ^{12}C/^{13}C ratio raises to ~ 6. These values are observed also in the following evolutionary phase (RHB and early AGB).

Astrophysics and Space Science is the original source of publication of this article. It is recommended that this article is cited as: *Astrophysics and Space Science* **265**: 189–190, 1999.
© 1999 *Kluwer Academic Publishers.*

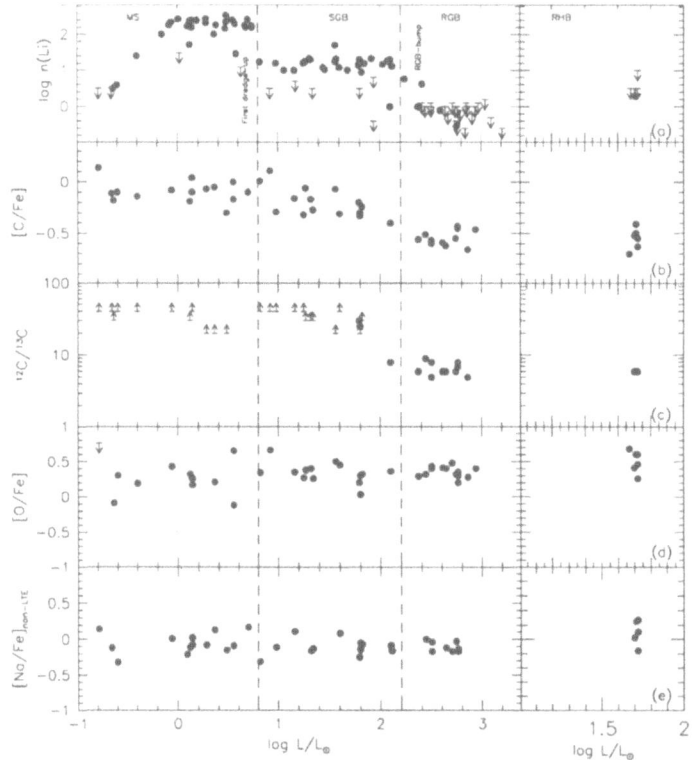

Figure 1. Abundances of light elements as a function of luminosity in field stars.

O and Na abundances in bright RGB and HB field stars are similar to those observed in unevolved stars. *Field stars do not display any signature* of the Na-O anticorrelation seen amongst globular cluster giants (see e.g. Kraft, 1994).

References

Charbonnel, C.: 1995, *Astrophys. J* **453**, L41.
Kraft, R.P.: 1994, *Publ. Astron. Soc. Pacific* **106**, 553.

ALUMINUM ABUNDANCES ON THE RED GIANT BRANCH: TESTING HELIUM MIXING AS A SECOND PARAMETER

R.M. CAVALLO

NASA/Goddard Space Flight Center, Greenbelt, MD, USA

Although globular cluster horizontal-branch (HB) morphology is determined primarily by the cluster metallicity, observations have shown that clusters with similar metallicities sometimes have very different HB's, indicating the existence of a second parameter. Sweigart (1997a, b) suggested that deep mixing on the red giant branch (RGB) may be a second parameter if He is mixed into the stellar envelope. A He-mixed RGB star would evolve onto the HB at a bluer color and brighter luminosity than an unmixed star. Cavallo *et al.* (1998) showed that Al is produced in large quantities only inside the H-burning shell where He is synthesized, so that Al makes a good tracer of He. Thus, a relationship should exist between the ratio of Al-rich to Al-normal stars on the RGB and the ratio of blue to red HB stars. Since He cannot be directly measured in cool giants, Al enhancements can be used as an indicator of He mixing.

In Figures 1a and 1b, we present the Al abundance data for the blue-HB cluster M13 and the red-HB cluster M3, both of which have [Fe/H] ~ -1.5. The distributions of the M13 and M3 giants are both bimodal with a ratio of 2:1 (Al strong to Al poor). The Al abundances in three M13 giants and all six M3 giants are from Cavallo and Nagar (1999, CN99) and were derived using the models of Kraft *et*

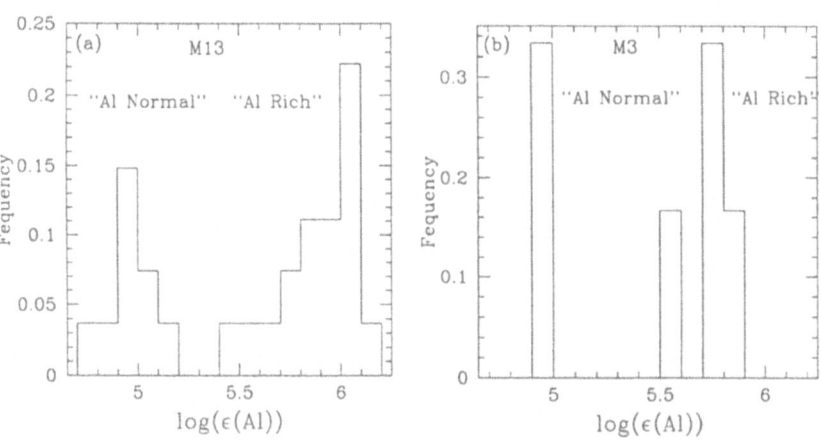

Figure 1. Al abundance distribution for (a) 27 M13 and (b) 6 M3 giants.

 Astrophysics and Space Science is the original source of publication of this article. It is recommended that this article is cited as: *Astrophysics and Space Science* **265**: 191–192, 1999.
© 1999 *Kluwer Academic Publishers.*

al. (1992, 1993, 1995). The the rest of the M13 data are from the Kraft papers and from Pilachowski *et al.* (1996). The similarity between the distributions is surprising given the difference between the two HB's, although the M3 sample is very small and its Al-rich stars are not as enhanced as the M13 giants.

Ferraro *et al.* (1998) used HST data to show that the empirical ratio of blue to red HB stars is 58:42 in M13; i.e., close to the ratio of Al-strong to Al-normal giants. No similar HST data exist for M3, although it does have a redder HB morphology. While the data might support the He-mixing hypothesis for M13, they suffer from two shortcomings: (1) the sample sizes of RGB stars with Al measurements is small, even for M13, and (2) some giants are likely to have been contaminated by an early generation of massive asymptotic-giant-branch (AGB) stars that evolved quickly before shedding their processed envelopes into the cluster. If the AGB stars are greater than $\sim 4~M_\odot$, it is likely that hot-bottom-burning products, such as Al, would be ejected along with *s*-process elements, such as Ba (Lattanzio *et al.*, 1997). Thus, looking for a relationship between Ba and Al, for example, might help to discriminate between stars that have been contaminated and those that are truly self-enriched; i.e., giants with both enhanced Al and Ba could be removed from the sample, making the connection between He mixing and Al clearer. That such a relationship might exist is suggested by the data of Ivans *et al.* (1999, this issue) who show that for 20 giants in M4 ([Fe/H] ~ -1) the Al and Ba abundances are both elevated on average. Unfortunately, they do not report on individual stars, so it is difficult to determine if a correlation exists or not. Such a relationship might, however, explain the M3 giants with Al enhancements.

References

Cavallo, R.M. and Nagar, N.: 1999, in preparation (CN99).
Cavallo, R.M., Sweigart, A.V. and Bell, R.A.: 1998, *Astrophys. J.* **492**, 575.
Ferraro, F.R., *et al.*: 1998, *Astrophys. J.* **500**, 311.
Kraft, R.P., *et al.*: 1992, *Astron. J.* **104**, 645.
Kraft, R.P., *et al.*: 1993, *Astron. J.* **106**, 1490.
Kraft, R.P., *et al.*: 1992, *Astron. J.* **109**, 2586.
Lattanzio, J.C., *et al.*: 1997, *Nuc. Phys. A* **621**, 435.
Pilachowski, C.A., *et al.*: 1996, *Astron. J.* **112**, 546.
Sweigart, A.V.: 1997a, *Astrophys. J.* **474**, L23.
Sweigart, A.V.: 1997b, in: A.G.D. Philip, J.W. Liebert and R. Saffer (eds.), *The Third Conference on Faint Blue Stars*, CUP, Cambridge, p. 3.

BLUE HORIZONTAL BRANCH STARS IN FOUR GLOBULAR CLUSTERS

A. BRAGAGLIA and C. CACCIARI
Osservatorio Astronomico di Bologna, via Ranzani 1, I-40127 Bologna, Italy

E. CARRETTA
Osservatorio Astronomico di Padova, vicolo Osservatorio 5, I-35122 Padova, Italy

F. CASTELLI
Osservatorio Astronomico di Trieste, via G.B. Tiepolo 11, I-34131 Trieste, Italy

1. The Problem

Several globular clusters with the same values of the primary parameters (age and chemical composition) show different horizontal branch (HB) morphologies, which may also be characterized by gaps or bimodal stellar distributions. Several parameters have been suggested as possibly responsible for this phenomenon, some of which can affect the physical parameters of the stars populating the HB.

Essential information on the above phenomena can be provided by the knowledge of the physical parameters (i.e. temperature and gravity) of HB stars. To derive these parameters we have collected low-resolution optical spectra of blue HB stars in some clusters that have extra-blue HB tails and gaps, monitoring in particular the hottest (and faintest) stars.

We have then fitted (with a χ^2-minimization algorithm) the calibrated spectral distributions of our program stars with the model atmospheres by Castelli (1997, NOVER), obtaining temperatures and gravities.

2. Our Sample: Preliminary Results

The low-resolution spectra ($R \sim 500 - 1000$, $\Delta\lambda \sim 3400 - 9000$ Å) have been acquired at the ESO 1.5 m telescope. We observed 4 globular clusters, selected because they all show extra-blue HB tails in spite of being more metal-rich than about [Fe/H] $= -1.5$: NGC288 (27 stars), NGC1904 (16 stars), NGC2808 (10 stars), NGC6752 (28 stars).

We did not obtain fair results for all the stars in our sample, for many different reasons (e.g. contamination from a close companion, non-gray light loss due to

 Astrophysics and Space Science is the original source of publication of this article. It is recommended that this article is cited as: *Astrophysics and Space Science* **265:** 193–194, 1999.
© 1999 *Kluwer Academic Publishers.*

differential atmospheric refraction, low S/N, etc.), but for about one half we get sufficiently good fits, i.e. a formal error of about 2% in T_{eff} and 0.2 in log g.

From the derived T_{eff}'s and gravities we may tentatively conclude the following:

(a) Most stars lie on the zero age HB (ZAHB), and there is no compelling evidence of evolution away from the ZAHB. The two stars which give a 'good' fit in NGC2808 and fall below the ZAHB are the faintest of our sample, and we don't think this result is robust enough to allow us to draw any conclusion.

(b) We see no evidence of the bimodal mass (i.e. gravity) distribution across the gap in NGC6752, that was found by Moehler *et al.* (1997). Our sample, however, does not reach the faintest end of the horizontal branch blue tail.

The detailed results, the full discussion and comparison with literature data (i.e. Crocker *et al.*, 1988 for NGC288; Caloi *et al.*, 1986; Cacciari *et al.*, 1995, and Moehler *et al.*, 1997 for NGC6752) will be given in a forthcoming paper, now in preparation.

References

Cacciari, C., Fusi Pecci, F., Bragaglia, A. and Buzzoni, A.: 1995, IUE observations of blue horizontal branch stars in the globular clusters M 3 and NGC 6752, *Astron. Astrophys.* **301**, 684.

Caloi, V., Castellani, V., Danziger, J., Gilmozzi, R., Cannon, R.D., Hill, P.W. and Boksenberg, A.: 1986, Optical and UV spectroscopy of blue horizontal branch stars in the globular cluster NGC 6752, *Mon. Not. R. Astron. Soc.* **222**, 55.

Castelli, F.: 1997, available at http://cfaku5.harvard.edu.

Crocker, D.A., Rood, R.T. and O'Connell, R.W.: 1988, Multiple populations on the horizontal branch. I – Observations in the (log g, log T_{eff})-diagram, *Astrophys. J.* **332**, 236.

Moehler, S., Heber, U. and Rupprecht, G.: 1997, Hot HB stars in globular clusters: physical parameters and consequences for theory III. NGC 6752 and its long blue vertical branch, *Astron. Astrophys.* **319**, 109.

THE MILDLY METAL-POOR GLOBULAR CLUSTER M4:
STAR-TO-STAR ABUNDANCE VARIATIONS IN 20 GIANTS

I.I. IVANS and C. SNEDEN

McDonald Observatory, University of Texas, Austin, USA

R.P. KRAFT

Lick Observatory, University of California, Santa Cruz, USA

Many low-metallicity globular clusters exhibit large star-to-star variations of C, N, O, Na, Mg, & Al abundances. In higher-metallicity clusters, the abundance swings are muted. Evidence of anticorrelations of O versus Na and Mg versus Al exists in classical northern-hemisphere clusters spanning a range of metallicities, $-0.8 \leq$ [Fe/H] ≤ -2.2. These and other anticorrelations, observed to be a function of giant branch position, are consistent with material having undergone proton-capture nucleosynthesis, presumably taking place in regions where the CNO-cycle is in operation, and brought to the surface by a deep-mixing mechanism. Distinctly bimodal distributions of cyanogen strengths at nearly all giant branch positions are observed in some clusters, including M4 (Norris, 1981). The metallicity of M4 ([Fe/H] ~ -1) places it among clusters in which the O versus Na and Mg versus Al anticorrelations are expected to be suppressed. Indeed, not much abundance variation has been observed in small samples of M4 giant stars (see e.g. Brown and Wallerstein, 1992). This puzzle led us to consider an abundance study of a large sample of bright giants in the mildly metal-poor globular cluster M4, the nearest, brightest, and one of the most accessible targets in which to study the CN-bimodal phenomenon.

We observed 20 M4 giant stars in 1997 using high resolution échelle spectrographs covering the effective wavelength range: $\lambda\lambda$ 5200 to 8800 Åon the 2.7-m at McDonald Observatory ($R \approx 60\,000$) and the 3.0-m at Lick Observatory ($R \approx 50\,000$).

Large and differential interstellar reddening due to the dark nebulosity in Scorpio-Ophiuchus does not permit a careful estimate of reddenings for individual stars; one cannot assume a strict $(B - V)$ versus Teff relationship. Instead, we obtained precise Teffs by employing two different spectroscopic techniques: (1) ratios of central depths of temperature sensitive absorption features (Gray, 1994) to initially estimate relative stellar Teffs and (2) as in previous Lick-Texas work, final Teffs were determined in conjunction with surface gravity and microturbulence constraints.

Astrophysics and Space Science is the original source of publication of this article. It is recommended that this article is cited as: *Astrophysics and Space Science* **265**: 195–196, 1999.
© 1999 *Kluwer Academic Publishers.*

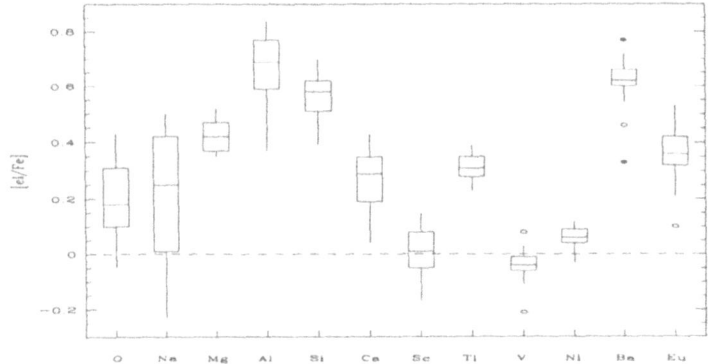

Figure 1. Boxplot of M4 Giant Star Abundances – A boxed horizontal line indicates the interquartile range [IQ] & median found for a particular element. Vertical tails extending from the boxes indicate the total range of abundances determined for each element, excluding outliers. Mild outliers (within 1.5 × IQ) are denoted by hollow circles (○) & severe outliers (from 1.5 to 3 × IQ) by filled circles (●). The dashed line represents the solar value for a particular elemental abundance.

The scatter about the mean (of elements not expected to be sensitive to proton-capture nucleosynthesis) compares well to those obtained in other high resolution cluster work and, as expected, no trend with Teff is observed for these non-volatiles. In individual M4 stars, anticorrelated abundances of N/Na/Al with C/O provide strong evidence for high-temperature proton-capture nucleosynthesis. We also observe increasing O-depletion/N-enhancement as stars ascend the RGB but combined $\log_\epsilon$ (C+N+O) remains ~ constant. Extremely low carbon isotope ratios ($^{12}C/^{13}C = 4.6 \pm 0.8$) indicate that the onset of mixing of the CNO-cycled material occurs at a luminosity lower than that of our observed stars. Elements Na, Mg, Si, & Ca are all enhanced with respect to the iron abundance, in approximate agreement with previous high resolution studies of M4 (Brown and Wallerstein, 1992). Al and Ba show uniform enhancements indicative of primordial enrichment. The analysis also confirms anomalous absorption properties of dust along line of sight to M4 ($E(V - K)/E(B - V) = R = 3.8 \pm 0.4$) that deviate from the normal law of interstellar extinction.

References

Brown, J.A. and Wallerstein, G.: 1992, *Astron. J.* **104**, 1818–1829.
Gray, D.F.: 1994, *Publ. Astron. Soc. Pacific* **106**, 1248–1257.
Norris, J.E.: 1981, *Astrophys. J.* **248**, 177–188.

STRÖMGREN CCD PHOTOMETRY OF GLOBULAR CLUSTERS

FRANK GRUNDAHL

Dominion Astrophysical Observatory, 5071 W. Saanich Road, Victoria, B.C, Canada

MICHAEL I. ANDERSEN

Nordic Optical Telescope, Ap. 474, E-38700 Santa Cruz de La Palma, Spain

Abstract. We present results from Strömgren $uvby$ photometry for the globular clusters M3 and M13, obtained from the Nordic Optical Telescope and the Danish 1.54 m telescope on La Silla obtained as part of a larger program to investigate cluster ages.

1. Results

Strömgren photometry offers the possibility for determining the effective temperature, chemical abundance ([Fe/H]) and gravity for F and G type stars. We are currently carrying out a large programme to study a number of globular clusters using CCD Strömgren photometry. So far we have obtained observations for more than 16 globular and five old open clusters. The first results for M13 have been published in Grundahl, VandenBerg and Andersen (1998). All observations for the clusters have been obtained with the Nordic Optical Telescope (NOT) on La Palma, and the Danish 1.54m telescope on La Silla.

From our M13 study the most surprising result was the finding that stars on the red giant branch (RGB) exhibit a very large scatter in the c_1 index at fixed color and luminosity. This is most likely caused by star–to–star variations in C, N, and/or O due to mixing. We have found this c_1 scatter to be present in clusters over the metallicity interval $-2.3 <$ [Fe/H] < -1.3. In M13 these variations in c_1 persists from the brightest giants to the base of the RGB, which indicates that mixing could occur at lower luminosities than predicted by canonical stellar theory.

In Figure 1 we show: a) the $(u - y, u)$ color–magnitude diagram for M13 obtained with NOT. The most striking feature of this diagram is the disagreement between the canonical ZAHB model (kindly provided by Don A. VandenBerg) and the observations. We see this feature in several other clusters with with a blue HB (Grundahl *et al.*, 1999, in preparation), but it is currently not clear what is causing this phenomenon. Furthermore it is worth noting that the coolest BHB stars in M13 now form a horizontal sequence in the CMD. This allows us to compare the difference in luminosity between the HB and (nearly flat) subgiant branch, which is age sensitive, and grows at a rate of 0.065 magnitudes per Gyr. We can use this

Astrophysics and Space Science is the original source of publication of this article. It is recommended that this article is cited as: *Astrophysics and Space Science* **265**: 197–198, 1999.
© 1999 *Kluwer Academic Publishers.*

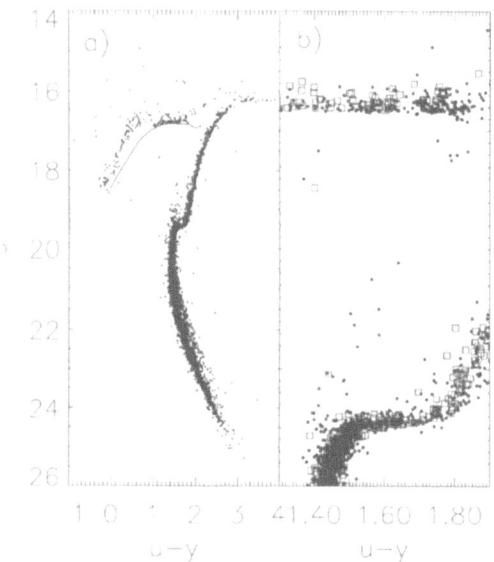

Figure 1. Left: The color–magnitude diagram for M13. Right: The CMD for M3 (*filled circles*) overplotted that of M13 (*open squares*). See text for details.

to compare to M3 in Figure 1b, which shows that the two clusters are nearly co-eval. A quantitative analysis yields an age difference between the two clusters of ∼ 0.5 Gyr which is inconsistent with age being the only second parameter varying between M3 and M13.

Reference

Grundahl, F., VandenBerg, D.A. and Andersen, M.I.: 1998, *Astrophys. J.* **500**, L179.

OLD STELLAR POPULATIONS IN NEARBY DWARF GALAXIES

ELINE TOLSTOY

European Southern Observatory, Garching bei München, Germany

Abstract. What can we learn from the somewhat arduous study of old stellar populations in nearby galaxies? Unless the nearby universe is subtly anomalous, it should contain a relatively normal selection of galaxies whose histories are representative of field galaxies in general throughout the Universe. We can therefore take advantage of our ability to resolve local galaxies into individual stars to directly, and accurately, measure star formation histories. The star formation histories are determined from numerical models, based on stellar evolution tracks, of colour-magnitude diagrams. The most accurate information on star formation rates extending back to the earliest epochs can be obtained from the structure of the main sequence. However, the oldest main sequence turnoffs are very faint, and it is often necessary to use the brighter, more evolved, populations to infer the star formation history at older times. A complete star formation history can be compared with the spectroscopic properties of galaxies seen over a large range of lookback times in redshift surveys. There is considerable evidence that the faint blue galaxies seen in large numbers in cosmological surveys are the progenitors of the late-type irregular galaxies seen in copious numbers in the Local Group, and beyond. We consider how the 'Madau-diagram', the star formation history of the Universe, would look if the Local Group were to be considered representative of the Universe as a whole.

1. Introduction

The study of resolved stellar populations provides a powerful tool to follow galaxy evolution directly in terms of physical parameters such as age (star formation history, SFH), chemical composition and enrichment history, initial mass function, environment, and dynamical history of the system. Photometry of individual stars in at least two filters and the interpretation of Colour-Magnitude Diagram (CMD) morphology gives the least ambiguous and most accurate information about variations in star formation within a galaxy back to the oldest stars. Some of the physical parameters that affect a CMD are strongly correlated, such as metallicity and age, since successive generations of star formation may be progressively enriched in the heavier elements. Careful, detailed CMD analysis is a proven, uniquely powerful approach (e.g., Tosi *et al.*, 1991; Tolstoy and Saha, 1996; Aparicio *et al.*, 1996; Mighell, 1997; Dohm-Palmer *et al.*, 1997, 1998; Hurley-Keller *et al.*, 1998; Gallagher *et al.*, 1998; Tolstoy *et al.*, 1998) that benefits enormously from the high spatial resolution of *HST* to the point that ground based CMD analysis is only worthwhile in ideal conditions beyond about the distance of the Magellanic Clouds.

 Astrophysics and Space Science is the original source of publication of this article. It is recommended that this article is cited as: *Astrophysics and Space Science* **265**: 199–206, 1999.
© 1999 *Kluwer Academic Publishers.*

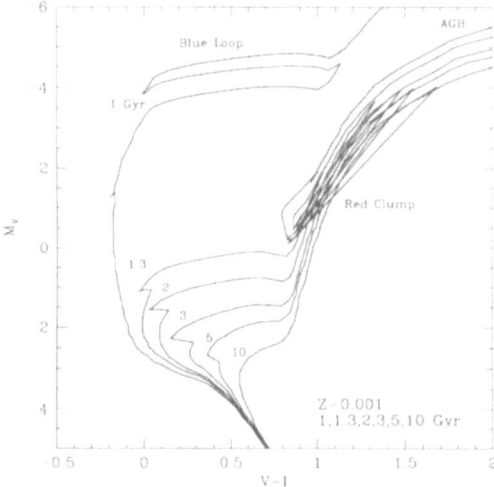

Figure 1. Isochrones (Bertelli *et al.*, 1994) for a single metallicity ($Z = 0.001$) and a range of ages, as marked in Gyr at the MSTOs. Isochrones were designed for single age globular cluster populations and are best avoided in the interpretation of composite populations, which can best be modeled using Monte-Carlo techniques (e.g. Tolstoy, 1996). They are used here for the purpose of illustration.

Because of the tremendous gains in data quality and thus understanding which have come from recent high quality CMDs of nearby galaxies it is now clearly worthwhile and fundamentally important to complete a survey of the resolved stellar populations of all the galaxies in our Local Group (LG). This will provide a uniform picture of the global star formation properties of galaxies with a wide variety of mass, metallicity, gas content etc. (e.g. Mateo, 1998), and will make a sample that ought to reflect the SFH of the Universe and give results which can be compared to high redshift survey results (e.g., Madau *et al.*, 1998). Initial comparisons suggest these different approaches do not yield the same results (Fukugita *et al.*, 1998), but the errors are large.

2. Colour-Magnitude Diagram Analysis

Much of our detailed knowledge of the SFHs of galaxies beyond 1 Gyr ago comes from the Milky Way and its nearby dSph satellites or from *HST* CMDs. To date, the limiting factors have been crowding and resolution limits for accurate stellar photometry from the ground. *HST* provides a unique opportunity to extend beyond our immediate vicinity and encompass the whole LG. To date *HST* has observed the resolved stellar populations in variety of nearby galaxies (e.g., dE, NGC 147, Han *et al*, 1997; Irr, LMC, Geha *et al.*, 1998; Spiral, M 31, Holland *et al.*, 1996; BCD, VII Zw 403, Lynds *et al.*, 1998; dI, Leo A, Tolstoy *et al.*, 1998; dSph, Leo I, Gallart *et al.*, 1998). For every LG galaxy at which *HST* has pointed at we have learnt something new and fundamentally important that was not discernable from ground based images, especially in the case of small dIs. The small dIs, like the dSph appear to exhibit a wide variety of SFHs. These results have affected our understanding of galaxy formation and evolution by demonstrating the importance of episodic star formation in nearby low mass galaxies. The larger galaxies in the LG have evidence of sizeable old halos, which appear to represent the majority of star formation in the LG by mass, although the problems distinguishing between effects of age and metallicity in a CMD result in a degree of uncertainty in the exact age distribution in these halos. It is important that detailed comparative studies of all galaxies in the LG are made in the future, including the M 31 and M 33 halo populations, to obtain a picture of the fossil record of star formation in galaxies of various types and sizes, and to identify both commonalities and differences in their SFH across the LG. In addition to a better understanding of galaxy evolution this will enable the comparison with cosmological surveys to be made more accurately.

Stellar evolution theory provides a number of clear predictions, based on relatively well understood physics, of features expected in CMDs for different age and metallicity stellar populations (see Figure 1). There are a number of clear indicators of varying star formation rates (*sfr*) at different times which can be combined to obtain a very accurate picture of the entire SFH of a galaxy.

Main Sequence Turnoffs (MSTOs): If we can obtain deep enough exposures of the resolved stellar populations in nearby galaxies we can obtain the *unambiguous age information that comes from the luminosity of MSTOs*. Along the Main Sequence itself different age populations overlie each other completely making the interpretation of the Main Sequence luminosity function complex, especially for older populations. However the MSTOs do not overlap each other like this and hence provide the most direct, accurate information about the SFH of a galaxy. MSTOs can clearly distinguish between bursting star formation and quiescent star formation, (e.g. Hurley-Keller *et al.*, 1998).

The Red Giant Branch (RGB): The RGB is a very bright evolved phase of stellar evolution, where the star is burning H in a shell around its He core. For a given metallicity the RGB red and blue limits are given by the young and old limits (respectively) of the stars populating it (for ages $\gtrsim$ 1 Gyr). As a stellar population ages the RGB moves to the red, for constant metallicity, the blue edge is determined by the age of the oldest stars. However increasing the metallicity of a stellar population will also produce exactly the same effect as aging, and also makes the RGB redder. This is the (in)famous age-metallicity degeneracy problem. The result is that if there is metallicity evolution within a galaxy, it impossible to uniquely disentangle effects due to age and metallicity on the basis of the optical colours of the RGB alone.

The Red Clump/Horizontal Branch (RC/HB): Red Clump (RC) stars and their lower mass cousins, Horizontal Branch (HB) stars are core helium-burning stars, and their luminosity varies depending upon age, metallicity and mass loss (Caputo *et al.*, 1995). The extent in luminosity of the RC can be used to estimate the age of

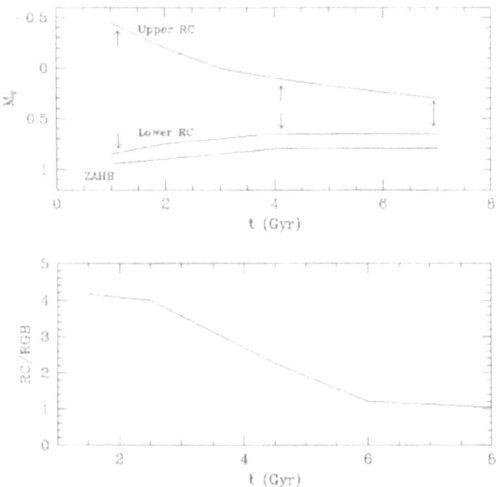

Figure 2. In the top panel are plotted the results of Caputo, Castellani and Degl'Innocenti (1995) for the variation in the *extent* in M_V magnitude of a RC with age, for a metallicity of $Z = 0.0004$. We plot the magnitude of the upper and lower edge of the RC versus age, in Gyr. We can thus clearly see that this extent is strong function of the age of the stellar population. Also plotted is M_V of the zero age HB against age. In the lower panel are plotted the results of running a series of Monte-Carlo simulations (Tolstoy, 1996) using stellar evolution models at $Z = 0.0004$ (Fagotto *et al.*, 1994) and counting the number of RC and RGB stars in the same part of the diagram, and thus we determine the expected ratio of RC/RGB stars versus age.

the population that produced it (Caputo *et al.*, 1995), as shown in the upper panel of Figure 2. The ratio, t_{RC}/t_{RGB}, is a decreasing function of the age of the dominant stellar population in a galaxy, and the ratio of the numbers of stars in the RC, and the HB to the number of RGB is sensitive to the SFH of the galaxy (Tolstoy *et al.*, 1998; Han *et al.*, 1997). Thus, the higher the ratio, N(RC)/N(RGB), the younger the dominant stellar population in a galaxy, as shown in the lower panel of Figure 2. This age measure is *independent of absolute magnitude and hence distance*, and indeed these properties can be used to determine an accurate distance measure on the basis of the RC (e.g. Cole, 1998). The presence of a large HB population on the other hand (high N(HB)/N(RGB) or even N(HB)/N(MS), is caused by a predominantly much older (> 10 Gyr) stellar population in a galaxy. The HB is the brightest indicator of very lowest mass (hence oldest) stellar populations in a galaxy.

The Extended Asymptotic Giant Branch (EAGB): The temperature and colour of the EAGB stars in a galaxy are determined by the age and metallicity of the population they represent (see Figure 3). However there remain a number of uncertainties in the comparison between the models and the data (Gallart *et al.*, 1994; Lynds *et al.*, 1998). It is very important that more work is done to enable a better calibration of these very bright indicators of past star formation events. In Figure 3 theoretical EAGB isochrones (Bertelli *et al.*, 1994) are overlaid on the HST CMD of a post-

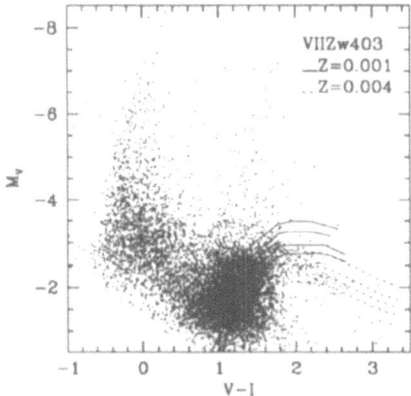

Figure 3. EAGB isochrones (Bertelli *et al.*, 1994) for metallicities, $Z = 0.001$ and $Z = 0.004$, are shown superposed on the observed CMD of VII Zw403 (Lynds *et al.*, 1998). For each metallicity the isochrones are for populations of ages 1.3, 2, 3, and 5 Gyrs, with the youngest isochrone being the brightest. This shows the potential discriminant between the age and metallicity of older populations, if the models could be better calibrated to a known SFH, e.g. for a nearby EAGB rich system like NGC 6822 where old MSTOs are observable.

starburst BCD galaxy VII Zw403, and we can see that a large population of EAGB stars is a bright indicator of a past high *sfr*, and the luminosity spread depends upon metallicity and the age of the *sfr*. That the RGB+AGB population of VII Zw403 looks so similar to NGC 6822 (Gallart *et al.*, 1994) is suggestive that dI and BCD galaxies can easily transform into each other on very short timescales.

3. The Connection to High Redshift

Star-forming, dI galaxies represent the largest fraction by number of galaxies in the LG, and it is clear from deep imaging surveys that this number count dominance appears to *increase* throughout the Universe with lookback time (Ellis, 1997). The large numbers of 'Faint Blue Galaxies' (FBG) found in deep imaging-redshift surveys appear to be predominantly intermediate redshift ($z < 1$, or a look-back time out to roughly half a Hubble time), intrinsically *small* late type galaxies, undergoing strong bursts of star formation (Babul and Ferguson, 1996). Thus we can assume that the dIs we see in the LG are a cosmologically important population of galaxies which can be used to trace the evolutionary changes in the *sfr* of the Universe with redshift. The 'Madau-diagram' (Madau *et al.*, 1998) uses the results of redshift surveys to plot the SFH of the Universe against redshift. It predicts that most of the stars that have formed in the Universe have done so at redshifts, $z \sim 1 - 2$. If it is correct, then the MSTOs from the most active period of star formation in the Universe will be easily visible as $7 - 9$ Gyr old MSTOs in the galaxies of the LG (e.g. Rich, 1998). Determining accurate SFHs for all the galaxies in the local Universe using CMD analysis provides an alternate route to and thus check upon the Madau-diagram.

Recent detailed CMDs of several nearby galaxies and self-consistent grids of theoretical stellar evolution models have transformed our understanding of galactic SFHs. Most of the dI CMDs to date suggest that the *sfr* was higher in the past, although the peak in the *sfr* has occured at relatively recent times as defined by Madau-diagram (the peaks occur at $z = 0.1 - 0.2$, within the first bin). The Mateo review of *all* LG dwarf galaxies (Mateo, 1998) and studies of M31 and our Galaxy (Renzini, 1998), on the other hand, suggest that the LG had its most significant peak in star formation >10 Gyr ago (i.e at $z > 3$), the epoch of halo formation. Many galaxies contain large numbers of RR Lyr variables (or HB) and/or globular clusters which can only come from a significant older population. It is possible that dI galaxies have quite different SFHs to the more massive galaxies. Thus although the small dI galaxies in the LG have been having short, often intense, bursts of star formation in comparatively recent times this is not representative of the majority of the star formation in the LG. However direct observations of the details of the oldest star forming episodes in any galaxy are limited at best. This is an area where advanced CMD analysis techniques have been developed (e.g. Tolstoy and Saha, 1996) and telescopes with sufficient image quality exist and the required deep, high quality imaging are observations are waiting to be made.

Figure 4 summarises what can currently be said about the SFH of the LG and how this compares with the Madau *et al.* (1998) and Shanks *et al.* (1998) redshift survey predictions. We have not included the dominant large galaxies in the LG, the Galaxy and M 31, but the SFH of the combined dwarfs is broadly consistent with what is known about the SFH of these large systems. They have, as far as we can tell, had a global *sfr* that has been gradually but steadily declining since their (presumed) formation epoch > 10 Gyr ago. There is currently no evidence for a particular peak in *sfr* around $7 - 9$ Gyr ago or any other time, as predicted by the Madau-diagram for either large galaxies or dwarfs. The dominant population by mass in the LG dwarfs are dE, if dIs are singled out a population with a star formation peak in the Madau-diagram range can be found. But at present the statistics are too limited to determine the typical fraction of old population in LG dIs. There is clearly a total mismatch between the SFH of the LG and the results from the redshifts surveys. This might hint at serious incompleteness problems in high redshift galaxy surveys, which appear to miss passively evolving systems in favour of small bursting systems.

The recent HST CMD results give much cause for optimism that we can hope to sort out in detail the SFH of all the different types of galaxies within in the LG if only HST would point at them occasionally. There is also great potential for ground based imaging using high quality imaging telescopes with large collecting areas, such as VLT is clearly going to be.

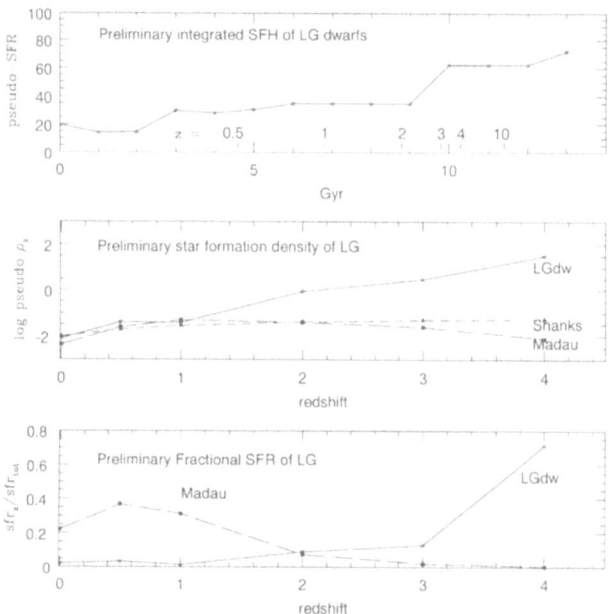

Figure 4. In the upper panel is a *rough* summation of the *sfr*s of the LG dwarf galaxies with time (data taken from Mateo, 1998) to obtain the integrated SFH of all the LG dwarfs. The redshifts corresponding to lookback times (for $H_0 = 50$, $q_0 = 0.5$). In the middle panel, a wild extrapolation is made; the assumption that the integrated SFH of the LG *dwarfs* in the upper panel is representative of the Universe as a whole. The resulting star formation density of the LG versus redshift is plotted using the same scheme as Madau *et al.* (1998) and Shanks *et al.* (1998), and these two models are also plotted and the LG curve is **arbitrarily**, and with a very high degree of uncertainty, normalised to the other two models. In the lowest panel the The LG dwarf *sfr* as a fraction of the total star formation integrated over all time is plotted versus redshift, and the Madau curve is also replotted in this form, for the volume of the LG. This highlights the totally different distibution of star formation with redshift found from galaxy redshift surveys and what we appear to observe in the stellar population of the LG.

References

Aparicio, A., *et al.*: 1996, *Astrophys. J. Lett.* **469**, 97.
Bertelli, G., *et al.*: 1994, *Astron. Astrophys. Suppl.* **106**, 275.
Caputo F., *et al.*: 1995, *Astron. Astrophys.* **304**, 365.
Cole, A.: 1998, *Astrophys. J. Lett.* **500**, L137.
Dohm-Palmer, R.C., *et al.*: 1997, *Astron. J.* **114**, 2514.
Dohm-Palmer, R.C., *et al.*: 1998, *Astron. J.* **116**, 1227.
Fagotto, *et al.*: 1994, *Astron. Astrophys. Suppl.* **104**, 365.
Fukugita, M., Hogan, C.J. and Peebles, P.J.E.: 1998, *Astrophys. J.* **503**, 518.
Gallagher, J.S., *et al.*: 1998, *Astron. J.* **115**, 1869.
Gallart C., *et al.*: 1994, *Astrophys. J. Lett.* **425**, 9L.
Gallart C., *et al.*: 1999, *Astrophys. J.*, in press (astro-ph/9811122).
Geha, *et al.*: 1998, *Astron. J.* **115**, 1045.
Han, M., *et al.*: 1997, *Astron. J.* **113**, 1001.

Holland, *et al.*: 1996, *Astron. J.* **112**, 1035.

Hurley-Keller, D., Mateo, M. and Nemec, J.: 1998, *Astron. J.* **115**, 1840.

Lynds, R., Tolstoy, E., O'Neil, E. and Hunter, D.A.: 1998, *Astron. J.* **116**, 146.

Madau, P., Pozzetti, L. and Dickinson, M.: 1998, *Astrophys. J.* **498**, 106.

Mateo, M.: 1998, *Annu. Rev. Astron. Astrophys.* **36**, 435.

Mighell, K.: 1997, *Astron. J.* **114**, 1458.

Tolstoy, E.: 1996, *Astrophys. J.* **462**, 684.

Tolstoy, E. and Saha, A.: 1996, *Astrophys. J.* **462**, 672.

Tolstoy, E., *et al.*: 1998, *Astron. J.* **116**, 1244.

Tosi, M., *et al.*: 1991, *Astron. J.* **102**, 951.

GLOBULAR CLUSTERS IN BLUE COMPACT GALAXIES AS TRACERS OF THE STARBURST HISTORY

G. ÖSTLIN

Institut d'Astrophysique de Paris, 98 bis Boulevard Arago, 75014 Paris, France

Abstract. Representing single stellar populations, globular clusters (GCs) are relatively easy to model, thus providing powerful tools for studying the evolution of galaxies. This has been demonstrated for the blue compact galaxy ESO338-IG04. GC systems in galaxies may be fossils of starbursts and mergers. Thus studies of GCs in the local universe may add to our understanding of the formation and evolution of galaxies and the distant universe.

1. Introduction – Globular Cluster Formation and Destruction

Globular clusters (GCs) are in general old stellar systems with masses 10^4 to $10^7 M\odot$, and are believed to be among the first ingredients to form in the process of galaxy formation. Thus understanding how GCs form is vital for understanding how galaxies form. Extragalactic GC systems in e.g. elliptical (E) galaxies often have bimodal colour distributions, indicating the presence of populations with different metallicity and/or age. Studies of merging galaxies, e.g. the 'Antennae' (Whitmore and Schweizer, 1995) and NGC 7252 (Miller *et al.*, 1997), have shown that these often contain young GC candidates in great numbers. These galaxies are believed to evolve into E galaxies as the merger remnants relax. The GC candidates in the more evolved mergers have redder colours indicating higher ages. Very young so-called 'super star clusters' (SSCs) have been found in many starburst galaxies, including dwarfs (e.g. Meurer *et al.*, 1995). The SSCs have properties which largely agree with those expected for young GCs, although the masses are quite uncertain, and are probably younger examples of the objects seen in the mergers. The triggering mechanism for the starbursts in the dwarfs that host SSCs is still an open question, but dwarf mergers are not ruled out. SSCs have also been found in the centres and circum-nuclear rings in giant barred spirals (e.g. Barth *et al.*, 1995; Kristen *et al.*, 1997) .

In conclusion young globular cluster candidates are found in extreme and energetic environments indicating that they can only be formed under special conditions. The coeval formation of many GCs will require extreme conditions, e.g. very high pressures and gas densities (Elmegreen and Efremov, 1997), conditions which are fulfilled in mergers. Moreover, bars trigger gas flows and circum nuclear rings may be created by dynamical resonances, enhancing the density. A newly

Astrophysics and Space Science is the original source of publication of this article. It is recommended that this article is cited as: *Astrophysics and Space Science* **265**: 207–210, 1999.
© 1999 *Kluwer Academic Publishers*.

formed GC will not automatically become an old GC (like the ones in our Galaxy)
as time goes by, but faces the risk of destruction and dissolution. Their ability to
survive depends on their interaction with their environment, but also on their IMF.
A galactic bar for instance is not a favourable place for GC survival since strong
shocks will easily disrupt many young GCs. The conditions in mergers, and in
particular dwarf galaxies may be more favourable for GC survival.

2. Young and Old GCs in ESO 338-IG04

ESO338-IG04 is a blue compact galaxy (BCG). A BCG is characterised by compact appearance and HII region like spectra indicating high star formation rates,
and in general low chemical abundances. Most BCGs are dwarf galaxies that, for
some reason, presently are undergoing starbursts. HST/WFPC2 images reveals that
ESO338-IG04 hosts a very rich population of compact star clusters counting more
than one hundred objects (Östlin *et al.*, 1998). The centre is crowded with young
SSCs, but in addition lots of intermediate age and old objects are found outside
the starburst region (there might well be some in the centre as well but they drown
in the light from young SSCs). Photometric modelling (using Salpeter and Miller-
Scalo IMFs) indicates masses in the range 10^4 to more than $10^7 M\odot$ and a wide
range of ages, from a few Myr up to 10 Gyr (see Figure 1). The spread in ages is
real and not caused by observational errors. Moreover, there are groupings in the
age distribution indicating that the cluster formation history has not been continuous. Most apparent is of course the numerous very young and luminous SSCs, but
perhaps even more interesting is the presence of one or two populations of massive
GCs with age 2 to 5 Gyr. This population, when corrected for objects below the
detection limit, shows that starburst progenitor had a specific frequency S_N (a
measure of the number of GCs relative to the host galaxy luminosity) comparable
to those of giant Es. In addition there are old ($\geq$ 10 Gyr) and 0.5 Gyr GCs present.
The dynamics and morphology of this galaxy indicates that a dwarf merger is responsible for the starburst (Östlin *et al.*, 1998, 1999). Thus we have a splendid low
luminosity ($M_B = -19$) counterpart of the giant mergers mentioned above. Even
if young GCs of course may disrupt or dissolve, the presence of intermediate age
GCs proves that GC formation is not a phenomenon that was isolated to the very
earliest days of galaxy formation. SSCs are often found in BCGs which suggest
that they might be a good laboratory for studying GC formation; and a closer look
for aged GCs may provide important information on previous bursts.

3. GCs as Starburst Fossils

We have seen that objects which are likely to be young GCs have been found in
galaxies which are starbursts and/or mergers. The presence of intermediate age

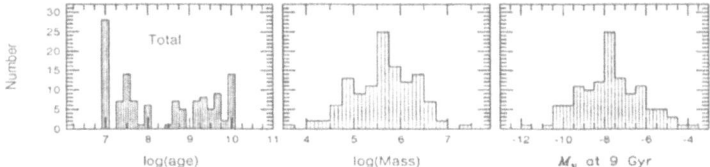

Figure 1. Modelled ages (*left*), masses (*centre*) and absolute *V* magnitudes when transforming all objects to an age of 9 Gyr (*right*). The results are based on a standard Salpeter IMF, but are very similar for a Miller-Scalo IMF (Östlin *et al.*, 1998a).

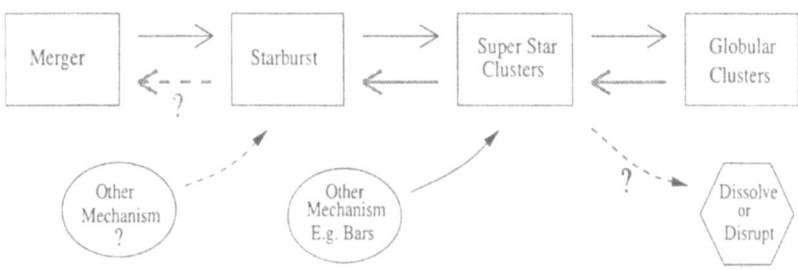

Figure 2. A sketch of the connection between GCs, SSCs, starbursts and mergers.

GCs in ESO338-IG04 and somewhat aged objects in the merger remnants suggest that at least a considerable fraction of young GCs will survive. It is a general property of starbursts to reveal the presence of SSCs when studied at high spatial resolution. R136, the central cluster in 30 Doradus, may *perhaps* be the closest example of a GC in the making. Even if not all SSCs become GCs, we can conclude that all newly formed GCs must have properties similar to SSCs. Moreover the coeval formation of many massive GCs would produce a starburst in itself. Let us illustrate this in a figure (Figure 2): A starburst leads to SSCs of which at least a fraction becomes GCs, and the reverse is also true, although SSCs may form also in bars, which however are hostile environments. Therefore the age distribution of the GC population can be used to infer the SSC formation, and thus starburst, history. It is also clear that mergers are capable of trigger starbursts, but we do not know yet if a merger is required for the occurrence of a global starburst in a galaxy. If that would be the case, GC populations could be used to trace the merger history of galaxies. Of course, the question marks in Figure 2 must be straightened out before GCs can be used as general probes.

Although GCs provide information on the starbursts history of galaxies, they do not necessarily tell us about the overall star formation history, afterall GC free galaxies exist. There is obviously two different modes of star formation: the 'violent' starburst mode which favours cluster formation and the 'quiet' mode that may still produce the bulk of the field stars in most galaxies (Van den Bergh, 1998). In nearby galaxies the star formation history (SFH) of the field population can be studied through deep colour magnitude diagrams (cf. Tolstoy this volume). In most galaxies however one has to infer the overall SFH from the integrated stellar population. A GC is much easier to model because it represents a true single coeval (on the scale of a few Myrs) stellar population. Spectroscopy can be used to investigate metallicities to circumvent the age-metallicity degeneracy. Thus GCs can serve as excellent probes for unveiling the starburst history in moderately distant galaxies. But also a more complete picture of the GC content in local galaxies would provide important information. This might be of importance for interpreting deep surveys which often are biased towards starburst galaxies. Depending on how strong the connection to mergers will prove to be, GC populations may also be useful for studying the morphological and number evolution of the galaxy population.

References

Barth, A., Ho, L., Filippenko, A. and Sargent, W.: 1995, *Astron. J.* **110**, 1009.

Elmegreen, B. and Efremov, Y.: 1997, *Astrophys. J.* **480**, 235.

Kristen, H., Jörsäter, S., Lindblad, P.O. and Boksenberg, A.: 1997, *Astron. Astrophys.* **328**, 483.

Östlin, G., Bergvall, N. and Rönnback, J.: 1998, *Astron. Astrophys.* **335**, 85.

Östlin, G., Amram, P., Bergvall, N., Masegosa, J. and Boulesteix, J.: 1999, *Astron. Astrophys. Suppl.*, **137**, 419.

Meurer, G., Heckman, T., Leitherer, Kinney, Robert and Garnett: 1995, *Astron. J.* **110**, 2665.

Miller, B., Whitmore, B., Schweizer, F. and Fall, M.: 1997, *Astron. J.* **114**, 2381.

van den Bergh, S.: 1998, *Astrophys. J.* **507**, L39.

Whitmore, B. and Schweizer, F.: 1995, *Astron. J.* **109**, 960.

IV – THE OLD DISK POPULATION

THE GALACTIC THICK DISK – STATUS

JOHN E. NORRIS

*Research School of Astronomy & Astrophysics, The Australian National University, Private Bag,
Weston Creek P.O., ACT 2611, Australia*

Abstract. The status of the Galactic thick disk is reviewed. Consideration of the recent literature
suggests that its vertical scale height and normalisation with respect to the thin disk remain uncertain
to within a factor two, with values reported in the ranges 750–1500 pc, and 0.02–0.13, respectively.
The bulk of the thick disk has kinematics $(\sigma_U, \sigma_V, \sigma_W) = (65, 54, 38$ km s$^{-1})$, and lags the thin
disk by some 40 km s^{-1}; differences of opinion exists as to whether kinematics change with distance
from the Galactic plane. The bulk of the thick disk has [Fe/H] ~ -0.6, with little or no evidence for
a vertical gradient. The question of gradients is critical for an understanding of thick disk cosmogony
and needs closer attention.

The reality of the so-called metal-weak thick disk (material having disklike kinematics and
[Fe/H] < -1.0) is also considered. The case for such material seems to be steadily growing: in
the range $-1.6 <$ [Fe/H] < -1.0, recent estimates suggest $\rho_{MWTD}/\rho_{Halo} \sim 0.1 - 0.3$.

While many workers regard the thick disk as a discrete entity, the caveat is made that this is a
sufficient condition, but not one necessarily required by the observations. Best practice requires that
both the discrete model and the alternative extended configuration be compared with observational
data to examine the relative likelihood of their relevance.

Recent theoretical advances are also discussed, together with the need for *in situ* measurements
of the thick disk away from the Galactic plane.

1. Introduction

In the Galactic context, the term 'thick disk' was introduced by Gilmore and Reid
(1983) to explain the observation that star counts towards the South Galactic Pole
were not well fit by a single exponential, but rather could be nicely represented
by two such components – one with scale height 300 pc, the second with scale
height 1350 pc and normalisation (at the Galactic plane) 0.02 that of the first. This
normalisation is an order of magnitude larger than that of the halo population* and
Gilmore and Reid suggested that the thick disk was most likely associated with
Intermediate Population II, and that it was a 'moderately metal poor population of
spheroid stars responding to the gravitational potential of the Galactic disk', as had
been suggested by van der Kruit and Searle (1981) to explain their observations of
edge-on spiral systems. Historians of science will note that this intermediate 'thick
disk' component was also present in the star counts of Yoshii (1982) towards the
North Galactic Pole.

---

* We use the term halo to refer to Population II material.

 Astrophysics and Space Science is the original source of publication of this article. It is recommended that this article is cited as: *Astrophysics and Space Science* **265**: 213–220, 1999.
© 1999 *Kluwer Academic Publishers.*

Work by Gilmore and Wyse (1985) and Wyse and Gilmore (1986) led to the suggestion that the thick disk had abundance and kinematic characteristics intermediate between that of the halo and thin disk populations ($\langle$[Fe/H]$\rangle \sim -0.6$, $\sigma_W \sim 60$ km s^{-1}, and asymmetric drift ~ -100 km s^{-1}), and laid the ground work for the suggestion that it was a discrete component, with origins quite different from those of the other components. Later investigations would support the abundance result, but lead to more moderate kinematics, with $\sigma_W \sim 40$ km s^{-1}, and asymmetric drift ~ -40 km s^{-1}(see e.g. Norris, 1993, and references therein).

The question of the discreteness or otherwise of the population is an important one, for while it provides a sufficient explanation of the star counts it is not a necessary one. An extended disk with a continuum of properties is also sufficient (e.g. Norris, 1987; Norris and Ryan, 1991). The issue is important in the context of disk cosmogony: if the thick disk is a discrete population one will be drawn towards explanations involving discrete events involving perhaps minor mergers, while if it is part of an extended disk configuration explanations involving dissipative processes might be favoured.

A number of reviews involving the thick disk exist in the literature. Some which the reader might find useful are those of Gilmore, Wyse and Kuijken (1989), Nemec and Nemec (1991a), Majewski (1993), and Norris (1996). The structure of the present discussion is as follows. §2 considers recent observational determinations of the parameters of the thick disk and the question of its discrete versus continuous nature, while in §3 we consider the status of the so-called metal-weak thick disk. Recent theoretical advances are addressed in §4. Finally, brief comments are made on a projected *in situ* study currently being undertaken with a view to more closely understanding the thicker parts of the disk.

2. Recent Determinations of Thick Disk Parameterisation

2.1. SCALE HEIGHT AND NORMALISATION

Table I compares recent determinations of scale height and normalisation with the original values of Gilmore and Reid (1983). It demonstrates that the scale height of the thick disk is known only to within a factor 2, and lies in the range 750–1500 pc. The relatively small quoted errors suggest that the uncertainty is due to systematic problems rather than internal ones. Similar statements pertain to the thick disk/thin disk normalisation. It seems reasonable to suggest to those who advocate that the thick disk is a discrete population that one should be able to determine its basic parameters to higher accuracy than this.

TABLE I

Recent determinations of thick disk scale height & normalisation

Author	Z_0 (pc)	ρ_{thick}/ρ_{thin}
Gilmore and Reid (1983)	1350	0.02
Reid and Majewski (1993)	1500	0.02
Bell (1996)	...	0.13 ± 0.03
Robin et al. (1996)	760 ± 50	0.056 ± 0.01
Buser et al. (1998)	1150 ± 150	0.054 ± 0.015
Phleps et al. (this volume)	1038 ± 66	...

2.2. KINEMATICS

As noted above, most workers agree that in the solar neighbourhood the thick disk has kinematics closer to those of the thin disk than to those of the halo, with perhaps the most relevant of the velocity ellipsoid determinations being that of Strömgren (1987) for material with $-0.79 < $ [Fe/H] $ < -0.70$: $(\sigma_U, \sigma_V, \sigma_W) = (65, 54, 38$ km s^{-1}), together with asymmetric drift -40 km s^{-1}. A remaining point of controversy concerns the question of possible gradients perpendicular to the galactic plane. Majewski (1992, Figure 32), on the one hand, finds such a gradient, with $V_{lag} \sim -18*Z$(kpc) -15 km s^{-1}, while Soubiran (1993) and Ojha et al. (1994), on the other, find no gradient and report the canonical value of -40 km s^{-1} up to 3 kpc from the plane.

2.3. METALLICITY

Table II compares recent abundance results with the initial values reported by Gilmore and Wyse (1985). Since abundances are notoriously difficult to determine with external accuracy better than ~ 0.20 dex it seems reasonable to suggest that all workers agree that the bulk of the thick disk has [Fe/H] ~ -0.6. If there is a vertical abundance gradient, it is small and lies within the errors of measurement.

2.4. DISCRETE VERSUS CONTINUOUS

During the past decade many workers have argued that there appears to be a distinct break between the properties of the thin and thick disks (see e.g. Carney et al., 1989; Nissen, 1990; Freeman, 1991; Edvardsson et al., 1993; Robin et al., 1996) supporting the concept of a discrete thick disk.

In contrast, Reid (1998) (considering the $(M_V, B-V)$ distribution of high proper motion stars with Hipparcos parallaxes and spectroscopically determined [Fe/H])

TABLE II

Recent thick disk abundance determinations

Author	$\langle$[Fe/H]$\rangle$	Vertical gradient
Gilmore and Wyse (1985)	-0.6	...
Carney *et al.* (1989)	-0.5	None
Gilmore, Wyse and Jones (1995)	-0.7	None
Bell (1996)	-0.36 ± 0.06	...
Robin *et al.* (1996)	-0.7 ± 0.2	None
Buser *et al.* (1998)	-0.6	0.1 ± 0.1 dex kpc^{-1}

supports the case of non-discreteness. He finds: 'the metallicity distribution inferred (from position in the colour-magnitude diagram) is more consistent with a continuum (the disk) than with a combination of two separate populations, the old disk and the thick disk'.

It is salutary to recall the admonition of Nemec and Nemec (1991b): 'In most models of the Galaxy the number of components is based on preconceived ideas or on a subjective assessment of the observational data. ... More objective ... procedures for estimating the number of components and goodness-of-fits tests ... for comparing competing models are available.' What is needed to decide between the various cases is an objective assessment of their relative ability to best represent the observational material. A noteworthy recent contribution to the debate is the work of Bell (1996), who undertook a maximum likelihood analysis of a complete sample of some 570 red giants towards $(l,b) = (90,0)$ and $(270,0)$ to determine whether a discrete or a continuous disk model better fitted the data. His conclusion was that the discrete representation provided a somewhat better fit. The only concern that the present author would have with the analysis is that only ~ 25 objects in the Bell sample have abundances in the critical regime [Fe/H] < -0.6. It would be well worthwhile if other larger datasets were subjected to this comparative type of analysis.

3. Metal-Poor Tail of the Thick Disk

Perhaps the major outstanding observational question concerning the thick disk is whether it contains a significant amount of material having [Fe/H] < -1.0, and if it does what is its kinematic signature. Norris, Bessell and Pickles (1985) argued from samples of metal-poor giants and dwarfs, and RR Lyrae variables that some 20% of objects having [Fe/H] < -1.0 had orbits with (disklike) eccentricities less that 0.4, which received some support from Carney and Latham (1986) and Allen, Poveda and Schuster (1991). The result was (justifiably) criticised by Twarog and Anthony-

Twarog (1994) and Ryan and Lambert (1995) on the grounds of systematic errors in the abundances of the giants. With this background the recent analysis of Chiba and Yoshii (1998) is then of considerable interest. Using better space velocities based on Hipparcos data, and abundances based on high-resolution spectroscopic analysis, they conclude that in the abundance range $-1.6 <$ [Fe/H] < -1.0 some 10% of material has disklike kinematics ($e < 0.4$), while in the range $-1.4 <$ [Fe/H] < -1.0 the proportion rises to 20%. Earlier reports in favour of the existence of a rather more substantial metal-weak tail to the disk include those of Beers and Sommer-Larsen (1995) and Rodgers and Roberts (1993)[*], while Layden (1995) had also reported $\rho_{MWTD}/\rho_{Halo} = 0.2 \pm 0.1$ for material having $-1.6 <$ [Fe/H] < -1.0.

One investigation still in progress is that of Nordström et al. (Andersen, 1995, private communication) who have obtained accurate abundances and kinematics of some 5600 stars in the solar neighbourhood. Some 27 of these have [Fe/H] < -1.0, and appear to have disklike kinematics. While the significance of these objects must await the analysis of possible abundance related selection effects, we note that they represent 0.005 of the sample, some five times the halo/disk ratio (~ 0.001) in the solar neighbourhood.

We conclude by drawing the reader's attention to the work of Chiba, Yoshii and Beers (1999) who have enlarged the dataset of Chiba and Yoshii (1998) discussed above. With the larger sample they are able to analyse the individual contribution of giants, dwarfs and horizontal branch stars to the metal-poor thick disk. It is of great interest that while they report substantial fractions ($\sim 0.2-0.3$) of metal-poor material for giants and dwarfs, they do not find a significant one for horizontal-branch stars. This behaviour needs to be confirmed: one possible explanation (first suggested in a related context by Morrison et al., 1990) is that it may result from age differences between the halo and metal-weak thick disk, with perhaps the thick disk being a few Gyr younger. Similar suggestions have been made by Norris and Green (1989) and Marquez and Schuster (1994). For completeness, however, we note that Martin and Morrison (1998) have also recently reanalysed the available RR Lyrae data and argue for a significant disklike component, with $\rho_{MWTD}/\rho_{Halo} = 0.26 \pm 0.06$ for material having $-1.6 <$ [Fe/H] < -1.0.

3.1. WHERE DOES THE METAL-WEAK THICK DISK FIT?

The question which then begs to be answered is whether the metal-weak thick disk is simply the non-gaussian tail of the thick disk, or a discrete population in its own right. Indeed, just how many components comprise the Galaxy's disk? Among those currently found in the literature we now have: the young disk, the intermediate disk, the old disk, the thin disk, the thick disk (identified as Intermediate Population II), the metal-weak thick disk; not to mention the extended disk.

---

[*] We exclude the work of Morrison, Freeman and Flynn (1990) from this list because of their use of the questionable abundances for the giants of Norris et al. (1985).

It is an interesting comment on the state of the subject that one eminent speaker at this meeting noted with some flippancy that he would not talk about 'the ten components of the Galactic disk', but would leave that to later speakers.

4. Theoretical Developments

Thick disks can be made in dissipative collapse models (Larson, 1976) and by the merger of small satellites with an existing disk system (Quinn, Hernquist and Fullagar, 1993). The former will generate gradients perpendicular to the Galactic plane, but while the model of Quinn *et al.* preserves existing ones it does not create any. Of particular interest in the present context is the prediction of the latter model that the asymmetric drift of a disk will not be significantly increased ($\Delta V_{lag} < 10$ km s^{-1}) by the merger with a 10% satellite. Thus, if the kinematic gradient reported by Majewski (1992) stands the test of time, it could not have been created in a merger of the type reported by Quinn *et al.*

Proponents of a discrete thick disk often cite the work of Quinn *et al.* in support of their position. It is thus important to examine the reliability and relevance of those models. While the recent work of Walker, Mihos and Hernquist (1996) finds similar results to those of Quinn *et al.*, that of Huang and Carlberg (1997), which undertakes simulations of mergers with low-density satellites (as opposed to the high-density ones used in the earlier works) reports rather different results. In particular, they find that minor mergers are more likely to cause disk tilting than thickening, that mergers with satellites of mass less than 20% of the existing disk produce no observable thickening, and in the case of a merger with a 30% satellite there is little observable thickening inside the half-mass radius, but great damage beyond this value. Huang and Carlberg argue that mergers with low-density satellites are more likely to have occurred in the case of the Milky Way, given what we know of its environment.

A second interesting development is the modelling by Noguchi (1998) of clumpy star-forming regions in an effort to explain the peculiar morphology of high-redshift galaxies. Beginning with a rotating ensemble of gaseous and dark matter particles, he finds that a galaxy forms on a timescale of 1 Gyr, and that its evolution is critically determined by the formation of distinctive clumps of mass 10^9 M$_\odot$. Dynamical friction causes the clumps to spiral towards the galactic center, creating a bulge, while during the process stars and gas scatter off the clumps, thickening the disk. The scale height of the old stellar component in the model he reported was 1.9 kpc.

To the present author at least, theory offers no definitive guidance. The results of Huang and Carlberg cause some problems for those who might argue that mergers provide a natural explanation of a discrete thick disk. That having been said, more detailed modelling, with substantial predictions, is required to make the case that a clumpy formation scenario for the Milky Way is the correct model for its formation.

5. Pending *In-Situ* Studies

The most limiting factor in efforts to understand the Galaxy has been our necessary reliance on solar-neighbourhood studies, where the important halo and thick disk stars represent only a very small minority of field stars. To date investigations away from the Galactic plane have been limited to small samples of only marginal statistical significance. This is now changing with the advent of large multifibre spectrographs on 4 m class telescopes. I conclude with a very brief description of such an investigation.

The 2dF facility of the Anglo-Australian Telescope permits one to simultaneously obtain intermediate resolution spectra ($\sim 2 - 4$ Å FWHM) of some 400 objects over a two degree field. This field coverage, together with the sensitivity of the system, is well matched to the investigation of main sequence turnoff objects a few kpc from the Galactic plane, and permits one for the first time to obtain significant numbers of objects well away for the contaminating effects of the thin disk. Gilmore, Wyse, Freeman, and the present author have initiated a programme to obtain samples of a few thousand stars in each of 5–6 fiducial directions (towards the Galactic poles, center and anticenter, and in directions of Galactic rotation) with a view to disentangling the populations discussed above, and to more strongly constraining the reality of otherwise of gradients. Hopefully, when projects such as this are completed, many of the remaining issues, such as those discussed above, will come closer to solution.

References

Allen, C., Poveda, A. and Schuster, W.J.: 1991, *Astron. Astrophys.* **244**, 280.
Bell, D.J.: 1996, *PhD Thesis*, Univ. of Illinois.
Beers, T.C. and Sommer-Larsen, J.: 1995, *Astrophys. J. Suppl.* **96**, 175.
Buser, R., Rong, J. and Karaali, S.: 1998, *Astron. Astrophys.* **331**, 934.
Carney, B.W. and Latham, D.W.: 1986, *Astron. J.* **92**, 60.
Carney, B.W., Latham, D.W. and Laird, J.B.: 1989, *Astron. J.* **97**, 423.
Chiba, M. and Yoshii, Y.: 1998, *Astron. J.* **115**, 168.
Chiba, M., Yoshii, Y. and Beers, T.C.: 1999, in: B.K. Gibson, T.S. Axelrod and M.E. Putnam (eds.), *The Third Stromlo Symposium: The Galactic Halo – Bright Stars & Dark Matter*, ASP, San Francisco, in press.
Edvardsson, B., Andersen, J., Gustafsson, B., Lambert, D.L., Nissen, P.E. and Tomkin, J.: 1993, *Astron. Astrophys.* **275**, 101.
Freeman, K.C.: 1991, in: B. Sundelius (ed.), *Dynamics of Disk Galaxies*, Göteborg Univ., Göteborg, p. 15.
Gilmore, G. and Reid, N.: 1983, *Mon. Not. R. Astron. Soc.* **202**, 1025.
Gilmore, G. and Wyse, R.F.G.: 1985, *Astron. J.* **90**, 2015.
Gilmore, G., Wyse, R.F.G. and Jones, J.B.: 1995, *Astron. J.* **109**, 1095.
Gilmore, G., Wyse, R.F.G. and Kuijken, K.: 1989, *Annu. Rev. Astron. Astrophys.* **27**, 555.
Huang, S. and Carlberg, R.G.: 1997, *Astrophys. J.* **480**, 503.
Larson, R.B.: 1976, *Mon. Not. R. Astron. Soc.* **176**, 31.

Layden, A.C.: 1995, *Astron. J.* **110**, 2288.

Majewski, S.R.: 1992, *Astrophys. J. Suppl.* **78**, 87.

Majewski, S.R.: 1993, *Annu. Rev. Astron. Astrophys.* **31**, 575.

Marquez, A. and Schuster, W.J.: 1994, *Astron. Astrophys. Suppl.* **108**, 341.

Martin, J.C. and Morrison, H.L.: 1998, *Astron. J.* **116**, 1724.

Morrison, H.L., Flynn, C. and Freeman, K.C.: 1990, *Astron. J.* **100**, 1191.

Nemec, J.M. and Nemec, A.F.L.: 1991a, *Publ. Astron. Soc. Pacific* **103**, 95.

Nemec, J. and Nemec, A.F.L.: 1991b, in: K. Janes (ed.), *The Formation & Evolution of Star Clusters*, ASP, San Francisco, p. 512.

Nissen, P.E.: 1990, in: R. Terlevich (ed.), *Elements and the Cosmos*, Cambridge University Press, Cambridge, p. 110.

Noguchi, M.: 1998, *Nature* **392**, 253.

Norris, J.: 1987, *Astrophys. J.* **314**, L39.

Norris, J.E.: 1993, in: G.H. Smith and J.P. Brodie (eds.), *The Globular Cluster–Galaxy Connection*, ASP, San Francisco, p. 259.

Norris, J.E.: 1996, in: H. Morrison and A. Sarajedini (eds.), *Formation of the Galactic Halo . . . Inside and Out*, ASP, San Francisco, p. 14.

Norris, J., Bessell, M.S. and Pickles, A.J.: 1985, *Astrophys. J. Suppl.* **58**, 463.

Norris, J. and Green, E.M.: 1989, *Astrophys. J.* **337**, 272.

Norris, J. and Ryan, S.G.: 1991, *Astrophys. J.* **380**, 403.

Ojha, D.K., Bienaymé, O., Robin, A.C. and Mohan, V.: 1994, *Astron. Astrophys.* **290**, 771.

Quinn, P.J., Hernquist, L. and Fullagar, D.P.: 1993, *Astrophys. J.* **403**, 74.

Reid, I.N.: 1998, *Astron. J.* **115**, 204.

Reid, N. and Majewski, S.R.: 1993, *Astrophys. J.* **409**, 635.

Robin, A.C., Haywood, M., Crezé, M., Ojha, D.K. and Bienaymé, O.: 1996, *Astron. Astrophys.* **305**, 125.

Rodgers, A.W. and Roberts, W.H.: 1993, *Astron. J.* **106**, 1839.

Ryan, S.G. and Lambert, D.L.: 1995, *Astron. J.* **109**, 2068.

Soubiran, C.: 1993, *Astron. Astrophys.* **274**, 181.

Strömgren, B.: 1987, in: G. Gilmore and B. Carswell (eds.), *The Galaxy*, Reidel, Dordrecht, p. 229.

Twarog, B.A. and Anthony-Twarog, B.J.: 1994, *Astron. J.* **107**, 1371.

van der Kruit, P.C. and Searle, L.: 1981, *Astron. Astrophys.* **95**, 116.

Walker, I.R., Mihos, J.C. and Hernquist, L.: 1996, *Astrophys. J.* **460**, 121.

Wyse, R.F.G. and Gilmore, G.: 1986, *Astron. J.* **91**, 855.

Yoshii, Y.: 1982, *Publ. Astron. Soc. Jpn.* **34**, 365.

IN SITU STUDY OF THE THICK DISK. PRELIMINARY RESULTS

D. KATZ, R. CAYREL, M.-F. COUPRY, M.-N. PERRIN and C. VAN 'T VEER
Observatoire de Paris, 61, av. de l'Observatoire, F-75014 Paris, France

C. SOUBIRAN
Observatoire de Bordeaux, BP89, F33270 Floirac, France

B. BARBUY
Universidade de São Paulo, C.P. 3386, São Paulo 01060-970, Brazil

O. BIENAYMÉ
CDS, Observatoire Astronomique de Strasbourg, 11 rue de l'Université, F-67000 Strasbourg, France

E. FRIEL
National Science Foundation, Division of Astronomical Sciences, 4201 Wilson blvd, Arlington VA 22230, USA

Abstract. We report the advancement of our chemical and kinematical study of the thick disk. The methods used to derive the stellar parameters are discussed and the preliminary results presented.

1. Introduction

There is a large variety of scenarios to explain the formation of the thick disk of our Galaxy: from continuous pressure supported collapse to dynamical friction (see Majewski, 1993, for a complete revue). But at the present time the uncertainties on the characteristics of the thick disk (scale length, scale height, metallicity distribution, velocity distribution, ...) are too large to discriminate between the models. So it is with the aim of reducing the error bars on those observables that we have undertaken, since 1991, an astrometric, photometric and spectroscopic survey of two fields: the first one in the direction of the north galactic pole and the second one in the direction $l = 4°$, $b = 47°$. Because the fields are located respectively near the clusters M3 and M5, we will refer to them as field M3 and field M5. Section 2 summarizes the observations. Section 3 describes the methods used in the determination of the stellar parameters and Section 4 presents the preliminary results.

Astrophysics and Space Science is the original source of publication of this article. It is recommended that this article is cited as: *Astrophysics and Space Science* **265:** 221–224, 1999.
© 1999 *Kluwer Academic Publishers.*

2. Observations

Proper motions have been derived for 4384 stars over 7 degrees in field M3 (Soubiran, 1992) and for 20000 stars over 15.5 degrees in field M5 (Ojha *et al.*, 1994). The samples are complete respectively to $V = 17$ and $V = 18$.

Three spectrographs have been used, each in a different interval of magnitude. The spectrographs ELODIE and CARELEC, mounted on the 193 cm telescope of the 'Observatoire de Haute-Provence', have been used respectively to observe stars with $12.0 \leq V \leq 13.5$ and $13.5 \leq V \leq 14.5$. Stars with $14.5 \leq V \leq 17.5$ were observed with the spectrograph MOS mounted on the 'Canada-France-Hawaii' telescope. A total of 600 spectra has been collected. Stars from the field M3 observed with ELODIE and CARELEC were selected by their reduced proper motion: $H = V + 5 \log \mu + 5$ and by their color (to restrict the observations to F, G and K stars). The stars from the field M5 observed with ELODIE and CARELEC were selected only by their color. No selection has been applied to the stars observed with MOS (except in apparent magnitude).

V, *B* and *I* Cousins photometry has been obtained during four observing runs at CFHT, Lowell, Hopkins and Kitt-Peak Observatories.

3. Data Reduction

From the 600 observed spectra, 230 have been fully reduced. The radial velocities where extracted from ELODIE's spectra using the ELODIE reduction software (Queloz, 1996) and from CARELEC using the softwares HALO and ETOILE. The atmospheric parameters T_{eff}, $\log g$ and [Fe/H] were estimated from the spectra by the softwares TGMET (Katz *et al.*, 1998; Soubiran *et al.*, 1998), HALO and ETOILE. The three programs rely on the least square comparison of an object spectrum with a library of spectra of standard stars.

Absolute magnitudes were derived from the isochrones of D. VandenBerg (1999) (private communication). The distances were calculated from the distance modulus and the three components of the velocities were determined from distances, radial velocities and proper motions. Figure 2 shows the distribution of stars as a function of $\log Z$ (distance perpendicular to the plane, in pc).

The orbital parameters (R_{min}, R_{max}, z_{min}, z_{max}, eccentricity) were calculated by M. Odenkirchen using the model of Galactic potential of C. Allen (Allen *et al.*, 1991).

4. Preliminary Results

The different populations (thin disk, thick disk and halo) have been separated using the plane [v_{rot}, [Fe/H]]. The criteria used to classify the stars are illustrated in Figure 3. *In fine* 130 stars were classified as thick disk stars.

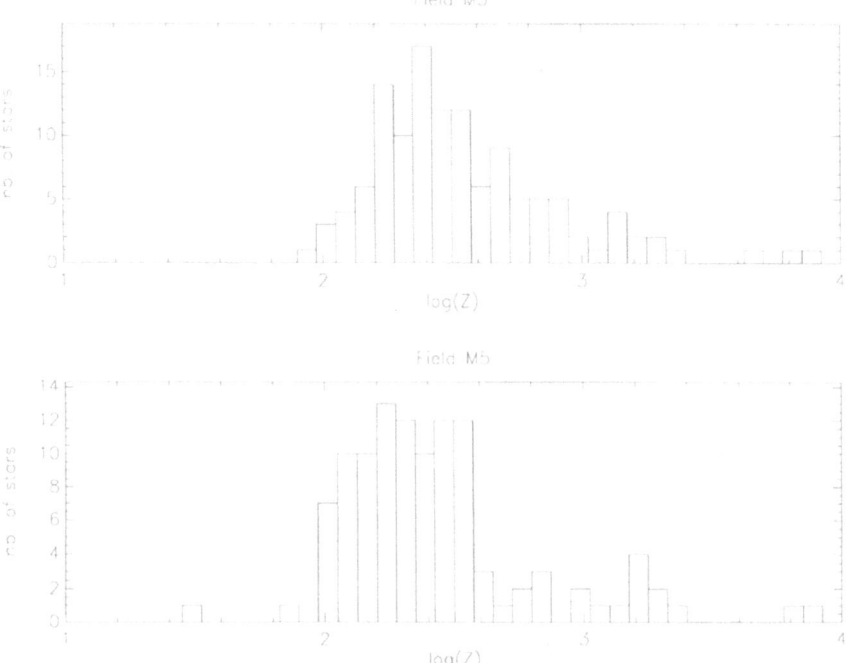

Figure 1. Distribution of stars in field M3 (top) and field M5 (bottom) as a function of log Z (distance perpendicular to the plane, in pc).

The analysis of the sample of thick disk stars has revealed two interesting subgroups. The first one is made of 7 stars with thick disk kinematics and very low metallicities ranging from −1.57 to −2.21. Stars with similar behaviour have already been reported by E. Morisson (Morrison, 1993) and by T. Beers and J. Sommer-Larsen (Beers *et al.*, 1995). The second subgroup is made of 12 stars with thick disk kinematics and near solar metallicities. The nature of these two subgroups is ambiguous because they exhibit properties of different populations: thick disk kinematics and halo or thin disk metallicities.

5. Conclusion

The preliminary results of the survey suggest that some stars can have 'thick disk kinematics' and at the same time chemical properties of the halo or of the thin disk. One way to investigate the true nature of those stars should be to study their age distribution which could be characteristic of a peculiar population. The ratio [O/Fe] can also give information on the history of the nucleosynthesis that took place before the birth of the stars.

Figure 2. [Fe/H] versus v_{rot} for the thick disk stars from M3 (top) and M5 (bottom). The symbols represent the different evolutionary stages: dwarf (square), turn-off (triangle) and evolved (circle).

Acknowledgements

We are very grateful to Michaël Odenkirchen for computing the orbits.

References

Allen, C. and Santillan, A.: 1991, *Rev. Mex. Astron. Astrofis.* **22**, 225.
Beers, T. and Sommer-Larsen, J.: 1995, *Astrophys. J. Suppl.* **96**, 175.
Katz, D., Soubiran, C., Cayrel, R., Adda, M. and Cautain, R.: 1998, *Astron. Astrophys.* **338**, 151.
Majewski, S.R.: 1993, *Annu. Rev. Astron. Astrophys.* **31**, 575.
Morrison, H.L.: 1993, *Astron. J.* **105**, 539.
Ojha, D.K., Bienaymé, O., Robin, A.C. and Mohan, V.: 1994, *Astron. Astrophys.* **290**, 771.
Queloz, D.: 1996, *ELODIE user's guide*, http://www.obs-hp/.
Soubiran, C.: 1992, *Astron. Astrophys.* **259**, 394.
Soubiran, C., Katz, D. and Cayrel, R.: 1998, *Astron. Astrophys. Suppl.* **133**, 221.

DYNAMICAL MODELS OF THE THICK DISC

O. BIENAYMÉ

Observatoire de Strasbourg, 11 rue de l'Université, F-67000, Strasbourg, France

Abstract. Thick discs are components intermediate between thin discs and stellar halos, they are also expected to gives clues concerning intermediate stages of the galaxy formation. Numerical models from N-body accretion to hydrodynamical collapse may reproduce part of kinematics or other observational facts. Consistent kinematic models of thick discs allow one to understand which kinematic features are intrinsic to thick discs and consequently do not depend really on the precise conditions of their formation. Establishing a dynamical continuity between the thick disc and the other stellar populations will provide the opportunity to discriminate between formation scenarios: -free-fall collapse, halo formed as a merger of dwarf galaxies or thick disc puff-up by galaxy harassment.

1. Introduction

The thick disc is an intermediate galactic component with intermediate properties for all the following characteristics: its stellar density distribution, its velocity moments (i.e. the rotational lag and the dispersion tensor), its metallicity abundances and the ages of the stars. These key observables and correlations show that the kinematics is correlated with the density structure and the age-metallicity groups. It shows that the thick disc is an intermediate step during the galactic formation.

Numerous scenarii of the galactic formation have been proposed. Simplifying, the two main schemes are the monolithic collapse (ELS, 1962) and formation by accretion (SZ, 1978) while the real formation processes must be a more complex combination of these two scenarii. Detail of scenarii shows that the thick disc is a key feature in order to understand the detail of galactic formation.

Some recent reviews are given by Freeman (1987), Gilmore *et al.* (1989). A very extensive review covering a wide range of thick disc formation scenarii is given by Majewski (1993). See also more recently Norris (1996).

2. Some Observational Evidences

Edge-on galaxies: Thick discs are identified in some external edge-on disc galaxies (van der Kruit and Searle, 1981; Shaw and Gilmore, 1989, 1990; de Grijs and Peletier, 1997). For such galaxies, the light profile perpendicular to the galactic plane shows an approximately double exponential profile. The radial scale length of these thick discs is larger than for the thin disc component and then the thick

 Astrophysics and Space Science is the original source of publication of this article. It is recommended that this article is cited as: *Astrophysics and Space Science* **265:** 225–230, 1999.
© 1999 *Kluwer Academic Publishers.*

disc is clearly visible and dominant in the outer part of such galaxies. If the thick disc scale lengths were much shorter than the thin disc scale lengths, they would be extremely difficult to identify. Though thick discs are apparently not universal in external edge-on disc galaxies, this could be just the result of an observational bias.

Star counts: The existence of an intermediate stellar population in our Galaxy have been recognized since a long time from the kinematic analysis of local stellar samples (see for instance the Vatican Conference on Stellar Populations, 1957). More recently star counts at high galactic latitude from intermediate (Gilmore and Reid, 1983; Robin *et al.*, 1996) to deep (Majewski, 1992) to ultra-deep (with the HST, Gould *et al.*, 1998; Mendez and Guzman, 1998) star counts allowed to determine the properties of the thick disc. Some degeneracy remains between the various published determinations of the local thick disc density and its scale height. However, the thick disc surface mass density, about 15–20 percent that of the thin disc, is more correctly determined. The existence of a thicker intermediate component between the thick disc and the halo could explain the variance of the published results ($h_z = 750$ to 1500 pc).

Star counts at high galactic latitude, combined with proper motions, enable us to determine the kinematic of the thick disc and allow a more accurate splitting of the populations. This is extremely clear for the halo component for which the rotational lag is so high that large proportion of halo stars are identified only from their proper motions (Soubiran, 1993; Ojha *et al.*, 1994ab, 1996; Spagna, 1996).

Local samples: High proper motion survey is an extremely efficient way to identify stars with thick disc or halo kinematics. However it introduces a strong kinematic bias that does not allow to define unambiguously (i.e. without complementary modelings of the thin disc populations) the link between the thin and thick discs: the respective local number density, the exact kinematics, a discontinuity between thin and thick discs ... .

Only the Beers *et al.* (1998) catalogue is sufficiently large about 2000 stars and based on non-kinematical criteria (low metallicity stars from a preliminary low dispersion survey). It will allow a determination (free from kinematics bias) of the local properties of the various stellar populations. It favors some continuity between the traditional thick disc and halo populations.

Other intermediate components exist and add some confusion: a bulge extension that could move up stars to the solar neighbourhood and explain the super metal rich stars observed in the solar neighbourhood (Grenon, 1998). Blue metal poor stars with a large rotational drift that could be remnants of an accreted dwarf galaxy (Preston *et al.*, 1994), or the A stars that could be remnants of a recently accreted galaxy (a third 'magellanic cloud').

3. Models

The kinematic signatures of a thick disc are its velocity lag relative to the rotational circular velocity, the three axes of the velocity ellipsoid and its vertical scale height. In fact these quantities are all related and are nearly a measure of the 'temperature' of this component. This is detailed in the next section, but first we detail the main signatures of various thick disc formation scenarii.

3.1. CHEMICODYNAMICAL MODELS

Chemico-dynamical models of galactic formation have been developed (Larson, 1976; Burkert *et al.*, 1992; Samland *et al.*, 1997) in order to follow the various step of the collapse of the primordial gas and the formation of the successive galactic stellar and ISM components.

In these models after a rapid collapse the gas remains for some time distributed in a thick component because the radiative cooling (free-free) is too long and prevents more flattening of the gas distribution. After a few hundreds of Myr, metal enrichment of the interstellar medium by stellar processing allows emission lines to be much more efficient than the free-free radiative cooling, and suddenly the collapse of the gas component occurs again, quickly forming a thin disc.

This scenario explains naturally a discontinuity between thin and thick discs. Due to a different time scale, these models do not attempt to explain the stellar halo formation.

However, in the inner part of the galaxy, we expect some continuity between the halo and the thick disc while in the most outer part the halo evolves independently from the disc.

3.2. DYNAMICAL HEATING

The accretion of a satellite in an inclined prograde orbit produces a dynamical heating of the disc (Quinn *et al.*, 1993). If such an event appears early when only 10 or 20 percent of the stellar disc is formed, this disc is thickened by a factor two or three and more in the outer part. The density scale length is also enlarged. Then the thin disc is formed after this early interaction.

Such scenario explains easily the discontinuity (for the density and the kinematics) between disc and thick discs, and the resulting thick disc appearance depends strongly on the exact accretion configuration. It also explains why thick discs do not seem to be universal. We expect an age and a chemical continuity between the old thin disc and the thick disc.

3.3. RESONANCE TRAPPING

It has been remarked (Sridhar and Tourma, 1996) that in slowly growing stellar disc, the change of the potential and the displacement of a 2:1 resonance through

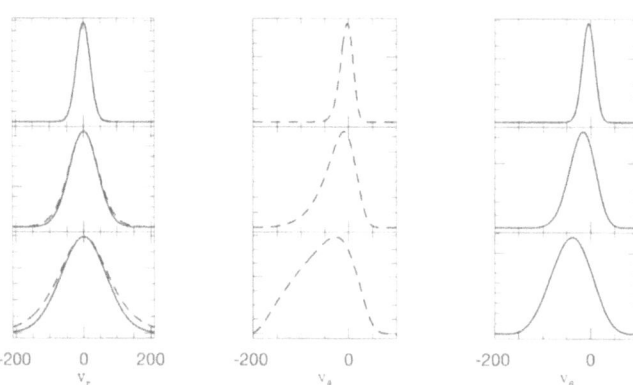

Figure 1. Marginal velocity distributions at solar position, for two models with short (full lines) and long (dashed lines) density scale lengths. Top to bottom with increasing velocity dispersions. Left: radial velocity distribution for both models, middle: tangential velocity distribution for model with $R_\Sigma = 2.5$ kpc, $R_\sigma = 7.5$ kpc, right: for model with $R_\Sigma = 1.8$ kpc, $R_\sigma = 15$ kpc.

the phase space modifies the distribution of orbits close to this resonance. While the resonance sweeps the phase space as the potential changes with time, a significant fraction of stars see their planar radial motion changed into vertical motion. It results that these stars formed a thick disc with stars having quasi mono-energetic vertical motions. These stars move quickly through the galactic plane, staying a long time at a maximum heigh showing a maximum density at about 1 kpc from the galactic plane and a much smaller density at $z = 0$ (similar to the high density shells phenomenon in some elliptical galaxies). In fact thick disc observational evidences are not really well explained by this mechanism that is probably efficient for only a small fraction of the disc stars. Discovery of such a stellar population would allow to strongly constrain the gravitational formation of the stellar disc.

4. Kinematic Modeling

A kinematic model can be used to describe most of the characteristics of disc stellar populations. The velocity dispersion, the asymmetric drift (or the rotational lag) and the thickness of a stellar disc are directly linked. Jeans equations allow to relate the moments while exact solutions of Boltzmann equation (distribution functions – d.f. – depending on the integral of motions) give a more complete description for a stationary potential. Shu (1969) has proposed such a two-integral of motions d.f. describing an axisymmetric exponential stellar disc. We have generalized this d.f. to independent density and kinematic scale lengths (Bienaymé and Séchaud, 1997) and to 3D (Bienaymé, 1999). For a locally relaxed stellar distribution, we expect that the radial velocity distribution is gaussian, the generalization of the Shu's models show that this is not necessarily true for the rotational velocity distribution (Figure 1). A very asymmetric v-distribution may be associated to a gaussian radial

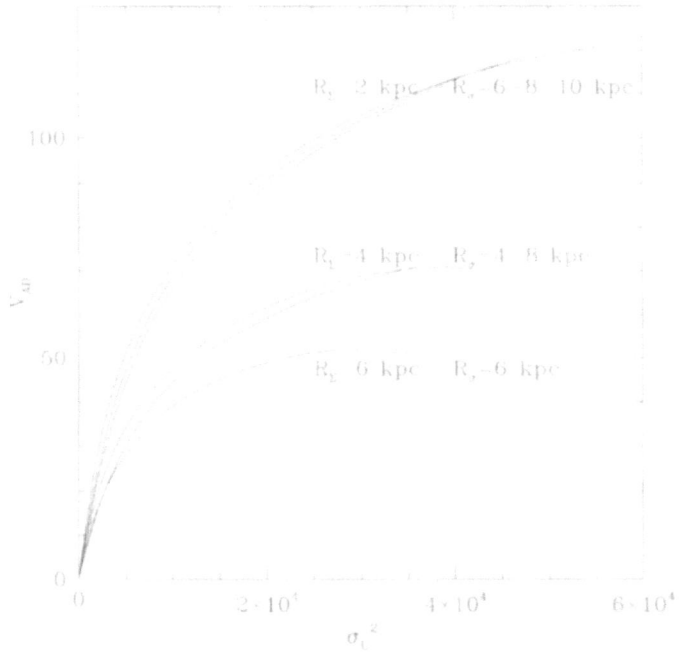

Figure 2. Asymmetric drift relations. Tangential velocity drift versus radial velocity variance for models with various density and kinematic gradients ($R_0 = 8.5$ kpc, $V_0 = 220$ km s^{-1}). Models with the largest density scale length present a saturation effect on the amplitude of the tangential velocity drift.

u-distribution. This means that splitting the kinematics in gaussian components is not a dynamically consistent approach. The mean drift increases with the velocity dispersion, but the rate depends on the scale lengths and, in the case of a large density scale length, the drift quickly reaches his maximum. This peculiar features implies that if the stellar discs is regularly heated, we could have an accumulation of a population having a maximum drift but corresponding to a large range of radial and vertical velocity dispersions (Figure 2). Analysing the v-velocity distribution without a dynamically consistent model can lead to a misinterpretion of the apparent discontinuity between thin and thick discs (for v-velocities).

5. Conclusion

Among the many models proposed for the origin of the thick disc, two models are favored: 1) a dynamical heating of an early disc by a satellite accretion 2) or a formation through a collapse probably independent from the formation of the present halo. There is a need to know exactly the gap or the link (discrete versus continuum) between old disc-TD and between TD-Halo. There is some apparently contradictory arguments favoring both scenario. The apparent non-universality of

TD in disc galaxies favors the early accretion scenario since it can explain the absence of TD in many S galaxies. The TD formation through collapse could explain the apparent kinematic-chemical continuity between TD-halo (Beers, 1998) and the apparent chemical discontinuity between thin disc-TD. But such a scenario should apply to spiral galaxies that should have all a thick disc.

References

Beers: 1998, see this proceedings.

Bienaymé, O.: 1999, *Astron. Astrophys.* **341**, 86.

Bienaymé, O. and Séchaud, N.: 1997, *Astron. Astrophys.* **323**, 781.

Burkert, A., Truran, J.W. and Hensler, G.: 1992, *Astrophys. J.* **391**, 651.

de Grijs, R. and Peletier, R.F.: 1997, *Astron. Astrophys.* **320**, L21.

Eggen, O.J., Lynden-Bell, D. and Sandage, A.: 1962, *Astrophys. J.* **136**, 748.

Freeman, K.C.: 1987, *Annu. Rev. Astron. Astrophys.* **25**, 603.

Gilmore, G. and Reid, N.: 1983, *Mon. Not. R. Astron. Soc.* **202**, 1025.

Gilmore, G., Wyse, R. and Kuijken, K.: 1989, *Annu. Rev. Astron. Astrophys.* **27**, 555.

Gould, A., Flynn, C. and Bahcall, J.: 1998, *Astrophys. J.* **503**, 798.

Grenon, 1998, see this proceedings

Larson, R.B.: 1976, *Mon. Not. R. Astron. Soc.* **176**, 31.

Majewski, S.R.: 1992, *Astrophys. J. Suppl.* **357**, 87.

Majewski, S.R.: 1993, *Annu. Rev. Astron. Astrophys.* **31**, 575.

Mendez, R.A. and Guzman, R.: 1998, *Astron. Astrophys.* **333**, 106.

Norris, J.E.: 1996, *ASP Conf. Ser.* **49**, 14.

Ojha, D.K., Bienaymé, O., Robin, A.C. and Mohan, V.: 1994a, *Astron. Astrophys.* **284**, 810.

Ojha, D.K., Bienaymé, O., Robin, A.C. and Mohan, V.: 1994b, *Astron. Astrophys.* **290**, 771.

Ojha, D.K., Bienaymé, O., Robin, A.C., *et al.*: 1996, *Astron. Astrophys.* **311**, 456.

Preston, G.W., Beers, T. and Shectman, S.A.: 1994, *Astron. J.* **108**, 538.

Quinn, P.J., Hernquist, L. and Fullagar, D.P.: 1993, *Astrophys. J.* **403**, 74.

Robin, A.C., Haywood, M., Crézé, M., *et al.*: 1996, *Astron. Astrophys.* **305**, 125.

Samland, M., Hensler, G. and Theis, CH.: 1997, *Astrophys. J.* **476**, 544.

Searle, L. and Zinn: 1978, *Astrophys. J.* **225**, 357.

Shaw, M.A. and Gilmore, G.: 1989, *Mon. Not. R. Astron. Soc.* **237**, 903.

Shaw, M.A. and Gilmore, G.: 1990, *Mon. Not. R. Astron. Soc.* **242**, 59.

Shu, F.H.: 1969, *Astrophys. J. Suppl.* **158**, 505.

Soubiran, C.: 1993, *Astron. Astrophys.* **274**, 181.

Spagna, A., *et al.*: 1996, *Astron. Astrophys.* **311**, 758.

Sridhar and Tourma: 1996, *Science* **271**, 973.

van der Kruit, P.C. and Searle, L.: 1981, *Astron. Astrophys.* **95**, 105.

van der Kruit, P.C. and Searle, L.: 1981, *Astron. Astrophys.* **95**, 116.

FAINT STARS AND THE STRUCTURE OF THE GALAXY

S. PHLEPS, K. MEISENHEIMER and C. WOLF
Max-Planck-Institut für Astronomie, Königstuhl 17, D-69117 Heidelberg, Germany

B. FUCHS and H. JAHREISS
Astronomisches Rechen-Institut, Mönchhofstr. 12–14, D-69120 Heidelberg, Germany

The *Calar Alto Deep Imaging Survey* (see Meisenheimer *et al.*, 1998) was mainly designed to search for emission line galaxies. We have identified a sample of about 400 faint stars ($R \geq 15$). The stars lie in three fields with Galactic coordinates $l = 85°, b = 45°, l = 175°, b = 45°$, and $l = 179°, b = -48°$ respectively, and have been separated from galaxies by a classification scheme based on photometric spectra and morphological criteria. The classification is reliable down to a magnitude limit of 23^{mag} in the CADIS R band. For each star we derive photometric parallaxes by using an empirical main-sequence color magnitude relation. We corrected the distance dependent incompleteness of the sample as follows: We divided the data into logarithmic distance bins. If $\hat{N}_j$ is the number of stars in the j th bin up to the corresponding color ($\equiv$ absolute magnitude) limit, $\hat{N}_{j-1}$ the number of stars in the previous bin up to the same limit, and N_{j-1} is the corrected number of stars in that bin, then the corrected number in the j th bin is $N_j = \hat{N}_j \cdot N_{j-1}/\hat{N}_{j-1}$. The first bin is assumed to be complete and can be used as a starting point, the other bins are corrected iteratively. The resulting density distribution as function of height above the midplane is shown in Figure 1. Since all stars are fainter than $R = 15^{mag}$ the nearest stars are still at a distance of about 200 pc, so we cannot determine the density at the midplane from the CADIS sample. The normalization points shown in Figure 1 (stars) have been derived by counting stars of the same colour as in our data set in the catalogue of nearby stars CNS4 (Jahreiß and Wielen, 1997). In a similar way we have determined the local density of halo stars using the data of Fuchs and Jahreiß (1998). The density distribution of the disk stars shows very clearly the contribution of the Galactic thick disk and can be fitted by a superposition of two exponentials with scaleheights $h_1 = (213 \pm 24)$ pc for the thin disk and $h_2 = (1038 \pm 66)$ pc for the thick disk components, respectively. The halo can be described best by a de Vaucouleurs law up to a distance of 24 kpc, but can be fitted as well by an – softened – inverse power law with exponent $\alpha = 4.15 \pm 0.16$. We derived an axial ratio for the halo of $a/c = (0.85 \pm 0.21)$.

The knowledge of the density distribution enables us to deduce the stellar luminosity function (SLF) free of systematic errors. We correct for incompleteness and Malmquist bias by dividing the number counts found in the absolute mag-

Astrophysics and Space Science is the original source of publication of this article. It is recommended that this article is cited as: *Astrophysics and Space Science* **265**: 231–232, 1999.
© 1999 *Kluwer Academic Publishers.*

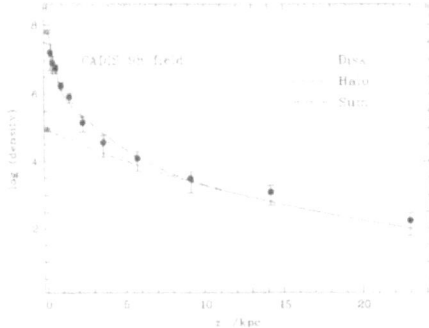

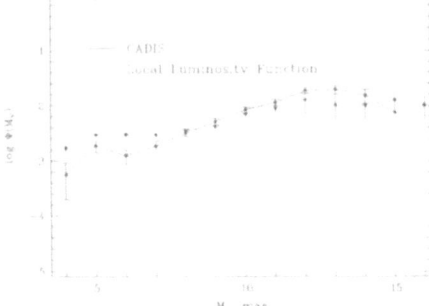

Figure 1. Density distribution.   *Figure 2.* Luminosity function.

nitude intervall $[M_V - 1/2, M_V + 1/2]$ by the maximum effective volume $V_{max}^{eff} = \omega \int_{r_{min}}^{r_{max}} v(r, l, b) r^2 \, dr$, where the integration limits are given by the distance moduli derived for the upper and lower limiting apparent magnitudes. $v(r, l, b)$ denotes the distribution function, normalised to unity at $z = 0$. As can be seen from Figure 2, the CADIS SLF is equal within statistical errors to the local SLF, which is based on HIPPARCOS parallaxes (Jahreiß and Wielen, 1997). It does *not* decline beyond $M_V = 12^{mag}$, as all other photometric SLFs do. The errors are very small at the faint end but larger at the bright end, complementary to the local SLF. Thus, the weighted mean of the CADIS and local SLFs, which has extremely small errors on both ends, can be regarded as the best determination of the SLF up to date.

References

Fuchs, B. and Jahreiß, H.: 1998, *Astron. Astrophys.* **329**, 81.
Meisenheimer, K., *et al.*: 1998, in: S. D'Odorico, A. Fontana and E. Giallongo (eds.), *The Young Universe, ASP Conf. Ser.* **146**, 134.
Jahreiß, H. and Wielen, R.: 1997, in: M.A.C. Perryman and P.L. Bernacca (eds.), *Hipparcos – Venice '97'*, ESA SP-402, p. 675.

WIDE BINARIES AMONG HIGH-VELOCITY AND METAL-POOR STARS

CHRISTINE ALLEN, M.A. HERRERA and A. POVEDA
Institute of Astronomy, Universidad Nacional Autónoma de México, México, D.F.

The properties of old disk and halo binaries are of interest for the understanding of the processes of formation and early dynamical evolution of the Galaxy. The luminosity function of the components of wide binaries and multiples, their mass function, the fraction of halo or old disk stars that are members of wide binaries, and the distribution of their separations are some of the basic properties that are poorly understood, mainly because of the paucity of known wide binaries among halo and old disk stars.

The present work is an attempt to ameliorate this situation. We have elaborated a list of 133 wide binaries mostly belonging to the halo or high-velocity disk, by searching for common-proper-motion companions to the high-velocity and metal-poor stars studied by Schuster and collaborators (1988, 1993). Based on Stromgren photometry, these authors have derived distances, metallicities and ages for their stars. Since each star has a large and well determined proper motion it was possible to compare it with that of the NLTT stars of its vicinity. In this way we were able to identify over 100 high-velocity and metal-poor common-proper-motion binary systems. Each system was carefully checked to avoid misidentifications, and in most of the cases distances were improved using the Hipparcos trigonometric parallaxes.

We have determined the distribution of angular separations for our wide binaries. Reliable distances are available for all of our systems, so this distribution can be converted into a separation distribution in AU. We find 11 systems that have projected separations in excess of 10000 AU, or 16 systems with expected semiaxes larger than 10000 AU; we note that their existence poses interesting dynamical problems.

The analysis of the frequency distribution of the expected major semiaxes shows that they closely follow Oepik's relation, namely, $f(a) \propto 1/a$, for $a \leq 10000$ AU. A comparison of the frequency distribution of major semiaxes in the present list of wide binaries with that of the wide binaries in the solar vicinity, shows that although both groups of binaries closely follow Oepik's law, they begin to depart from it at very different separations. In the solar vicinity we found the wide *younger* pairs to follow Oepik's law for major semiaxes of up to 8000 AU, whereas the wide *older* pairs follow this law only up to 2400 AU (Poveda *et al.*, 1997). This difference was interpreted as due to disruptive effects caused by massive perturbers which, with the passage of time, tend to eliminate the widest systems. It is worth noting that

Astrophysics and Space Science is the original source of publication of this article. It is recommended that this article is cited as: *Astrophysics and Space Science* **265**: 233–234, 1999.
© 1999 *Kluwer Academic Publishers.*

a frequency distribution of Oepik's form has been found also for very young stars (see Allen *et al.*, 1997 and references therein).

In the present group of wide binaries, the departure from Oepik's law occurs at about 10000 AU. This much larger value can be interpreted as a result of the smaller fraction of their lifetimes that the binaries spend in the galactic disk, where the volume density of the perturbers is highest, as well as to their large space velocities; both effects will tend to decrease the rate of energy exchanges of the binary with external perturbers, and hence to preserve the widest systems. Even when these systems finally become unbound, their motions will follow parallel tracks for a long time, causing them to be identified as common-proper-motion pairs.

Since many systems also have known radial velocities, space velocities for them were be determined, and galactic orbits computed and characterized. The secondaries of these wide binaries are interesting in themselves, since they represent a sampling of the very faint end of the main sequence of old disk and halo stars.

Acknowledgements

Our sincere thanks go to G. Cordero, A. Nigoche and A. Hernández.

References

Allen, C., Poveda, A. and Herrera, M.A.: 1997, in: Docobo *et al.* (eds.), *Visual Double Stars: Formation, Dynamics and Evolutionary Tracks*, Dordrecht, Kluwer, p. 133.

Poveda, A., Allen, C. and Herrera, M.A.: 1997, in: Docobo *et al.* (eds.), *Visual Double Stars: Formation, Dynamics and Evolutionary Tracks*, Dordrecht, Kluwer, p. 191.

Schuster, W.J. and Nissen, P.E.: 1988, *Astron. Astrophys. Suppl.* **73**, 225.

Schuster, W.J., Parrao, L. and Contreras, M.E.: 1993, *Astron. Astrophys. Suppl.* **97**, 951.

LOCAL STARS AS TRACERS OF GALACTIC EVOLUTION

B. NORDSTRÖM, J. ANDERSEN and E.H. OLSEN

Niels Bohr Institute for Astronomy, Physics, and Geophysics, Juliane Maries Vej 30, DK-2100 Copenhagen, Denmark

R. FUX, M. MAYOR, N. MOWLAVI and F. PONT

Observatoire de Genève, 51 Chemin des Maillettes, CH-1290, Switzerland

Abstract. We describe the overall properties of a new catalogue of metallicities, ages, and galactic orbits for a large, complete sample of F and G dwarfs in the solar neighbourhood. Based on a magnitude-limited sample of $\sim 14\,000$ stars, it is volume-complete to ~ 40 pc. Together with the astrophysical parameters of direct relevance to models of the evolution of the disk, it will contain the basic photometric, astrometric, and radial velocity data from which they are derived. Information on duplicity is also included. The full exploitation of the data will require a lengthy analysis, in particular to assess the degree of completeness of subsamples of stars of different population types. An early result on the effects of diffusion of galactic orbits in the disk – essential for understanding the scatter in the age-metallicity diagram and estimating the birth radius of stars – is briefly illustrated.

1. Introduction

The solar neighbourhood serves as a fundamental testbed for models of the evolution of the disk of our Milky Way Galaxy. Yet, our knowledge of the properties of the stars in the solar neighbourhood has long been based on incomplete data for heterogeneous and biased samples of stars. As described by, e.g., Strömgren (1987) and Nordström *et al.* (1997b), our group has long been engaged in remedying this situation by providing homogeneous metallicities, ages, kinematical data, and duplicity information for a large, complete sample of nearby F and G dwarfs in the whole sky. The massive observational effort to obtain photometry and radial velocities has been completed, and the newly-released HIPPARCOS data allow to expand and consolidate the results in various respects. A catalogue listing the key astrophysical parameters as well as the basic data from which they are derived is being prepared for immediate publication.

Key questions that can be addressed anew from this material include the following:

- What are the ages and other properties of the thin and thick disk?
- Is there a real Age-Metallicity Relation in the solar neighbourhood?
- What causes the scatter in the Age Metallicity Diagram?
- What is the real Abundance Distribution Function for long-lived stars in the solar neighbourhood ("G dwarf problem")?

Astrophysics and Space Science is the original source of publication of this article. It is recommended that this article is cited as: *Astrophysics and Space Science* **265**: 235–241, 1999.
© 1999 *Kluwer Academic Publishers.*

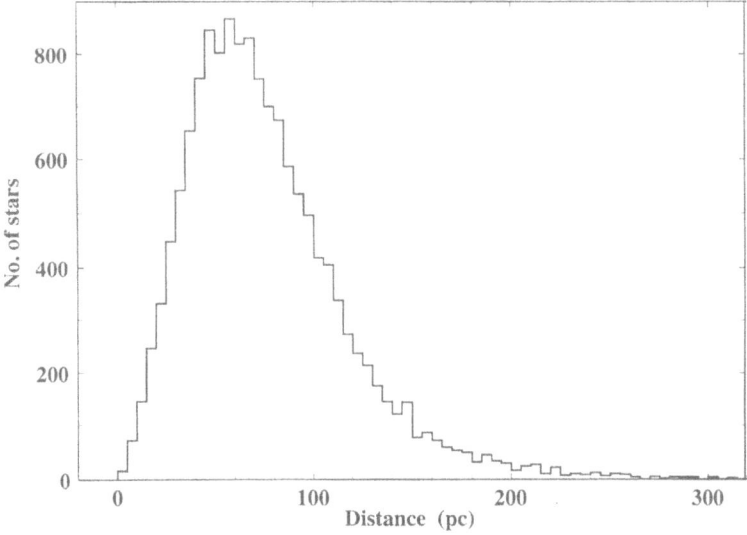

Figure 1. Distribution of the sample vs. distance from the Sun.

–   What are the cause and amount of the heating of the disk?

–   What are the IMF and star formation history of the disk?

The discussion of these issues will occupy a series of future papers. In particular, the proper assessment of duplicity, completeness and other selection corrections for the subgroups of stars that need to be defined in each case is an important and highly non-trivial matter. Here we outline the main characteristics of the material and illustrate a simple test of one of the fundamental questions in its interpretation.

2. The Sample and the Data

The motivations for conducting an inventory of the solar neighbourhood in order to define large, homogeneous, complete, and kinematically unbiased samples of nearby F and G dwarf stars have been detailed, e.g. by Strömgren (1987) and Nordström *et al.* (1997b). Sample selection is based on the all-sky $uvby$-H_β photometry by Olsen (1994 and earlier papers), which allows the determination of distances, metallicities and individual ages. PPM proper motions (Bastian and Röser, 1991–1993), improved with modern positions from the Carlsberg Automatic Meridian Circle (Carlsberg, 1991–1994), reach an accuracy at least equal to that of the HIPPARCOS proper motions now available for many of the stars (ESA, 1997). Finally, multiple radial-velocity observations for most stars have been obtained in both hemispheres with the CORAVEL scanners; radial velocities for ~600 mostly fast-rotating stars north of declination −40° were determined at the Center for Astrophysics (Nordström *et al.*, 1997c).

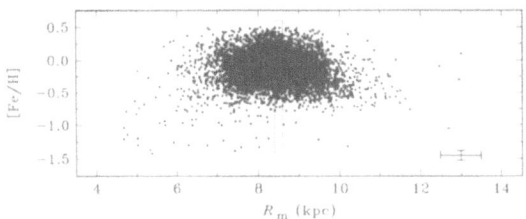

Figure 2. Distribution of mean galactocentric orbital radii of the stars vs. [Fe/H]. At the present time, the stars are found inside the box shown.

Altogether, the sample contains just over 14,000 stars with *uvby* photometry, from which [Fe/H] has been derived for nearly 11,000 and ages for nearly 10,000 stars. Kinematical data are available for some 12,000 stars, and nearly 8,000 stars have complete data of all three types. The distribution of the stars with distance from the Sun is shown in Figure 1.

The photometric metallicities for the sample agree well with those determined for a subsample of F and early G dwarfs by Edvardsson *et al.* (1993, E93), and the photometrically determined distances have been verified using the available parallaxes from the HIPPARCOS satellite. Using our distances, proper motions and radial velocities, space motions (U, V, W) have been computed, and from these in turn, the main parameters of the galactic orbits of the individual stars have been determined by direct integration in an analytical axisymmetric gravitational potential reproducing the observed rotation curve of the disk (see Fux, 1997). Figure 2 shows the mean galactocentric distances R_m for ~8,500 individual stars vs. their iron abundance; as seen, the small local volume we have observed contains stars from a large range of distances in the disk; the term "local" sample is certainly to be used with large grains of salt.

Ages have been determined (by NM) from the photometry for the F and early G stars by direct comparison with newly-calculated models from the latest Geneva series for the range of metallicities encountered here. These models, which include convective core overshooting for masses above 1.3 $M_\odot$ have been extensively tested on open cluster data and found to match real stars very closely (see, e.g. Nordström *et al.*, 1997a).

3. A Sample Application: How Much Orbital Diffusion?

Many of the correct answers to the questions outlined in the Introduction require that the relative frequencies of different types of star can be determined precisely, a process that requires careful evaluation of the effects of differences in absolute magnitude, kinematic properties, and duplicity in selected subsamples of stars. Some fundamental questions can, however, be addressed immediately from a "clean" subset of the catalogue from which binaries of all kinds (visual and spectroscopic)

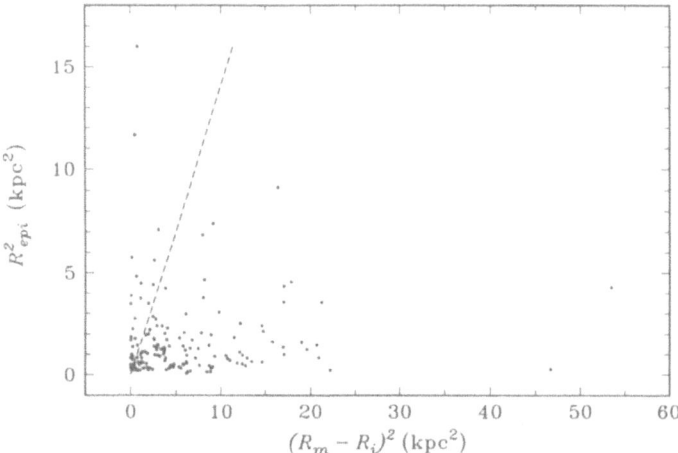

Figure 3. Epicyclic orbital radius vs. change in mean galactic orbital radius from R_i as defined by WFD (both squared) for the E93 stars. Dashed line: Predicted correlation.

have been eliminated. One of these is the scatter in the age-metallicity diagram and the dependence of key element ratios with galactocentric distance which were discussed by E93 (their Figures 31 and 20) - results of key importance for our understanding of star formation and chemical enrichment in the disk.

It was assumed in E93 that the mean radius, R_m of the present galactic orbit of a star is a reliable, if approximate, indicator of its birthplace. This view has been challenged by Wielen *et al.* (1996; WFD), who find that the scattering of stellar orbits by massive objects in the disk (giant molecular clouds, massive black holes) essentially erases all memory of the initial orbital radius. Instead, they propose a model in which the initial orbital radius R_i of a star can be computed from its age and metal abundance only, assuming a single-valued age-metallicity relation combined with a constant galactic radial abundance gradient. Using model parameters based on the data by E93, they find good agreement with the observations. Clearly, the underlying processes of star formation and chemical enrichment, the dynamical evolution of the disk, and the basis on which the observational data can be interpreted are radically different in the E93 and WFD scenarios.

It is, however, a fundamental consequence of the WFD theory that the changes in R_m due to collisions with massive objects must, on average, be accompanied by a similar increase in the eccentricity or, equivalently, epicyclic radius R_{epi} of the orbit: These random collisions must inject energy in both the mean and epicyclic orbital motions. More specifically, the WFD theory predicts that $< (\Delta R_{epi})^2 > / < (\Delta R_m)^2 >= 1.4$ and $< (\Delta R_{epi})^2 > = 0.50$ kpc^2 Gyr^{-1}. Assuming that orbits are initially circular with radius R_i, $\Delta R_m = R_m - R_i$ and $\Delta R_{epi} = R_{epi}$.

Figures 3 and 4 show these relations for the stars in E93, i.e. using exactly the same data and definitions as used by WFD; the predicted relations are shown

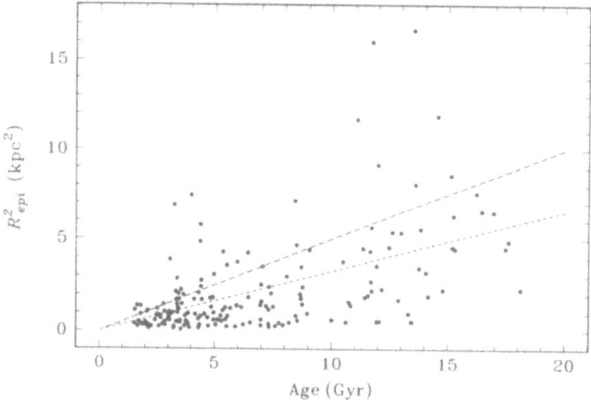

Figure 4. Epicyclic orbital radius squared vs. age for the E93 stars. Long-dashed line: Predicted relation (WFD); short-dashed line: Fit to the data (see text).

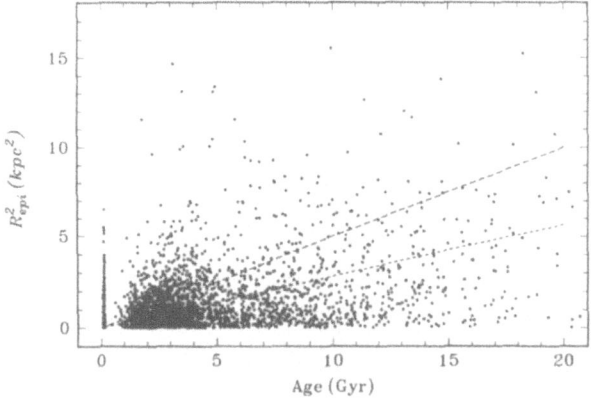

Figure 5. Same as Figure 4, but for the 'clean' subset of the new sample (see text).

as long-dashed lines. Nothing like the predicted correlation is discernible in Figure 3. In Figure 4, a direct fit to the observed points (short-dashed line) gives $< (\Delta R_{epi})^2 > = 0.33\pm0.02$ kpc^2 Gyr^{-1} only. As the E93 sample has a strong metallicity bias, these fits might be misleading. Therefore, we have repeated them with the new sample, using only single stars with very well-determined ages (better than 0.15 dex; 3473 stars). The lack of any correlation in Figure 3 appears even more striking with the larger sample, and the fit to the epicyclic motion as a function of time (Figure 5) gives $< (\Delta R_{epi}^2) > = 0.28\pm0.01$ kpc^2 Gyr^{-1}, indistinguishable from the fit in Figure 4, but incompatible with the prediction.

We conclude from Figure 3 that R_i appears unrelated to the dynamical history of the stars, and thus is not a useful indicator of their birthplaces. Figures 4 and 5 would appear to indicate that orbital diffusion does operate at the level of perhaps 1 kpc at the solar age, as expected given the known existence of giant molecular

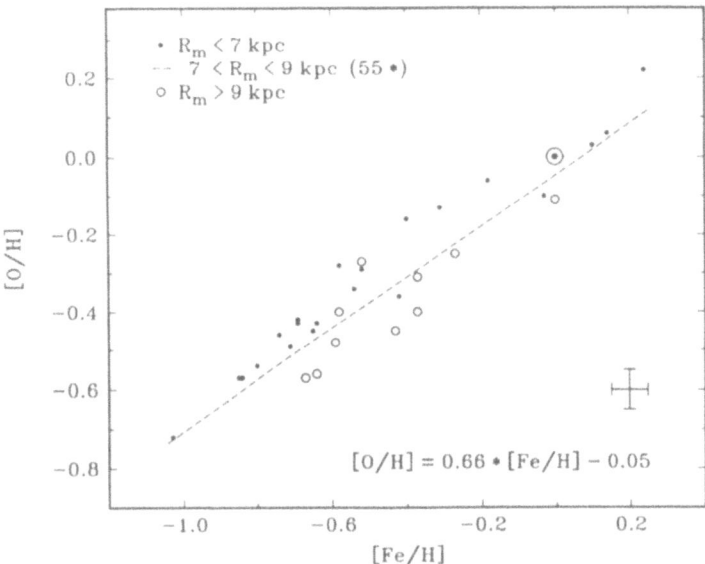

Figure 6. Oxygen vs. iron abundances for subgroups of the stars of E93 inside and outside the solar circle. The dashed line shows the fit for stars on the solar circle.

clouds in the disk. But the large effects required to completely erase the memory of the birth radius of the star and explain the scatter in the age-metallicity diagram appear incompatible with the observations.

This conclusion is strengthened by the dependence of the [O/H] ratio of the E93 stars on R_m as well as on [Fe/H] (Figure 6), a result that would appear incomprehensible if stars were scattered through such large distances in the disk as suggested by the WFD theory. R_m thus appears to remain a useful average indicator of galactic origin in discussions of the evolution of the disk using stars with computed galactic orbits. The bar recently discovered in the Milky Way would thus seem to be too short-lived to have seriously disturbed the majority of stellar orbits at the solar radius (Gummersbach *et al.*, 1998; but see Grenon (this volume) for a contrasting view).

4. Summary and Outlook

The catalogue of data as described above is being prepared for publication in printed and electronic form and should be in press by the time this preliminary report appears. Its contents, in particular ages and metal abundances, but also the intermediate parameters based on the actual observations which lead to these final results (effective temperatures, log g, distances, etc.) will be based on up-to-date calibrations. However, as recently re-emphasized e.g. by Nordström *et al.* (1997a), the transformations between observed colour indices and fundamental atmospheric parameters are indeed the main current limitation in the application of stellar model computations to real stars.

The accuracy of these transformations is limited by the current incompleteness in stellar model atmosphere and synthetic spectrum calculations (convection theory, non-LTE effects, completeness and atomic data for absorption lines, etc.). It is to be hoped and expected that much progress will be made in this area over the next few years, and some of the derived parameters will need revision. We trust that our careful sample selection and the observational parameters we have chosen to obtain will facilitate such updates and ensure the catalogue a long and useful life in studies of our Milky Way as a prototype of the distant galaxies.

Acknowledgements

The work described here involves many hundreds of observing nights over many years at the Danish 0.5m and 1.5m telescopes at ESO, La Silla, Chile, as well as on several northern telescopes. We are grateful to the many colleagues who have collaborated in this multi-year effort, including several CORAVEL observers and David Latham and Robert Stefanik at the Harvard-Smithsonian Center for Astrophysics. Financial support throughout this project from the Carlsberg Foundation, the Danish Natural Science Research Council, the Danish Board for Astronomical Research, and the Smithsonian Institution (to BN and JA), and from the Swiss Fund for Scientific Research is gratefully acknowledged.

References

Bastian, U. and Röser, S.: 1991–93, *PPM Star Catalogues*, Vols. I–IV, Astron. Rechen-Institut Heidelberg, Germany.

Carlsberg Meridian Catalogue La Palma: 1991–94, *Observations of positions of stars and planets*, Vol. 5–8. Copenhagen University Observatory, Royal Greenwich Observatory and Real Instituto y Observatorio de la Armada en San Fernando.

Edvardsson, B., Andersen, J., Gustafsson, B., Lambert, D.L., Nissen, P.E. and Tomkin, J.: 1993, *Astron. Astrophys.* **275**, 101 (E93) .

ESA: 1997, *The Hipparcos and Tycho Catalogue*, ESA SP-1200.

Fux, R.: 1997, *Ph.D. Thesis no. 2954*, Geneva Observatory, pp. 147–148 .

Gummersbach, C.A., Kaufer, A., Schäfer, D.R., Szeifert, T. and Wolf, B.: 1998, *Astron. Astrophys.* **338**, 881.

Nordström, B., Andersen, J. and Andersen, M.I.: 1997a, *Astron. Astrophys.* **322**, 460.

Nordström, B., Olsen, E.H., Andersen, J., Mayor, M. and Pont, F.: 1997b, in: A. Burkert, D. Hartmann and S. Majewski (eds.), *The History of the Milky Way and its Satellite System, ASPC* **112**, 145.

Nordström, B., Stefanik, R.P., Latham, D.W. and Andersen, J.: 1997c, *Astron. Astrophys. Suppl.* **126**, 21–30.

Olsen, E.H.: 1994, *Astron. Astrophys. Suppl.* **104**, 429; and **106**, 257.

Strömgren, B.: 1987, in: G. Gilmore and R.F. Carswell (eds.), *The Milky Way Galaxy*, Reidel, Dordrecht, p. 299.

Wielen, R., Fuchs, D. and Dettbarn, C.: 1996, *Astron. Astrophys.* **314**, 438 (WFD).

EIGHT GYR MINIMUM DISK AGE FROM HIPPARCOS

CHRIS FLYNN
Tuorla Observatory, Piikkiö, Finland

RAUL JIMENEZ
ROE, Edinburgh, UK

EIRA KOTONEVA
Tuorla Observatory, Piikkiö, Finland

Extant methods for determining the age of the Galactic disk range from the cooling of old disk white dwarfs (e.g. Liebert *et al.*, 1989; Bergeron *et al.*, 1997; Oswalt *et al.*, 1996), individual ages of evolved F stars (e.g. Edvardsson *et al.*, 1993) to isotope dating (e.g. Butcher, 1987; Morell *et al.*, 1992). These methods indicate a disk age in the range of 9 to 12 Gyr. We present here a *minimum age estimate* of 8 Gyr based on the red edge of the distribution of giants and sub-giants in the colour magnitude diagram (CMD) of nearby stars measured by the Hipparcos satellite. This feature is traced out by the oldest, metal rich disk stars, and can thus be used derive a minimum disk age (Janes, 1975). This work is described in full in Jimenez *et al.* (1998).

Ideally, metallicities of stars along the red edge would be needed in order to constrain the disk age by comparison with theoretical isochrones, but in cool subgiants of this type metallicities are not yet available, although they should be practical within a few years. Metallicities are available for cool stars on the main sequence and on the giant branch, and these at least can be used to check the positions of our isochrones. Figure 1 shows (Padua) isochrones of ages 8, 11, 13 and 15 Gyr for [Fe/H] = 0.0 and [Fe/H] = +0.3. G and K dwarfs of the same metallicities are shown by solid symbols (from a sample by Flynn and Morell 1997). The position of the reddest solar metallicity red clump giants (which mark the first ascent giant branch in metal rich stars) from the sample of Høg and Flynn (1998) is shown as a large square. The available giant branch and main sequence stars match the isochrones well, indicating that the transformations of the theoretical isochrones to the CMD are satisfactory. The [Fe/H] = +0.3 isochrones set a lower limit to the disk age. Almost all the stars lie to the left of the the 8 Gyr isochrone, leading to our *minimum disk age* of 8 Gyr. If there were significant numbers of stars more metal rich than [Fe/H] = +0.3 in the disk, then this age limit would decrease (by circa 1 Gyr per 0.1 increase in [Fe/H]). However an abundance analysis of complete samples of disk stars indicates that [Fe/H] = +0.3 is an effective upper limit for samples of this size; see poster by Kotoneva and Flynn (this conference). Some

Astrophysics and Space Science is the original source of publication of this article. It is recommended that this article is cited as: *Astrophysics and Space Science* **265**: 243–244, 1999.
© 1999 *Kluwer Academic Publishers.*

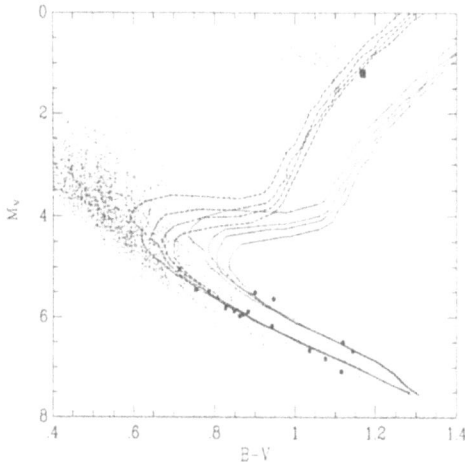

Figure 1. CMD of the nearby stars observed by Hipparcos compared to isochrones (from left to right) of 8, 11, 13 and 15 Gyr for [Fe/H] = 0.0 (dashed lines) and [Fe/H] = +0.3 (dotted lines). Nearby stars with accurate abundances in the ranges −0.1 < [Fe/H] < +0.1 (squares) and +0.2 < [Fe/H] < +0.4 (circles) are shown to check the positioning of the isochrones.

stars do lie redward of the 8 Gyr solar abundance isochrone; these stars could be as old as 11 to 13 Gyr if they are of solar metallicities. Good spectroscopic abundances of stars along the red edge of the subgiant region would allow us to estimate the disk age itself rather than just its lower limit.

References

Bergeron, J., Ruiz, M.T. and Leggett, S.K.: 1997, *Astrophys. J. Suppl.* **108**, 339.
Butcher, H.R.: 1987, *Nature* **328**, 127.
Edvardsson, B., Andersen, J., Gustafsson, B., Lambert, D.L., Nissen, P.E. and Tomkin, J.: 1993, *Astron. Astrophys.* **275**, 101.
Morell, O., Källander, D. and Butcher, H.R.: 1992, *Astron. Astrophys.* **259**, 543.
Flynn, C. and Morell, O.: 1997, *Mon. Not. R. Astron. Soc.* **286**, 617.
Høg and Flynn, C.: 1998, *Mon. Not. R. Astron. Soc.* **294**, 28.
Janes, K.: 1975, *Astrophys. J. Suppl.* **29**, 161.
Jimenez, R, Flynn, C. and Kotoneva, E.: 1998, *Mon. Not. R. Astron. Soc.* **299**, 515.
Liebert, J., Dahn, C. and Monet, D.: 1989, in: *White dwarfs; Proceedings of IAU Colloquium* **114**, Springer-Verlag, pp. 15–23.
Oswalt, T.D., Smith, J.A., Wood, M.A. and Hintzen, P.: 1996, *Nature* **382**, 692.

AGE–METALLICITY RELATION AND STAR FORMATION HISTORY OF THE GALACTIC DISK

HELIO J. ROCHA-PINTO and WALTER J. MACIEL
Instituto Astronômico e Geofísico, São Paulo, Brazil

JOHN SCALO
The University of Texas at Austin, Austin, USA

CHRIS FLYNN
Tuorla Observatory, Piikkiö, Finland

1. Introduction

In this work, we have determined the age–metallicity relation (AMR) and star formation rate (SFR) for the Galactic disk. The sample was selected from the surveys of chromospheric activity in solar-type stars by the Mount Wilson Group, and includes 730 stars (see details in Rocha-Pinto and Maciel, 1998, hereafter RPM).

2. Results and Discussion

The size of the sample for the building of AMR is 552 stars. We have eliminated from the sample all stars more distant than 80 pc and all very active stars. The metallicities are photometric, and for the active and very active stars they have been corrected to allow for the m_1 deficiency caused by the chromospheric activity (see RPM). We have found very good agreement with other AMRs in the literature. During the lifetime of the disk, the mean [Fe/H] has increased from circa -0.50 dex, 14 Gyr ago, to 0.12 dex today. We find small metallicity dispersion around the mean AMR although this is probably an artifact of our method.

A subsample with 319 stars was used to find kinematical constraints related to age and [Fe/H]. For these stars, we calculated the spatial velocities U, V, W. The velocity dispersion ellipsoid increases steeply with age during the first 3 Gyr, saturating after 6 Gyr in $(\sigma_U, \sigma_V, \sigma_W) = (55 \pm 5, 40 \pm 10, 27 \pm 5)$ km s^{-1}.

The chromospheric age distribution is transformed into star formation history through the application of three corrections: a volume correction, a stellar evolution correction and a scale height correction. Figure 1 shows the results. We have used the same nomenclature introduced by Majewski (1993), namely bursts A, B and

Astrophysics and Space Science is the original source of publication of this article. It is recommended that this article is cited as: *Astrophysics and Space Science* **265**: 245–246, 1999.
© 1999 *Kluwer Academic Publishers*.

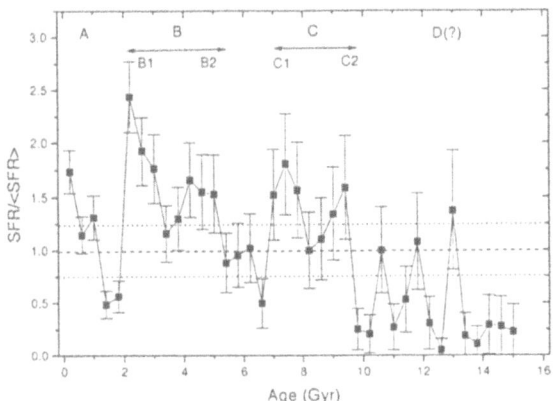

Figure 1. SFR history of the Galactic disc. The dotted lines indicate the 2 σ variations around a constant SFR for a sample of this size.

C. We have run around 3000 simulations to test the statistical significance of the bursts we have found. From the comparison with the results of the simulations, we are able to say that the SFR history we have found for the Milky Way disk is not constant at a confidence level greater than 98%.

We have calculated how this SFR history would be seen at several redshifts, for the case of a flat universe. Our Galaxy does not show enhanced star formation episodes at intermediate redshifts, as is seen in recent determinations of the cosmic SFR. A possible explanation for this difference is that the star formation rate we found is only valid for the disk. The cosmic SFR should reflect mainly the star formation history of the central parts of the galaxies, from where the main part of the ultraviolet light originates.

Acknowledgements

H.J.R-P. and W.J.M. acknowledge support by FAPESP and CNPq; J.S. acknowledges support by NASA; C.F. acknowledges the Finnish Academy.

References

Majewski, S.R.: 1993, *Annu. Rev. Astron. Astrophys.* **31**, 575.
Rocha-Pinto, H.J. and Maciel, W.J.: 1998, *Mon. Not. R. Astron. Soc.* **298**, 332 (RPM).

NEARBY STARS AND THE HISTORY OF THE GALACTIC DISK

H. JAHREISS, B. FUCHS and R. WIELEN

Astronomisches Rechen-Institut, Mönchhofstraße 12-14, D-69120 Heidelberg, Germany

A first version of the *Fourth Catalogue of Nearby Stars* (CNS4) has been completed now and is publicly available on the internet (http://www.ari.uni-heidelberg.de/aricns/). The construction of the catalogue by emulating Hipparcos data (ESA, 1997) as well as the principal results on the luminosity function of nearby stars, their kinematics etc. are described in detail by Jahreiß and Wielen (1997) and Jahreiß *et al.* (1998). In this *note* we report on two recent extensions of the catalogue.

The Hipparcos Survey made an extension of the distance limit to 50 pc possible for nearby stars having $M_V \leq 4^m$ without any loss of completeness. For most of these stars radial velocities could be collected from the literature in order to determine reliable space velocities.

This indeed provided an eightfold increase of the initially rather small sample size for the young ($\leq 1 \cdot 10^9$ yr) CM groups (formed according to the position of the stars in the CM diagram) as well as for the 'old' K giants. Consequently, the kinematical parameters of the corresponding age groups became more reliable. Especially, the former odd behaviour of the CM group with age $1 \cdot 10^9$ yr is now changed and the total velocity dispersion fits now much better to the age-velocity relation shown in Figure 1 where the full line represents a diffusion process with constant diffusion coefficient (Wielen, 1977).

For 206 main-sequence stars with $0.5 \leq B - V \leq 1.0$ the S-measurements (Duncan *et al.*, 1991; Henry *et al.*, 1996) of the stellar Ca II H and K line intensity were transformed (Noyes *et al.*, 1984) to log R'_{hk}, a measure of the chromospheric emission (CE). Apart from six high-velocity stars showing anomalously strong CE emission the remaining stars were divided in seven different CE groups. For each CE group a mean age could be determined from the relative sample size under the assumption of a constant rate of star formation. The resulting total velocity dispersions are shown as open squares in Figure 1. It is evident that apart from the youngest and oldest group the remaining CE groups fit rather satisfactory to the data of the unrelated HK groups formed according to the chromospheric emission of the Mc Cormick K and M dwarfs or the CM groups of early type main sequence stars that were defined by different B–V intervals.

In Figure 1 the colour-magnitude diagram of the 228 'old' K giants within 50 pc is shown. Besides 4 giants (open triangles) belonging to the Hyades cluster, clump giants (open circles) show up very distinctly in the CMD at $M_V \approx 0.7^m$ in the range $B - V = 0.8^m$ to 1^m. Interestingly, the 40 clump giants have velocity dispersions,

 Astrophysics and Space Science is the original source of publication of this article. It is recommended that this article is cited as: *Astrophysics and Space Science* **265**: 247–248, 1999.
© 1999 *Kluwer Academic Publishers.*

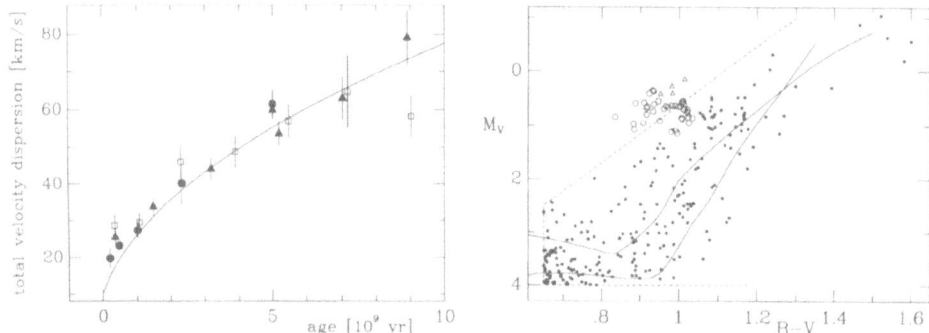

Figure 1. Left panel: Total velocity dispersion versus age. Full circles: CM groups; full triangles: HK groups of the Mc Cormick K and M dwarfs; open squares: CE groups of main-sequence stars $(0.5 \leq B - V \leq 1.0)$.

Right panel: CM-diagram of 288 K giants in the solar neighbourhood. Dotted lines indicate the old giant branch window in the CMD. The full lines indicate the giant branch of the old open clusters M 67 and NGC 188, respectively.

$\sigma_U, \sigma_V, \sigma_W = (34, 23, 12 \text{ km s}^{-1})$ which are much lower than that of the remaining 197 K giants, $\sigma_U, \sigma_V, \sigma_W = (42, 33, 23 \text{ km s}^{-1})$. This seems to indicate that the clump giants are much younger than the other 'old' K giants.

References

Duncan, D.K., Vaugham, A.H., Wilson, O.C., *et al.*: 1991, *Astrophys. J. Suppl.* **76**, 383.

Henry, T.J., Soderblom, D.R., Donahue, R.A. and Baliunas, S.L.: 1996, *Astron. J.* **111**, 439.

ESA: 1997, *The Hipparcos and Tycho Catalogues*, ESA SP-1200.

Jahreiß, H. and Wielen, R.: 1997, in: M.A.C. Perryman and P.L. Bernacca (eds.), *Hipparcos-Venice '97'*, ESA SP-402, p. 675.

Jahreiß, H., Fuchs, B. and Wielen, R.: 1998, in: P. Brosche, W.R. Dick, O. Schwarz and R. Wielen (eds.), *The Message of the Angles. Astrometry 1798–1998*, Reviews of Modern Astronomy, *Astron. Ges.*, in press.

Noyes, R.W., Hartmann, L.W., Baliunas, S.L., *et al.*: 1984, *Astrophys. J.* **279**, 763.

Wielen, R.: 1977, *Astron. Astrophys.* **60**, 263.

ABUNDANCE RATIOS IN METAL-RICH HALO AND THICK-DISK STARS

P.E. NISSEN

Institute of Physics and Astronomy, University of Aarhus, DK-8000 Aarhus C, Denmark

Abstract. Recent determinations of precise abundance ratios for nearby halo and thick disk stars in the metallicity range $-1.3 <$ [Fe/H] < -0.5 have revealed a significant cosmic spread in the abundances of oxygen, magnesium, sodium, nickel, s-process and r-process elements relative to iron. Possible explanations of these variations are reviewed. In particular, it is discussed if the differences in abundance ratios are correlated with the kinematics of the stars, and hence can be used to identify stellar populations in the Galaxy.

1. Introduction

Abundance studies of stars with metallicities around [Fe/H] $\simeq -1.0$ are not very frequent in literature – partly because extreme metal-poor halo stars have been considered more interesting and partly because both halo stars and thick disk stars with [Fe/H] $\simeq -1.0$ are relatively rare. Such 'mildly' metal-poor stars are, however, very interesting. They represent the end of the evolution of the Galactic halo and the beginning of the thick disk. Recent large photometric and spectroscopic surveys of high velocity main-sequence stars have shown a considerable overlap in the metallicity range $-1.3 <$ [Fe/H] < -0.5 between halo stars having a slow Galactic rotation velocity, $V_{rot} \simeq 0$ km s^{-1}, and thick disk stars with $V_{rot} \simeq 170$ km s^{-1} (Nissen and Schuster, 1991; Schuster *et al.*, 1993, and Carney *et al.*, 1996). As reviewed in the following, studies of the two populations in the 'overlap' metallicity range have revealed an interesting cosmic scatter in the abundance ratios of elements formed by different nucleosynthesis processes. The interpretation of these new data is still unclear, but potentially the variations in the abundance ratios may provide new information on the formation and evolution of the halo and the thick disk as well as the relation between these two components of the Galaxy.

The differences in abundance ratios found in the works reviewed are quite small – typically 0.1 to 0.3 dex – so high precision is needed to detect these differences. In this connection it is important to distinguish between *absolute* and *differential* values of abundance ratios. Absolute values are subject to many error sources: e.g. uncertainties in the gf-values of the absorption lines, errors in the atmospheric models, and possibly non-LTE effects. This means that the accuracy of the absolute abundance ratios are seldom better than 0.1 dex and often worse. Differential

Astrophysics and Space Science is the original source of publication of this article. It is recommended that this article is cited as: *Astrophysics and Space Science* **265**: 249–256, 1999.
© 1999 *Kluwer Academic Publishers*.

abundance ratios for a group of stars having a small range in effective temperature, surface gravity and overall metal abundance may, however, be determined with a precision approaching 0.02 dex in the abundance ratios (Nissen and Schuster, 1997; Jehin *et al.*, 1999). The trick is to determine the abundance ratios of two elements from absorption lines, which have about the same dependence on T_{eff} and log g. As an example, the Ti/Fe ratio is determined from Ti I and Fe I lines, whereas Eu/Fe is determined from Eu II and Fe II lines. In the latter case ionized lines are used because no neutral Eu lines are available. Weak lines (on the linear part of the curve of growth) should be used in order to be independent of uncertain broadening parameters, although it may also be possible to use strong lines if the damping constants are sufficiently well known (Anstee *et al.*, 1997). In the case of weak lines the precision is limited by the error in the equivalent width measurements, so high signal-to-noise ($S/N \simeq 200$) and high resolution ($R \simeq 60\,000$) spectra are needed.

2. The Ages of Thick Disk and Metal-Rich Halo Stars

Before discussing the abundance ratios it is appropriate with a short digression to the ages of thick disk and halo stars, because Hipparcos parallaxes have made it possible to derive fairly precise relative ages of the two populations in their over-lapping metallicity range. It is of considerable interest for the abundance discussion to see if one of the populations is significantly older than the other.

Figure 1 shows a plot of M_V vs. log T_{eff} for all stars from Nissen & Schuster (1991) with $-1.0 <$ [Fe/H] < -0.7 having Hipparcos parallaxes (ESA, 1997) with errors less than 10%. T_{eff} is calculated from $b - y$ using the IRFM calibration of Alonso *et al.* (1996). [Fe/H] is derived from the m_1 index with the calibration of Schuster and Nissen (1989). The absolute magnitude M_V follows directly from the visual magnitude and the parallax. None of stars are significantly reddened or affected by absorption according to $E(b - y)$ as determined from β and $b - y$. Furthermore, all known binary stars have been excluded.

Most stars in Figure 1 have thick-disk kinematics. Their velocity components in the direction of Galactic rotation have a mean value of $V_{rot} = 165$ km s^{-1} with a dispersion of ± 40 km s^{-1}. The remaining four stars have halo kinematics with $V_{rot} < 50$ km s^{-1}. As can be seen from the comparison with the new alpha-enhanced isochrones of VandenBerg (1997) corresponding to the the mean metallicity of the sample, most stars have ages around 14 Gyr. The two younger stars could be thin disk stars contaminating the sample. CD-57 1633 is a halo star with a low (solar) [α/Fe] (Nissen and Schuster, 1997). Its age should actually be estimated from [α/Fe] $= 0.0$ isochrones, which means that it is older than it appears from Figure 1. The few stars to the right of the 16 Gyr isochrone may be undetected binaries. Clearly, the sample should be investigated in more detail with respect to chemical composition and binarity, but it seems rather safe to conclude that both

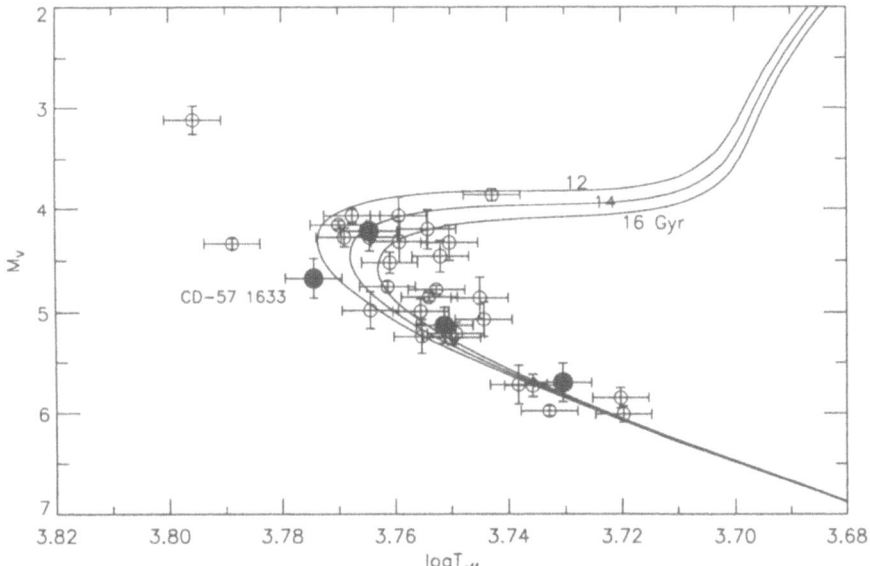

Figure 1. The $M_V - \log T_{eff}$ diagram of stars from Nissen and Schuster (1991) with $-1.0 < $ [Fe/H] < -0.7 having Hipparcos parallaxes with $\sigma(\pi)/\pi < 0.1$. Filled circles: halo stars. Open circles: thick-disk stars. Error bars correspond to the error of the parallax and to ± 70 K. The isochrones shown are from VandenBerg (1997) and correspond to stellar models with [Fe/H] $= -0.83$, [α/Fe] $= +0.3$, and $Y = 0.241$. They have been shifted with $\Delta T_{eff} = -70$ K in order to fit the group of six unevolved stars.

the thick disk and the metal-rich halo are very old populations, and in particular that the thick disk is not significantly younger than the metal-rich halo.

3. Abundance Ratios

In this section we briefly discuss three papers dealing with abundance ratios of main sequence and subgiant stars in the metallicity range $-1.3 < $ [Fe/H] < -0.5 and T_{eff} in the range from about 5500 K to 6500 K. All papers are characterized by determinations of abundance ratios with high internal precision and by finding significant cosmic dispersions in various element ratios at a given [Fe/H].

3.1. EDVARDSSON *et al.* (1993)

This paper deals with 189 disk stars selected to be about evenly distributed in [Fe/H] from -1.0 to $+0.3$. The sample contains both thin and thick-disk stars. Only one star, HD 148816 with [Fe/H] $= -0.74$, has halo kinematics.

As seen from Figure 2 the differences in [α/Fe] in the metallicity range $-0.8 < $ [Fe/H] < -0.4 are correlated with the mean orbital Galactocentric distance, R_m.

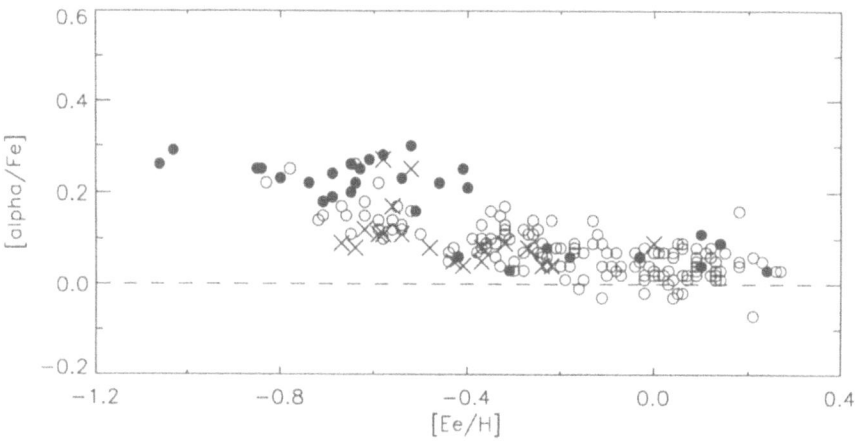

Figure 2. [alpha/Fe] vs. [Fe/H] for the Edvardsson *et al.* (1993) stars with 'alpha' referring to the mean abundance of Mg, Si, Ca and Ti. Stars shown with filled circles have a mean Galactocentric distance in their orbits, $R_m < 7$ kpc. Open circles refer to stars with $7 < R_m < 9$ kpc, and × to stars with $R_m > 9$ kpc.

Stars with large Galactic orbits, $R_m > 9$ kpc, have smaller [α/Fe] values than stars with $R_m < 7$ kpc. This correlation may be due to a decline of the star formation rate with Galactocentric distance, i.e. supernovae of Type Ia start contributing with iron at a lower [Fe/H] in the outer parts of the Galaxy than in the inner parts. Alternatively, it could be that the differences in [α/Fe] is due to an overlap between thick-disk and thin-disk stars in the $-0.8 <$ [Fe/H] < -0.5 interval, the thick disk stars having high values of [α/Fe] and the younger thin disk stars having lower values. This is the explanation favored by Fuhrmann (1998). His suggestion is in fact supported by the kinematics and ages of the Edvardsson *et al.* stars with [Fe/H] < -0.4. The 23 stars with $R_m < 7$ kpc have a velocity dispersion perpendicular to the Galactic plane $\sigma(W) = \pm 45$ km s^{-1} and a mean age of 12 Gyr, typical for the thick disk, whereas the 12 stars with $R_m > 9$ kpc have $\sigma(W) = \pm 28$ km s^{-1} and a mean age of 7 Gyr, corresponding to the old thin disk.

3.2. NISSEN AND SCHUSTER (1997)

In this paper two groups of stars were selected from the V_{rot}-[Fe/H] diagram of Schuster *et al.* (1993). Stars in the first group (the halo stars) have $V_{rot} < 50$ km s^{-1}, whereas stars in the second group (the disk stars) have $V_{rot} > 150$ km s^{-1}. In both groups the stars are distributed from -1.3 to about -0.5 in [Fe/H].

As seen from Figure 3 the halo stars show a range in [O/Fe] and [Mg/Fe] from about 0.0 to +0.3. The other alpha-elements (Si, Ca and Ti) have a similar range and show excellent correlation with oxygen and magnesium. [Na/Fe] has a relatively large and [Ni/Fe] a small variation. Both ratios are well correlated with

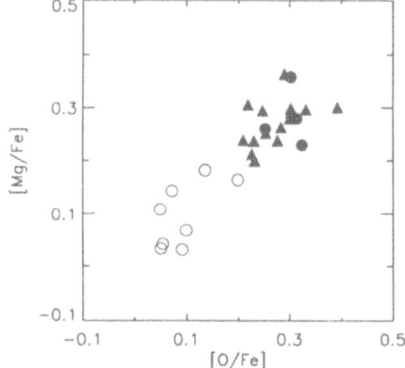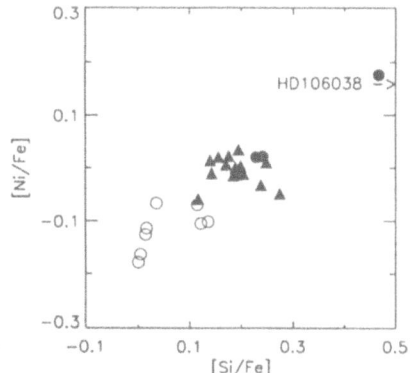

Figure 3. [Mg/Fe] vs. [O/Fe] and [Ni/Fe] vs. [Si/Fe] from Nissen and Schuster (1997). Thick disk stars are shown with filled triangles and halo stars with filled or open circles. The star HD 106038 is peculiar by being enhanced in the abundances of Li, Si, Ni and the s-process elements by a factor of 2 to 4 ([Si/Fe] = 0.57).

[α/Fe]. These variations are mainly due to the presence of 8 halo stars (shown by open circles in Figure 3), which tend to have larger Galactic orbits than the 4 halo stars with the same abundance ratios as the disk stars. Although this correlation of abundance ratios with orbital parameters (R_{max} and z_{max}) could be due to small number statistics, it suggests that the halo stars with 'anomalous' abundance ratios have formed in the outer parts of the halo or have been accreted from dwarf galaxies with a chemical evolution history different from that of the inner halo and the disk.

Nissen and Schuster (1997) interpreted the 'low-alpha' halo stars as being accreted from dwarf galaxies like the Magellanic Clouds for which several models (Gilmore and Wyse, 1991; Tsujimoto *et al.*, 1995; Pagel and Tautvaišienė, 1998) predict a solar α/Fe ratio around [Fe/H] $= -1.0$ as a consequence of an early star formation burst followed by a long dormant period. Gilmore and Wyse (1998) have, however, questioned this hypothesis. They point out that the 'low-alpha' stars are on orbits of such low periGalactic distances (< 1 kpc) that they could not be accreted from dwarf galaxies of known types. Instead they suggest that the 'low-alpha' stars comes from self-enriching transient regions sustaining star formation slowly enough to incorporate the ejecta from Type Ia supernovae at a low value of [Fe/H]. They further point out that the systematic difference in [α/Fe] between the thick disk and the majority of the halo stars suggests that the halo and the thick disk are *dis*-connected, very old components of the Galaxy. This interpretation is consistent with previous inferences based on angular momentum considerations (Wyse and Gilmore, 1992), and with the fact that the metal-rich halo and the thick disk have about the same old age (Section 2).

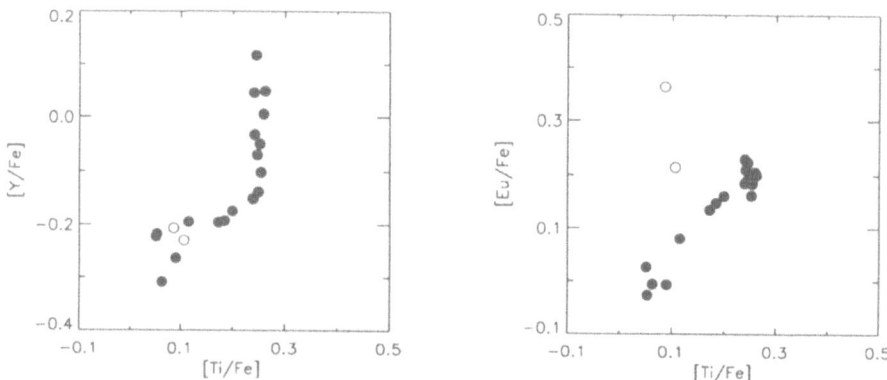

Figure 4. [Y/Fe] vs. [Ti/Fe] and [Eu/Fe] vs. [Ti/Fe] from Jehin *et al.* (1999). The two stars with clear halo kinematics are shown with open circles.

3.3. JEHIN *et al.* (1999)

In this paper 21 stars situated in the turnoff region of the HR-diagram and having $-1.3 <$ [Fe/H] < -0.7 are studied with a special emphasis on the *s*- and *r*-process elements. Most of the stars have thick-disk kinematics, but at least two stars are halo stars having $V_{rot} < 0$ km s^{-1}.

Figure 4 shows a plot of [Y/Fe] and [Eu/Fe] vs. [Ti/Fe] with data taken from Tables 6 and 7 of Jehin *et al.* Whereas the *r*-process element Eu closely follows the alpha-element Ti (except for the two deviating halo stars), the *s*-process element Y shows a two-branch behaviour. For about half of the stars, corresponding to the low values of [Ti/Fe], Y is well correlated with Ti. The remaining stars, however, have a constant (maximum) value of [Ti/Fe] and a range of [Y/Fe] values. Jehin *et al.* see no correlation between the abundance ratios and the kinematics of the stars and suggest an entirely new explanation of the variations – the so-called EASE (Evaporation/Accretion/Self-Enrichment) scenario. All the stars are assumed to be born in globular clusters undergoing chemical evolution. First the cluster is enriched with alpha-elements from Type II SNe and later with *s*-process elements from AGB stars. Early disruption of the cluster then leads to stars on the lower branch of the [Y/Fe]-diagram, whereas later disruption or 'evaporation' of stars leads to the vertical branch consisting of stars with different amounts of accreted *s*-process elements.

4. Conclusion

Precise determinations of abundance ratios in metal-rich halo stars and thick disk stars have revealed interesting variations and correlations between the abundances of elements formed by different nucleosynthesis processes. At a given metallicity, say [Fe/H] $= -1.0$, [O/Fe], [Mg/Fe], [Si/Fe], [Ca/Fe] and [Ti/Fe] as well as

[Eu/Fe] have a range corresponding to nearly 0.3 dex. The s-process elements and sodium show an even larger range, i.e. close to 0.5 dex. Also among the iron-peak elements there are variations; [Ni/Fe] shows a range of 0.2 dex with a very good correlation to e.g. [Na/Fe].

Edvardsson et al. (1993) and also Nissen and Schuster (1997) suggest that the variations in [α/Fe] are due to different star formation rates in the regions, where the stars have been born. The high value of [α/Fe] corresponds to pure Type II SNe nucleosynthesis with a normal (i.e. solar-neighborhood) IMF (Nissen et al., 1994), whereas the low (solar) value corresponds to incorporation of Fe from Type Ia supernovae (Tsujimoto et al., 1995). As the Type Ia SNe occur with a delay of about 1 Gyr, the [Fe/H] at which these supernovae start contributing with iron will depend on the star formation rate. Furthermore, there seems to be a correlation between [α/Fe] and the orbital parameters of the stars suggesting that the 'low-alpha' stars have been formed in the outer parts of the Galaxy, or – in the case of the halo stars – accreted from dwarf galaxies.

Jehin et al. (1999) suggest a completely different scenario according to which field halo and thick disk stars are formed in globular clusters with self-enrichment and (later) disruption and/or evaporation of stars. Note, that in this scenario Type Ia SNe play no role. The proto globular clusters start with a near solar ratio of [α/Fe] and are then enriched with the products of Type II SNe and AGB stars. Although this 'EASE' scenario provides a nice explanation of the two branches seen in the [Y/Fe] – [Ti/Fe] diagram (Figure 4) it fails to justify the assumption that the proto globular clusters are formed with a near solar [α/Fe] value. Furthermore, the exclusion of Type Ia supernovae as contributors to the chemical evolution of the globular clusters does not agree with currently accepted models (e.g. Matteucci and François, 1992; Yoshii et al., 1996) according to which Type Ia SNe start contributing with iron after a delay of about 1 Gyr, i.e. simultaneously with or even before AGB stars begin to contribute with s-process elements.

References

Alonso, A., Arribas, S. and Martínez-Roger, C.: 1996, *Astron. Astrophys.* **313**, 873.

Anstee, S.D., O'Mara, B.J. and Ross, J.E.: 1997, *Mon. Not. R. Astron. Soc.* **284**, 202.

Carney, B.W., Laird, J.B., Latham, D.W. and Aguilar, L.A.: 1996, *Astron. J.* **112**, 668.

Edvardsson, B., Andersen, J., Gustafsson, B., Lambert, D.L., Nissen, P.E. and Tomkin, J.: 1993, *Astron. Astrophys.* **275**, 101.

ESA: 1997, *The Hipparcos and Tycho Catalogues*, ESA SP-1200.

Fuhrmann, K.: 1998, *Astron. Astrophys.* **338**, 161.

Gilmore, G. and Wyse, R.F.G.: 1991, *Astrophys. J.* **367**, L55.

Gilmore, G. and Wyse, R.F.G.: 1998, *Astron. J.* **116**, 748.

Jehin, E., Magain, P., Neuforge, C., Noels, A., Parmentier, G. and Thoul, A.A.: 1999, *Astron. Astrophys.* **341**, 241.

Matteucci, F. and François, P.: 1992, *Astron. Astrophys.* **262**, L1.

Nissen, P.E. and Schuster, W.J.: 1991, *Astron. Astrophys.* **251**, 457.

Nissen, P.E. and Schuster, W.J.: 1997, *Astron. Astrophys.* **326**, 751.

Nissen, P.E., Gustafsson, B., Edvardsson, B. and Gilmore, G.: 1994, *Astron. Astrophys.* **285**, 440.

Pagel, B.E.J. and Tautvaišienė, G.: 1998, *Mon. Not. R. Astron. Soc.* **299**, 535.

Schuster, W.J. and Nissen, P.E.: 1989, *Astron. Astrophys.* **221**, 65.

Schuster, W.J., Parrao, L. and Contreras Martínez, M.E.: 1993, *Astron. Astrophys. Suppl.* **97**, 951.

Tsujimoto, T., Nomoto, K., Yoshii, Y., Hashimoto, M., Yanagida, S. and Thielemann, F.-K.: 1995, *Mon. Not. R. Astron. Soc.* **277**, 945.

VandenBerg, D.A.: 1997, in: T.R. Bedding, A.J. Booth and J. Davis (eds.), *Fundamental Stellar Properties, IAU Symp.* **189**, Kluwer, p. 439.

Wyse, R.F.G. and Gilmore, G.: 1992, *Astron. J.* **104**, 144.

Yoshii, Y., Tsujimoto, T. and Nomoto, K.: 1996, *Astrophys. J.* **462**, 266.

A GRID OF METAL-POOR MODEL STELLAR ATMOSPHERES FOR STARS BORN IN THE EARLY GALAXY

C. VAN 'T VEER-MENNERET, D. KATZ and R. CAYREL
Observatoire de Paris, 61 avenue de l'Observatoire, F75014 Paris, France

C. SOUBIRAN
Observatoire de Bordeaux, BP89, F33270 Floirac, France

1. Introduction

In the frame of an extensive study of the formation and chemical evolution of stars belonging to the galactic halo and to the intermediate population II, grids of model atmospheres have been built for a large range of metallicities. The Kurucz codes and opacity tables were used (Kurucz, 1993) to compute metal-poor atmospheres. Hereafter the procedure will be described and new results will be shown for the intermediate population II star 85 Pegasi.

2. Description and Use of the Grids

The automation of Kurucz' codes on UNIX station has been achieved, and allows to compute grids of models, fluxes, Balmer line profiles, spectra and colours. Following the results of Fuhrmann *et al.* (1993, 1994, 1997), and of van 't Veer-Menneret and Mégessier (1996, VM), we have admitted as established that *the line profiles of the Balmer series for stars with convective transfer of radiation are all together well represented using models constructed with the MLT parameter* $\alpha = 0.5$. This finding seems to be valid for metal-rich, solar composition and metal-poor stars, in the range of effective temperatures (T_{eff}) 5000 K–8750 K.

3. Method and Results

With our own observational material and reduction tools, we came to the same conclusions as Fuhrmann *et al.* (1993, 1994, 1997) i. e.: *BLPs are effective indicators of stellar atmosphere structure, so long as* Hα *and the following lines are jointly interpreted, and constitute a direct evidence of the depth stratification of the*

 Astrophysics and Space Science is the original source of publication of this article. It is recommended that this article is cited as: *Astrophysics and Space Science* **265:** 257–258, 1999.
© 1999 *Kluwer Academic Publishers.*

TABLE I

The second row gives the results of Axer *et al.* (1994)

[M/H]	T_{eff}	log g	v_t	[Fe/H]	[Ca/Fe]	[Si/fe]	[Ti/Fe]	[Ni/Fe]
−0.8	5600 K	4.6	0.5	−0.65	+0.3	+0.3	+0.3	0.0
	5524	4.61	1.79	−0.86				

atmosphere. First of all, we derive T_{eff} from $H\alpha$ and find the only model fitting both $H\alpha$ and $H\beta$. Then metallicity and gravity are derived by iterative method through abundance analysis. In Table I, the first column gives metallicity of the ODFs used including enhancement of 'α elements', the others give the parameters derived from a detailed abundance analysis of 85 Peg.

The well known multiple system 85 Peg presents the anomaly that the secondary component is slightly more massive than the primary, while it is 3 magnitudes fainter. The difference with Axer *et al.* in the value of [Fe/H] is obviously due to the difference in the value of the microturbulence v_t. An interesting result is the enrichment with respect to iron of the 3 α elements represented in the spectral range studied.

4. Conclusion

The new results to be emphasized are: (i) in the isochrone grid computed by Lebreton *et al.* (1997), the position of 85 Peg, slightly modified, is in a better agreement with its new metallicity; (ii) the enrichment in α elements by +0.3 with respect to iron.

Acknowledgement

We acknowledge R.L. Kurucz for his generous distribution of his codes and data, and F. Castelli for her efficient help to understand them.

References

Fuhrmann, K., Pfeiffer, M., Frank, C., Reetz, J. and Gehren, T.: 1997, *Astron. Astrophys.* **323**, 909.
Kurucz, R.L.: 1993, *CD-ROM* 13, 14.
Van 't Veer-Menneret, C. and Mégessier, C. (**VM**): 1996, *Astron. Astrophys.* **309**, 879.
Axer, M., Fuhrmann, K. and Gehren, T.: 1994, *Astron. Astrophys.* **291**, 8.
Lebreton, Y., *et al.*: 1997, *ESA Symp., 'Hipparcos-Venice '97'*, ESA SP-402, pp. 379.

PRESSURE BROADENING COEFFICIENTS FOR STELLAR APPLICATIONS

A. SPIELFIEDEL and N. FEAUTRIER

DAMAP and URA812 du CNRS, Observatoire de Paris, 92195 Meudon Cedex, France

1. Context of the Study

The precise determination of star gravity from the wing intensity of collisionally broadened lines may be limited by the quality of the calculation of the line parameters and specifically of the line broadening constants pertaining to collisions with hydrogen atoms (Cayrel *et al.*, 1996). Among the CaI triplet lines 6102, 6122, 6162 Å the strongest line (6162 Å) is considered as a good gravity indicator (Edvarsson, 1988).

For the relevant temperature range (4000 K–8000 K), the relative mean velocity of the hydrogen and calcium atoms is small, thus one can consider that a transient CaH molecule is formed in the collisional process. The typical parameters for the calculations are the perturber density $N_H \cong 10^{12}$ cm^{-3} and the temperature $T \cong$ 5000 K.

2. Results

Calculations of the broadening constants require the determination of the potential energy curves of Ca+H correlated with the atomic levels of the Ca transition, and collision calculations. Using the CaH potential curves of Chambaud and Lévy (1989), we have calculated all the elements of the collisional S-matrix. The line broadening and shift are expressed in terms of these matrix elements. Since the kinetic energies are relatively small compared to the potential energy variations during the collision, the collision process is best described in a quantum mechanical approach. The lower Ca^{3}P level is split into a triplet by the fine structure interaction which must be included in the calculation since we know from its order of magnitude that it will affect significantly the collision dynamics and thus the line broadening constants.

As expected (Spielfiedel *et al.*, 1991), the fine structure effects are important as one can see in table 1 for $T = 5000$ K. A comparison of $\frac{W_{ij}}{N_P}$ for the ij line with the Van der Waals broadening coefficient W_{VDW} (constant for all the lines) shows that

 Astrophysics and Space Science is the original source of publication of this article. It is recommended that this article is cited as: *Astrophysics and Space Science* **265**: 259–260, 1999.
© 1999 *Kluwer Academic Publishers.*

TABLE I

Broadening coeffi-
cients. Units are 10^{-8}
$cm^3 \, s^{-1}$

ij	$\frac{W_{ij}}{N_P}$	$\frac{W_{ij}}{W_{VDW}}$
21	5.16	2.44
11	4.71	2.22
01	6.20	2.92

this usual approximation underestimates the broadening constants by more than a factor of two.

3. Comments and Consequences

An analysis of the differences between the broadening coefficients for the three lines of the multiplet shows that they are mainly due to the particular shape of the potential curves and particularly to the mid and long range part of the $^2\Sigma^+$ molecular states which exhibit large perturbations due to the ionic Ca^+H^- configuration (Chambaud and Lévy, 1989). The Van der Waals approximation which assumes a R^{-6} long range potential cannot take into account such perturbation. As a consequence, we can expect large effects for all the lines for which the $^2\Sigma^+$ states have a major contribution, in particular for the 6712.7 Å ($^1P_1 - {}^1D_2$) line since only doublet molecular states contribute to the broadening.

References

Cayrel, R., Faurobert-Scholl, M., Feautrier, N., Spielfiedel, A. and Thévenin, F.: 1996, On the use of Ca I triplet lines as luminosity indicators, *Astron. Astrophys.* **312**, 549–552.

Chambaud, G. and Lévy, B.: 1989, CaH potential curves: a simple theoretical treatment of intershell effects, *J. Phys. B: At. Mol. Opt. Phys.* **22**, 3155–3165.

Edvarsson, B.: 1988, Spectroscopic surface gravities and chemical composition for 8 nearby single subgiants, *Astron. Astrophys.* **190**, 148–166.

Spielfiedel, A., Feautrier, N., Chambaud, G. and Lévy, B.: 1991, Collision broadening of the 4s4p($^3P^o$)-4s5s(^3S) line of calcium perturbed by hydrogen and collision induced transitions among the 4s4p($^3P^o$) states, *J. Phys. B: At. Mol. Opt. Phys.* **24**, 4711–4721.

THE CHEMICAL EVOLUTION OF CARBON IN THE GALACTIC DISK

T. KARLSSON, B. EDVARDSSON, B. GUSTAFSSON, E. OLSSON and N. RYDE

Uppsala Astronomical Observatory, Box 515, SE-751 20 Uppsala, Sweden

1. Observations of Disk Stars, Analysis and Results

A subsample of 80 F and G Disk dwarf stars was selected from Edvardsson *et al.* (1993, *Astron. Astrophys.* **275**, 101) and observed with the ESO 1.4 m CAT telescope. The carbon abundance was determined using the weak [C I] line at 8727 Å, which is presumably less sensitive to errors in the model atmospheres and other changes in the model parameters than the abundance.

We used a strict synthetic spectrum analysis to determine the abundances and found the data to satisfy the relation $[C/Fe] = (-0.17 \pm 0.03) \times [Fe/H] + (0.065 \pm 0.008)$. This leads to an increasing [C/O] ratio with metallicity due to the significantly steeper slope of the [O/Fe] ratio (Edvardsson *et al.*, 1993). The total statistical error is ~ 0.06 dex in [C/H].

2. Possible Carbon Production Sites

The stellar origin of carbon became clear with the discovery of the Triple Alpha process together with Hoyle's prediction of a resonance state at 7.65 MeV of carbon. But which type of stars is the most efficient carbon producer? A number of sources have been suggested: (1) Supernovae type II, (2) Supernovae type Ia, (3) Novae, (4) Very massive stars such as Wolf-Rayet stars (through radiatively driven winds), (5) Intermediate- and low-mass stars (through winds and PNe).

The first three sources can not explain the observed evolution of [C/O]. Either, the events are too rare given their predicted carbon production (*SNe Ia, novae*) or they produce a nearly constant and too low [C/O] ratio with metallicity (*SNe II*).

WR stars could contribute through mass loss, particularly in their WC stage. The yields are suggested to be strongly dependent on metallicity due to the effect of radiatively driven winds, exposing and blowing away the carbon shells before the carbon is ignited.

Intermediate- and low-mass stars, undergoing dredge-ups and high mass loss during their AGB and PNe phases, could also contribute substantially to the total amount of carbon. The effect on the carbon enrichment in the Galaxy would then

 Astrophysics and Space Science is the original source of publication of this article. It is recommended that this article is cited as: *Astrophysics and Space Science* **265**: 261–262, 1999.
© 1999 *Kluwer Academic Publishers.*

be time dependent (reflecting the relatively long time-scale of the low-mass stars) rather than metallicity dependent.

3. A Test to Deduce the Origin of Carbon

Wolf-Rayet stars and intermediate-/low-mass stars, may both qualitatively reproduce the observed [C/O] slope. We will, instead of depending on chemical evolution modelling, propose a model-independent test to decouple the sources and determine the origin of carbon in our Galaxy.

The test is built on the fact that the two factors behind the [C/O] increase – metallicity and time-scale – are interrelated in different ways in different galaxies. The [C/O] slope in the Disk maps the build-up of carbon. Comparing with the corresponding slope for stellar systems of equal age but different metallicity, i.e. where only the metallicity-dependent source contributes to the slope, we are able to determine the significance of the time-dependent source. This is the low- and intermediate-mass stars, the upper mass limit being dependent on how far back in time we map the Disk. If the [C/O] slope in the Disk is steeper, the lower mass stars have produced a significant amount of carbon, while equal slopes suggest WR stars to be the main contributor.

In practice, we compare the build-up of carbon as a function of metallicity in the Galaxy with H II regions in dwarf irregular galaxies assuming that the irregulars are old stellar systems. The slope of the Disk stars in the [C/O]–[O/H] diagram is not steeper than the corresponding slope for the dwarf irregulars. This is interpreted as a result of carbon being produced predominantly by massive stars, in particular during their WC stage, and not in the AGB or PN stage of lower mass stars.

4. Conclusions

Using our accurate determinations of carbon abundances in local Disk solar-type dwarfs and comparing the evolution of [C/O] in the Disk with that of dwarf irregular galaxies, we conclude that the main contributor to carbon production in our Galaxy is metal-rich WC stars.

A complete report of this study is given in Gustafsson *et al.* (1999), *Astron. Astrophys.*, in press. See also astro-ph/9811303.

SIMULTANEOUS SOLUTIONS OF STELLAR PARAMETERS (T_{eff}, [Fe/H]) FROM SYNTHETIC $UVBY$ STRÖMGREN PHOTOMETRY

E. LASTENNET

Queen Mary & Westfield College, Astronomy Unit, Mile End Road, London E1 4NS, UK

T. LEJEUNE, P. WESTERA and R. BUSER

Astronomisches Institut der Universität Basel, Venusstr. 7, CH-4102 Binningen, Switzerland

1. Introduction

The comprehensive knowledge of fundamental parameters of single stars is the basis of the modelling of star clusters and galaxies. Most fundamental stellar parameters of the individual components in SB2 eclipsing binaries are known with very high accuracy. Unfortunately, while masses and radii are well determined, the temperatures strongly depend on photometric calibrations. In the present work, we have used an empirically-calibrated grid of theoretical stellar spectra (BaSeL models) for simultaneously deriving homogeneous effective temperatures and metallicities from observed data. For this purpose, we have selected 20 binary systems (40 stars) for which we have $uvby$ Strömgren photometry with estimated errors (see Lastennet *et al.*, 1998 for details).

2. An Example of Simultaneous (T_{eff},[Fe/H]) Determination

To compute synthetic colours from the BaSeL models, we need effective temperature (T_{eff}), surface gravity ($\log g$), and metallicity ([Fe/H]). Consequently, given the observed colours (namely, $b - y$, m_1, and c_1), we are able to derive T_{eff}, $\log g$, and [Fe/H] from a comparison with model colours. As the surface gravities can be derived very accurately from the masses and radii of the stars in our working sample, only two physical parameters have to be derived (T_{eff} and [Fe/H]). This has been done by minimizing the χ^2-functional, defined as

$$\chi^2(T_{eff}, \text{[Fe/H]}) = \sum_{i=1}^{n} \left[\left(\frac{\text{colour(i)}_{\text{syn}} - \text{colour(i)}}{\sigma(\text{colour(i)})} \right)^2 \right],$$

where n is the number of comparison data, colour(1) $= (b - y)_0$, colour(2) $= m_0$, and colour(3) $= c_0$. The best χ^2 is obtained when the synthetic colour, colour(i)$_{\text{syn}}$,

 Astrophysics and Space Science is the original source of publication of this article. It is recommended that this article is cited as: *Astrophysics and Space Science* **265**: 263–264, 1999.
© 1999 *Kluwer Academic Publishers*.

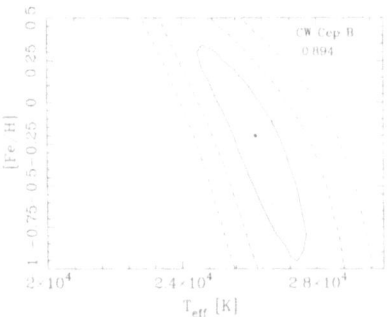

Figure 1. Simultaneous solutions of T_{eff} and [Fe/H] for CW Cep B. Best fit (*filled dot*) and 1-σ (*solid line*), 2-σ (*dashed line*), and 3-σ (*dot-dashed line*) confidence levels are also shown. Previous estimates of T_{eff} from Clausen and Giménez (1991) are indicated as vertical dotted lines.

is equal to the observed one. Finding the central minimum value χ^2_{min}, we form the χ^2-grid in the (T_{eff}, [Fe/H])-plane and compute the boundaries corresponding to 1 σ, 2 σ, and 3 σ respectively, as shown in Figure 1 for one of the 40 stars.

The range of T_{eff} values that we derive for CW Cep B agrees well with previous estimates (as indicated by the vertical dotted lines), and its metallicity is compatible with those of the Galactic disk stars. As a general trend for the whole sample, it is worth noticing that our T_{eff} ranges do not provide estimates systematically different from previous ones. However, the confidence regions show that most previous estimates of T_{eff} are optimistic, and that [Fe/H] should not be neglected in such determinations. The effects of surface gravity and interstellar reddening have also been carefully studied. Moreover, comparisons for 8 binaries of our working sample with Hipparcos parallaxes show good agreement with the most reliable parallaxes. Full details and critical discussions are given in Lastennet *et al.* (1998).

References

Clausen, J.V. and Giménez, A.: 1991, *Astron. Astrophys.* **241**, 98.
Lastennet, E., Lejeune, T., Westera, P. and Buser, R.: 1998, *Astron. Astrophys.*, accepted (astro-ph/9811103).

THE DISK POPULATIONS IN THE [MG/H]-[FE/MG] PLANE

KLAUS FUHRMANN

Universitäts-Sternwarte, Scheinerstrasse 1, D-81679 München, Germany

Abstract. The disk populations' metal abundance degeneracy is shown to be considerably relaxed in a two-dimensional presentation of their chemical properties. As such, the metallicities of a sample of nearby F- and G-stars are given in terms of magnesium abundances along the abscissa and iron-to-magnesium abundance ratios perpendicular to that. In combination with stellar age estimates and kinematics the disk populations turn out to be fairly well separable in this *abundance plane*, which in turn allows to address a number of important issues on the Milky Way's history.

1. Introduction

The thick disk of the Milky Way Galaxy is and has always been an elusive member of its main constituents. The intermediate character of this population in view of kinematics and metallicity repeatedly caused doubts as to whether this is indeed a distinct entity on the stage of Galactic evolution or merely part of the high and low endings of rather ill-understood distribution functions of halo and disk stars. As a particular example one may refer to the so-called metal-weak thick-disk stars, some of which reach down to at least [Fe/H] ~ -1.6, i.e. considerably deep into the halo star regime. With respect to the formation and age of the thick-disk population a good deal of our understanding persists in the dark. It is however important to note that a number of observations lend support to a very old population, presumably not much different from that of the Galactic halo (e.g. Marquez and Schuster, 1994), and there is now also growing evidence that the disk populations are separated in age by a gap in star formation (e.g. Gratton *et al.*, 1996).

Many excellent reviews of the thick disk exist in the literature and many updates of the most recent results can be found in this volume, to which we refer here for convenience. Our own focus will be on the separability of thick- and thin-disk stars by means of chemistry, kinematics and age estimates of nearby F- and G-type stars.

2. Observations and Analyses

The data presented here consist of spectroscopic observations obtained in 1995–1998 at the Calar Alto Observatory in Spain. The fiber optics Cassegrain échelle spectrograph FOCES (Pfeiffer *et al.*, 1998) was employed with a spectral resolution up to $\lambda/\Delta\lambda \sim 60000$.

Astrophysics and Space Science is the original source of publication of this article. It is recommended that this article is cited as: *Astrophysics and Space Science* **265**: 265–268, 1999.
© 1999 *Kluwer Academic Publishers.*

The basic stellar parameters of our F- and G-star sample are derived from the Balmer lines (T_{eff}), the iron ionization equilibrium and/or the wings of the Mg Ib lines (log g), and the profile analyses of iron and magnesium lines for the metallicity ([Fe/H]), microturbulence (ξ_t) and abundance ratio ([Fe/Mg]). Details are given in Fuhrmann (1998), where most of the derived stellar parameters are tabulated as well. Here we only briefly recall the basic features of our analysis, namely its differential character with respect to the Sun, the use of purely spectroscopic tracers for the basic stellar parameters, the full wavelength coverage and low level stray-light ($\sim$ 1%) provided by FOCES, the high spectral resolution that enables detailed line profile analyses, and, in particular, the reference to the accurate Hipparcos parallaxes that provides the robust framework for almost every aspect. The investigated F- and G-stars belong to the main-sequence and turnoff region, with only a small number of subgiants. The accuracy of the atmospheric parameters is characterized by ΔT_{eff} $\sim$ 80 K, Δ log g $\sim$ 0.1 dex, Δ[Fe/H] $\sim$ 0.07 dex, and Δ[Fe/Mg] $\sim$ 0.05 dex. In combination with the kinematics, and with age estimates from evolutionary tracks, we now proceed with the separability of the thin- and thick-disk stars.

3. The Disk Populations in the [Mg/H]-[Fe/Mg] Plane

One of the main characteristics of the thick-disk population is that its metallicity distribution function – which may peak at [Fe/H] $\sim$ -0.6 – shows considerable overlap with the halo and thin-disk stars. We have already mentioned the metal-weak thick-disk stars, but there is also a noticeable common region towards high metallicities, as thin-disk objects reach down to at least [Fe/H] $\sim$ -0.6.

In view of this abundance overlap along the [Fe/H] 'coordinate' the introduction of an additional chemical component by means of e.g. an appropriate *abundance ratio* may rectify this degeneracy. Thus a 2-d presentation of the chemical properties of stars may establish a better definition of the individual stellar populations. We therefore arrange our program stars in Figure 1 in a [Mg/H]-[Fe/Mg] plane, i.e. we refer to the α-element magnesium as the primary abundance indicator and compare the relative enrichment of iron with respect to this element. The opposite perspective, [Mg/Fe] vs. [Fe/H], is the more common version in the literature, which we also include for comparison. Note that with respect to the stellar ages the data in Figure 1 are depicted such that small circles correspond to young stars. The principal importance of the α-element abundance ratios is based on the well-known fact that α-chain nuclei are predominantly produced in massive stars and rapidly distributed via type II SNe, whereas the bulk of iron comes from SN Ia, i.e. on considerably longer timescales. Thus the observational fact that many metal-poor stars are overabundant by $\sim$ $+0.4$ dex in the α-elements is a very useful constraint to our understanding of the evolution of the Galaxy.

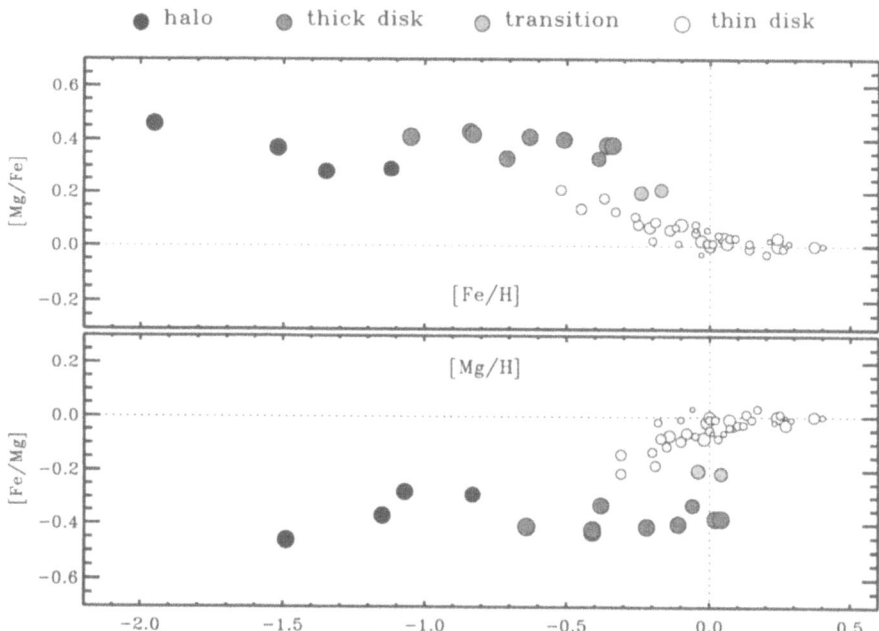

Figure 1. Abundance ratios of the α-element magnesium for the sample of nearby F- and G-stars. Circle diameters are in proportion to the stellar ages. Different *stellar populations* are given with various grayscale symbols as illustrated in the legend on top, and are based on chemistry, kinematics and age informations.

Keeping in mind that the sample in Figure 1 is yet rather small, the displayed data may be summarized as follows:

- the stars of the thin-disk population (light circles) are fairly well separable from thick-disk objects by means of their relative positions in the abundance plane, the mean rotational velocities with a differential lag of $\langle \Delta V_{rot} \rangle \sim -70$ km s^{-1} for the identified thick-disk members, and the finding that all thin-disk stars are younger than ~ 9 Gyr

- the stars of the thick disk are throughout old objects, most of them exceed 12 Gyr. Hence, the thick disk is most evidently the precursor population of the thin disk. A merger scenario like that discussed in Quinn, Hernquist and Fullagar (1993) is thereby very unlikely

- from the latter item it is also clear that the thick disk's pre-enrichment in metallicity provides a simple answer to the well-known G-dwarf problem (cf. e.g. Gilmore and Wyse, 1986)

- the data of Figure 1 give no permission to distinguish between thick-disk stars and halo stars. For this purpose we made use of the data in the [Fe/H]-V_{rot} diagram of Schuster, Parrao and Contreras Martínez (1993)

- the abscissa in Figure 1 is not a useful timescale: 72 Her and HD 221830, the two most metal-rich thick-disk stars, have an age close to $\sim$ 14 Gyr, whereas the most metal-poor thin-disk stars are at best half as old
- the metallicity distribution of thin-disk stars spans a range of at least $-0.6 \leq$ [Fe/H] $\leq +0.4$. Thick-disk stars show a considerable abundance overlap and can even exceed the solar magnesium abundance
- there is at best a weak age-metallicity relation for the stars of the thin disk. In particular, old thin-disk stars, such as μ Her or 31 Aql, can be very metal-rich
- the metal-poor thin-disk stars at intermediate [Fe/Mg] values may be interpretable in terms of a delayed infall of processed thick-disk material. At the high side of the thin-disk metallicities the restriction to solar [Fe/Mg] ratios argues against a short-term starburst of, say, less than 10^8 years at the onset of the thin disk and provides valuable constraints on the relative occurrence and/or yields of SN II vs. SN Ia
- the relative distribution of the disk populations in Figure 1 implies the existence of a star formation gap with considerable enrichment of iron, before the thin disk came into play. The two transition stars ρ CrB and HR 7569 at intermediate [Fe/Mg] values are also intermediate in their ages ($\sim$ 9.5 $-$ 10 Gyr), and thus provide a firm upper limit for the age of the thin disk.

References

Fuhrmann, K.: 1998, *Astron. Astrophys.* **338**, 161.
Gilmore, G. and Wyse, R.F.G.: 1986, *Nature* **322**, 806.
Gratton, R., *et al.*: 1996, *ASP Conf. Ser.* **92**, 307.
Marquez, A. and Schuster, W.J.: 1994, *Astron. Astrophys. Suppl.* **108**, 341.
Pfeiffer, M.J., *et al.*: 1998, *Astron. Astrophys. Suppl.* **130**, 381.
Quinn, P.J., Hernquist, L. and Fullagar, D.P.: 1993, *Astrophys. J.* **403**, 74.
Schuster, W.J., Parrao, L. and Contreras Martínez, M.E.: 1993, *Astron. Astrophys. Suppl.* **97**, 951.

ABUNDANCE GRADIENTS IN THE OUTER GALACTIC DISK

W.J. MACIEL and C. QUIREZA
IAG/USP, São Paulo, Brazil

Abstract. Radial abundance gradients of the element ratios O/H, Ne/H, S/H, and Ar/H are determined for a sample of disk planetary nebulae, emphasizing the behaviour of the gradients at large galactocentric distances.

1. Introduction

The existence of radial abundance gradients is firmly established, both in the Galaxy and in other spirals. The gradients can be derived for several abundance ratios from photoionized nebulae (HII regions and planetary nebulae) and young stars (cf. Maciel, 1997). These gradients have some important consequences on chemical evolution models, particularly considering their *magnitudes*, *spatial* and *temporal* variations in the galactic disk.

Planetary nebulae (PN) play a distinct role in the solution of these problems. In the present work, gradients of the element ratios O/H, Ne/H, S/H, and Ar/H are determined for a sample of disk PN, emphasizing the behaviour of the gradients at large galactocentric distances.

2. The Data

The present investigation is based on a sample of disk PN generally classified as Type II objects. The basic sample is that of Maciel and Köppen (1994) and Maciel and Chiappini (1994), supplemented by some new objects with abundances derived by the IAG/USP group or from the recent literature. As a consequence, our new sample forms the largest database of galactic PN with reliable abundances and distances ever to be considered in order to estimate radial abundance gradients.

3. Results and Discussion

We have obtained plots of the O/H, Ne/H, S/H and Ar/H ratios as a function of the galactocentric distance R, and derived linear and second order fits to the data. The main conclusions by Maciel and Köppen (1994) are maintained, in the sense that

 Astrophysics and Space Science is the original source of publication of this article. It is recommended that this article is cited as: *Astrophysics and Space Science* **265:** 269–270, 1999.
© 1999 *Kluwer Academic Publishers.*

for $R \leq R_0 = 7.6$ kpc a radial gradient is obtained averaging $d \log(\text{X}/\text{H})/dR \simeq -0.06$ dex kpc^{-1}, in excellent agreement with the well known O/H gradient of about -0.07 dex kpc^{-1} derived from HII regions. Moreover, the new results suggest some flattening for $R > R_0$, particularly for $R \geq 10$ kpc, in agreement with the recent results for HII regions by Vílchez and Esteban (1996).

Open cluster stars generally produce consistent results with HII regions, as also cepheids and supergiants. Data for Be stars are somewhat contradictory, but a recent work by Smartt and Rolleston (1997) suggests a similar gradient as seen in the HII region and PN data near R_0. For larger R, there may be no flattening for B stars, which could be explained by the temporal variations of the gradients suggested by Maciel and Köppen (1994).

Our results can be used to constrain some of the recent chemical evolution models of the Galaxy. Classical models (Chiappini et al., 1997) predict flatter gradients than observed, but are consistent with the observed flattening at the outer Galaxy. The multiphase models (cf. Mollá et al., 1997) predict some temporal flattening of the gradients, in contrast with the suggestion by Maciel and Köppen (1994). The predicted magnitudes of the gradients are similar, and even steeper than observed, but the models predict a steepening for large R, in contrast with the results of the present work. Probably the most promising theoretical models are the so-called chemodynamical models (cf. Samland et al., 1997). Their predictions are in good agreement with the gradients from photoionized nebulae, both regarding the magnitude of the gradients and their space variations. Since the application of these models to galaxies like our own is still in its infancy, it is expected that more detailed models will be able to account also for the time behaviour of the abundance variations in the near future.

Acknowledgements

This work was partially supported by CNPq and FAPESP.

References

Chiappini, C., Matteucci, F. and Gratton, R.: 1997, *Astrophys. J.* **477**, 765.
Maciel, W.J.: 1997, in: H.J. Habing and H.J.G.L.M. Lamers (eds.), *IAU Symp. 180*, Kluwer, Dordrecht, p. 397.
Maciel, W.J. and Chiappini, C.: 1994, *Astrophys. Space Sci.* **219**, 231.
Maciel, W.J. and Köppen, J.: 1994, *Astron. Astrophys.* **282**, 436.
Mollá, M., Ferrini, F. and Díaz, A.I.: 1997, *Astrophys. J.* **475**, 519.
Samland, M., Hensler, G. and Theis, C.: 1997, *Astrophys. J.* **476**, 544.
Smartt, S.J. and Rolleston, W.R.J.: 1997, *Astrophys. J.* **481**, L47.
Vílchez, J.M. and Esteban, C.: 1996, *Mon. Not. R. Astron. Soc.* **280**, 720.

OPEN CLUSTERS AS A RECORD OF THE PAST

E.D. FRIEL

*Division of Astronomical Sciences, National Science Foundation, Arlington, VA, and
Astronomy Department, Boston University, Boston, MA, USA*

Abstract. The Galactic open cluster population has long been used as a probe of the structure of the Galactic disk and a timeline for studying its evolution. With ages that range up to 12 billion years and positions that span a large range of Galactocentric distances, the open clusters provide a broad sample with which to investigate issues such as the history of star formation in the Galaxy, the chemical evolution of the disk, and the competing influences of cluster formation and disruption that mold the properties of the current cluster population.

1. Introduction

The open cluster population presents us with fossil traces of a complex history, which we must unravel from the limited clues left behind. As we have seen at this meeting, viewing the globular cluster and halo field populations in this way has been productive, and it is only recently that we have acquired the tools that allow us to gain similar insights using the open cluster population. We now have much larger samples to work with and the large format CCDs and multi-object spectrographs now available are well suited to the study of the extended fields of open clusters.

I can highlight only some of the areas that will be most productive for study in the coming years, and where some of the most interesting questions remain. I will begin with a sketch of some basic observations, or fossil indications, that we have from the open cluster population. Following the long tradition of work in stellar populations, we will look at the distribution of the open clusters in terms of their location, ages, kinematics, and abundances.

2. Cluster Spatial Distribution

When we plot the spatial distribution of the open clusters as projected onto the Galactic plane, we see immediately a striking difference between the distributions for young versus old clusters. Young clusters are found throughout the disk, while old clusters, those with ages roughly the age of the Hyades or older (about 700 million years), are only found more than 7 kpc from the Galactic center. By contrast,

 Astrophysics and Space Science is the original source of publication of this article. It is recommended that this article is cited as: *Astrophysics and Space Science* **265**: 271–278, 1999.
© 1999 *Kluwer Academic Publishers.*

the disk globular clusters are found in the inner disk, in regions where we do not find old open clusters.

This tendency for the old open clusters to avoid the inner regions of the Galaxy was noted decades ago, even with small samples, and has only become more pronounced as more clusters are discovered (van den Bergh and McClure, 1980; Phelps *et al.*, 1994). It can no longer be attributed solely to observational selection effects, for we have searched for clusters in these regions, and those we find are all relatively young.

The observed distribution must be controlled by the competing effects of formation and destruction, in some delicate balance that produces the characteristics of the remnant population we sample today. Our efforts at understanding these fossil clusters then become attempts to quantify these competing effects. We might consider how the dependence on initial conditions such as cloud properties, infall or accretion events, or the intrinsic cluster properties like cluster mass and central densities, or cluster orbital characteristics combine to produce a population of clusters that we see today. On the other hand, we know that clusters are destroyed over time through both internal forces of mass loss through dynamical evolution and stellar evolution, and by the external forces of interactions with the galactic tidal field or with molecular clouds. To fully understand the clusters as we now see them, and to use them as probes of galactic structure and evolution, we need to understand in a quantitative sense the relative importance of these effects and how they might change through the lifetime of the Galaxy.

3. Cluster Age Distribution

Star clusters provide us with a reasonably well established relative timeline for our investigations, a benefit over many field star studies. The cumulative age distribution for the entire population of open clusters shows evidence for two populations with different characteristic lifetimes: a large population of clusters with typical lifetimes of 200 million years, and a much smaller, but substantial, population of clusters with typical lifetimes of 3 to 5 billion years (Janes and Phelps, 1994).

We will concentrate on these 'old' clusters to probe the earliest stages of the disk. There are now almost 100 clusters with ages greater than the Hyades. In the past few years many groups have contributed to increasing this sample and determining cluster properties (e.g. Carraro, Tosi, Phelps, Geisler, among others). There are continuing attempts to add to this sample through systematic surveys of the roughly 1200 open clusters that have been cataloged (Phelps, 1998) and to improve our understanding of the old clusters known (Bragaglia *et al.*; Carraro, this volume).

A histogram of the age distribution of these clusters, as shown in Figure 5 of Friel (1995), raises a number of intriguing questions: Is there a continuity in ages from the globular clusters to the open clusters? Are there open clusters that can been seen as 'transition' objects from the halo to the disk? How old IS the oldest open cluster? What is it about these oldest clusters that makes them survive to ages approaching, or perhaps overlapping the globular clusters? Are they special? And if so, in what ways?

4. Kinematics of Old Disk Clusters

We turn, next, to the kinematics of the old disk clusters, in an attempt to understand how the clusters' orbital characteristics affect their survivability, perhaps in keeping them away from the objects that would act to destroy them or in revealing characteristics of their formation. It has been known for some time, from radial velocities alone, that the old open clusters are a rapidly rotating population, with a rotation speed that is similar to the old disk field stars, somewhat more rapid than the thick disk field or the disk globular clusters. There are also some clusters which appear, from their radial velocities, to be on unusual orbits (Scott *et al.*, 1995).

We would like to go beyond the radial velocities to understand cluster orbits. Unfortunately, the sample we have to work with is quite limited. Early work on a few classic clusters was recently expanded to a sample of 5 by Carraro and Chiosi (1994). Their motivation was primarily an investigation of how the place of formation affected the radial abundance gradient, and they concluded that the clusters experienced limited excursions in galactocentric radius, and thus their motion had little effect on the shape of the abundance gradient over time.

We continued this work several years ago with some exploratory calculations by Finlay *et al.* (1995). This work included two more clusters for a total of 7 old clusters but, unlike previous studies, included the effect of uncertainties in the observed parameters (proper motion, distance, and radial velocity). Finlay also performed a careful comparison between open and globular cluster populations. He found that the open clusters generally show limited excursions in galactocentric radius, and stay in the outer disk, outside 8 kpc. As Table I shows, most clusters have low orbital eccentricities, of about 0.1. However, the effect of observational uncertainty is appreciable for some clusters, making it impossible to extrapolate back to a place of formation with any confidence.

The old clusters generally stay in the outer disk, where we now see them, thereby avoiding the giant molecular clouds in the inner disk. However, there are several puzzles that remain. The clusters also experience many passages through the disk, so we still must ask how they can survive for so long. In addition, the sample now contains a very unusual cluster – NGC 6791, with an orbital eccentricity of 0.3, similar to that of disk globular clusters. At perigalacticon, NGC 6791 approaches within 4.6 kpc of the galactic center, inhabiting a region of the Galaxy where we currently see no open clusters of similar large age.

E.D. FRIEL

TABLE I

Orbital parameters for selected old open clusters

Cluster	Age (Gyr)	R_{gc}(peri) (kpc)	eccen	P(R) (Myr)	P(z) (Myr)	No. of Z passages
NGC 752	2	8.8	0.10	200 ± 5	85 ± 5	50
NGC 2158	3	12.4	0.13	300 ± 4	150 ± 60	40
NGC 2420	4	9.7	0.12	230 ± 20	150 ± 30	50
NGC 2506	5	10.8	0.11	380 ± 140	150 ± 40	70
M 67	5	7.8	0.08	170 ± 10	90 ± 30	110
NGC 188	8	9.2	0.13	220 ± 10	140 ± 20	120
NGC 6791	9	4.6	0.33	420 ± 120	130 ± 30	140

To explain the survivability of an ancient cluster like NGC 6791, other effects must come into play. A natural guess might be the mass of the cluster, since NGC 6791 is among the most massive of the open clusters known today.

We can turn to N-body simulations, to quantify the dependence of survivability on mass, and to look at how the various effects of orbital location, mass, and internal structure are related in controlling the survivability of the cluster.

Previous N-body studies of open clusters focused on understanding the dynamical evolution of clusters on circular orbits in the disk at the solar radius (Terlevich, 1987; de la Fuente Marcos, 1995, 1996, 1997). Working from Finlay's orbital calculations, Hsu *et al.* (1996) took a new approach: to perform N-body simulations for individual clusters on their known orbits. She found, not surprisingly, that the clusters undergo a significant amount of mass loss over their lifetimes. For NGC 6791, in a calculation lasting for just 1 Gyr, the cluster lost roughly 50% of its mass. This cluster has an age of roughly 9 Gyr, so at this rate of dynamical mass loss, and at a current day estimated mass of at least 4000 $M_\odot$ (Kaluzny and Udalski, 1992), we can extrapolate to an initial mass of close to 10^5 to 10^6 $M_\odot$. The older clusters are not only remnants of a larger initial population, but are shades of their former selves. NGC 6791 has survived long enough for us to see it, even in the inner disk, because it was, and still is, so massive. A look at our Galaxy 10 Gyr ago may well have revealed massive, young clusters, almost as massive as those seen now in nearby galaxies. The present day open cluster system reflects the effect of the environment on the intrinsic mass distribution of the cluster population.

Combining Hsu *et al.*'s results with those of Terlevich and de la Fuente Marcos, we can make some general observations. Small, poorly populated clusters dissolve primarily from internal dynamical effects on the order of a few 10^8 years, and explain the majority of the open cluster population. For these clusters, we see no trend in their rate of dissolution with location in the Galaxy. For richer, more massive clusters, the galactic tidal effects become significant, moderated by the cluster mass. Intermediate mass clusters ($\sim$ 500 to 1000 $M_{\odot}$) can survive several billion years, if they reside in the outer Galactic disk (e.g. NGC 752, NGC 3680). Clusters that survive for more than a few Gyr must be either very much more massive than 'normal' (e.g. NGC 6791) or must stay exclusively in the outer disk (e.g. NGC 188), or both.

This preliminary N-body work could profitably be pursued further to better define the relative importance of these effects. We also need proper motion studies of more clusters, particularly the oldest ones and those whose radial velocities hint at unusual behavior. I understand that some new proper motion work on particularly interesting old clusters is underway by Majewski and Phelps (Phelps, 1998).

5. Nucleosynthesis in the Galactic Disk

The open clusters have long been used as tracers of abundance gradients in the disk. Since the early work by Janes in the 1970's, others have found general agreement in the existence and magnitude of the trend. Samples have increased to 40 to 60 objects, over a wide range of ages and distances, and have used a variety of methods for abundance determinations. Most investigators have found gradients of -0.06 to -0.09 dex kpc^{-1} over a range of distances from 7 to 16 kpc from the Galactic center (Friel, 1995).

However, a recent analysis by Twarog *et al.* (1997) questioned this interpretation and raised some interesting issues. Their sample of 63 clusters spans all ages in the disk with abundances coming primarily from DDO photometry, but supplemented by spectroscopic values from Friel and Janes (1993), transformed to the DDO metallicity scale. Their analysis indicates no strong evidence for a linear relationship, but instead, strong statistical evidence for an abrupt discontinuity in the abundance profile in the disk at 10 kpc.

We have, in the meantime, been working on a larger sample of spectroscopic data for old clusters, and revising (and improving) the metallicity calibration that was used in our previous study (Friel and Janes, 1993; Friel and Tavarez, 1998). The first results from that larger sample and new, improved calibration are shown in Figure 1.

This sample of old clusters shows no sign of an abrupt discontinuity in the abundance distribution at 10 kpc. Our sample contains only clusters Hyades-aged or older. It is based on moderate resolution spectroscopic data that helps us eliminate non-members from the samples. A formal linear regression to these data yields a slope of -0.06 ± 0.01 dex kpc^{-1}, which is shallower than we have found previously, but within the range of values found by other investigators. The shallower slope is due to changes in the abundance scale adopted, primarily by using more recent high-resolution abundance determinations for metallicity calibrators. The difference with the Twarog et al result arises from differences in the placement of several key clusters in this figure. This sensitivity to the parameters for a small

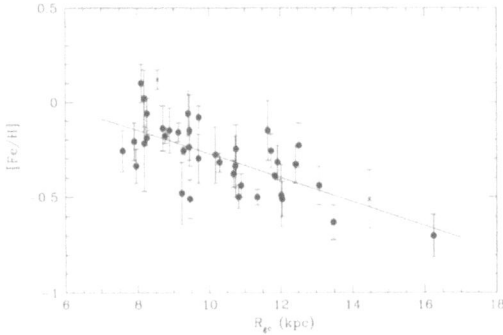

Figure 1. Radial abundance gradient.

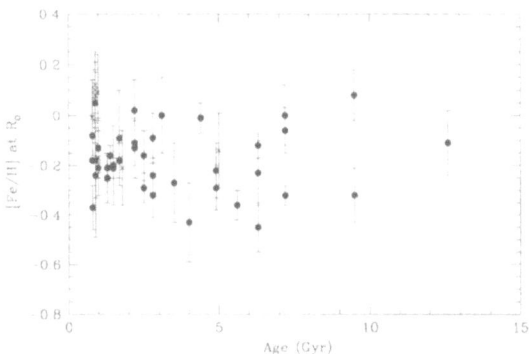

Figure 2. Age-metallicity relationship

number of clusters suggests we should be cautious about how we interpret the fine details of the abundance trend with position in the Galaxy.

As in previous studies, this general trend shows no dependence on age, indicating that the abundance 'gradient' as traced by the open clusters has not evolved significantly over the lifetime of the disk. A plot of mean cluster abundance as a function of age (Figure 2) shows no dependence. Abundances plotted here have been normalized for the gradient found earlier, to give the metallicities the cluster would have at the solar position. The dispersion in this diagram is about 0.15 dex, and appears to be independent of age, though the number of clusters at the oldest ages is still limited.

This basic result, that there is no apparent relationship between the cluster age and its metallicity, has been clear from many studies. It is also consistent with the results from recent field star studies (Nordstrom and Andersen, this volume).

It also indicates that there had to be prompt initial enrichment of the disk, at least locally, where these clusters formed. It lends support to, and calls for theories of galactic chemical evolution that include infall of un-enriched material, and/or radial flows, or the sequential stellar enrichment proposed by Van den Hoek and De Jong (1997).

To make progress now, we must move beyond the low resolution data that yield these global trends to detailed abundance analyses for the oldest of the disk clusters. This work is essential to confirm the relatively high metallicities of the oldest open clusters, and, with the determination of the [O/Fe] and [α/Fe] ratios, to ensure that the ages of the open and globular clusters are on the same scale, and to look at the evolution of the oxygen, α- and r-process elements in the early disk.

Some of this work has begun (Brown *et al.*, 1996; Tautvaisiene *et al.*, this volume). A recent paper by Peterson and Green (1998), finds a metallicity of [Fe/H] = +0.4 dex for NGC 6791 from high resolution data of a blue horizontal branch star in the cluster. L. Fullton is leading an effort to determine detailed abundance profiles for the very oldest open clusters and key disk globular clusters (Barrett *et al.*, 1996; Kramer *et al.*, 1996).

6. Conclusions and Future Work

In this short space I have outlined only some of the highlights of efforts at using open clusters as probes of the Galactic disk. These clusters hold great promise, and as we look forward to the availability of new telescopes and instruments, we can hope for further insights into the fossils of the old open clusters. To take advantage of these new facilities, we will need to continue our efforts to identify more old clusters, and to refine our determinations of their basic properties. Some fundamental data remain to be collected or measured, such as cluster proper motions. Clusters remain our best probes of many aspects of stellar dynamics, and we can look toward increasingly complete surveys of radial velocities, which, when combined with improved theoretical investigations and N-body simulations, can further elucidate the relative importance of dynamical and tidal effects on the cluster population. Finally, we can expect high-resolution spectroscopic work to provide us with the detailed understanding of nucleosynthesis and processing in the early disk to move beyond the rough global patterns we have found thus far. We have much to look forward to.

Acknowledgements

Giusa and Roger Cayrel have, in many ways, been my astronomical 'parents'. They sponsored me while I worked with them in Paris, they have included me in a broad range of their remarkable research, and they have generously shared data, resources, experience, and, most importantly, their unique scientific wisdom. This simple paper, given as part of a celebration of their extraordinary careers, seems a small token with which to thank them. But I do so sincerely. I must also thank the students whose work I featured in this contribution – Jarod Finlay, Lauren Hsu, Elizabeth Barrett, and Dan Kramer – and my colleague Maritza Tavarez, whose work on open clusters has been a source of pleasure and stimulation over the years.

References

Barrett, E., Kramer, D., Friel, E.D., Fullton, L. and Balachandran, S.: 1996, *Bull. Am. Astron. Soc.* **28**, 1367.

Brown, J., Wallerstein, G., Geisler, D. and Oke, J.B.: 1996, *Astron. J.* **112**, 1551.

Carraro, G. and Chiosi, C.: 1994, *Astron. Astrophys.* **288**, 751.

Finlay, J., Noriega-Crespo, A., Friel, E.D. and Cudworth, K.: 1995, *Bull. Am. Astron. Soc.* **27**, 1437.

Friel, E.D.: 1995, *Annu. Rev. Astron. Astrophys.* **33**, 381.

Friel, E.D. and Janes, K.A.: 1993, *Astron. Astrophys.* **267**, 75.

Friel, E.D. and Tavarez, M.: 1998, in preparation.

de la Fuente Marcos, R.: 1995, *Astron. Astrophys.* **301**, 407.

de la Fuente Marcos, R.: 1996, *Astron. Astrophys.* **308**, 141.

de la Fuente Marcos, R.: 1997, *Astron. Astrophys.* **322**, 764.

Hsu, L., Noriega-Crespo, A. and Friel, E.D.: 1996, *Bull. Am. Astron. Soc.* **28**, 1366.

Kaluzny, J. and Udalski, A.: 1992, *Acta. Astron.* **42**, 29 .

Janes, K.A. and Phelps, R.L.: 1994, *Astron. J.* **108**, 1773.

Kramer, D., Barrett, E., Friel, E.D., Fullton, L. and Balachandran, S.: 1996, *Bull. Am. Astron. Soc.* **28**, 1363.

Peterson, R. and Green, E.: 1998, *Astrophys. J. Lett.*, in press.

Phelps, L.: 1998, private communication.

Phelps, R.L., Janes, K.A. and Montgomery, K.A.: 1994, *Astron. J.* **107**, 1079.

Scott, J.E., Friel, E.D. and Janes, K.A.: 1995, *Astron. J.* **109**, 1706.

Terlevich, E.: 1987, *Mon. Not. R. Astron. Soc.* **224**, 193.

Twarog, B., Ashman, K.M. and Anthony-Twarog, B.: 1997, *Astron. J.* **114**, 2556.

van den Bergh, S. and McClure, R.D.: 1980, *Astron. Astrophys.* **80**, 360.

van den Hoek, L.B. and de Jong, T.: 1997, *Astron. Astrophys.* **318**, 231.

SEQUENCES OF NEARBY OPEN CLUSTERS WITH HIPPARCOS

N. ROBICHON

Sterrewacht Leiden, Leiden, The Netherlands

Y. LEBRETON and F. ARENOU

DASGAL, CNRS URA 335, Observatoire de Meudon, Meudon Cedex, France

Accurate mean Hipparcos parallaxes of the Pleiades, Praesepe and Coma Ber have been used to compare their observationnal main sequences with theoretical ZAMS in the colour-magnitude diagram. The aim of this study is to test the agreement between Hipparcos parallaxes and the models, the observed metallicities and the relations $(T_{eff}, B - V)$ and (T_{eff}, BC).

Cluster	$m - M$	σ_{m-M}	[M/H]	$\sigma_{[Fe/H]}$	#	$E(B - V)$
Coma	4.70	0.04	−0.048	0.012	20	0.00
Pleiades	5.36	0.06	−0.112	0.025	62	0.04
Praesepe	6.28	0.12	+0.170	0.010	94	0.00

Cluster mean parameters are given in the table above. Distance moduli are derived from mean cluster parallaxes computed using Hipparcos intermediate data (Robichon *et al.*, 1999). The global metallicities $[M/H]$ have been computed by Grenon (1998) from Geneva photometry of single stars of spectral type in the range F4-K3. The number # of stars used are also indicated in the table. Reddenings are taken from the Lyngå catalogue.

Internal structure models are from Lebreton (1999), with $\Delta Y / \Delta Z = 2.2$. $(T_{eff}, B - V)$ empirical calibrations from Alonso et al. (1996) and bolometric corrections from theoretical grids of Bessel *et al.* (1998) are used to transform the theoretical tracks into the $((B - V)_0, M_V)$ plane. Comparisons has been carried out with other empirical or theoretical calibrations. Discrepancies exist in the $B - V$ relations which can reach 0.05 mag, while bolometric correction calibrations are in good agreement within 0.05 mag.

The theoretical ZAMS of the three clusters are superposed, in the figure above, with the cluster members extracted from the 'Base Des Amas' (Mermilliod, 1995). The observed sequences of Praesepe and the Pleiades are in good agreement with theoretical ZAMS in the range $0.4 < B - V < 0.6$ within the error bars of the Hipparcos distance moduli. In the range $0.6 < B - V < 0.8$, the observed sequences are systematicaly 0.1–0.15 magnitude brighter. This could be a signature of problems in the $(T_{eff}, B - V)$ or (T_{eff}, BC) relations or in the adopted atmosphere models adopted in the theoretical ZAMS in this range of temperature.

 Astrophysics and Space Science is the original source of publication of this article. It is recommended that this article is cited as: *Astrophysics and Space Science* **265**: 279–280, 1999.

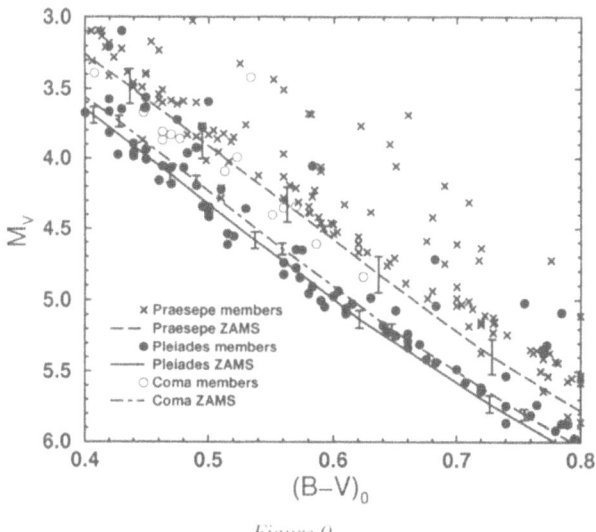

Figure 0.

The sequences of Coma Berenices differ by about 0.15 mag in M_V at a given $B-V$. Such a difference cannot be explained only by the error bar on the Hipparcos distance modulus (0.04 mag). Without drastically decreasing the helium content, the most likely explanation would be a true metallicity 0.1 dex higher than the value given by Grenon (1998).

Looking only at these three clusters, the positions of their main sequences in the HR diagram do not completely agree with their Hipparcos distance moduli. In order to better understand the source of the discrepancies observed, more constraints are needed and the study is to be extended to the twenty or so clusters closer than 500 pc with a known Hipparcos distance (Robichon *et al.*, 1998).

References

Alonso, A., Arribas, S. and Martinez-Roger, C.: 1996, *Astron. Astrophys.* **331**, 873.
Bessell, M.S., Castelli, F. and Plez, B.: 1998, *Astron. Astrophys.* **333**, 231.
Grenon, M.: 1998, private communication.
Lebreton: 1999, in preparation.
Mermilliod, J.C.: 1995, in: D. Egret and M.A. Albrecht (eds.), *Information and On-Line Data in Astronomy*, Kluwer, p. 127.
Robichon, N., Arenou, F., Mermilliod, J.-C. and Turon, C.: 1999, *Astron. Astrophys.*, submitted.

OLD OPEN CLUSTERS: CONSTRAINTS ON DISK AGE AND EVOLUTION

A. BRAGAGLIA and M. TOSI

Osservatorio Astronomico di Bologna, via Ranzani 1, I-40127 Bologna, Italy

G. MARCONI

Osservatorio Astronomico di Roma, Via Osservatorio 5, I-00040 Monte Porzio, Roma, Italy

1. Why Open Clusters?

Open clusters are found in different regions of the disk, cover a large interval in age (a few Myr to $\sim$ 10 Gyr), and in metallicity ([Fe/H] $\geq$ -1.0); moreover, their ages and distances can be measured more accurately than for any single field star. Hence, they offer information on the age of the Galactic disk (on relative and absolute ages, on the possible time continuity between halo and disk, etc.), and on the evolution of the disk metallicity (both in space and time).

Any attempt to define reliable age or metallicity rankings must be based on the study of a large sample of clusters homogeneously treated (see e.g. Janes and Phelps, 1994; Carraro and Chiosi, 1994; Friel, 1995). To this end we have started to analyze a sample of clusters at various Galactic locations, and covering a large range in ages and metallicities. We obtain metallicity, reddening, distance, and age from photometric observations (using the synthetic colour-magnitude method, see e.g. Tosi *et al.*, 1991).

We use a numerical code for CMD simulation based on stellar evolutionary tracks (taken from different groups, to study the effects of the adoption of different models), and taking into account theoretical and observational uncertainties (e.g. photometric errors, incompleteness, binary fraction). We derive the best combination of metallicity, reddening, distance, and age by comparing the observed and the synthetic CMDs, on the basis of: a) morphology (e.g. shape of the MS, SGB, RGB; position of the red clump and of the TO; gaps), and b) population ratios (distribution in colour and magnitudes, e.g. the LFs).

We have already analyzed several clusters with ages from about 0.1 to about 10 Gyr: NGC7790 (Romeo *et al.*, 1989), NGC2506 (Marconi *et al.*, 1997), Be 21 (Tosi *et al.*, 1998), NGC6253 (Bragaglia *et al.*, 1997), NGC2243 (Bonifazi *et al.*, 1990), Cr261 (Gozzoli *et al.*, 1996). Work is in progress for Be 22, To 2, Mel 71, NGC2660, NGC2849 (all intermediate-old), and NGC6603 (young).

 Astrophysics and Space Science is the original source of publication of this article. It is recommended that this article is cited as: *Astrophysics and Space Science* **265:** 281–282, 1999.
© 1999 *Kluwer Academic Publishers.*

2. Preliminary Conclusions

(a) Open clusters indicate that the disk age is larger than about 8–9 Gyr, and possibly reaches 12 Gyr. There are several clusters older than about 7 Gyr (e.g. Cr 261, NGC6791, NGC188), and Be 17 is $\simeq$ 12 Gyr old (Phelps, 1997; but see also Carraro *et al.*, 1998, who give for it only 9 Gyr).

(b) Results based on our sample are qualitatively consistent with the standard evolutionary scenarios of a negative radial abundance gradient and a positive time-metallicity relation.

Our sample is still too small to generalize to the whole cluster system: we intend to obtain new photometric data, and reanalyze data from literature. Moreover, more accurate metallicities are needed and we are planning high-resolution spectroscopy of selected old open clusters.

References

Bonifazi, A., Fusi Pecci, F., Romeo, G. and Tosi, M.: 1990, CCD photometry of Galactic open clusters. II – NGC 2243, *Mon. Not. R. Astron. Soc.* **245**, 15.

Bragaglia, A., Tessicini, G., Tosi, M., Marconi, G. and Munari, U.: 1997, UBVRI CCD photometry of the old open cluster NGC 6253, *Mon. Not. R. Astron. Soc.* **284**, 477.

Carraro, G. and Chiosi, C.: 1994, The Galactic system of old open clusters: age calibration and age-metallicity relation, *Astron. Astrophys.* **287**, 761.

Carraro, G., Ng, Y.K. and Portinari, L.: 1998, On the Galactic disc age-metallicity relation, *Mon. Not. R. Astron. Soc.* **296**, 1045.

Friel, E.D.: 1995, The Old Open Clusters Of The Milky Way, *Annu. Rev. Astron. Astrophys.* **33**, 381.

Gozzoli, E., Tosi, M., Marconi, G. and Bragaglia, A.: 1996, CCD photometry of the old open cluster Collinder 261, *Mon. Not. R. Astron. Soc.* **283**, 66 .

Janes, K.A. and Phelps, R.L.: 1994, The galactic system of old star clusters: The development of the galactic disk, *Astron. J.* **108**, 1773.

Marconi, G., Hamilton, D., Tosi, M. and Bragaglia, A.: 1997, Old open clusters: UBGVRI photometry of NGC 2506, *Mon. Not. R. Astron. Soc.* **291**, 763.

Phelps, R.L.: 1997, Berkeley 17: The Oldest Open Cluster?, *Astrophys. J.* **483**, 826.

Romeo, G., Bonifazi, A., Fusi Pecci, F. and Tosi, M.: 1989, CCD photometry of galactic open clusters. I – NGC 7790, *Mon. Not. R. Astron. Soc.* **240**, 459 .

Tosi, M., Greggio, L., Marconi, G. and Focardi, P.: 1991, Star formation in dwarf irregular galaxies – Sextans B, *Astron. J.* **102**, 951.

Tosi, M., Pulone, L., Marconi, G. and Bragaglia, A.: 1998, Old open clusters: the interesting case of Berkeley 21, *Mon. Not. R. Astron. Soc.* **299**, 834.

OLD OPEN CLUSTER AND THE AGE OF THE GALACTIC DISK

GIOVANNI CARRARO

Department of Astronomy, Padova University, Vicolo dell'Osservatorio 5, I-35122 Padova, Italy

In this paper we analyse the Color–Magnitude Diagrams (CMDs) of six very old open clusters (namely: NGC 188, NGC 6791, Collinder 261, Melotte 66, Berkeley 39 and Berkeley 17, Table 1) to determine carefully their age (Carraro *et al.*, 1998).

TABLE I

Derived properties of the open clusters sample

Cluster	Z	$E(B-V)$	$E(V-I)$	$(m-M)_o$	τ (Gyr)	$\Delta\tau$ (Gyr)
NGC 188	0.017	0.13	0.16	11.25	6	1
NGC 6791	0.030	0.13	0.17	13.00	9	1
Collinder 261	0.014	0.30	0.36	12.10	7	1
Melotte 66	0.006	0.20	0.25	13.30	4	1
Berkeley 39	0.010	0.18	0.22	12.95	6	1
Berkeley 17	0.008	0.65	0.75	12.15	8	1

We use updated photometry and spectroscopy, and study these CMDs by means of isochrone fitting for the exact cluster metallicity derived from Padova stellar evolutionary models (Girardi *et al.* 1998).

We find that the ages of these clusters are confined between 4 and 9 Gyr, Melotte 66 and NGC 6791 being the youngest and the oldest one, respectively. Berkeley 17, insofar believed the oldest open cluster (Phelps, 1997) has an age around 8 Gyr.

When using open clusters, the age of the Galactic Disk turns out to be not less than 9 Gyr, but surely lower than the Galactic Halo.

References

Carraro, G., Girardi, L. and Chiosi, C.: 1999, Is the Galactic Disk older than the Halo?, *Mon. Not. R. Astron. Soc.* **309**, 430.

Girardi, L., Bressan, A., Bertelli, G. and Chiosi, C.: 1999, in press.

Phelps, R.L.: 1997, Berkeley 17: the oldest open cluster?, *Astrophys. J.* **483**, 826.

Astrophysics and Space Science is the original source of publication of this article. It is recommended that this article is cited as: *Astrophysics and Space Science* **265**: 283–284, 1999.
© 1999 *Kluwer Academic Publishers.*

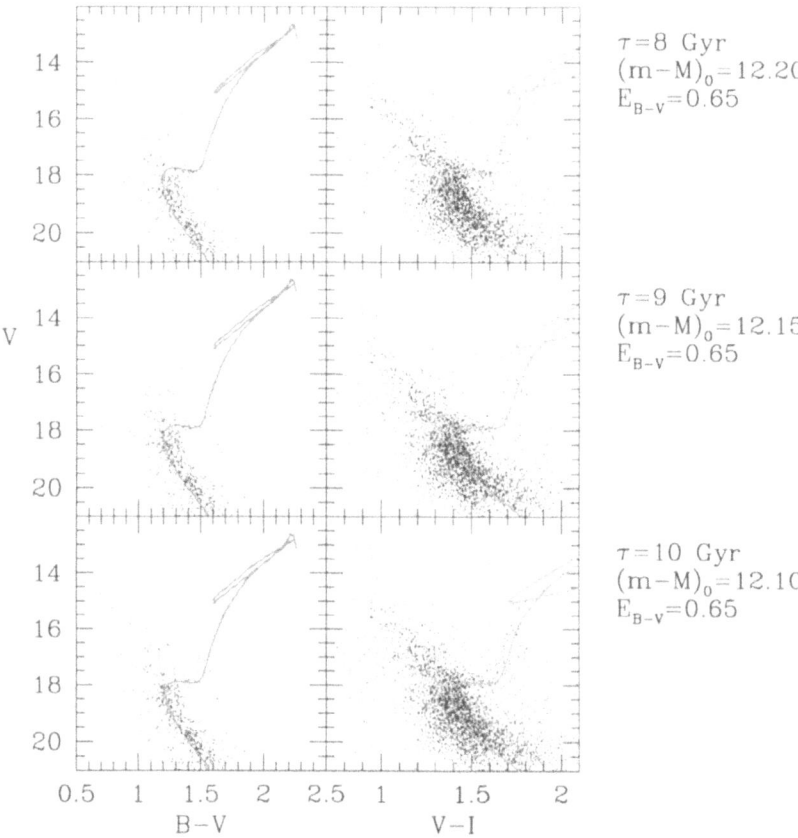

Figure 1. The CMD of Berkeley 17 from Phelps 1997. Overimposed are isochrones for the set of parameters listed on the left.

M 67: AN EXHIBITOR OF METAL ABUNDANT CORE HELIUM-BURNING STARS

G. TAUTVAIŠIENĖ

Institute of Theoretical Physics and Astronomy, Goštauto 12, Vilnius 2600, Lithuania

I. TUOMINEN and I. ILYIN

Astronomy Division, University of Oulu, P.O. Box 300, 90401 Oulu, Finland

1. Introduction

The old galactic open cluster M 67 is a cornerstone in understanding stellar evolution. An advantage that cluster members have to be identical, except of mass and evolutionary state, is most efficient for the analysis of changes in mixing-sensitive abundances. However, abundances of carbon and nitrogen, particularly sensitive indicators of stellar evolution, have been investigated from high-resolution spectra only in three giants of M 67 (Brown, 1985). Results obtained for 19 giants from moderate-resolution spectra (Brown, 1987) are suspected to have a systematic offset (c.f. Charbonnel *et al.*, 1998). High-resolution analyses are very scarce for the oxygen abundances: Griffin (1975) has measured the [O I] line at 6300 Å in one star, Cohen (1980) has analyzed four stars, but the weaker line of [O I] at 6363 Å was used. In the paper by Cohen (1980), a low value of [Fe/H] = –0.39 for M 67 has been received. This is the same paper that gave a rather low [Fe/H] value for M 71, which was later increased by 0.5 dex. In the same paper by Cohen (1980), abundances of some other elements in M 67 look very extraordinary. The ratios of [Mg/Fe] reach –0.8 dex at the same time [Si/Fe] being enhanced by about 0.6 dex, [Ba/Fe] are approximately equal to –0.4 dex while for a very similar element lanthanum [La/Fe] ≈ +0.6 dex.

2. Observations and Analysis

High-resolution spectra of six core He-burning 'clump' stars and three giants in M 67 were observed on the Nordic Optical Telescope with the SOFIN échelle spectrograph (Tuominen *et al.*, 1992). The 2nd optical camera ($R \approx 60\,000$) and the 3rd optical camera ($R \approx 30\,000$) were used. All spectra were exposed to a $S/N \geq 100$. Reductions of the CCD images were made with the *3A* software package (Ilyin 1995).

Astrophysics and Space Science is the original source of publication of this article. It is recommended that this article is cited as: *Astrophysics and Space Science* **265:** 285–286, 1999.
© 1999 *Kluwer Academic Publishers.*

The spectra were analyzed using a differential model atmosphere technique. The *Eqwidth* and *Synthetic Spectrum* program packages, kindly made available from the Uppsala Astronomical Observatory, were used to calculate equivalent widths of lines and synthetic spectra. A net of model atmospheres has been computed with an updated version of the *MARCS* and was supplied by Bengt Edvardsson, Uppsala.

The forbidden lines of [C I] at 8727 Å and [O I] at 6300 Å were used for the analysis.

3. Results

Abundances of 25 chemical elements were determined in atmospheres of six clump stars and three giants in M 67. The main results of our detailed analysis are the following. The mean [Fe/H] = −0.03 ± 0.01. Relative abundances of oxygen, α-process elements and other heavy elements to iron are very homogeneous and consistent with those observed in dwarfs of the galactic disk, carbon being by about 0.1 dex underabundant and nitrogen and sodium being by about 0.2 dex overabundant. In comparison to the red giant branch stars in our sample, the [C/N] ratio of the clump stars as a group is reduced by 0.13 dex and the [Na/Fe] ratio is increased by 0.10 dex. These results indicate that an extra-mixing may occur in the core He-burning stars which produces an additional alteration of the surface C, N and Na abundances.

References

Brown, J.: 1985, *Astrophys. J.* **297**, 233.
Brown, J.: 1987, *Astrophys. J.* **317**, 701.
Charbonnel, C., Brown, J.A. and Wallerstein, G.: 1989, *Astron. Astrophys.* **332**, 204.
Cohen, J.G.: 1980, *Astrophys. J.* **241**, 981.
Ilyin, I.V.: 1995, *Acquisition, Archiving and Analysis (3A) Software Package and User's Manual*, Observatory, University of Helsinki.
Griffin, R.: 1975, *Mon. Not. R. Astron. Soc.* **171**, 181.
Tuominen, I.: 1992, *NOT News* **5**, 15.

LINKING FIELD METAL-POOR STARS AND GLOBULAR CLUSTERS: THE EASE SCENARIO

E. JEHIN, P. MAGAIN, C. NEUFORGE, A. NOELS, G. PARMENTIER and A. THOUL
Institut d'Astrophysique et de Géophysique, Université de Liège, 5 avenue de Cointe, 4000 Liège, Belgium

1. Introduction

We have analysed high resolution and high signal-to-noise spectra of 21 mildly metal-poor stars ([Fe/H] ~ -1) (Jehin *et al.*, 1998, 1999). We studied the correlations between the relative abundances of 16 elements, with a special emphasis on the neutron-capture ones. This analysis reveals the existence of two sub-populations of field metal-poor stars, namely Pop IIa and Pop IIb, which differ by the behaviour of the s-process elements versus the α and r-process elements. We suggest a scenario for the formation of these metal-poor stars, which relates the origin of these stars to the evolution of globular clusters. According to this scenario, thick disk and field halo stars were born in globular clusters from which they escaped, either during an early disruption of the cluster (Pop IIa) or through a later disruption or an evaporation process (Pop IIb).

2. Observations

When looking at correlation diagrams for typical r-process elements versus α elements* we observe a one-to-one correlation between them. All points are located on a single straight line with a slope of about 1, ending with a clumping at the maximum value of [α/Fe]. On the other hand, the correlation between an s-process element and an α-element shows two distinct behaviours. About half of the stars show a correlation with a slope smaller than 1, whereas the other stars exhibit a constant (and maximum) value for [α/Fe] and increasing values of [s/Fe], starting at the maximum value reached by the first group of stars. When we add points coming from stars in other studies (Zhao and Magain, 1991; Edvardsson *et al.*, 1993; Nissen and Schuster, 1997), we observe a change in the behaviour of the correlation of *s* versus α elements at a metallicity [Fe/H] ≈ -0.6. Low-metallicity

---

* Due to a lack of space in these proceedings, we refer the reader to our paper (Jehin *et al.*, 1999) for the figures.

Astrophysics and Space Science is the original source of publication of this article. It is recommended that this article is cited as: *Astrophysics and Space Science* **265**: 287–288, 1999.
© 1999 *Kluwer Academic Publishers.*

stars follow the relation described above (a 'two-branches diagram'). However, this relation does not apply to higher-metallicity stars, which scatter in the upper-left corner of the diagram.

3. The EASE Scenario

We suggest a scenario for the formation and origin of the metal-poor stars, which provides an explanation for the relation described by the two-branches diagram. This scenario provides a link between the field halo stars (FHS) and globular clusters (GCs):

(1) FHS were born in (proto-) GCs;

(2) GCs have undergone a chemical evolution; their evolution can be separated in two distinct phases;

(3) during the first phase, massive (first-generation) stars in the proto-GC cloud evolve, ending their lives as supernovae and ejecting α elements and r-process elements into the ISM. A second generation of stars forms out of this enriched ISM. The proto-GC can become unstable and get disrupted, the lower-mass stars forming Pop IIa;

(4) during the second phase, if the GC has survived, intermediate mass stars reach the AGB, ejecting s elements into the ISM through stellar winds or superwinds events; the matter released in the ISM by the AGB stars is accreted by the lower mass stars, enriching those stars in s elements;

(5) some of these low mass stars evaporate from the GCs or get dispersed in the halo when the GC is disrupted; they form Pop IIb FHS.

Acknowledgements

This work has been supported by the Pole d'Attraction Interuniversitaire p4/05 (SSTC, Belgium).

References

Edvardsson, B., Andersen, J., Gustafsson, B., Lambert, D.L., Nissen, P.E. and Tomkin, J.: 1993, *Astron. Astrophys.* **275**, 101.

Jehin, E., Magain, P., Neuforge, C., Noels, A. and Thoul, A.A.: 1998, *Astron. Astrophys.* **330**, L33.

Jehin, E., Magain, P., Neuforge, C., Noels, A., Parmentier, G. and Thoul, A.: 1999, *Astron. Astrophys.* **341**, 241.

Nissen, P.E. and Schuster, W.J.: 1997, *Astron. Astrophys.* **326**, 751.

Zhao, G. and Magain, P.: 1991, *Astron. Astrophys.* **244**, 425.

STELLAR POPULATIONS IN THE NEAR INFRARED

RICARDO P. SCHIAVON

Departamento de Astronomia, CNPq/Observatório Nacional, Rua General José Cristino, 77, Rio de Janeiro 20921-400, Brazil

1. Introduction

A new grid of high resolution synthetic stellar spectra for population synthesis is presented. The spectra were computed in the $\lambda\lambda 6000 - 10200$ Å interval, employing an updated and comprehensive list of atomic and molecular lines (for details, see Schiavon and Barbuy, 1999). The behaviour of spectral indices contained in the spectral region under study, such as the NaI 'doublet' (~ 8190 Å), the FeH Wing-Ford band (~ 9900 Å) and TiO bands, as a function of stellar atmospheric parameters has been studied by Schiavon *et al.* (1997a), Schiavon *et al.* (1997b) and Schiavon and Barbuy (1999), respectively. In this contribution the first results of the study of integrated synthetic spectra of single-aged stellar populations (SSPs) are presented.

2. Age, Metallicity and IMF Effects on Integrated Spectra

In Figure 1, integrated synthetic spectra are shown as a function of the fundamental parameters of SSPs. These spectra were computed by combining the isochrones of Bertelli *et al.* (1994), Baraffe *et al.* (1998) and the T_{eff}-scale from Lejeune *et al.* (1998). Metallicity is the leading parameter in shaping the integrated spectra. The effects due to age are weaker. Changes in IMF have an important influence only on a few gravity sensitive lines.

The NaI index defined by Faber and French (1980) has its sensitivity to metallicity influenced by a TiO bandhead on the red continuum window of the index (for details, see Schiavon, Barbuy and Bruzual, 1999, in preparation). It has been proposed by many authors (Delisle and Hardy, 1992 and references therein) that the observed gradient of the NaI index towards galaxy centers can be explained by the corresponding well known metallicity gradient. Our results suggest that IMF effects can contribute significantly to the observed gradients. More detailed observations are required in order to tackle this problem.

Astrophysics and Space Science is the original source of publication of this article. It is recommended that this article is cited as: *Astrophysics and Space Science* **265:** 289–290, 1999.
© 1999 *Kluwer Academic Publishers.*

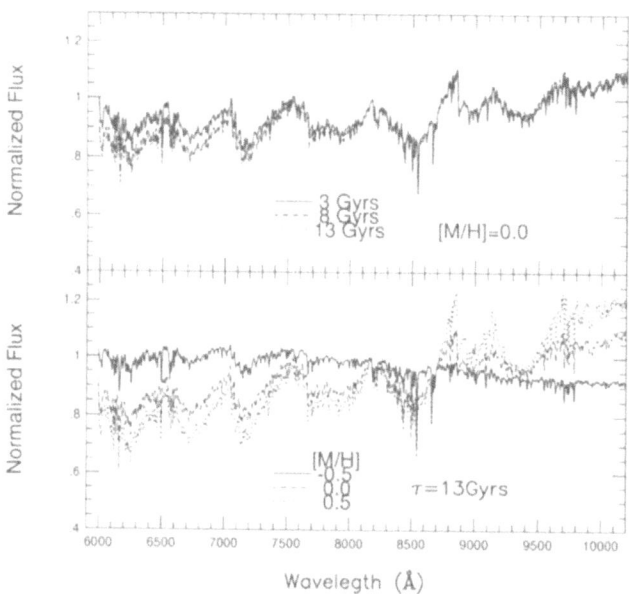

Figure 1. Integrated synthetic spectra of SSPs computed for a set of ages and metallicities, employing a Salpeter IMF.

The CaII triplet ($\lambda\lambda$8498, 8542, 8662 Å), although being a strong stellar gravity indicator, is not sensitive to the contribution of low-mass stars, being IMF-insensitive. Most of the metallicity-sensitive indices are weakly sensitive to age. Thus, red-NIR indices can help solving the age-metallicity degeneracy. A full account of the response of spectral indices age, metallicity and IMF variations will be given elsewhere (Schiavon, Barbuy and Bruzual, 1999, in preparation). The connection of our spectral library with the evolutionary population synthesis code of Bruzual and Charlot (1999, in preparation) is currently under way.

References

Baraffe, I., Chabrier, G., Allard, F. and Hauschildt, P.H.: 1998, *Astron. Astrophys.* **337**, 403.
Bertelli, G., Bressan, A., Chiosi, C., Fagotto, F. and Nasi, E.: 1994, *Astron. Astrophys. Suppl.* **106**, 275.
Delisle, S. and Hardy, E.: 1992, *Astron. J.* **103**, 711.
Faber, S.M. and French, H.B.: 1980, *Astrophys. J.* **235**, 405.
Schiavon, R.P., Barbuy, B., Rossi, S.C.F. and Milone, A.: 1997a, *Astrophys. J.* **479**, 902.
Schiavon, R.P., Barbuy, B. and Singh, P.D.: 1997b, *Astrophys. J.* **484**, 499.
Schiavon, R.P. and Barbuy, B.: 1999, *Astrophys. J.*, in press.

CARBON STARS IN NEARBY LOCAL GROUP GALAXIES

M. AZZOPARDI

IGRAP, Observatoire de Marseille, 2, place Le Verrier, F-13248 Marseille Cedex 04, France

Abstract. Well pronounced molecular bands displayed by the spectra of carbon stars make their detection possible in relatively distant galaxies. However, so far, extensive surveys for this kind of object have only been made in the Galactic halo galaxies, mainly in the Magellanic Clouds and in the dwarf spheroidals. We review the carbon star surveys of these systems with special emphasis on low luminosity carbon stars that have been found in the Small Magellanic Cloud and recently in the Fornax dwarf galaxy.

1. Introduction

Asymptotic Giant Branch (AGB) stars are valuable objects for the study of the morphology, stellar evolution, and kinematics of galaxies. In addition, the ratio of the number of carbon (C) stars to the number of late-type M stars, and possibly the faint end of the C star luminosity function, may provide information on the metallicity of the parent system. The presence of pronounced molecular bands in the envelopes and the atmospheres of the AGB stars have made their photometric detection possible as far away as NGC 2403 (Hudon *et al.*, 1989).

The excess carbon with regard to oxygen (C/O > 1) in C stars allows their detection through their CN and Swan C_2 bands. Due to their relative proximity, the Magellanic Clouds and the Galactic dwarf spheroidals have, for a long time, been the subject of extensive slitless spectroscopic surveys for C stars. Note that papers on C star populations in nearby galaxies have recently been published by the author in the proceedings of the IAU Symposia No. 177 (The Carbon Star Phenomenon), No. 190 (New Views of the Magellanic Clouds) and No. 192 (The Stellar Content of Local Group Galaxies). Hence, this review paper has to be understood as an up to date compilation of the previous contributions.

2. AGB Star Survey Methods

Slitless spectroscopic surveys for C and late M-type stars have been successfully carried out in the Magellanic Clouds and the Galactic dwarf spheroidal galaxies. Photographic plates obtained either with Schmidt telescopes equipped with low-dispersion objective prisms or at the prime focus of 4-m class telescopes equipped

 Astrophysics and Space Science is the original source of publication of this article. It is recommended that this article is cited as: *Astrophysics and Space Science* **265**: 291–299, 1999.
© 1999 *Kluwer Academic Publishers.*

with wide-field correctors and very low dispersion transmission gratings (grism or grens) have been employed in extensive surveys of these systems. Occasionally, suitable intermediate to broad bandpass filters have been used to restrict the instrumental spectral range and minimize the number of overlaps. At present, the same observing technique, using CCD cameras behind focal reducers at the Cassegrain focus of 4-m class telescopes, is carried out in an effort to identify fainter C stars that have previously eluded photographic surveys.

In systems more distant than the Galactic halo dwarf spheroidals, transmission grating surveys are ineffective and an imaging technique must be used for C star surveys. By imaging with intermediate-band filters that monitor CN and TiO band strength, C stars are easily identified and distinguished from M-stars. As the imaging technique does not disperse the light, it works well in crowded fields and goes to fainter magnitudes. Based on the earlier photometric system of Palmer and Wing (1982), intermediate-band photometric systems have been defined by Richer *et al.* (1985) and Cook and Aaronson (1989) and used to survey NGC 205 (Richer *et al.*, 1984) as well as M 33, NGC 6822, IC 1613 and WLM (Cook *et al.*, 1986). Subsequent spectroscopy has demonstrated that the imaging technique identifies C stars with a high degree of success (for example, see the study of M 31 by Brewer *et al.* (1996) and papers quoted therein).

In transmission grating surveys C stars are identified mainly through their near-infrared CN-band blends at 7945, 8125 and 8320 Å. However, they can also be found by means of their Swan C_2 bands at 4737, 5165 and 5636 Å in the blue-green spectral domain, a detection technique pioneered by Sanduleak and Philip (1977). As shown by McCarthy (1987), comparing the independent C star surveys carried out in selected Small Magellanic Cloud (SMC) fields by Blanco *et al.* (1980) and by Westerlund *et al.* (1986), Westerlund and associates' technique is more sensitive to the bluer C stars (including R-, CH-, BaII- and J-type stars) than the near-infrared technique which favours detection of the redder and cooler classical N-type C stars. Hence, in the visible domain, these two surveys techniques are complementary. However, both techniques miss luminous C stars with strong stellar winds responsible for high mass-loss rates, which can only be detected at infrared or radio wavelengths and consequently, are dedicated targets for surveys such as 2MASS and DENIS.

3. Carbon Stars in the Magellanic Clouds

The readily recognized spectral features described above, and the proximity of the Magellanic Clouds (MCs) have contributed significantly to the success of C star surveys in these galaxies for the past four decades.

3.1. CARBON STARS IN THE MAIN BODY OF THE MCS

The discovery of nine C star candidates in the Large Magellanic Cloud (LMC) was first reported by Westerlund (1960). An extension of the photometric (R, I) and near-infrared objective-prism surveys ($\approx$ 2100 Å mm^{-1} dispersion at the A-band) ($I > 13$) to the major part of the LMC (62.5 sq. deg.) led to the identification of an additional 302 C star candidates (Westerlund 1964; Westerlund *et al.* 1978). Meanwhile, the CTIO blue-green Curtis-Schmidt telescope objective-prism search (1360 Å mm^{-1} dispersion at Hγ) by Sanduleak and Philip (1977) of part of the area previously surveyed by Westerlund led to the detection of 474 LMC C star candidates of which 399 were newly discovered objects. This result shows clearly that the strength of the Swan C$_2$ bands more than compensates the loss of flux by C stars in the blue-green spectral domain. Subsequent spectroscopic and photometric studies by Crabtree *et al.* (1976) and Richer *et al.* (1979) of selected survey candidates confirmed their carbon star nature and gave information on the physical properties of LMC C stars.

Deeper ($I > 17$) near-infrared grism surveys (2350 Å mm^{-1} and 1700 Å mm^{-1} dispersions at the A-band) for C and late M-type stars in selected areas of the Magellanic Clouds (52 circular LMC fields of 0.12 deg^2 each, and 28 circular SMC fields of 0.12 deg^2 each plus 9 square fields of 0.38 deg^2 each) were carried out by Blanco *et al.* (1980) using a photographic camera equipped with a Schott RG 630 filter at the prime focus of the CTIO 4.0-m reflector. Their observations resulted in the detection of 1045 C star candidates in the LMC and 860 in the SMC in $\approx$ 6.2 and $\approx$ 6.8 sq. deg. total fields, respectively (Blanco *et al.*, 1980; Blanco and McCarthy, 1983, 1990). Star-count isopleths, based on the integration of the C star surface densities of the selected sample areas, led Blanco and McCarthy (1983) to estimate the total number of cool C stars as 11000 in the LMC and 2900 in the SMC. Also, study of the M-giant distribution led these authors to investigate the ratio of the frequencies of C stars and M giants i.e. (number of C stars)/(number of giants of type M6 or later) in the Magellanic Clouds. They found that this ratio is 2.2 ± 0.1 in the LMC and appears to be constant throughout the galaxy; for the SMC, the corresponding ratio is 19.2 ± 0.8 in the central regions and decreases continuously outwards to reach 4.7±0.4 in its periphery. From these results Blanco and McCarthy stated that if C/M is related to metallicity alone (and not age) no significant metallicity gradient exists in the LMC, while an increase of metallicity towards the SMC external regions is indicated at the time (3–5 Gyr ago) when the red-giant progenitors formed.

Later on, an extensive photographic slitless spectroscopic survey for field C stars in the SMC was achieved by us at the prime focus of the ESO 3.6-m telescope. We used the wide field corrector equipped with a blue-green grism (2200 Å mm^{-1} dispersion) in combination with a Schott GG 435 filter and IIIa-J plates in order to restrict the instrumental spectral range to a useful domain (4300–5300 Å) and minimize the number of spectrum overlaps. The main body of the SMC (except its western regions) was covered by 13 partially overlapping fields of 0.78 deg^2 each. Adopting the Swan C_2 bands at 4737 and 5165 Å for C star selection criteria we (Rebeirot *et al.*, 1993) found 1707 SMC C star candidates. The overall surface distribution of the SMC C stars shows that they belong to a spheroidal-like system, in fairly good agreement with the smoothed red light SMC isophots (see de Vaucouleurs and Freeman, 1972).

3.2. CARBON STARS IN THE OUTER HALOES OF THE MCS

Studies of the dynamical interactions between the components of the Galaxy-LMC-SMC system lead to a better understanding of the origin of the Magellanic Stream and of the morphological disturbances shown by the SMC and to a certain extent by the LMC. Also, the need for an accurate determination of the total mass of the Magellanic Clouds has driven searches for suitable kinematical tracers in the outer regions of these galaxies. The discovery of intrinsically bright CH stars in the outer halo of the LMC (Hartwick and Cowley, 1988; Feast and Whitelock, 1992) has emphasized the usefulness of the C-type stars for dynamical investigations. Consequently, extensive surveys for C stars in the outer haloes of the Magellanic Clouds, including the inter-Cloud region, and subsequent spectroscopy for kinematical studies have appeared in the past few years.

The UK Schmidt Telescope (UKST) objective-prism (800 Å mm^{-1} dispersion at $\approx$ 4300 Å) by Morgan and Hatzidimitriou (1995) of the outer regions of the SMC resulted in the detection of 1185 objects in an area of about 220 sq. deg. on the sky. When coupling this result with that obtained by Rebeirot *et al.* (1993) for the central parts, Morgan and Hatzidimitriou estimated that the total number of C stars in the SMC would be $\simeq$ 3060. Radial velocities of a selected sample of 71 C stars belonging to the outer regions of the SMC allowed Hatzidimitriou *et al.* (1997) to calculate the mean heliocentric velocity of these stars as 149.3 ± 3.0 km s^{-1} with a velocity dispersion of 25.2 ± 2.1 km s^{-1}. Subsequently, they determined the mass of the SMC as 1.2 10^9 $M\odot$.

A similar approach was used by Demers *et al.* (1993) and Kunkel *et al.* (1997a) to investigate the inter-Cloud region and the Magellanic Cloud haloes using photometric data obtained from UKST B_J and R plates of 19 fields. Their investigations resulted in the discovery of over one thousand C stars. Examination of the kinematics of the LMC periphery, using radial velocities of 759 C stars, led Kunkel *et al.* (1997b) to determine the LMC rotation curve from 3 to 12 kpc and the mass of this system (more than 90%), contained within a 5 kpc radius, to be 6.2 ± 0.9 10^9 $M\odot$.

4. Carbon Stars in the Galactic Dwarf Spheroidal Galaxies

The recent interest in looking for dwarf systems, satellites of more massive galaxies, has increased the number of known Galactic dwarf spheroidal galaxies to 9, namely Carina, Draco, Fornax, Leo I, Leo II, Sagittarius, Sculptor, Sextans and Ursa Minor. Intermediate systems between globular clusters and elliptical galaxies, their low surface brightness and lack of very luminous stars make them difficult to detect. Owing to the fact that their colour-magnitude diagrams generally display well-defined asymptotic giant branches (except for Sextans), as well as their proximity, have fostered searches for red giants in these galaxies. The present census shows that Fornax contains by far the largest number (104) of C stars among the Galactic halo dwarf spheroidal galaxies. However, note that Fornax has a comparable normalized C star number as a function of parent galaxy luminosity that the Magellanic Clouds and most of the other Galactic dwarf spheroidals (see Azzopardi *et al.*, 1999).

4.1. THE FORNAX SYSTEM

Spectroscopy of seven objects by Aaronson and Mould (1980), selected in the full colour range of the 66 very red stars discovered by Demers and Kunkel (1979) using B and V plates obtained with the CTIO 1.5-m telescope, revealed the presence of the first five C stars in the Fornax dwarf spheroidal. From these results, Aaronson and Mould claimed that 'for $B - V$ greater than ~ 2.1, the giant branch of Fornax may consist entirely of C stars'. As a matter of fact, the subsequent near-infrared grism searches for late-type stars, carried out by Blanco and McCarthy (Frogel *et al.*, 1982) at the prime focus of the CTIO 4-m reflector (one field having 23′ diameter) and by Westerlund (Richer and Westerlund, 1983) at the prime focus of the ESO 3.6-m telescope (three 16′ diameter fields), increased considerably the number of C star candidates known in this galaxy. The ESO survey (Westerlund *et al.*, 1987; Lundgren, 1990) provided a list of 48 C star candidates, 23 of them being newly identified objects.

Our CFHT blue-green slitless spectroscopic searches, using the Swan C_2 bands as C star selection criteria, more than doubled the number of known C stars in Fornax. Our first survey was photographic and used the wide-field corrector and a grens giving 2000 Å mm^{-1} dispersion. We obtained two plates with the same one sq. deg. field of view centred on the nominal centre of the Fornax system. This resulted in the detection of 42 C star candidates. Subsequent slit- and multislit-spectroscopy with various ESO telescopes allowed us to confirm the carbon star nature of 29 objects, 12 of them only being inside the boundaries of the previous near-infrared grism fields. Then, the C star population of the Fornax system was revisited using the Multi-Object Spectrograph (MOS) equipped with an interference filter ($\lambda_0 = 4850$ Å; $\Delta\lambda = 1025$ Å) and a prism providing ≈ 800 Å dispersion at 4800 Å. In five partially overlapping fields (CCD camera 2048 × 2048 pixel, 15 μm/pixel) of 10′ × 10′ covering the central regions of Fornax, several new C star candidates were found. Subsequent multi-object spectroscopy with the Meudon ESO Fiber Object Spectrograph (MEFOS) led us to identify 29 additional C stars, that had escaped previous photographic surveys due to their relative faintness. So, adding the two very red stars found by Demers and Kunkel (which lie outside the region covered by the transmission grating surveys) afterwards classified as C stars by Aaronson and Mould (1980), and taking into account that the C star nature of four candidates detected by the near-infrared grism searches was not confirmed by our medium resolution spectroscopy, the total number of Fornax C stars is presently 104. One can consider the C star census in Fornax virtually complete in the regions covered by the CFHT/MOS survey.

4.2. THE SAGITTARIUS SYSTEM

A number of C stars have been identified in the region of Sagittarius, the most recently discovered and nearest Galactic dwarf spheroidal. However, owing to its large extent on the sky and position, a serious risk of confusion with stars from the Galactic bulge occurs, hence radial velocities are mandatory to prove the membership of any C star detected in the direction of the Sagittarius galaxy. The first four C stars belonging to this system were found by Ibata *et al.* (1995). Then a set of 26 objects (whose magnitudes and colours are consistent with Sagittarius C stars) were selected in a 3×3 degree field by Whitelock *et al.* (1996) from a colour-magnitude diagram and objective-prism spectra provided by the UKST. One additional C star, comparable to those found by Azzopardi *et al.* (1991), has been identified by Ng and Schultheis (1997) inside field #3 of the Palomar-Groningen Survey at the outer edge of the Sagittarius system. In this connection, one has to drawn attention to the suggestion by Ng (1997), that at least some of the 34 C stars discovered by Azzopardi *et al.* (1991) in eight high transparency fields in the direction of the Galactic centre likely originated from Sagittarius and are not members of the Galactic bulge.

4.3. OTHER SYSTEMS

The Sculptor system was searched for C stars by Blanco and McCarthy (Frogel *et al.*, 1982) and by Richer and Westerlund (1983) using the same technique as used for the Fornax dwarf galaxy. These surveys led to the detection of three C star candidates. We re-identified these objects plus five other candidates; the C star nature of all these objects was subsequently confirmed by our slit spectroscopy (Azzopardi *et al.*, 1986).

Spectroscopy of six objects selected as red giant stars in a colour-magnitude diagram of Leo I by Aaronson *et al.* (1983) resulted in the discovery of the first C star belonging to this system. Our CFHT grens survey and subsequent multi-object spectroscopy with EFOSC at the ESO 3.6-m reflector (Azzopardi *et al.*, 1986) led to the detection of 19 C stars, including the one previously found in this system. Using the same method Aaronson *et al.* (1983) identified four C stars in Leo II. These stars were rediscovered on our CFHT grens plates in addition to two more C star candidates.

A systematic blue-green UKST objective-prism survey led Mould *et al.* (1982) to identify seven C stars in the Carina galaxy, including the two objects (selected as the brightest possible members of this system from a colour-magnitude diagram) previously found spectroscopically by Cannon *et al.* (1981). Using our grism detection technique at the ESO 3.6-m telescope we re-identified six of Mould and associates' C stars and also detected three additional C star candidates. All nine objects were subsequently confirmed as C stars by our medium resolution spectroscopy.

Near-infrared grism surveys at the prime focus of the KPNO 4-m telescope resulted in the detection of three C stars in the Draco system (Aaronson *et al.*, 1983) and only one (object V335 in the van Agt (1967) list) in the Ursa Minor dwarf galaxy (Aaronson *et al.*, 1983). All these objects, but no more, were re-identified by our CFHT blue-green grens surveys (Azzopardi *et al.*, 1986).

Note that no C stars have yet been found in Sextans and, unlike other dwarf spheroidal galaxies, this system appears to have no AGB stars.

5. Low-Luminosity Carbon Stars

The relation between the carbon abundance and a colour-equivalent inferred from the spectrophotometry of most C stars in the catalogue of Rebeirot et al. (1993) allowed us to select a sample ($\approx$ 5% of the set) of faint and relatively blue objects. Subsequent medium resolution spectroscopy and JHK photometry of 50 of these with the ESO 3.6-m telescope led Westerlund *et al.* (1992, 1995) to discover SMC C stars at the highest temperature and lowest luminosity ever found in an extragalactic system. The faint-end of the C star luminosity function derived from the JHK photometry of 161 stars using the method of Wood *et al.* (1983) was found to be $M_{bol} \approx -1.4$ mag (assuming an SMC absorption-free distance modulus $(m - M)_0 = 18.8$).

JHK photometry of about 85% of the Fornax dwarf galaxy C star population at the ESO/MPI 2.2-m telescope equipped with IRAC 2 led to an almost complete C star luminosity function. The absolute bolometric magnitudes, obtained by the same method that we used for the SMC, led us to identify low-luminosity objects as faint as $M_{bol} \approx -1.2$ mag (taking $(m - M)_0 = 21.0$ for the Fornax absorption-free distance modulus). These objects spectroscopically resemble the faintest C stars we have found in the SMC, and in many ways in the Galactic bulge (Westerlund *et al.*, 1991).

Deep slitless spectroscopy frames covering the inner region of the Leo II galaxy with EFOSC mounted on the ESO 3.6-m telescope equipped with a prism (1500 Å mm^{-1} dispersion or 22.5 Å pixel^{-1} at λ4770) and an interference filter ($\lambda_0 = 4880$ Å; $\Delta\lambda = 1030$ Å) and subsequent medium resolution spectroscopy with MOS at CFHT led us to identify two additional, potentially low-luminosity C stars. However, an attempt to find very faint C stars in the Carina system using the same detection technique at the ESO New Technology Telescope (NTT) led to a null result. Using the same selection criteria, we are now looking for faint C stars in the LMC from the spectrophotometry of about 1200 objects identified in seven selected fields which have been observed, as for the SMC, with the wide-field corrector and the blue-green grism at the prime focus of the ESO 3.6-m telescope. Work is in progress, but so far we have found no C stars fainter than $M_{bol} = -3.3$ mag.

These preliminary results as well as our present firm discovery of low-luminosity C stars in two slightly metal-poor galaxies corroborate the theoretical predictions about the dependence of the mean C star luminosity on metallicity leading to the expectation that, for two systems with similar star-formation histories, the mean luminosity of C stars would be lower in the more metal-deficient system. Low-luminosity C star formation is still a question under discussion and a number of scenarios to explain how these objects form have been suggested (Westerlund *et al.*, 1991, 1992, 1995; van Eck *et al.*, 1998). For instance, according to Frantsman (1997) the low-luminosity C stars may be in the early AGB evolutionary stage and should be 'formed as a result of the mass transfer in close binary systems while the primary is a C star on the thermally-pulsing AGB stars'. Deeper surveys in other nearby extragalactic systems are needed to better determine the intrinsic luminosity of the faintest C stars and investigate the possible C star formation processes able to produce C stars at low luminosities.

References

Aaronson, M. and Mould, J.: 1980, *Astrophys. J.* **240**, 804.

Aaronson, M., Olszewski, E.W. and Hodge, P.W.: 1983, *Astrophys. J.* **267**, 271.

van Agt, S.L.Th.J.: 1967, *Bull. Astron. Inst. Netherlands* **19**, 275.

Azzopardi, M., Lequeux, J. and Westerlund, B.E.: 1986, *Astron. Astrophys.* **161**, 232.

Azzopardi, M., Lequeux, J., Rebeirot, E. and Westerlund, B.E.: 1991, *Astron. Astrophya. Suppl.* **88**, 265.

Azzopardi, M., Muratorio, G., Breysacher, J. and Westerlund, B.E.: 1999, in: R.D. Cannon and P.A. Whitelock (eds.), *IAU Symp.* **No. 192**, ASP Conf. Series **144**.

Blanco, V.M. and McCarthy, M.F.: 1983, *Astron. J.* **88**, 1442.

Blanco, V.M. and McCarthy, M.F.: 1990, *Astron. J.* **100**, 674.

Blanco, V.M., McCarthy, M.F. and Blanco B.M.: 1980, *Astrophys. J.* **252**, 133.

Brewer, J.P., Richer, H.B. and Crabtree, D.R.: 1996, *Astron. J.* **112**, 491.

Cannon, R.D., Niss, B. and Norgaard-Nielsen, H.U.: 1981, *Mon. Not. R. Astron. Soc.* **196**, 1P–5P.

Cook, K.H. and Aaronson, M.: 1989, *Astron. J.* **97**, 923.

Cook, K.H., Aaronson, M. and Norris, J.: 1986, *Astrophys. J.* **305**, 634.

Crabtree, D.R., Richer, H.B. and Westerlund, B.E.: 1976, *Astrophys. J.* **203**, L81–L85.

Demers, S. and Kunkel, W.E.: 1979, *Publ. Astron. Soc. Pacific* **91**, 761.

Demers, S., Irwin, M.J. and Kunkel, W.E.: 1993, *Mon. Not. R. Astron. Soc.* **260**, 103.

van Eck, S., Jorissen, A., Udry, S., Mayor, M. and Pernier, B.: 1998, *Astron. Astrophys.* **329**, 971.

Feast, M.W. and Whitelock, P.A.: 1992, *Mon. Not. R. Astron. Soc.* **259**, 6.

Frantsman, Ju.: 1997, *Astron. Astrophys.* **319**, 511.

Frogel, J.A., Blanco, V.M., McCarthy, M.F. and Cohen, J.G.: 1982, *Astrophys. J.* **252**, 133.

Hartwick, F.D.A. and Cowley, A.P.: 1988, *Astrophys. J.* **334**, 135.

Hatzidimitriou, D., Croke, B.F., Morgan, D.H. and Cannon, R.D.: 1997, *Astron. Astrophys. Suppl.* **122**, 507.

Hudon, J.D., Richer, H.B., Pritchet, C.J., Crabtree, D., *et al.*: 1989, *Astron. J.* **98**, 1265.

Ibata, R.A., Gilmore, G. and Irwin, M.J.: 1995, *Mon. Not. R. Astron. Soc.* **277**, 781.

Kunkel, W.E., Irwin, M.J. and Demers, S.: 1997a, *Astron. Astrophys. Suppl.* **122**, 463.

Kunkel, W.E., Demers, S., Irwin, M.J. and Albert, L.: 1997b, *Astrophys. J.* **488**, L129–L132.

Lundgren, K.: 1990, *Astron. Astrophys.* **233**, 21.

McCarthy, M.F.: 1987, in: M. Azzopardi and F. Matteucci (eds.), *Stellar Evolution and Dynamics of the Outer Halo of the Galaxy*, ESO Conference Proceedings, No. 27, 203.

Morgan, D.H. and Hatzidimitriou, D.: 1995, *Astron. Astrophys. Suppl.* **113**, 539.

Mould, J., Cannon, R.D., Aaronson, M. and Frogel, J.A.: 1982, *Astrophys. J.* **254**, 500.

Ng, Y.K.: 1997, *Astron. Astrophys.* **328**, 211.

Ng, Y.K. and Schultheis, M.: 1997, *Astron. Astrophys. Suppl.* **123**, 115.

Palmer, L.G. and Wing, R.F.: 1982, *Astron. J.* **87**, 1739.

Rebeirot, E., Azzopardi, M. and Westerlund, B.E.: 1993, *Astron. Astrophys. Suppl.* **97**, 603.

Richer, H.B. and Westerlund, B.E.: 1983, *Astrophys. J.* **264**, 114.

Richer, H.B., Crabtree, D.R. and Pritchet, C.J.: 1984, *Astrophys. J.* **287**, 138.

Richer, H.B., Olander, N. and Westerlund, B.E.: 1979, *Astrophys. J.* **230**, 724.

Richer, H.B., Pritchet, C.J. and Crabtree, D.R.: 1985, *Astrophys. J.* **298**, 240.

Sanduleak, N. and Philip, A.G.D.: 1977, *Publ. Warner Swasey Obs.* **2**, 105.

de Vaucouleurs, G. and Freeman, K.C.: 1972, *Vistas in Astronomy* **14**, 163.

Westerlund, B.E.: 1960, *Uppsala Astron. Obs. Ann.* **4**, No. 7.

Westerlund, B.E.: 1964, in: F.J. Kerr and A.W. Roggers (eds.), *IAU/URSI Symp.* **20**, Australian Academy of Science, Canberra, Australia, 239.

Westerlund, B.E., Azzopardi, M. and Breysacher, J.: 1986, *Astron. Astrophys. Suppl.* **65**, 79.

Westerlund, B.E., Azzopardi, M., Breysacher, J. and Rebeirot, E.: 1991, *Astron. Astrophys.* **244**, 367.

Westerlund, B.E., Azzopardi, M., Breysacher, J. and Rebeirot, E.: 1992, *Astron. Astrophys.* **260**, L4–L6.

Westerlund, B.E., Azzopardi, M., Breysacher, J. and Rebeirot, E.: 1995, *Astron. Astrophys.* **303**, 107.

Westerlund, B.E., Edvardsson, B. and Lundgren, K.: 1987, *Astron. Astrophys.* **178**, 41.

Westerlund, B.E., Olander, N., Richer, H.B. and Crabtree, D.R.: 1978, *Astron. Astrophys. Suppl.* **31**, 61.

Whitelock, P.A., Irwin, M. and Catchpole, R.M.: 1996, *New Astronomy* **1**, 57.

Wood, P.R., Bessel, M.S. and Fox, M.W.: 1983, *Astrophys. J.* **272**, 99.

V – THE BULGES

A BRIEF HISTORY OF STAR FORMATION AND CHEMICAL ENRICHMENT IN THE BULGE OF THE MILKY WAY

JAY A. FROGEL

The Ohio State University, Department of Astronomy, Columbus, OH, USA

Abstract. Observations of the stellar content of the bulge of the Milky Way can provide critical guidelines for the interpretation of observations of distant galaxies, in particular for understanding their stellar content and evolution. In this brief overview I will first highlight some recent work directed towards measuring the history of star formation and the chemical composition of the central few parsecs of the Galaxy. These observations point to an episodic history of star formation in the central region with several bursts having occurred over the past few 100 Myr (e.g. Blum *et al.*, 1996b). High resolution spectroscopic observations by Ramírez *et al.* (1998) of luminous M stars in this region yield a near solar value for [Fe/H] from direct measurements of iron lines. Then I will present some results from an ongoing program by my colleagues and myself the objective of which is the delineation of the star formation and chemical enrichment histories of the central 100 parsecs of the Galaxy, the 'inner bulge'. From new photometric data we have concluded that there is a small increase in mean [Fe/H] from Baade's Window to the Galactic Center and deduce a near solar value for stars in the central region. For radial distances greater than 1° from the Galactic Center we fail to find a measurable population of stars that are significantly younger than those in Baade's Window. Within 1° we find a number of luminous M giants that most likely are the result of a star formation episode not more than one or two Gyr ago.

1. Introduction

We care about the bulge of the Milky Way both because of what it tells us about the formation and evolution of our own galaxy and because its structure and stellar content are often used as a proxies in the study of other galactic bulges and elliptical galaxies. Thus, in the spirit of this meeting, it serves as a vital link between the near and the far, between the present and the past.

Buried within the Galactic bulge is the center of the Galaxy, a region which, on a small scale, has some properties in common with luminous AGNs and starburst galaxies. Because of its proximity, it can be studied in greater detail than any other such galactic nucleus. On the other hand, between us and it lie clouds of interstellar dust of great enough optical depth that the average visual extinction is about 30 magnitudes. Thus it is only at infrared and longer wavelengths that the central few arc minutes of the Galaxy can be studied. Indeed, out to a radius of about 2° (and much farther if observing on or close to the major axis) the visual extinction is still great enough that optical observations are difficult to impossible along most lines of sight.

 Astrophysics and Space Science is the original source of publication of this article. It is recommended that this article is cited as: *Astrophysics and Space Science* **265**: 303–309, 1999.
© 1999 *Kluwer Academic Publishers.*

This review of our current state of knowledge of the star formation and chemical enrichment history of the Galactic bulge will be brief. I will make no attempt to cite the many research papers that have appeared over the past decade or so that deal with these topics. The papers that are referred to, though, do contain extensive references to the literature. Furthermore, the review will be restricted to stars and their optical and infrared photospheric radiation.

Simply put, the history of star formation can be traced by a survey of either hot blue stars or cool red ones. The former, which will primarily be massive main sequence stars, are effective tracers of the most recent epoch of star formation. The latter will not only be effective tracers of the most recent epoch – the late-type supergiants – but will also trace out older epochs of star formation via the presence of luminous AGB stars. This review will deal primarily with surveys of the cool stellar population in the central part of the Milky Way so that star formation can be investigated over a broader period of time. The next section, though, will briefly consider some work on the hot stellar component in the immediate vicinity of the Galactic Center itself.

2. The Central Few Arc Minutes Of The Galaxy

Krabbe *et al.* (1995) have reported on an extensive survey of the central few arc seconds of the Galaxy. They identified more than 20 luminous blue supergiants and Wolf-Rayet stars in a region not more than a parsec in radius. The inferred masses of some of these stars approaches 100 $M_\odot$. From this they conclude that between 3 and 7 Myr ago there was a burst of star formation in the central region. They also identified a small population of cool luminous AGB stars from which one can conclude that there was significant star formation activity a few 100 Myr ago as well.

Blum *et al.* (1996a) carried out a K-band survey of the central 2 arc minutes of the Galaxy and drew renewed attention to the presence of a significant excess of luminous stars ($K_0 < 6$) when compared to a typical old stellar population such as is found in Baade's Window, for example. Most of these stars were found by Blum and others to be M stars, presumably a mixture of supergiants and AGB stars. However, as Blum *et al.* (1996a,b) pointed out, the distinction between an M supergiant and a luminous M-type AGB stars cannot be made on the basis of luminosity alone since there is a two magnitude range in which the luminosities of the two very different class of stars overlap (see Figure 5 of Blum *et al.*, 1996b). But assigning stars to one class or the other is critical in deciphering the star formation history of this region. With K-band spectra, though, it becomes straightforward to make this distinction for almost all cases (Figure 1 of Blum *et al.*, 1996b). As first quantified by Baldwin *et al.* (1973) M-type supergiants can be easily distinguished from ordinary giants of the same temperature (or color) via the strengths of the H_2O and CO absorption bands in K-band spectra.

Blum *et al.* (1996b) analyzed K-band spectra for a representative sample of 19 of the luminous stars identified in their survey area. Only 3 of these stars were found to be supergiants; one of these is the well known IRS 7. The remainder are AGB stars, some of which could be long period variables as well. From the spectra and the multi-color photometry they were able to calculate effective temperatures and bolometric luminosities for the stars. With the assumption that the abundance of the stars they observed is comparable to that of disk stars in the solar neighborhood, they were able to estimate ages for the stars from a comparison with stellar interior models. Rather than continuous star formation, they concluded that there have been multiple epochs of star formation in the central few parsecs of the Galaxy. The most recent epoch, less than 10 Myr ago, corresponds with that found by Krabbe *et al.* (1995). Blum *et al.* also identified significant periods of star formation as having occurred about 30 Myr, between 100 and 200 Myr, and more than about 400 Myr in the past. The majority of stars they observed are associated with the oldest epoch of star formation.

What about the abundances of stars in the central few parsecs? Ramírez *et al.* (1998) have obtained high resolution K band spectra for 10 M giants in this region and did a full spectral synthesis analysis of them. They were able to measure a true [Fe/H] with their observations of iron lines and thus remove any ambiguity that could arise by inferring [Fe/H] from measurements of elements that are often used as proxies (e.g. Mg or Ca). For these 10 stars they derive a mean [Fe/H] of 0.0 with a dispersion no larger than their uncertainties, about ± 0.2 dex. Their mean value is a few tenths of a dex greater than the mean [Fe/H] determined for Baade's Window K giants (Sadler *et al.*, 1996; McWilliam and Rich, 1994). While it may be surprising that the mean [Fe/H] at the Galactic Center is not super-solar, the small increase in the mean value of [Fe/H] compared with Baade's Window is consistent with estimates for the gradient in [Fe/H] in the inner Galactic bulge (Tiede *et al.*, 1995; Frogel *et al.*, 1999). On the other hand, a non-detectable dispersion in [Fe/H] stands in contrast to a dispersion that is more than an order of magnitude in size for the K giants in Baade's Window (Sadler *et al.*, 1996; McWilliam and Rich, 1994). It is, however, consistent with the lack of dispersion found for the M giants in Baade's Window (Frogel and Whitford, 1987; Terndrup *et al.*, 1991).

The fact that [Fe/H] is near solar at the Galactic Center with a star formation rate per unit mass at least at present is considerably in excess of the solar neighborhood value suggests that the rate of chemical enrichment has been quite different at the two locations.

The difference in the measured dispersions between K and M giants remains to be explained. In Baade's Window there is no detectable population of K giants with luminosities great enough to place them near the top of the giant branch (DePoy *et al.*, 1993). At the same time, it is generally thought that in a stellar population most of whose stars have [Fe/H] greater than -1.0, nearly all K giants eventually evolve into M giants. Thus the observed dispersions should be similar for the two groups. That they are not could imply that estimates for evolutionary rates and lifetimes near the upper end of the RGB and AGB are wrong. It could also point to problems with the analysis of the M giants, although in the case of the Ramírez *et al.* work this seems unlikely since the underlying principles of their analysis are basically the same as that employed for the optical studies of the K giants.

3. The Inner Galactic Bulge

Now we turn our attention to the inner 3° of the Galactic bulge. This region, which is interior to Baade's Window, will be referred to as the inner Galactic bulge. With the 2.5 meter duPont Telescope at Las Campanas Observatory I have obtained JHK images of 11 fields within the inner bulge, three of which are within 1° of the Galactic Center. The two questions that are being addressed are: What is the abundance of the stellar population in this region and is there any evidence that a detectable component of the population is relatively young, i.e. significantly younger than globular clusters? To answer the question about stellar abundances my collaborators and I are taking two independent approaches. The first is based on the finding of Kuchinski *et al.* (1995) that the giant branch of a metal rich globular cluster in a K, JK color magnitude diagram is linear over 5 magnitudes and has a slope that is proportional to its optically determined [Fe/H]. Results from this part of the study, based on the LCO data, will be summarized here. The second approach, which is expected to give a more detailed and precise answer to the abundance question, and is based on the analysis of K-band spectra obtained at CTIO of about one dozen M stars in each of 11 fields. This is a work in progress.

We have used two indicators to test for the presence of intermediate age stars in the bulge (i.e. an age not more than a few Gyr as opposed to closer to 10 Gyr). The first is a determination of the luminosity of the brightest stars on the giant branch of each of the fields. A sign of a relatively young age would be if there were stars brighter than those found in Baade's Window. The second indicator involves a comparison of the properties of long period variables in the bulge with their counterparts in Galactic globular clusters.

3.1. ABUNDANCES IN THE INNER GALACTIC BULGE

The best 'fixed reference point' in any attempt to determine abundances within the inner bulge is the determination by McWilliam and Rich (1994) of the mean abundance of a sample of K giants in Baade's Window based on high resolution spectroscopy. They found a mean [Fe/H] of about –0.2. A similar result was found by Sadler et al.(1996) based on spectroscopy of several hundred K giants in Baade's Window. Furthermore, both of these independent analyses agreed that the spread in [Fe/H] in Baade's Window was considerably greater than an order of magnitude and could be as large as two orders of magnitude. Observations of Baade's Window M giants, on the other hand, both in the near IR and of red TiO bands (e.g.

Frogel and Whitford, 1987; Terndrup *et al.*, 1991) consistently pointed to a greater than solar abundance with no measurable dispersion. The independent estimate of [Fe/H] for the Baade's Window giants based on the near-IR slope method (Tiede *et al.*, 1995) differed from the previous determinations in that they found an [Fe/H] close to the value based on the optical spectra of K giants.

The near-IR survey of inner bulge fields has yielded color-magnitude diagrams that, except for the fields with the highest extinction, reach as faint as the horizontal branch. Thus, with data for the entire red giant branch above the level of the HB we can apply the technique developed by Kuchinski *et al.* (1995) which derives an estimate for [Fe/H] based on the slope of the RGB above the HB. Although the calibration of this technique is based on observations of globular clusters, the applicability of the method to stars in the bulge was demonstrated by Tiede *et al.* (1995) in their analysis of stars in Baade's Window. Although we were able to estimate, statistically, the reddening to each field, the method itself is reddening independent since it depends only on a slope measurement. Based on 7 fields on or close to the minor axis of the bulge at galactic latitudes between $+0.1°$ and $-2.8°$ we derive a dependence of $\langle[Fe/H]\rangle$ on latitude for b between $-0.8°$ and $-2.8°$ of -0.085 ± 0.033 dex/degree. When combined with the data from Tiede *et al.* we find for $-0.8° \leq b \leq -10.3°$ the slope in $\langle[Fe/H]\rangle$ is -0.064 ± 0.012 dex/degree, somewhat smaller than the admittedly crude value derived by Minniti *et al.* (1995). An extrapolation to the Galactic Center predicts [Fe/H] $= +0.034 \pm 0.053$ dex, in close agreement with the result of Ramírez *et al.* (1998). Also in agreement with Ramírez *et al.*, we find no evidence for a dispersion in [Fe/H]. Details of this work are in Frogel *et al.* (1999).

Analysis of the K-band spectra of the brightest M giants in each of the fields surveyed is nearing completion; the results appear to be consistent with those based on the RGB slope method, namely, an [Fe/H] for Baade's Window M giants close to the McWilliam and Rich value but with little or no gradient as one goes into the central region. Also, little or no dispersion in [Fe/H] within each field is visible in the spectroscopic data. Further work on the calibration of these data must be done before definitive conclusions can be drawn.

In summary, several independent lines of evidence point to an [Fe/H] for stars within a few parsecs of the Galactic Center of close to solar. The gradient in [Fe/H] between Baade's Window and the Center is small – not more than a few tenths of a dex. Exterior to Baade's Window there is a further small decline in mean [Fe/H] (e.g. Terndrup *et al.*, 1991; Frogel *et al.*, 1990; Minniti *et al.*, 1995). It remains to be seen whether this gradient arises from a change in the mean [Fe/H] of a single population or a change in the relative mix of two populations, one relatively metal rich and identifiable with the bulge, the other relatively metal poor and more closely associated with the halo. Support for the latter interpretation is found in the survey of TiO band strengths in M giants in outer bulge fields by Terndrup *et al.* (1990) for which they found a bimodal distribution. McWilliam and Rich (1994) proposed an explanation based on selective elemental enhancements as to why earlier abundance estimates of bulge M giants seemed to consistently yield [Fe/H] values in excess of solar. What still remains to be understood is why even recent measurements of the M giant abundances do not reveal any evidence for an intrinsic dispersion in [Fe/H] in any given field. Finally, an issue that needs further investigation is the degree to which the indirect methods used for getting at [Fe/H] are influenced by selective element enhancements.

3.2. STELLAR AGES IN THE INNER GALACTIC BULGE

If a stellar population has an age significantly younger than 10 Gyr, say not more than a few Gyr, then stars on the AGB can reach luminosities several magnitudes brighter than they would in an older population. After correction for extinction we noted that our fields closest to the Galactic Center had significant numbers of bright, red stars. With the stars in Baade's Window as a guide we chose a reddening corrected K magnitude of 8.0 as the limit to the brightest magnitude obtainable in an old population and counted the number stars in each surveyed field brighter than this relative to the number in a predefined, fainter magnitude interval. We found that at radial distances greater than 1.3° the ratio was constant with a value equal to that for Baade's Window. On the hand, for the fields closer to the center than 1.0° this ratio was significantly larger, implying the presence of a relatively young population of stars, not more than a couple of Gyr old. This is consistent with Blum *et al.*'s work on the inner few arc minutes of the bulge. Details are in Frogel *et al.* (1999)

The second test we applied to see if there is evidence for a young population in the Galactic bulge was to compare the luminosities and periods of bulge long period variables (LPVs) with those found in globular clusters (Frogel and White-lock, 1998). For LPVs of the same age, those with greater [Fe/H] will have longer periods. LPVs with longer periods also have higher mean luminosities. In the past, claims have been made for the presence of a significant intermediate age population of stars in the bulge based on the finding of some LPVs with periods in excess of 500–600 days. It is necessary, however, to have a well defined sample of stars if one is going to draw conclusions based on the rare occurrence of one type of star. The M giants in Baade's Window are just such a well defined sample (e.g. Frogel and Whitford, 1987). Frogel and Whitelock (1998) presented a detailed comparison of LPVs in the bulge and in metal rich globular clusters. They demonstrated that with the exception of a few of the LPVs in Baade's Window with the longest periods, the distribution in bolometric magnitudes of the LPVs from the two populations over-lap completely. Furthermore, because of the dependence of period and luminosity on [Fe/H] and the fact that there has been no reliable survey for LPVs in globulars with [Fe/H] > −0.25, while a significant fraction of the giants in Baade's Window have [Fe/H] > 0.0 (McWilliam and Rich, 1994), the brightest Baade's Window LPVs can be understood as a result of this higher [Fe/H] compared with globular clusters.

Finally, observations with the Infrared Astronomical Satellite (IRAS) at 12 μm were used to estimate the integrated flux at this wavelength from the Galactic bulge as a function of galactic latitude along the minor axis. Galactic disk emission was removed from the IRAS measurements with the aid of a simple model. These fluxes were then compared with predictions for the 12 μm bulge surface brightness based on observations of complete samples of optically identified M giants in minor axis bulge fields (Frogel and Whitford, 1987; Frogel et al., 1990). No evidence was found for any significant component of 12μm emission in the bulge other than that expected from the optically identified M star sample plus normal, lower luminosity stars. Since these stars are themselves fully accountable by an old population, the conclusion from this study was, again, that most of the Galactic bulge has no detectable population of stars younger than those in Baade's Window, i.e. younger than an age comparable to that of globular clusters.

References

Baldwin, J.R., Frogel, J.A. and Persson, S.E.: 1973, *Astrophys. J.* **184**, 427.
Blum, R.D., Sellgren, K. and Depoy, D.L.: 1996a, *Astrophys. J.* **470**, 864.
Blum, R.D., Sellgren, K. and Depoy, D.L.: 1996b, *Astron. J.* **112**, 1988.
DePoy, D.L., Terndrup, D.M., Frogel, J.A., Atwood, B. and Blum, R.: 1993, *Astron. J.* **105**, 2121.
Frogel, J.A.: 1998, *Astrophys. J.* **505**, 659.
Frogel, J.A., Terndrup, D., Blanco, V.M. and Whitford, A.E.: 1990, *Astrophys. J.* **353**, 494.
Frogel, J.A., Tiede, G.P. and Kuchinski, L.E.: 1999, *Astron. J.*, submitted.
Frogel, J.A. and Whitelock, P.A.: 1998, *Astron. J.* **116**, 754.
Frogel, J.A. and Whitford, A.E.: 1987, *Astrophys. J.* **320**, 199.
Krabbe, A., Genzel, R., et al.: 1995, *Astrophys. J. Lett.* **447**, L95.
Kuchinski, L., Frogel, J.A., Terndrup, D.M. and Persson, S.E.: 1995, *Astron. J.* **109**, 1131.
McWilliam, A. and Rich, R.M.: 1994, *Astrophys. J. Suppl.* **91**, 749.
Minniti, D., Olszewski, E. and Rieke, M.: 1995, *Astron. J.* **110**, 1686.
Ramírez, S., Sellgren, K., Carr, J.S., Balachandran, S., Blum, R. and Terndrup, D.: 1998, in: H. Falcke, A. Cotera, W. Duschl and F. Melia (eds.), *ASP Conf. Ser.*, The Central Parsecs, in press.
Sadler, E.M., Rich, R.M. and Terndrup, D.M.: 1996, *Astron. J.* **112**, 171.
Terndrup, D.M., Frogel, J.A. and Whitford, A.E.: 1990, *Astrophys. J.* **357**, 453.
Terndrup, D.M., Frogel, J.A. and Whitford, A.E.: 1991, *Astrophys. J.* **378**, 742.
Tiede, G.P., Frogel, J.A. and Terndrup, D.M.: 1995, *Astron. J.* **110**, 2788.

THE CHEMICAL EVOLUTION OF THE GALACTIC BULGE

F. MATTEUCCI

Dipartimento di Astronomia, Università di Trieste, Via G.B. Tiepolo 11, 34100 Trieste and
SISSA/ISAS, Via Beirut 2-4, 34014 Trieste, Italy

D. ROMANO

Osservatorio Astronomico di Trieste, Via G.B. Tiepolo 11, 34100 Trieste and
SISSA/ISAS, Via Beirut 2-4, 34014 Trieste, Italy

Abstract. In this paper we review the chemical evolution models for the Galactic bulge: in particular, we discuss the predictions of models as compared with the available abundance data and infer the mechanism as well as the time scale for the formation of the Galactic bulge. We show that good chemical evolution models reproducing the observed metallicity distribution of stars in the bulge predict that the $[\alpha/Fe] > 0$ over most of the metallicity range. This is a very important constraint indicating that the bulge of our Galaxy formed at the same time and even faster than the inner Galactic halo. We also discuss predictions for the evolution of light elements such as D and ^7Li and conclude that the D astration should be maximum due to the high star formation rate required for the bulge whereas the evolution of the abundance of Li should be similar to that observed in the solar neighbourhood, but with an higher Li abundance in the interstellar medium at the present time.

1. Introduction

Several questions concerning the formation and the evolution of the Galaxy are still unanswered:
- did the Galaxy form **inside-out** or **outside-in** ?
- did the bulge form **before** or **after** the halo?
- did the bulge form **before** or **after** the disk?
- did the bulge form by **mergers** of bulgeless spirals?

In the last few years, as reviewed by Wyse and Gilmore (1992), different scenarios for the bulge formation have been proposed:
- accretion of extant stellar systems which by dynamical friction eventually settle in the center of the Galaxy. Negligible star formation (SF) *in situ*.
- The bulge forms from gas that has accumulated at the center of the Galaxy (either from gas rich mergers or simply reflecting initial conditions) and evolves independently of the other Galactic components with either i) rapid SF or ii) slow SF.
- The bulge forms from gas supply determined by inflow of metal enriched gas from the thick-disk or the halo and this process can be either (a) rapid or (b) slow.

Astrophysics and Space Science is the original source of publication of this article. It is recommended that this article is cited as: *Astrophysics and Space Science* **265:** 311–318, 1999.
© 1999 *Kluwer Academic Publishers.*

– The bulge forms from gas supply determined by inflow of metal enriched gas from the disk, this in turn is determined by the star formation timescale in the disk.

The study of the chemical evolution can help in answering these questions and deciding which of the above mentioned scenarios is the most likely to occur. In particular, the study of abundance ratios represent a powerful tool to understand the formation and the evolution of the Galaxy. In this paper we want to review the predictions about abundances and abundance ratios by means of chemical evolution models of the Galactic bulge as compared with the observational data.

Let us start from the observations: Rich (1988) obtained the metallicity distribution of K giants in the Galactic bulge from low resolution spectroscopy. He found a range of $-1.0 \leq$ [Fe/H] $\leq +1.0$ and a peak of [Fe/H] at roughly twice the solar value. He used a metallicity indicator which was a mixture of Fe and Mg (Faber indices). Geisler and Friel (1992) using Washington photometry confirmed Rich's results. Later, McWilliam and Rich (1994) with higher resolution and S/N data obtained an average [Fe/H] $=\simeq -0.15$ dex. Concerning abundance ratios the first paper was by Barbuy and Grenon (1990) who found an average [O/Fe] $= +0.2$ dex for some super metal rich stars which they attributed to the bulge. McWilliam and Rich (1994) found also overabundances of Mg and Ti of $+0.3$ dex over the full [Fe/H] range whereas for Si and Ca the inferred trend seems to follow that observed for the solar neighbourhood stars with solar ratios occurring above [Fe/H] ~ -0.2 dex. These authors pointed out that oxygen measurements are quite difficult and no clear trend was found for this element. More recently, Sadler *et al.* (1996) found an average [Fe/H] $= -0.11\pm0.04$ dex for a sample of M and K giants in the bulge. They also found that stars with [Fe/H] < 0 have [Mg/Fe] $\sim +0.2$ dex whereas stars more metal rich do not show a Mg overabundance. Therefore, the situation looks still unclear although recent data by Barbuy (this conference) seem to show overabundances for most of the α-elements observed in stars belonging to two bulge globular clusters.

In order to understand the bulge formation we should also look at the inferred stellar ages: Terndrup (1988) derived old ages ($\tau_{turn-off} \sim 11 \pm 3$ Gyr) from C-M diagrams of M giants. He concluded that bulge stars formed almost simultaneously and that there has been no star formation during the last 5 Gyr. Lee (1992) studied the bulge horizontal branch stars and concluded that the oldest stars in the bulge are even older than the oldest stars in the halo by 1.3 ± 0.3 Gyr. More recently, Ortolani *et al.* (1995) concluded that bulge globular clusters are as old as halo globular clusters. Therefore, although Holtzman *et al.* (1993), on the basis of HST data of bulge stars, concluded that there is a significant number of intermediate age (< 10 Gyr) stars, there is a general consensus in attributing old ages to most of the bulge stars.

From the theoretical point of view several models have been presented in the last years: the first attempt to model the chemical evolution of the bulge were made by adopting the simple closed- box model (Rich, 1988) and the conclusion was it fits very well the metallicity distribution of K giants with an effective yield of 1.8 $Z_\odot$. Then, the first detailed chemical evolution model for the Galactic bulge was that of Matteucci and Brocato (1990). In this model the bulge was assumed to form by a fast collapse of primordial material and to suffer a rapid *in situ* star formation with an initial mass function (IMF) flatter than in the solar vicinity. This was required by fitting the K-giant metallicity distribution found by Rich (1988). This model predicted $[\alpha/\text{Fe}] > 0$ for most of the bulge stars. This was due to the assumption of a fast evolution for the bulge leading the gas to be fastly enriched in Fe by essentially supernovae of type II. In fact, type Ia supernovae originate from white dwarfs in binary systems and their chemical enrichment is delayed relatively to the enrichment from type II supernovae which originate in massive stars.

Hensler *et al.* (1993) presented chemo-dynamical models where the bulge does not evolve separately from the other Galactic components and form from pre-enriched gas from the disk. However, no predictions about abundance ratios were presented.

Molla and Ferrini (1995) adopted an IMF like in the solar neighbourhood and a timescale of bulge formation relatively long ($\tau_{Bulge} = 1$ Gyr). They predicted, as expected, an evolution for the $[\alpha/\text{Fe}]$ ratios quite similar to that in the solar vicinity with $\langle [\text{O/Fe}] \rangle_{Bulge} \sim -0.15$ dex.

In this paper we will discuss in detail the results obtained in a very recent paper by Matteucci *et al.* (1998) who proposed again for the bulge a fast *in situ* star formation from primordial gas, the same forming the Galactic halo and collapsing faster in to the center.

2. The Model for the Bulge

This model (Matteucci *et al.*, 1998) is an extension to the bulge of the two-infall model of Chiappini *et al.* (1997). The model assumes that the halo and thick-disk formed during a relatively fast infall episode with a global timescale of ~ 1 Gyr whereas the thin-disk is supposed to have accumulated on large timescales increasing from the innermost to the outermost regions (from ~ 0.5 Gyr to ~ 15 Gyr with a timescale of 8 Gyr at the solar circle) from mostly extragalactic gas. The bulge is assumed to have formed on a free-fall time ($\sim 10^8$ years) as the most central part of the spheroid. The evolution of the halo and thick-disk is therefore disentangled from that of the thin-disk whereas the bulge is considered to form out of the same infalling gas forming the halo but accumulating faster in the center. The gas out of which the bulge forms is primordial, but a small enrichment would not affect the final results. The star formation rate in the Bulge is assumed to be the fastest among the various Galactic components, therefore the bulge is the first component to form in the *inside-out* formation of the Galaxy.

The star formation rate is assumed to be proportional to the gas density:

$$\left(\frac{d\sigma}{dt}\right)_* = \nu\sigma_{gas}^{1.5} \tag{1}$$

where ν varies in the different Galactic components: $\nu_{Bulge} = 20$ Gyr^{-1}, $\nu_{Halo} = 2$ Gyr^{-1}, $\nu_{Thin-disk} = 1$ Gyr^{-1}.

2.1. THE STELLAR NUCLEOSYNTHESIS

The model follows the evolution of the abundances of light and heavy elements: H, D, ^{3}He, ^{4}He, ^{7}Li, C, N, O, Ne, Mg, Si, S, Ca, Fe and others. The adopted nucleosynthesis prescriptions are:

– *Low and intermediate mass stars*: van den Hoek and Groenewegen (1997) (standard case) with $M_{up} = 8\ M_\odot$ (limiting mass for the formation of a C-O degenerate core). These stars are responsible for the production of ^{14}N, C and some s-process elements. The predicted ^{14}N is partly primary and partly secondary, generally in agreement with previous results of Renzini and Voli (1981).

– *Massive stars*: Woosley and Weaver (1995). Massive stars are responsible for the production of α-elements and part of the Fe. Some N is also produced in massive stars, however the *primary/secondary* origin of N in massive stars is still not clear. Several calculations suggest that it is possible to produce primary N in low metallicity massive stars and this production is crucially dependent on the treatment of convection. Here we assume N to be only secondary.

– *SN Ia*: Nomoto *et al.* (1984). These supernovae are responsible for most of the Fe production if a Salpeter-like IMF is assumed and traces of elements from C to Si.

Concerning the light elements we assume: deuterium is only destroyed and goes into ^{3}He. Prescriptions for ^{3}He are the same as in Tosi *et al.* (1998).

Lithium is assumed to be partly destroyed and partly produced in stars. We consider as Li producers: massive AGB stars, supernovae II and Carbon-stars (see Matteucci *et al.*, 1995 for more details).

3. Main Results

In order to fit the metallicity distribution of Bulge stars by Mc William and Rich (1994) (see Figure 1) we had to assume that the IMF in the bulge must be flatter ($x_{Bulge} = 1.1 - 1.35$) than in the solar vicinity where a Scalo (1986) IMF seems more appropriate (Chiappini *et al.*, 1997)

Moreover, the bulge formation should have been faster than the solar neighbourhood formation and even the inner halo formation, of the order of a free-fall time ($\tau_{Bulge} \sim 10^8$ years). This model, which best reproduces the observed distribution,

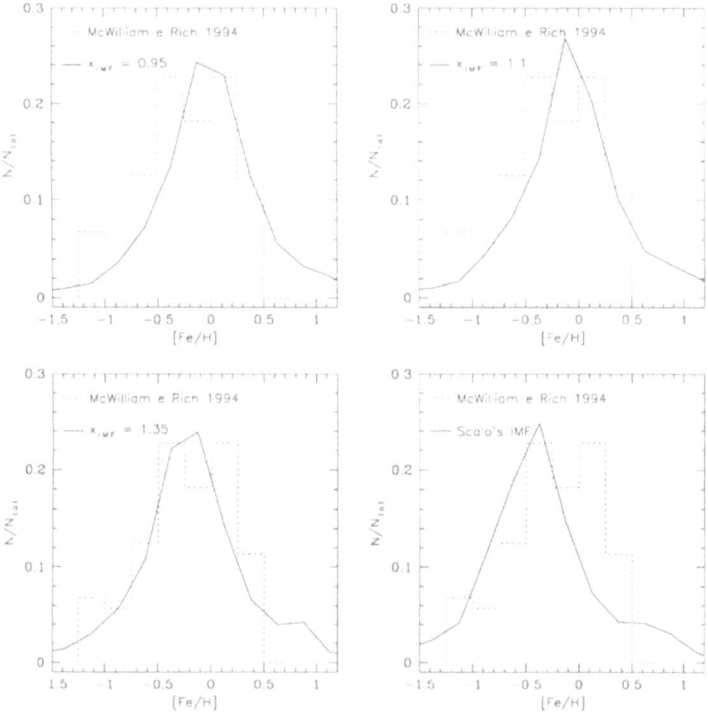

Figure 1. The predicted stellar metallicity distribution for the bulge stars under different assumptions about the IMF compared with the observed distribution by McWilliam and Rich (1994)

predicts [α/Fe]> 0 for most of the [Fe/H] range. This is in agreement with the predictions of Matteucci and Brocato (1990) who studied just Mg and Fe. In particular, our results show (see Figure 2) that the α-elements can be divided into two groups: i) O, Ne, Mg and ii) Si, S, Ca. The elements of group i) indicate a maximum [O,Ne,Mg/Fe] $\sim$ +0.5 − +0.8 dex at low metallicities and show a continuous slowly decreasing trend over the whole [Fe/H] range, whereas the elements of group ii) show an almost constant overabundance relative to Fe over almost the whole [Fe/H] range, of the order of [Si,S,Ca/Fe] $\sim$ +0.1 − +0.4 dex. It is worth noticing that the flatter IMF in the bulge than in the rest of the Galaxy implies that $[α/Fe]_{Bulge} > [α/Fe]_{Halo}$ and $[^{13}C, N/Fe]_{Bulge} < [^{13}C,N/Fe]_{Halo}$. This is due to the larger fraction of massive stars in the IMF assumed for the bulge and to the fact that massive stars are mostly responsible for the production of α-elements whereas low and intermediate mass stars are mostly responsible for the production of ^{13}C and ^{14}N. In the same figure we show for comparison the predicted abundance ratios versus [Fe/H] in the solar neighbourhood region. These predictions are the same as in Chiappini *et al.* (1997) with the exception of ^{14}N, ^{12}C and ^{13}C since here we used the yields of van den Hoek and Groenewegen (1997) instead of those of Renzini and Voli (1981).

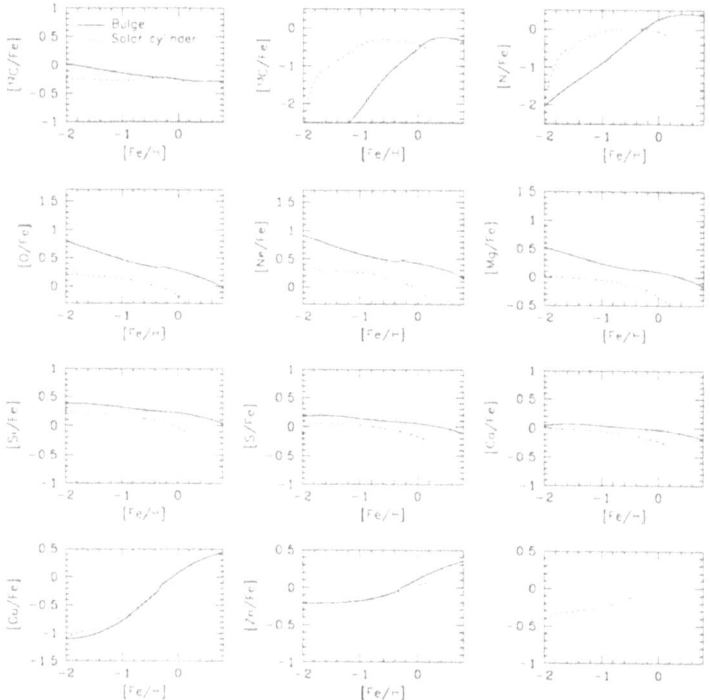

Figure 2. Predicted behaviours of [el/Fe] vs. [Fe/H] for the solar vicinity and for the bulge in the framework of the self-consistent model for the whole Galaxy presented here.

Deuterium should be completely destroyed in the interstellar medium in the Bulge due to the short duration of its formation process by accretion of primordial material and to the fact that inside stars D is completely destroyed. On the other hand, in the solar vicinity we expect to observe now in the ISM a D abundance which is a factor of two lower than the primordial one and this is because this Galactic region formed by slow infall of primordial material and the star formation rate was less efficient than assumed for the bulge. For the same reasons for which D is destroyed the predicted abundance of He should instead be high, namely $\langle Y \rangle \sim 0.35$, in agreement with predictions by Renzini (1994). In Figure 3 we show the prediction of our model for Log (Li/H) vs. [Fe/H] in the bulge. The predicted trend is similar to that in the solar vicinity (see Matteucci *et al.*, 1995), the only difference being the larger Li abundance achieved at the present time in the bulge. From our results it is clear that a high metallicity bulge stars can not have a 'primordial' Li abundance as claimed by Minniti *et al.* (1998) for a MS bulge star (source star of the MACHO microlensing effect 97BLG45), which is shown in the figure, unless we suppress most of the stellar Li producers. More probably the observed Li abundance is the result of Li depletion in a high metallicity star.

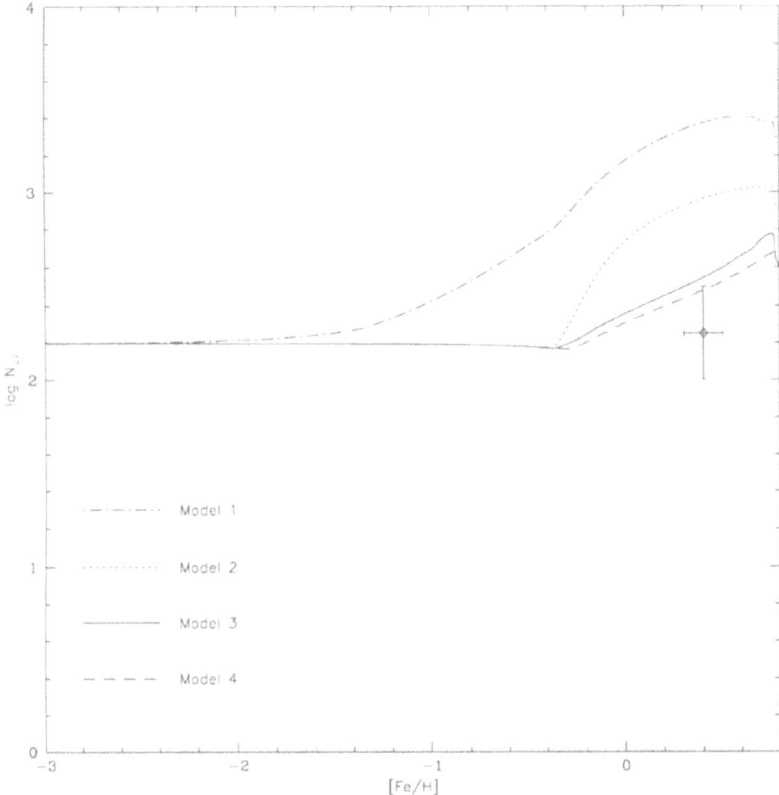

Figure 3. The predicted log (Li/H) vs. [Fe/H] for the bulge under different assumptions about stellar Li-producers: Model 1 assumes AGB stars, supernovae and C-stars, Model 2 assumes only AGB and C-stars, Model 3 assumes only AGB and C-stars but with a smaller contribution from AGB stars, and finally Model 4 is like Model 3 but with a flatter IMF ($x_{IMF} = 1.1$). The data point of Minniti *et al.* (1998) is shown for comparison.

4. Conclusions

We have presented the results of a model for the chemical evolution of the Galactic bulge and shown predictions for several chemical abundances. We have compared the model predictions with the available data (the stellar metallicity distribution for bulge stars by McWilliam and Rich 1994) and derived the following conclusions: the bulge formed quite fast ($\tau_{Bulge} \sim 10^8$ years), faster than the solar vicinity ($\tau_{SN} \sim 8$ Gyr) and the inner halo ($\tau_{Halo} \sim 2$ Gyr). Therefore, we favor the scenario of a fast formation out of primordial gas with star formation *in situ*. In order to reproduce the exact shape of the stellar metallicity distribution in the bulge, in particular the position of the peak, an IMF flatter than that suggested for the solar vicinity (Scalo, 1986) is required. If the IMF in the bulge is flatter than in the rest of the Galaxy then we should expect that $[\alpha/\text{Fe}]_{Bulge} > [\alpha/\text{Fe}]_{Halo}$

although the overabundance of the α-elements depends mainly on the timescales of bulge formation. For this reason, observing these abundance ratios represents a very powerful tool to infer the mechanism of formation of the bulge and its age. In fact, based on our model we can conclude that: if $[\alpha/\text{Fe}]_{Bulge} > 0$ then $\tau_{Bulge} < 1$ Gyr, otherwise if $[\alpha/\text{Fe}]_{Bulge} < 0$ then $\tau_{Bulge} > 1$ Gyr, due to the relative timescales for the formation of α-elements and Fe. More and more data on stars in the Galactic bulge are necessary to assess this point although very recent results (Barbuy, this conference) seem to confirm our predictions.

References

Barbuy, B. and Grenon, M.: 1990, in: B.J. Jarvis and D.M. Terndrup (eds.), *Bulges of Galaxies*, ESO/CTIO Workshop, p. 83.
Chiappini, C., Matteucci, F. and Gratton, R.: 1997, *Astrophys. J.* **477**, 765.
Geisler, D. and Friel, E.D.: 1992, *Astron. J.* **104**, 128.
Hensler, G., Samland, M., Theis, C. and Burkert, A.: 1993, in: G. Hensler *et al.* (eds.), *Panchromatic View of Galaxies-their Evolutionary Puzzle*, Editions Frontieres, p. 341.
Holtzman, J.A., Light, R.M., Baum, W.A., *et al.*: 1993, *Astron. J.* **106**, 1826.
Lee, Y.W.: 1992, *Astron. J.* **104**, 1780.
Matteucci, F. and Brocato, E.: 1990, *Astrophys. J.* **365**, 539.
Matteucci, F., Romano, D. and Molaro, P.: 1998, *Astron. Astrophys.* **341**, 458.
Matteucci, F., D'Antona, F. and Timmes, F.X.: 1995, *Astron. Astrophys.* **303**, 460.
McWilliam, A. and Rich, R.M.: 1994, *Astrophys. J. Suppl.* **91**, 749.
Minniti, D., Vandehei, T., Cook, K.H., Griest, K. and Alcock, C.: 1998, *Astrophys. J.* **499**, L175.
Molla, M. and Ferrini, F.: 1995, *Astrophys. J.* **454**, 726.
Nomoto, K., Thielemann, F.K. and Yokoi, K.: 1984, *Astrophys. J.* **286**, 644.
Ortolani, S., Renzini, A., Gilmozzi, R., *et al.*: 1995, *Nature* **377**, 701.
Renzini, A.: 1994, *Astron. Astrophys.* **285**, L5.
Renzini, A. and Voli, M.: 1981, *Astron. Astrophys.* **94**, 175.
Rich, R.M.: 1988, *Astrophys. J.* **95**, 828.
Sadler, E.M., Rich, R.M. and Terndrup, D.M.: 1996, *Astron. J.* **112**, 171.
Scalo, J.M.: 1986, *Fundam. Cosmic Phys.* **11**, 1.
Terndrup, D.M.: 1988, *Astron. J.* **96**, 884.
Tosi, M., Steigman, G., Matteucci, F. and Chiappini, C.: 1998, *Astrophys. J.* **498**, 226.
van den Hoek, L.B. and Groenewegen, M.A.T.: 1997, *Astron. Astrophys. Suppl.* **123**, 305.
Woosley, S.E. and Weaver, T.A.: 1995, *Astrophys. J. Suppl.* **101**, 181.
Wyse, R.F.G. and Gilmore, G.: 1992, *Astron. J.* **104**, 144.

ABUNDANCES IN INDIVIDUAL STARS OF THE GALACTIC BULGE

B. BARBUY

IAG/USP, São Paulo, Brazil

Abstract. Element ratios in two stars of NGC 6553 and one star in NGC 6528 reveal that α-elements are enhanced relative to Fe. The metallicity [Fe/H] $\approx$ -0.6, but taking into account the overabundances of several elements, the overall metallicity is close to solar.

1. Introduction

The question we would like to answer is if the bulge stellar population is old as predicted in models of galaxy formation 'inside-out' such as proposed by Larson (1990), or if it is young, disk-like, resulting from the dynamical evolution of a bar (Raha *et al.*, 1992; Spergel *et al.*, 1996; Pfenniger, 1993). A third possibility proposed by Wyse *et al.* (1997) is the formation of bulges and thick disk by accretion of dwarf galaxies.

If the bulge is formed rapidly in the beginning of the Galaxy life, a fast enrichment by SNs II at early times would result in a signature through [α-elements/Fe] enhanced. If the bulge is disk-like, then one would expect [α-elements/Fe] = 0.0.

In order to try to solve this question we are carrying out a program of spectroscopic determination of abundances in bulge stars.

2. Previous Studies

In fact very few studies are available concerning abundances in the bulge. Low resolution spectra obtained by Rich (1988) indicated a mean abundance for bulge stars of [Fe/H] = +0.25. Sadler *et al.* (1996) derived [Mg/Fe] $\approx$ +0.3 from measurementes of the Mg_2 index in about 400 stars. Idiart *et al.* (1996) derived [Mg/Fe] = +0.45 from an integrated spectrum of Baade's Window.

McWilliam and Rich (1994, hereafter MR94) presented a high-resolution spectroscopic study of 12 bulge K giants of Baade's Window. Their main results are: [Fe/H] = -0.25, [Mg/Fe] = [Ti/Fe] = +0.3, [Ca/Fe] = [Si/Fe] = 0.0, [O/Fe] = 0.0 and [Eu/Fe] = slightly enhanced.

Chemical evolution models for the bulge were presented by Matteucci and Brocato (1990), Matteucci *et al.* (1998) and Matteucci (1998, this meeting), where a inside-out model for the formation of the Galaxy is adopted. An increased star

Astrophysics and Space Science is the original source of publication of this article. It is recommended that this article is cited as: *Astrophysics and Space Science* **265**: 319–326, 1999.
© 1999 *Kluwer Academic Publishers.*

TABLE I

Stellar parameters for the program stars

star	T_{eff}	log g	[Fe/H]
NGC6553:II-85	4000	0.7	−0.5
NGC6553:III-3	4000	0.8	−0.6
NGC6528:I-2	3600	1.0	−0.6

formation efficiency, of a factor 10 relative to the solar neighbourhood is assumed, and a timescale for the bulge formation of about 10^7 yr is obtained. The main result is an enhancement of α elements, with the exception of calcium: [Ca/Fe] $\approx$ 0.0 as given in Matteucci *et al.* (1998), in very good agreement with observations.

3. The Bulge Globular Clusters NGC 6553 and NGC 6528

The bulge globular clusters are test probes of the bulge configuration and chemical abundances. In the last years we have been carrying out studies of globular clusters in the bulge, through imaging and spectroscopy.

In Barbuy *et al.* (1998) we gathered the results obtained for all clusters within 5° of the Galactic center. We concluded that the bulge is flattened and extended from the center to 4.5 kpc from the Sun. Also, the metallicity distribution of bulge globular clusters is very similar to that of bulge field stars (McWilliam and Rich, 1994). In Ortolani *et al.* (1995) it has been revealed that NGC 6528 and NGC 6553 should be nearly coeval with the halo clusters. Also, the luminosity function of these clusters is very similar to that of field stars in Baade's Window, confirming that the clusters and bulge field form a same stellar population, making the clusters ideal laboratories for bulge studies.

In the present paper we present the first results on detailed abundances of these clusters.

4. Analysis of Individual Stars

We have carried out a detailed analysis of the stars II-85 and III-3 (identifications given in Hartwick 1975) of NGC 6553 (Barbuy *et al.*, 1999a) and star 2 of ring 1 (identification given in van den Bergh and Younger, 1979) of NGC 6528 (Barbuy *et al.*, 1999b).

A major step in the analysis of bulge stars is to obtain their temperatures, for which derivation precise of colours and reddening are necessary. For such, we have obtained VI colours with the *Hubble Space Telescope* – HST (Ortolani *et al.*, 1995)

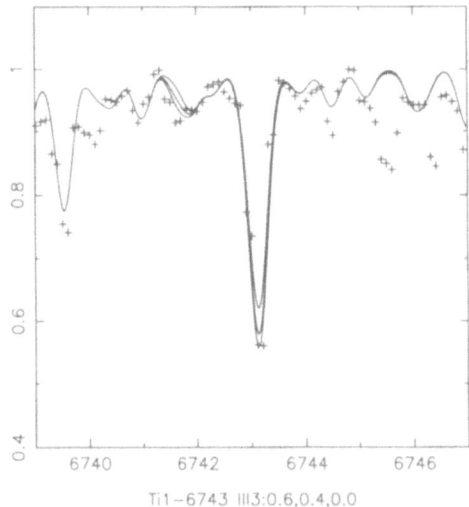

Figure 1. NGC 6553:III-3 – TiIλ6743 Å line computed with [Ti/Fe] = 0.0, +0.4, +0.6.

and JK colours using the IRAC2 detector at the 2.2 m telescope of the *European Southern Observatory* – ESO (Guarnieri *et al.*, 1998). Reddening corrections were obtained from Colour-Magnitude Diagrams (CMDs), using the colour difference between the RGB colour of the clusters and that of 47 Tuc. E(V–I) for NGC 6528 and NGC 6553 were derived in (Barbuy *et al.* (1999a,b) and Guarnieri *et al.* (1998) – see also discussions on reddening in Bruzual *et al.*, 1997). For the colour-temperature relations we have adopted the Bessell *et al.* (1998) calibrations.

High resolution échelle spectra were obtained at the ESO 3.6 m telescope using the CASPEC spectrograph. The reductions were carried out using the MIDAS package for échelle spectra, and equivalent widths of a large list of lines were measured using IRAF.

All calculations of curves-of-growth and spectrum synthesis were carried out employing atmospheric models by Plez *et al.* (1992). Curves-of-growth of FeI and FeII were built using the code RENOIR by M. Spite, wherefrom (a) excitation equilibrium was checked by verifying that FeI lines of different excitation potential give the same abundance; (b) ionization equilibrium was used by imposing that the FeI and FeII curves-of-growth give the same abundance; (c) the [Fe/H] value for the analysed stars were derived from the FeI curves-of-growth. The stellar parameters determined for the sample stars are listed in Table I.

5. Abundance Ratios

Elemental abundances were derived by computig synthetic spectra line-by-line, adopting in all cases laboratory oscillator strengths (Fuhr *et al.*, 1988; Martin *et*

B. BARBUY

TABLE II

Abundance ratios in NGC 6553, NGC 6528, and bulge K giants (MR94)

[X/Fe]	NGC6553	NGC6528	MR94
[Fe/H]	−0.55	−0.6	−
Na	+0.65	−	+0.20
O	−	+0.35	+0.03
Mg	+0.33	+0.40	+0.35
Al	+0.50	−	+0.58
Si	+0.35	−	+0.14
Ca	+0.32	0.0	+0.14
Ti	+0.60	+0.60	+0.37
Y	−0.20	−	+0.05
Zr	−0.40	−	−0.53
Ba	−0.10	−	+0.20
La	+0.13	−	+0.09
Eu	+0.0	−	+0.26

al., 1988; Wiese *et al.*, 1969, MR94). In the spectrum synthesis calculations, molecular lines of TiO, CN, C_2 and MgH were included; note that TiO bands become important for the cooler stars.

In Table II are given the resulting abundance ratios for a mean of two stars in NGC 6553 (see more details in Barbuy *et al.*, 1999a), one star in NGC 6528 (Barbuy *et al.*, 1999b – work in progress) and a mean of results for 12 bulge K giants by MR94.

In Figure 1 is shown the TiI λ6743 Å line in NGC6553:III-3 and in Figures 2 and 3 are shown the [OI]λ6300 and λ6363 Å lines in NGC 6528:I-2. Note that in NGC 6553, due to a low radial velocity, the [OI] lines are masked by the [OI] sky lines and several telluric lines.

6. Discussion

Let us discuss the results for NGC 6553 for which the whole set of abundances is available: a metallicity of [Fe/H] = −0.55 combined to an enhancement of α-elements results in $[Z/Z_\odot] = -0.08$, i.e. essentially a solar overall metallicity. For comparison, we computed Z for 47 Tuc adopting abundances by Brown *et al.* (1990) and Wallerstein *et al.* (1992), wherefrom $[Z/Z_\odot] = -0.32$ is obtained. This means that, recalling Figure 1b of Ortolani *et al.* (1995), an important conclusion can be drawn: the bulge clusters NGC 6528 and NGC 6553 have essentially the

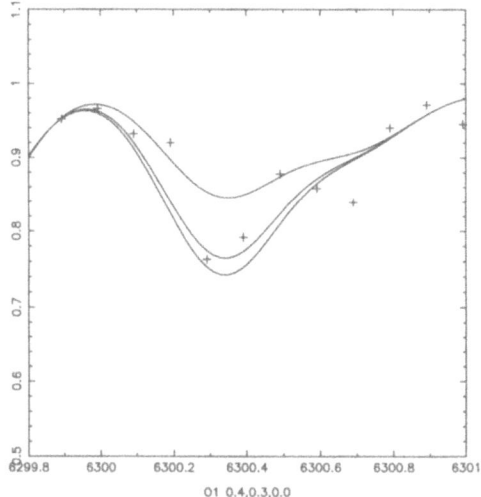

Figure 2. NGC 6528:I-2 – [OI]λ6300 Å line computed with [O/Fe] = 0.0, +0.3, +0.4.

same [Fe/H] as 47 Tuc and consequently the same age given that they show the same magnitude difference between turn-off and horizontal branch, and the difference in RGB morphology is due to a more pronounced enhancement of α-elements in NGC6528/NGC6553.

7. Other Samples?

In this section we would like to call the attention to the sample of Super-metal-rich (SMR) stars in the solar neighbourhood, described in Grenon (1998). Small subsamples of these stars were analysed in Barbuy and Grenon (1990) and Castro *et al.* (1997) – these two analyses show some discrepancies and it would be important to extend the sample and compare such results with 'in situ' bulge stars as those discussed in the previous sections.

8. Why So Few Spectroscopic Abundance Determinations?

The reason why so few spectroscopic abundance determinations are available for bulge stars stems from observational limitations: (a) the brightest stars in clusters have $V \approx 15.5$, therefore at the limit with 4m telescopes for high resolution studies; (b) CMDs are necessary beforehand in order to have the colours and reddening – high image quality is necessary for such crowded regions and CMDs were only obtained in the recent years. Besides, until recently there were no adequate colour-temperature relations for cool metal-rich stars, and the first reliable tables

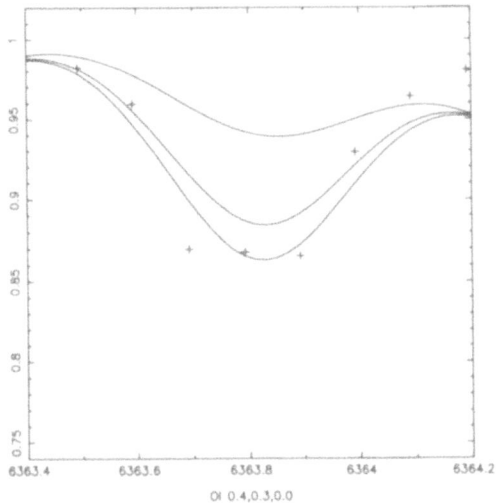

Figure 3. NGC 6528:I-2 – [OI]λ6363 Å line computed with [O/Fe] = 0.0, +0.3, +0.4.

for temperatures $T_{\text{eff}} < 4000$ K are now available with the work by Bessell *et al.* (1998) and Lejeune *et al.* (1998).

Probably the best sequence of work is to obtain CMDs, then low resolution spectra for a selection and finally high resolution spectra of selected individual stars. The same can be said about stellar populations in Local Group Galaxies, for which the first adequate CMDs are becoming available only now (see IAU Symp. 192 – Whitelock and Russell, 1998).

9. Conclusions

A first high resolution detailed abundance analysis of individual stars in metal-rich globular clusters was carried out. The clusters NGC 6528 and NGC 6553 show [Fe/H] ≈ -0.6 combined to a strong enhancement of α-elements, and for NGC 6553 we obtain $[Z/Z_\odot] = -0.08$. It is interesting to note that 47 Tuc show essentially a similar [Fe/H] = -0.7 (Zinn and West, 1984) and overall metallicity $[Z/Z_\odot] = -0.32$.

Two main important conclusions can be drawn:

(a) the inner bulge clusters are not peculiarly metal-rich in terms of [Fe/H], but show a strong enhancement of α-elements, which results in an overall metallicity essentially solar. This might well be the case of ellipticals and bulges of spiral galaxies.

(b) given that these clusters have [Fe/H] very similar to that of 47 Tuc, and that they show the same magnitude difference between turn-off and horizontal branch (Ortolani *et al.*, 1995), they are as old as 47 Tuc – this was not clear in Ortolani *et al.* (1995), since at that time we supposed that the [Fe/H] of NGC6553/NGC6528 should be higher. In order to obtain the precise age, it would be interesting to carry out stellar evolution calculations for the detailed abundances found for NGC 6553/NGC6528.

Our results show that the bulge of stellar populations in the Galactic bulge seem to be old, and their chemical enrichment proceeded in a fast timescale at early times. This lends support to an old age for galactic spheroids and to a high star formation rate in the early universe ($z < 3$) (Renzini, 1998).

Acknowledgements

This work was partially supported by CNPq and FAPESP.

References

Barbuy, B., Bica, E. and Ortolani, S.: 1998, *Astron. Astrophys.* **333**, 117.

Barbuy, B. and Grenon, M.: 1990, in: B.J. Jarvis and D.M. Terndrup (eds.), *Bulges of Galaxies, ESO Conf. and Workshop Proceedings* **35**, 83.

Barbuy, B., Renzini, A., Ortolani, S., Bica, E. and Guarnieri, M.D.: 1999a, *Astron. Astrophys.* **341**, 539.

Barbuy, B., Renzini, A., Ortolani, S. and Bica, E.: 1999b, *Astrophys. J.*, in preparation.

Bessell, M.S., Castelli, F. and Plez, B.: 1998, *Astron. Astrophys.* **333**, 231.

Brown, J.A., Wallerstein, G. and Oke, J.B.: 1990, *Astron. J.* **100**, 1561.

Brown, J.A. and Wallerstein, G.: 1992, *Astron. J.* **104**, 1818.

Bruzual A.G., Barbuy, B., Ortolani, S., Bica, E., Cuisinier, F., Lejeune, T. and Schiavon, R.: 1997, *Astron. J.* **114**, 1531.

Castro, S., Rich, R.M., Grenon, M. Barbuy, B. and McCarthy, J.: 1997, *Astron. J.* **114**, 376.

Guarnieri, M.D., Ortolani, S., Montegriffo, P., Renzini, A., Barbuy, B., Bica, E. and Moneti, A.: 1998, *Astron. Astrophys.* **331**, 70.

Fuhr, J.R., Martin, G.A. and Wiese, W.L.: 1988, Atomic Transition Probabilities: Iron through Nickel, *Journal of Physical and Chemical Reference Data* **17**, Suppl. No. 4.

Gallino, R., *et al.*: 1998, *Astrophys. J.* **497**, 388.

Grenon, M.: 1998, in: *Highlights of Astronomy*, in press.

Harris, W.E.: 1996, *Astron. J.* **112**, 1487.

Hartwick, F.D.A.: 1975, *Publ. Astron. Soc. Pacific* **87**, 77.

Idiart, T., Freitas Pacheco, J.A. and Costa, R.D.D.: 1996, *Astron. J.* **111**, 1169.

Larson, R.B.: 1990, *Publ. Astron. Soc. Pacific* **102**, 709.

Lejeune, T., Cuisinier, F. and Buser, R.: 1997, *Astron. Astrophys. Suppl.* **130**, 65.

Martin, G.A., Fuhr, J.R. and Wiese, W.L.: 1988, Atomic Transition Probabilities: Scandium through Manganese, *Journal of Physical and Chemical Reference Data* **17**, Suppl. No. 3.

Matteucci, F. and Brocato, E.: 1990, *Astrophys. J.* **365**, 539.

Matteucci, F.: 1998, this meeting.

Matteucci, F., Romano, D. and Molaro, P.: 1999, *Astron. Astrophys.* **341**, 458.

McWilliam, A. and Rich, R.M.: 1994, *Astrophys. J. Suppl.* **91**, 749 (MR) .

Ortolani, S., Renzini, A., Gilmozzi, R., Marconi, G., Barbuy, B., Bica, E. and Rich, R.M.: 1995, *Nature* **377**, 701.

Pfenniger, D.: 1993, *IAU Symp.* **153**, 387.

Plez, B., Brett, J.M. and Nordlund, Å.: 1992, *Astron. Astrophys.* **256**, 551.

Raha, N., Sellwood, J.A., James, R.A. and Kahn, F.D.: 1992, *Nature* **352**, 411.

Renzini, A.: 1998, astro-ph/9801209, in: S. D'Odorico, A. Fontana, E. Giallongo (eds.), *The Young Universe, ASP Conf. Ser.* **146**, 298.

Rich, R.M.: 1988, *Astron. J.* **95**, 828.

Sadler, E.M., Terndrup, D.M. and Rich, R.M.: 1996, *Astron. J.* **112**, 117.

Spergel, D.N., Malhotra, S. and Blitz, L.: 1996, in: D. Minniti and H.-W. Rix (eds.), *Spiral Galaxies in the Near-IR*, Springer-Verlag, pp. 128.

Whitelock, P. and Cannon, R.: 1998, *IAU Symp.* **192**, in press.

Wiese, W.L., Martin, G.A. and Fuhr, J.R.: 1969, *Atomic Transition Probabilities: Sodium through Calcium*, NSRDS-NBS 22.

Wyse, R., Gilmore, G. and Franx, M.: 1997, *Annu. Rev. Astron. Astrophys.* **35**, 637.

Zinn, R. and West, M.J.: 1984, *Astrophys. J. Suppl.* **55**, 45.

CHEMICAL ABUNDANCES OF NEWLY DISCOVERED PLANETARY NEBULAE IN THE GALACTIC BULGE

R.D.D. COSTA and W.J. MACIEL

Instituto Astronômico e Geofísico – USP, C.P. 3386, 01060-970 São Paulo, SP, Brazil

1. Introduction

The galactic bulge is a structure whose properties have been deeply studied recently. Its connection with other spheroidal structures as globular clusters or elliptical galaxies has been also examined under many aspects. However, many questions concerning its chemical and dynamical evolution, as well as the kinematics of their components, are still open. In particular, the study of bulge Planetary Nebulae (PNe) brings important information on the evolution of light element abundances (He,O,N,Ne,Ar,S), which are associated to the evolution of intermediate mass stars.

While a strong effort has been made in the last years to identify new PNe in the bulge (Acker *et al.*, 1992; Kohoutek, 1994), the determination of their chemical abundances is mostly to be made. Extensive works like those of Webster (1988), Acker *et al.* (1991) and Cuisinier *et al.* (1998) cover only a fraction of the known PNe of the bulge. We present here the first results of a long term project aimed to derive chemical abundances of recently discovered PNe in the galactic bulge, in order to enlarge the database of chemical abundances needed to study the chemical evolution of this structure. Object names on the table refer to Kohoutek (1994).

2. Observations and Data Reduction

All the objects were observed (at least twice) at the Pico dos Dias Observatory (LNA/CNPq) in Brazil, using a Cassegrain spectrograph attached to the 1.60 m telescope. Data reduction followed the standard procedure of bias, dark and faltfield corrections, extraction of the spectrum, wavelength calibration, and flux calibration through spectrophotometric standard stars observed each night. The table below summarizes our results.

Astrophysics and Space Science is the original source of publication of this article. It is recommended that this article is cited as: *Astrophysics and Space Science* **265:** 327–329, 1999.
© 1999 *Kluwer Academic Publishers.*

TABLE I

Chemical abundances of Bulge PNe

Object	He/H	$\varepsilon(O)$	$\varepsilon(N)$	$\varepsilon(S)$	$\varepsilon(Ar)$	log(N/O)
K5-3	0.144	8.25	7.57	6.80	6.06	−0.678
K5-4	0.130	8.61	7.88	7.35	6.33	−0.737
K5-8	0.234	8.00	6.80	7.09	5.77	−1.195
K5-11	0.213	8.36	8.00	6.49	6.58	−0.362
K5-13	0.211	8.43	8.40	6.73	5.73	−0.031
K5-16	0.178	8.34	8.47	7.00	6.23	0.121

3. Discussion

Our data were combined with those derived by Cuisinier *et al.* (1998) (CMAK), and with a sample of planetary nebulae from the galactic disk. Results show that the CMAK sample points to bulge PNe more abundant in Ar, S, O than their disk counterparts. Our sample, that covers weaker objects, indicates however that small abundances can also be found in bulge objects. As these elements are not produced by nucleosynthesis in the progenitor stars, their abundances reflect the composition of the interstellar medium at the epoch of the progenitor formation, so, the spread in abundances is consistent with the scenario of star formation in the bulge beginning in early phases of the Galactic history and spanning over a wide range of metallicities.

On the other hand, He and N abundances are an indication of the progenitor mass once these elements are mainly produced by stellar nucleosynthesis. In this case, behavior of bulge and disk objects are about the same, since progenitors have the same spread in mass and then in evolution time. Helium abundances are, on the average, higher for bulge than for disk objects, which was already noted by Ratag (1992).

Acknowledgement

This work was partly supported by FAPESP and CNPq.

References

Acker, A., Raytchev, B., Köppen, J. and Stenholm, B.: 1991, *Astron. Astrophys. Suppl.* **89**, 237.
Acker, A., Cuisinier, F., Stenholm, B. and Terzan, A.: 1992, *Astron. Astrophys.* **264**, 217.
Cuisinier, F., Maciel, W.J., Acker, A. and Köppen, J.: 1998 in: B. Barbuy *et al.* (eds.), *Science with Gemini*, IAG-USP/UFSC, p. 220.

Kohoutek, L.: 1994, *Ast. Nach.* **315**, 235.
Ratag, M.A., Pottasch, S.R., Dennefeld, M. and Menzies, J.W.: 1992, *Astron. Astrophys.* **255**, 255.
Webster, L.: 1988, *Mon. Not. R. Astron. Soc.* **230**, 377.

THE KINEMATICS AND ORIGIN OF SMR STARS

M. GRENON

Observatoire de Genève, 1290 Sauverny, Switzerland

Abstract. Ages and kinematics are revised for SMR stars ([M/H] >+0.30) common to NLTT and HIPPARCOS catalogues. The origin of MR to SMR stars appears heterogeneous: part of them, of intermediate age, were formed few kpc inside the solar orbit, from slowly enriched gas, whereas the oldest SMRs were born early, closer to the galactic center where the metal-enrichment had been fast. Perturbations by the galactic bar explain the peculiar motions of local SMRs as well as their outwards migration.

1. Introduction

Although SMRs were identified many years ago, Grenon (1972), as an old inner galactic population their birthplace remained uncertain. Their kinematics and chemistry suggested an origin close to or inside the bulge. SMRs high metallicity could have resulted either from a prompt enrichment in the bulge or from a delayed enrichment in the innermost thin disc. Their maximum metallicity, i.e. [M/H] = +0.6 dex, exceeds by far the maximum [M/H] reached now in the solar vicinity in recently formed stars. Constraints on time-scales for the gas metal-enrichment are provided by abundance anomalies. An early high-dispersion analysis by Barbuy *et al.* (1990) showed high [O/Fe] ratios, suggesting a prompt enrichment controlled by SN II supernovae. According to their HR-diagram, extreme SMRs are all very old whereas metal-rich stars (MR) and SMRs, with [M/H] = +0.15 to +0.35, form a mixed population, either very old or of intermediate age, with less eccentric orbits. The approach followed to disentangle the age, metallicity and birthplace, is summarized in the next sections.

2. The Observational Data

Although SMRs with [M/H] in the range +0.3 to +0.6 represent 4% of stars in the solar vicinity (Grenon, 1989), the nearby stars sample from Gliese and Jahrheiss is definitely too small to describe the early history of the Galaxy, namely when local halo and bulge counterparts are addressed. Our MRs and SMRs form a subsample of a set of 10054 proper-motion stars from the NLTT catalogue, proposed for the HIPPARCOS mission. From the subset of 7900 stars having astrometric results

Astrophysics and Space Science is the original source of publication of this article. It is recommended that this article is cited as: *Astrophysics and Space Science* **265**: 331–336, 1999.
© 1999 *Kluwer Academic Publishers.*

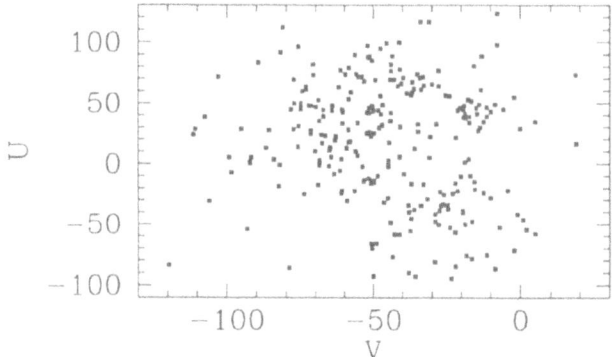

Figure 1. The U, V plane for MR and SMRs in the [M/H] range +0.25 to +0.55 dex. (Velocity components not corrected for solar motion).

from this mission, 5500 were measured in Geneva photometry to obtain estimates of T_{eff}, [M/H] and Mv. The photometric σ[M/H] is $0.05 - 0.07$ dex for G and K stars. Space velocities were computed for 3000 stars with known radial velocities from literature or mainly from CORAVEL measurements.

3. The Kinematics of MRs and SMRs

The (U, V) plane for stars within 125 pc and with [M/H] $= +0.25$ to $+0.55$, is shown in Figure 1. This metallicity range includes the metal-rich tail of the Hyades supercluster around $V - 18$ and $U + 40$. Older stars are scattered across the (U, V) plane. A trend to have either $V < -50$ and positive U, or $V > -50$ and negative U is noticed. Local SMRs with $V < -50$ show a mean outwards motion of $+20$ km s^{-1} (corrected for solar motion). If computed in an axisymmetric potential, the smallest pericentric distances Rp are close to 3 kpc. For extreme SMRs the minimum V component is around -120 with small $|U|$, an indication that their orbits barely extend outside the solar orbit.

In the V range $-100, -40$, there is an overlap between thick disc (TD) and SMR populations. The main difference, in addition to the metallicity, is the smaller scatter in $U : \sigma U = 72$ and 39 km s^{-1}, and especially in $W : \sigma W = 50$ and 18 km s^{-1} for TD and SMR respectively.

In Figure 2, where Zmax is plotted against [M/H] for all stars with orbital excentricity $e > 0.25$, the SMR population appears as the flattest component of the local old disc. The mean Zmax of 219 SMRs is only 0.16 kpc, a result in apparent contradiction with classical age versus σW relations, since SMRs are among the oldest disc stars. The origin of this anomaly is related to the orbital diffusion mechanism, see Section 6.

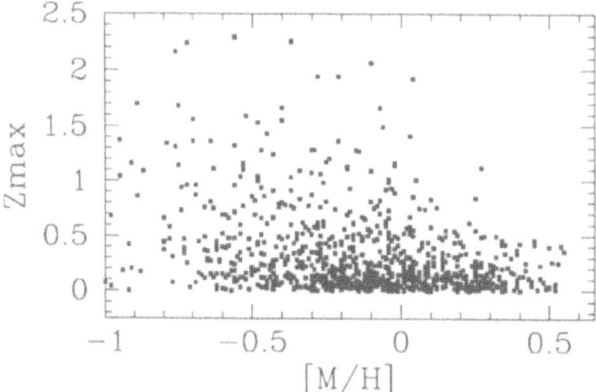

Figure 2. The maximum distance to the galactic plane of NLTT stars with projected orbital eccentricity larger than 0.25.

4. The Ages of SMRs

Ages of MR and SMR stars are inferred from HR-diagrams constructed with T_{eff} from Geneva photometry and Mv from HIPPARCOS parallaxes for narrow ranges in [M/H]. Stars with errors on parallaxes less than 15% are the only ones considered here. The diagram for mild metal-rich stars shows an age mixture: 0.7 Gyr for the Hyades generation, 3–4 Gyr for an intermediate age group and about 10 Gyr for the bulk of old stars. The classical MR star μ Leo belongs to the Hyades supercluster with (U, V, W) +38, -19, -16 and [M/H] $= +0.27 +/- 0.06$. This red giant, in core-He burning phase, cannot be compared directly to bulge SMR giants because of its higher mass and also because its high metallicity results from the secular metal enrichment of the thin disc.

The HR-diagram of SMRs with [M/H] in the range +0.25 to +0.55, Figure 3, contains essentially very old stars plus few blue stragglers around T_{eff} 6000 K and Mv 4.0. The age of the oldest group is very model dependent when derived from the turn-off T_{eff}. A densely populated unevolved main sequence, well above the Hyades sequence in the Mv/T_{eff} diagram, allows a correct fit of the isochrones. The luminosity of the horizontal part of the subgiant branch is nearly independent on C,N,O abundances and on the amount of He sedimentation. It leads to ages of 9 to 10 Gyr for the old SMRs. The red turn-off, around G7 IV, and the mis-classification at low resolution as K giants, explain why true SMRs are generally missed in programs using F and G stars to monitor the galactic chemical evolution.

5. Time-Scales for the Inner Galactic Metal-Enrichment

From a first set of SMRs, of heterogeneous ages, Barbuy *et al.* (1990) found an overabundance in Oxygen, [O/Fe] $= +0.3$, suggesting an early and prompt

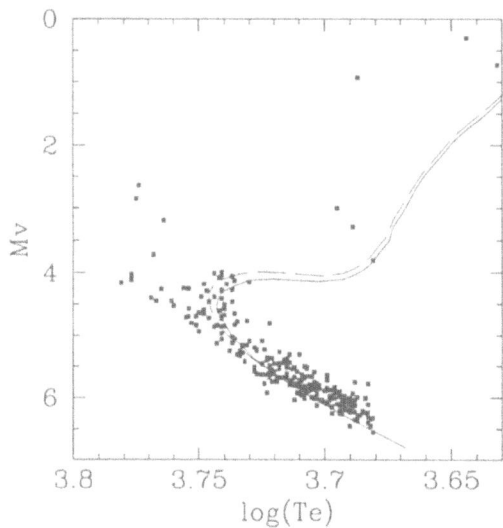

Figure 3. The HR-diagram for 257 stars with [M/H] = +0.25 to +0.55. Isochrones from Maeder and Meynet, with $Z = 0.04$, correspond to ages of 8.9 and 10.0 Gyr.

metal enrichment controlled by SN II. A set of 9 SMRs from our NLTT sample, kinematically very old, was analysed more recently by Castro *et al.* (1997). Part of them turned out to be the most metal-rich stars identified up to now in the disc. The Geneva photometric [M/H] scale was therewith confirmed. The Oxygen overabundance found by Barbuy *et al.* was partly infirmed. When extrapolating the trend [O/Fe] versus [Fe/H], observed in the solar vicinity by Edvardsson *et al.* (1993), to higher [Fe/H], the old SMRs show, at given [Fe/H], an excess in [O/Fe] of +0.1 dex at most. From this mild anomaly, if any, we conclude that the high metallicity is due to nucleosynthesis by both SN II and SN I supernovae.

A s-process elements deficiency, shown by the ratio [Ba/Fe] = −0.3 to −0.5, is a possible indication that AGB stars were not yet playing a significant nucleosynthetic role when SMRs were formed. The maximum metallicity was then reached in the inner disc in less than 1-2 Gyr. Our unique case with a clear r-process anomaly is G 161-29 where [Eu/Fe] = +0.5 and [Fe/H] = +0.1. In this star, the abundance distribution seems to result purely from nucleosynthesis by SN II. The time-scale for the metal-enrichment in the early bulge could then be as short as 0.7 Gyr. G 161–29 is for the moment the only local counterpart to some peculiar bulge stars observed *in situ* by McWilliam and Rich (1994).

6. The SMRs Birthplaces

If SMRs were formed close to the galactic center, the present pericentric distances of those seen now close to the Sun, are as large as 3 to 5 kpc. A scattering mechan-

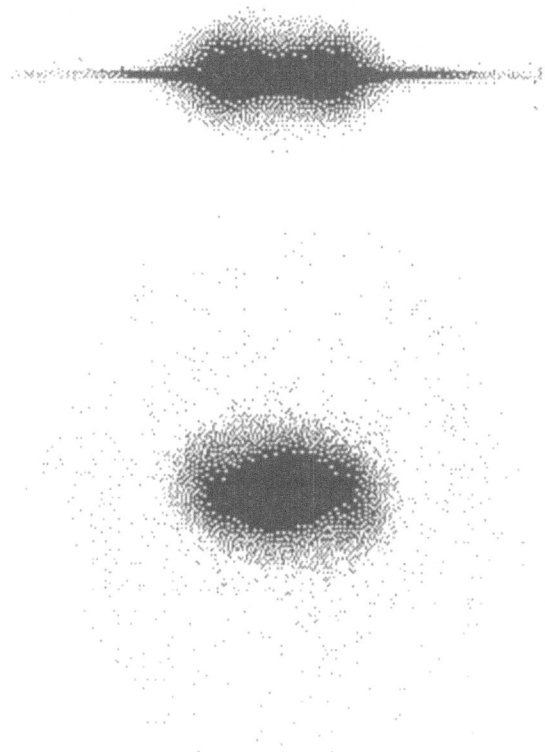

Figure 4. The re-distribution of stars, born in a ring of radius 2 +/ − 0.5 kpc, after a 4 Gyr evolution in a barred galaxy similar to the Galaxy. The size of the disc's minor axis is 11 kpc.

ism, able to increase the angular momentum but not significantly the scale height, had to be identified. Early mergers appear to be unefficient since the σU, σV, σW increase are as small as 8 to 10 km s^{-1} per impact, with little angular momentum change. Stochastic encounters with molecular clouds are also unable to produce significant outwards migrations.

Bar effects on the dynamics and kinematics of galaxies are discussed by Martinet *et al.* (this volume). Bars introduce orbital chaos and allow rather fast inwards and outwards migrations. SMRs streaming motions were explained as a local effect of the galactic bar by Raboud *et al.* (1998). In order to understand the SMRs origin, 3-D simulations were performed with Fux (1997) models, following the evolution of inner galactic populations formed before or after the bar formation, starting with various initial kinematical conditions. When stars occupy initially an inner ring of radius smaller than the bar size, most of them migrate towards the galactic center, get vertical energy and finally merge in the bulge peanut structure, see Figure 4. They become kinematically similar to the older bulge populations. The net result is a bulge growth at the expenses of the inner thin disc.

A tiny fraction of stars gets radial energy without significant increase in σW. A thin disc forms and expands outwards up to the solar orbit and and beyond it. The initial vertical scale height is nearly preserved through the expansion process. In a similar way, an inner flattened halo would produce a local thick disc. The very small scale height of local SMRs suggests the thin disc close to the bulge rather than the ellipsoidal bulge as the probable birthplace of old SMRs.

References

Grenon, M.: 1972, in: G. Cayrel *et al.* (eds.), *L'Age des Etoiles*, pp. LV 1–6.

Grenon, M.: 1989, *Astrophys. Space Sci.* **156**, 29–37.

Barbuy, B. and Grenon, M.: 1990, in: Jarvis *et al.* (eds.), *Bulges of Galaxies*, Garching, ESO/CTIO Workshop, pp. 83–87.

Castro, S., Rich, R.M., Grenon, M., Barbuy, B. and Mc Carthy, J.K.: 1997, *Astron. J.* **114**, 376–387.

Fux, R.: 1997, *Astron. Astrophys.* **327**, 983–1003.

Raboud, D., Grenon, M., Martinet, L., Fux, R. and Udry, S.: 1998, *Astron. Astrophys.* **335**, L61–64.

AGE AND METALLICITY GRADIENTS IN THE GALACTIC BULGE

A differential study using HST/WFPC2 observations

SOFIA FELTZING and GERRY GILMORE

Institute of Astronomy & RGO, UK

Abstract. We test for age and metallicity gradients in the Galactic Bulge between the two low extinction windows Baade's window ($l = 1°.1 \ b = -4°.8$) and Sagittarius-I ($l = 1°.3 \ b = -2°.7$). We derive a metallicity difference of ≤ 0.2 dex between BW and SGR-I window. This corresponds to a metallicity gradient of 1.3 dex kpc^{-1}. This steep gradient is reconcilable with the existence of a short scale length inner component to the Bulge, most likely that prominent in the NIR, which perhaps forms a separate entity superimposed on the larger, optical Bulge. Through number counts in colour-magnitude diagram around the turnoff for several fields and clusters, extending over a range in galactic longitude we find no evidence for a significant young stellar population in the Galactic Bulge. Previous suggestions of the existence of a significant young stellar population have most likely incorrectly identified the foreground disk stars.

1. Introduction

The nature and origin(s) of galactic bulges are key aspects of any galaxy formation model (Silk and Wyse, 1993). However, even the present-day properties of bulges in spiral galaxies are not well known (Wyse, Gilmore and Franx, 1997). The existence of smooth and/or discontinuous gradients in age and/or metallicity in the stellar population(s) in the Galactic Bulge can help to discriminate between different scenarios.

The most common approach to determine the properties of the Bulge is to study the so called bulge globular clusters (see e.g. Ortolani *et al.*, 1995; Minniti, 1996; Zinn, 1996 and Barbuy *et al.*, 1998). While primarily motivated by observational convenience, the rationale behind this approach is that these clusters may be valid tracers of the stellar population(s) of the Bulge (see however Zinn, 1996 and Harris, 1998). Formation scenarios relevant to this approach include the possibility that the Bulge has been assembled from numerous such clusters and these are the last surviving of the clusters form a system associated with the Bulge rather than with the rest of the spheroidal component(s) of the Galaxy. However, some of the clusters used, e.g. Ter7, have recently been shown to be associated with the satellite dwarf galaxy Sgr dSph, rather than with the Galaxy itself. It is therefore desirable to observe the stars in the galactic Bulge directly. Direct studies of the Bulge are difficult due to the severe crowding towards the central regions of the Galaxy and the large, patchy, reddening along the line of sight.

 Astrophysics and Space Science is the original source of publication of this article. It is recommended that this article is cited as: *Astrophysics and Space Science* **265**: 337–340, 1999.
© 1999 *Kluwer Academic Publishers.*

HST/WFPC2 images allows us to study the turn-off and main-sequence stars both in the Bulge and in the clusters associated with it. Ortolani *et al.* (1995) observed two clusters, NGC6553 and NGC6528, with HST/WFPC2 and found them to have ages comparable to the halo globular clusters, from this they inferred that the Galactic Bulge population is of the same age as the halo. Vallenari *et al.* (1996) obtained deep HST/WFPC1 photometry of another region in Baade's window (BW). Combining this with ESO/NTT data they found a significant young stellar population which they identified with the Bar, as well as an old stellar population. The old stellar population was found to contribute not more than 30% to the total stellar content observed. Similarly, Holtzman *et al.* (1993) found a dominant intermediate age population from analysis of HST/WFPC1 images.

2. The Metallicity and Age of the Bulge

Having established the parameters for the clusters (see Feltzing and Gilmore for a full discussion), we determine the properties of the Bulge stellar population(s) in BW relative to those of the globular cluster. Then, by comparing the colour-magnitude diagram (CMD) of BW with that of SGR-I we quantify any systematic offsets or gradients in age and metallicity in the field population(s).

2.1. THE BULGE METALLICITY GRADIENT

In order to check a possible metallicity gradient, we compare the CMDs of the BW and SGR-I field by constructing ridge lines from the CMDs and moving these according to their respective reddening. This is a particularly useful method since we are not very vulnerable to incompleteness at the faint end of the ridge lines, the CMDs being comparably complete.

The ridge lines were moved according to their respective reddening. If we attribute all the resulting difference between the ridge lines to metallicity it would imply a difference of 0.5 dex in metal abundance, which corresponds to 3.2 dex kpc^{-1}. However, the relative reddening between the fields may well be in error since they are taken from separate studies and derived by different methods. We thus test two further hypotheses. First, there is no significant metallicity gradient, but there is an age gradient. This led to and age gradient of $\sim$ 6 Gyr over 0.14 kpc a result we consider implausible. Finally, we conservatively assign as much as is possible of the apparent difference in the CMDs to reddening. There is still a residual systematic difference between SGR-I and BW, though of much reduced amplitude. Our best (uncertain) estimate of the difference is $\leq$ 0.25 dex. We conclude that there is an abundance gradient between BW (at projected distance from the centre of 550 pc) and SGR-I, and projected distance 412 pc (assuming a distance of 8 kpc for the Bulge). The amplitude is $\leq$ 0.2 dex, corresponding to $\leq$ 1.3 dex kpc^{-1}.

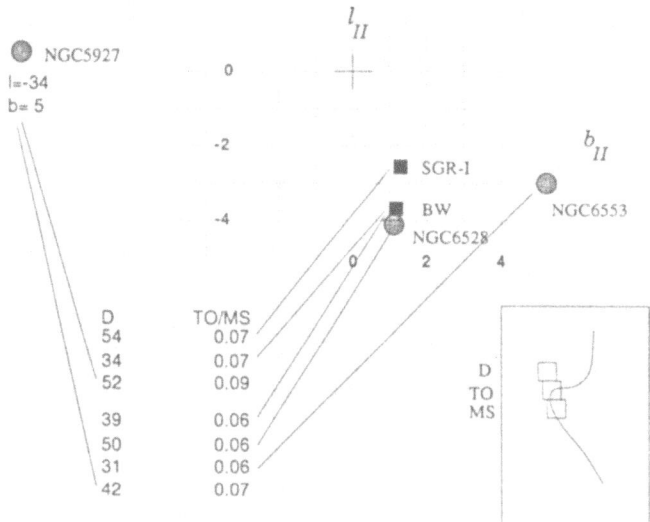

Figure 1. Sketch showing the positions of the fields and clusters. The positions of the 'windows' in which we perform number counts are shown in the schematic CMD that is inset on the right hand side. In the table we give the number counts in the disk bin and the relative numbers for turn-off and main sequence. The three first refer to deep observations and the four last to shallow observations (around 100s) (see Feltzing and Gilmore).

2.2. THE AGE OF THE BULGE

The turnoff in CMDs is sensitive to age and metallicity, but also to contamination by foreground stars. Rather then assuming the stars around the turnoffs in the CMDs of Bulge fields indicate a substantial young Bulge population, we test the possibility that they are foreground disk contamination. We define three 'windows' in our CMDs in which we count the number of stars. The placing of the windows was decided upon by inspecting the CMDs of SGR-I and NGC6528, corresponding to stars above the turn-off (young bulge and/or foreground disk) (D), at the turn-off of an old population (TO) and the main sequence (MS), see Figure 1.

The relative numbers between stars that are suspected to be foreground disk stars and stars that are thought to belong to the Bulge are constant between the fields. The absolute number of stars identified in the disk box indicate that we are seeing true disk stars and not true bulge stars in this part of the diagram since the numbers do not change with l. The COBE Bulge has a scale length of $\sim 1°$ and hence the number counts would change significantly over the area studied, in particular the counts in the observations of NGC5927 and NGC6553 would be much lower than the counts for NGC6528, BW and SGR-I. The ratio of disk to main sequence and turn-off stars for NGC6528 suggests that for this cluster the CMD is significantly affected by field stars.

The spatial distribution of the number counts in I_{814} in the four deep fields are consistent with results from simple simulations using the E2 model of the galactic bulge in Dwek *et al.* (1995).

3. Conclusions

We conclude that the Galactic Bulge have a mean metallicity equal to that of the old disk and that there is evidence for a metallicity gradient.

Furthermore, we conclude that the apparent evidence for a significant young Bulge population in our CMDs is an artifact of disk contamination and that there is no evidence in these HST/WFPC2 data for an age difference between the Bulge and the globular cluster system.

References

Barbuy, B., Bica, E. and Ortolani, S.: 1998, *Astron. Astrophys.* **333**, 117.

Binney, J., Gerhard, O. and Spergel, D.: 1997, *Mon. Not. R. Astron. Soc.* **288**, 365.

Dwek, R.G., *et al.*: 1995, *ApJ* **445**, 716.

Feltzing, S. and Gilmore, G.: *Astron. Astrophys.*, submitted.

Harris, W.E.: 1998, in: D. Zaritsky (ed.), *Galactic Halos: A UC Santa Cruz Workshop, ASP Conf. Ser.* **136**, 33.

Holtzman, J.A., Light, R.M., Baum, W.A., Worthey, G., *et al.*: 1993, *Astron. J.* **106**, 1826.

Minniti, D.: 1996, *Astrophys. J.* **459**, 175.

Ortolani, S., Renzini, A., Gilmozzi, R., *et al.*: 1995, *Nature* **377**, 701.

Silk, J. and Wyse, R.: 1993, *PhR.* **231**, 293.

Vallenari, A., Chiosi, C., Bertelli, G. and Ng, Y.K.: 1996, in: R. Gredel *et al.* (eds.), *The Galactic Centre, ASP Conf. Ser.* **102**, 320.

Zinn, R.: 1996, in: H. Morrison and A. Sarajedini (eds.), *Formation of Galactic Halo ... Inside and Out, ASP Conf. Ser.* **92**, 211.

Wyse, R., Gilmore, G. and Franx, M.: 1997, *Annu. Rev. Astron. Astrophys.* **35**, 637.

THE SUPER METAL RICH COMPONENT OF THE GALAXY

M.L. MALAGNINI
Dipartimento di Astronomia, Università degli Studi di Trieste, Via Tiepolo 11, I-34131, Trieste, Italy

C. MOROSSI
Osservatorio Astronomico di Trieste, Via G.B. Tiepolo, 11, I-34131 Trieste, Italy

A. BUZZONI
Osservatorio Astronomico di Brera, Via Brera, 28, I-20121 Milano, Italy

M. CHÁVEZ
Instituto Nacional de Astrofísica, Optica y Electrónica, CP 72000 Puebla, Mexico

Abstract. We present the results obtained by comparing mid-resolution stellar spectra of super metal rich candidates with synthetic spectra computed in the wavelength range 4850–5400 Å. Atmospheric parameters, derived by using the flux fitting method, are illustrated for a sample of representative stars. The final aim of the project is the definition of a fully consistent metallicity scale for SMR stars.

1. Introduction

Facing the extensive evidence for enhanced chemical evolution among stellar populations in the Milky Way and in external galaxies, we have undertaken a project based on a full comparison between observed and theoretical spectra. We aim at a quantitative definition of the metallicity scale at supersolar regimes in order to better constrain the chemical evolution of the Galaxy bulge and provide a useful interpretative tool for population synthesis studies in external galaxies.

The main requirements of the method are (i) a suitable observational database of stars of super solar metallicity, and, for each star, (ii) a set of reliable atmospheric parameters, namely effective temperature, surface gravity, and overall metallicity (T_{eff}, log g, [M/H]).

2. The Observational Database

We selected a sample of about 200 stars of spectral types from F to M as candidate super metal rich (SMR) stars ([Fe/H] > 0.0) on the basis of the current estimates of metal abundance reported in the literature (Chavez *et al.*, 1996).

Observations were taken during three runs at the 2.12 m telescope of the INAOE 'G. Haro' Observatory in Cananea (Mexico). The Böller and Chivens spectro-

Astrophysics and Space Science is the original source of publication of this article. It is recommended that this article is cited as: *Astrophysics and Space Science* **265:** 341–344, 1999.
© 1999 *Kluwer Academic Publishers.*

TABLE I

Atmospheric parameters for 4 representative stars

HD	Sp. Type	V	B − V	T_{eff}	log g	[Fe/H]	Ref. (Cayrel et al., 1997)
30652	F6V	3.19	0.45	6300		+0.16	75
30652				6462		+0.18	153
30652				6462	4.5	-0.69	266
30652				6380	4.4	+0.02	294
				6371	**4.15**	**+0.06**	**Present work**
83951	F3V	6.14	0.36	6720	4.0	+0.14	502
83951				6789	4.11	-0.02	565
83951				6840	4.00	+0.03	653
				6810	**4.08**	**+0.16**	**Present work**
182572	G8IV	5.16	0.77	5478		+0.42	154
182572				6000		+0.51	173
182572				5663	4.26	+0.50	188
182572				5727	4.13	+0.44	254
182572				5727	4.60	+0.39	270
182572				5663	4.0	+0.21	294
182572				5380	3.92	+0.15	574
				5448	**3.96**	**+0.13**	**Present work**
187691	F8V	5.10	0.56	6146	4.4	+0.10	315
187691				6146	4.4	+0.13	328
187691				6146	4.4	+0.14	376
187691				6146	4.14	+0.09	624
				6079	**4.14**	**+0.02**	**Present work**

graph was working in the range 4600–5500 Å at mid–resolution (2.5 Å FWHM at 5000 Å). So far we gathered about 250 good quality (S/N in the range 50–200) spectra for 139 stars.

3. Atmospheric Parameters

The catalogue by Cayrel et al. (1997), contains 5946 [Fe/H] determinations from high resolution spectroscopic analyses for 3247 stars. The atmospheric parameters for 110 stars of our sample are listed in this Catalogue. Unfortunately, high

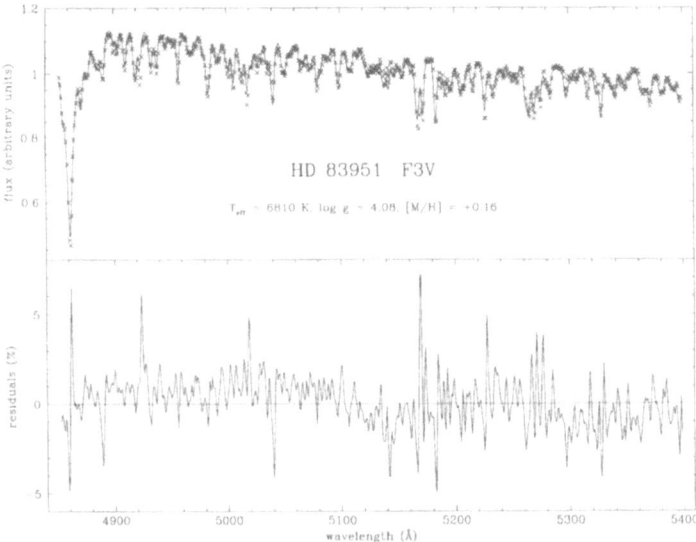

Figure 1. Upper panel: comparison between observations (×) and synthetic fluxes (solid line); Lower panel: percentage residuals.

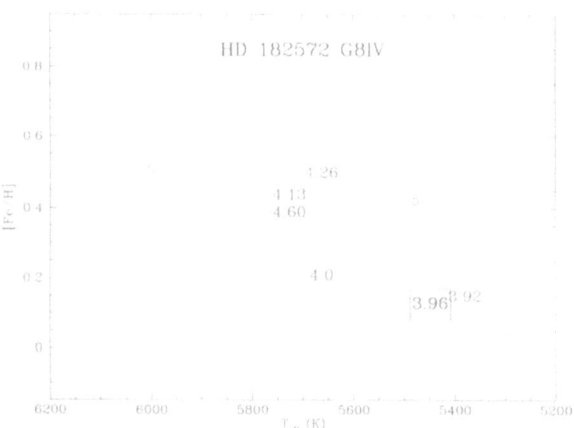

resolution observations in the literature cover, with very few exceptions, a limited wavelength range, thus preventing self–consistent determinations of the complete set of atmospheric parameters. To ensure homogeneity and completeness, we chose to derive the set of atmospheric parameters for each star by using our own observations and a grid of ad hoc computed synthetic spectra (Chavez *et al.*, 1997).

We applied the flux fitting method basically described in Malagnini and Morossi (1983), modified in order to derive simultaneously the three parameters (T_{eff}, log g, [M/H]) from mid resolution spectra instead of spectral energy distributions.

A sample of our results, for four representative stars of different spectral type, is illustrated in Table 1 where our fiducial atmospheric parameters are compared with the available estimates from the Cayrel *et al.* (1997) catalog according to different sources in the literature.

The case of HD 83951 (an F3V star) is reported in more detail in Figure 1. A fair agreement is achieved between original observations and synthetic fluxes with a typical scatter in the fit on the order of $\pm 2\%$ at $1 - \sigma$.

A recognized major problem when deriving individual estimates of metallicity, temperature and gravity from stellar spectra is a sort of 'degeneracy' of these fiducial distinctive parameters. An example of the interplay of the 3 parameters is illustrated in Figure 2; in particular, any safe estimate of $[Fe/H]$ is prone to the uncertainties affecting the determination of T_{eff} and log g. To disentangle the problem we should rely on the simultaneous determination of the atmospheric parameters when matching the synthetic fluxes over the whole wavelength range of the observations.

This work is in progress and the final results for the whole stellar sample will be presented in a forthcoming paper (Malagnini *et al.*, 1999).

Acknowledgements

Partial support from Università di Trieste (60% grant) is acknowledged.

References

Cayrel de Strobel, G., Soubiran, C., Friel, E., Ralite, N. and François, P.: 1997, *Astron. Astrophys. Suppl.* **124**, 299.
Chavez, M., Buzzoni, A., Malagnini, M.L. and Morossi, C.: 1996, in: A. Buzzoni, A. Renzini and A. Serrano (eds.), *Fresh Views of Elliptical Galaxies, ASP Conf. Ser.* **86**, 249.
Chavez, M., Malagnini, M.L. and Morossi, C.: 1997, *Astron. Astrophys. Suppl.* **126**, 1.
Malagnini, M.L. and Morossi, C.: 1983, *Statistical Methods in Astronomy*, ESA SP–201, 27.

OLD, LOW-MASS, METAL-RICH (SMR) STARS

G. CAYREL DE STROBEL and Y. LEBRETON
Dasgal, Observatoire de Paris-Meudon, F-92195 Meudon Cedex, France

C. SOUBIRAN
Observatoire de Bordeaux, BP21, F-33270 Floirac, France

E.D. FRIEL
Departement of Astronomy, Boston University and NSF Division of Astronomical Sciences, Arlington, VA 22230, USA

Abstract. The authors of this paper try to disentangle the many problems arisen from a new enlarged sample of nearby low-mass, metal-rich-stars. These stars have reliable absolute magnitudes, deduced from Hipparcos parallaxes, precise bolometric corrections, effective temperatures and metal abundances from high resolution detailed spectroscopic analyses. Their ages have been derived from a grid of isochrones calculated with up to date physics. The main goal of this paper is to determine the ages of the slightly evolved SMR stars. Among those with well determinated ages about 80% of them have intermediate ages of (2 to 5 Gyr), but only 20% have ages of 8 Gyr or more. Nevertheless, the existence of very old metal-rich stars is confirmed.

1. Introduction

The many not yet resolved problems concerning the kinematics, dynamics, place of birth, chemical compositon and age of the so-called SMR stars request experts in very wide domains of physics. The appellation 'super-metal-rich' for stars, which, once analysed spectroscopically in detail, exceed only by small amounts the metallicity of the Sun is surely unappropriate. It was introduced in the late sixties, and was based on photometric results which have displayed a metal rich group (with overabundances up to a factor of 10 with respect to the Sun) of late type stars in the solar neighbourhood. Since the paper by Cayrel de Strobel (1987) the number of SMR stars analysed in detail has tripled, (from 29 to 98). This number is still not breathtaking and represent only the 3% of all the stars contained in the [Fe/H] Catalogue (1996 Edition). It shows that in the immediate vicinity of the Sun SMR stars are rare, but, their study is essential for the determination of the upper limit of metal enrichment in stars and for the better comprehension of the nucleosynthesis processess which have acted in supernova explosions and in AGB stars. The discovery by Withford and Rich (1983) of a group of metal rich giants ([Fe/H] up to +0.6 dex) in the bulge of our Galaxy supported the existence in the Galaxy of a strong metal rich population of stars. Is this population spread over the whole

Astrophysics and Space Science is the original source of publication of this article. It is recommended that this article is cited as: *Astrophysics and Space Science* **265**: 345–352, 1999.
© 1999 *Kluwer Academic Publishers.*

Galaxy, or only concentraded in the Bulge ? and if existing, what is the span of its age ? the span of its metallicity ? Are some members of this population passing through the nearby solar neighbourhood in such a way that they are bright enough that high resolution, high S/N spectroscopic techniques may be applyed to them? Recently new high resolution, high S/N spectroscopic observations have enlarged the sample of nearby G and K disk-stars, with metallicities equal or higher than those of the Hyades: [Fe/H] = +0.14 ± 0.05, as the conventional threshold for the super-metal rich (SMR) stars. Two groups of metal rich stars have been displayed: the first is composed by low-mass G and K dwarfs and subgiants, the second by giants. The first group reflects the chemical composition in which these stars are formed, whereas the metal enrichment of the giants could, at least for some of them, reflect nucleosynthesis produced in their center. All along this paper our attention will be focussed on about a hundred of unevolved or slightly evolved SMR stars belonging to the first group. In the limits of our 8 pages, mainly we shall discuss the ages of the new sample of 98 SMR stars. In Section 2 we present in Table I the most intersting stars, all carefully analysed in detail. In Section 3 the observational HR diagram of the whole sample will be displayed in Figure 1 and discussed. The ages of the stars in connection with their metallicity are critically discussed in Section 4. A short conclusion will end this rapid overview on nearby SMR stars.

2. Spectroscopic Search for SMR Stars: A New Enlarged Sample

The '1996 Edition' of the Catalogue of [Fe/H] determinations by Cayrel de Strobel *et al.* (1997) and six recent papers by, Porto de Mello (1996), Feltzing and Gust-afsson (1997), Castro *et al.* (1997), Flynn and Morell (1997), Fuhrmann (1998) and Gustafsson *et al.* (1998) have allowed to increase the number of SMR, spec-troscopically analysed, stars.

In Table I we present a sub-sample of 22 of the most interesting stars of our sample of 98 SMR stars. We shall publish elsewhere (*Astron. Astrophys. Suppl.*) the complete list of nearby, spectroscopic analysed, SMR stars. Table I is divided in three parts: results from spectroscopic detailed analyses, ([Fe/H], log T_{eff}), results from Hipparcos astrometry (M_{bol} and perigalactic distance), from the comparison between an observational and a theoretical (log T_{eff}, M_{bol}) diagram (age). Many of the stars in Table I have been analysed in detail more than once, and we thought that it is interesting to publish individually the results on [Fe/H] and on T_{eff} of the different authors. Very recently two papers by Feltzing and Gustafsson (1997) and Castro (1997 *et al.*) have largely debated on 'The most metal rich stars in the Galactic Disk' (Feltzing, 1997), and have combined abundance ratios of trace elements with kinematical data of the most metal rich stars in the solar neighbour-hood. For instance, Barbuy and Grenon (this Conference) have found that SMR dwarfs with very excentric orbits contained more oxygen than what is expected from standard models of galactic chemical evolution of the Disk. In this Conference

TABLE I

Stellar parameters of a sample of SMR stars: the 7 columns give: identification, V, [Fe/H], log T_{eff}, bol. mag., age in Gyr, perigalactic distance in Kpc. For some stars, results of [Fe/H] and Mbol of two authors, or more are given: YC stands for Yves Chmielewski, LD for Licio Da Silva, BE for Bengt Edwardsson, SF for Sofia Feltzing, RG for Raffaele Gratton, KF for Klaus Fuhrmann, CN for Corinne Neuforge, JH for John Hearnshaw, BC for Bruce Campbell, and GC for Giusa Cayrel.

HD, name	V	[Fe/H]	log T_{eff}	M_{bol}	Age Gyr	Peri. Galac. Dist kpc
126614	8.8	0.55	3.74	4.546	–	5.345
138776	8.74	0.48	3.756	4.562	4.0	5.669
99109	9.10	0.45	3.732	5.079	–	4.827
115589	9.57	0.44	3.732	5.333	–	5.157
182572 31Aql (JH)	5.17	0.44	3.756	4.206	5.0	5.657
182572 (SF)	5.17	0.36	3.759	4.195	9.0	5.657
182572 (LD)	5.17	0.36	3.750	4.195	9.0	5.657
160691 μ Ara	5.12	0.41	3.748	4.164	9.0	8.417
75732 ρ^1 Cnc	5.94	0.40	3.727	5.327	–	7.522
171999 A	8.34	0.40	3.725	5.333	(11.5)	4.139
68988	8.20	0.37	3.775	4.315	(1.00)	6.366
10780	5.63	0.36	3.734	5.504	–	7.850
134987	6.47	0.36	3.766	4.376	3.00	6.527
110010	6.99	0.35	3.776	4.182	2.10	7.625
136442	5.54	0.35	3.681	3.075	4.3	–
85503 μ Leo (RG)	3.88	0.34	3.657	0.273	(0.6)	7.516
85503 (GC)	3.88	0.31	3.634	0.176	(0.6)	7.516
1461	6.47	0.33	3.773	4.578	–	6.748
32147	6.21	0.28	3.665	6.090	–	8.089
144585 (SF)	6.32	0.27	3.766	3.962	4.7	7.562
144585 (BE)	6.32	0.23	3.766	3.960	4.7	7.562
128621 α CenB (YC)	1.35	0.26	3.726	5.565	–	8.479
128621 (CN)	1.35	0.27	3.716	5.553	–	8.479
128620 α CenA (YC)	–0.01	0.22	3.763	4.282	4.0	8.492
128620 (CN)	–0.01	0.22	3.763	4.284	4.0	8.492
128620 (LD)	–0.01	0.23	3.766	4.287	3.5	8.492
128620 (BE)	–0.01	0.15	3.757	4.271	5.0	8.492
190360 (JH)	5.71	0.26	3.748	4.620	8.0	6.522
190360 (KF)	5.73	0.24	3.747	4.638	8.0	6.522
26846 39EriA (BC)	4.87	0.21	3.663	0.389	0.7	6.601
26846 39EriB (BC)	8.57	0.19	3.766	4.506	–	6.601
217014 51 Peg (KF)	5.46	0.20	3.763	4.468	3.0	7.172
217014 51 Peg (BE)	5.46	0.06	3.760	4.458	4.0	7.172

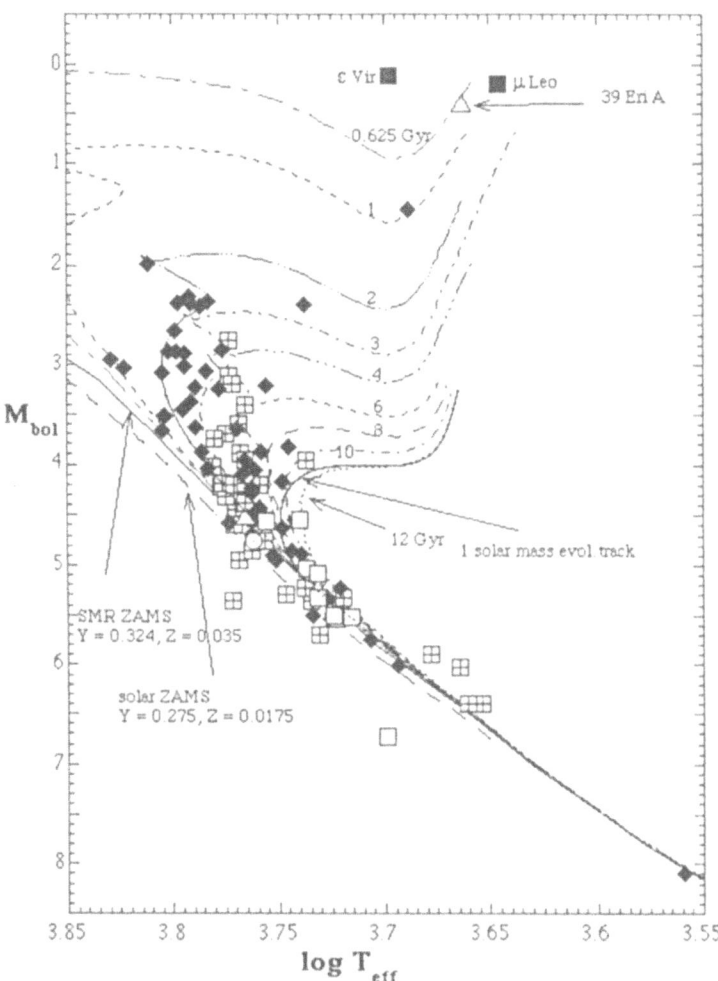

Figure 1. SMR-composition theoretical HR diagram (Lebreton1998): $Y = 0.324$, $Z = 0.0346$, $\ell/Hp = 1.64$, with an overplot of 5 samples of SMR dwarfs and subgiants: crossed squares from Feltzing and Gustafsson(1997), filled diamonds: from [Fe/H] Catalogue (Cayrel de Strobel, Soubiran *et al.*, 1997), open squares from Castro *et al.* (1997), filled squares: the giants ε Vir and μ Leo, open triangles: the binary 39 Eri, and the Sun with its usual symbol: $\odot$.

the poster by Soubiran gives an insight on the Kinematics and Orbits of SMR stars. The ages of these stars have not yet been discussed: this is what we are undertaking in the next section.

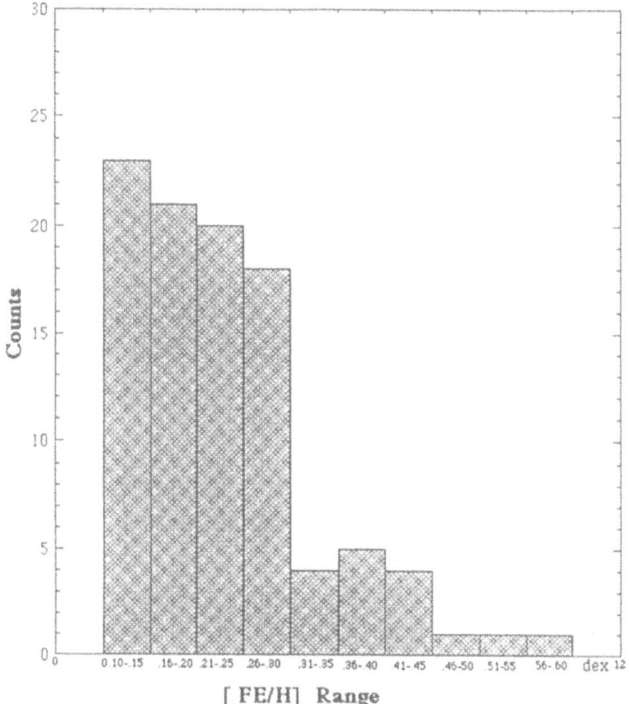

Figure 2. Metallicity – histogram of a sample of 98 SMR stars in the range: [Fe/H] = +0.1 to +0.5 dex.

3. Comparison of the Observational ($\log\ T_{eff}$, M_{bol}) Diagram with a Theoretical Diagram Computed for a SMR Chemical Composition

We present in Figure 1 the observational ($\log\ T_{eff}$, M_{bol}) diagram. We have used Hipparcos parallaxes (ESA, 1997) as the basis for M_{bol} determinations of the program stars (column 5) and we have taken the bolometric corrections of Alonso *et al.* (1996). The observational ($\log\ T_{eff}$, M_{bol}) diagram is compared to a metal rich grid of stellar structure models composed by a metal rich and a metal normal ZAMS, 9 isochrones (from the Hyades age (0.625 Gyr), to the oldest Disk-star ages, 10 Gyr), and 1 solar mass, metal rich, evolutionary track. The grid has been calculated by Lebreton (1998) who has derived the value of He for her models from a solar $\Delta Y / \Delta Z = 2.8$ enrichment relation:

$$Y_{SMR} = 0.324,\ Z_{SMR} = 0.035,\ \ell/Hp = 1.64$$
$$Y_{\odot} = 0.275,\ Z_{\odot} = 0.0175,\ \ell/Hp = 1.64.$$

The observational points in Figure 1 derive chiefly, from three sets of stars. Two of them have been observed and analyzed in detail by Feltzing and Gustafsson (1997) and by Castro *et al.* (1997). The last set has been built up by SMR stars listed in the [Fe/H] Catalogue (Cayrel *et al.*, 1997). For clarity, error bars have not been superimposed in Figure 1, but an estimate of the mean standard errors on effective temperature and bolometric magnitude of the program stars is: $\sigma T_{eff} = \pm 50$ K, and $\sigma M_{bol} = \pm 0.15$ mag.

The whole sample of SMR stars is plotted in Figure 1. Of these stars, 30% are not, or not evolved enough, to allow their 'evolutionary' age estimation, but circumscribe quite well the theoretical SMR ZAMS. As regards the evolved stars in Figure 1, their age was estimated on the best fitting isochcrone, or interpolated between the two nearest isochrones. Column 6 of Table I contains the estimated age of the evolved stars. One star in our sample, the nearby K dwarf HD17925, shows a very strong lithium line in its spectrum. This strong Li line indicates that it must be very young.

We shall comment on four interesting cases:

i) The effective temperatures of 31 Aql, a well known slightly evolved SMR star, as obtained by three different authors, (cf. Table I) involve for this star ages from 5 to 9 Gyrs, but note that, in this large brackett of age, 31 Aql remains an old SMR star.

ii) If one attributes the effective temperature of Neuforge-Verheecke, (1995) to α Cen B, and not that of Chmielewski *et al.* (1992), the star is placed, right, on the metal rich ZAMS.

iii) The SMR binary, 39 Eri, (Cayrel de Strobel, 1991) is composed by a giant and a dwarf. The dwarf fits very well the theoretical SMR ZAMS, and the Giant, 39 EriA, fits well the 0.625 isochrone. Its follow, that the age of the dwarf 39 Eri B is also known, and that the stellar system is very young.

iv) A new carefull detailed analysis by Fuhrmann (1998) shows that the star 51 Peg with a planetary companion, discovered by Mayor and Queloz (1995), is metal rich.

4. The Age Versus [Fe/H] Relation for SMR Field Stars

A metalllicity and an age histogram are shown in Figures 2 and 3, respectively. They have been constructed with the help of columns 3 and 6 of Table I. The bins representing strong metal rich stars are poorly populated in the histogram of Figure 2, as well as those representing old metal rich stars in that of Figure 3. Seven stars of our sample are older than 8 Gyr. If, the interruption of SMR star formation around 7 Gyr is real or not, and if we are faced to a first burst of SMR stars around 10 Gyr requests more observations. This burst would fit the model of chemical evolution of the Galaxy elaborated by Rama (1991). Figure 4 reproduces the [Fe/H] versus Age relation of the sample. This figure shows that the SMR stars have been formed since the existence of the Thick and Thin Disk, but, the support of observational material is too small to substantiate the idea that the SMR phenomenon was more active in the past than it is now.

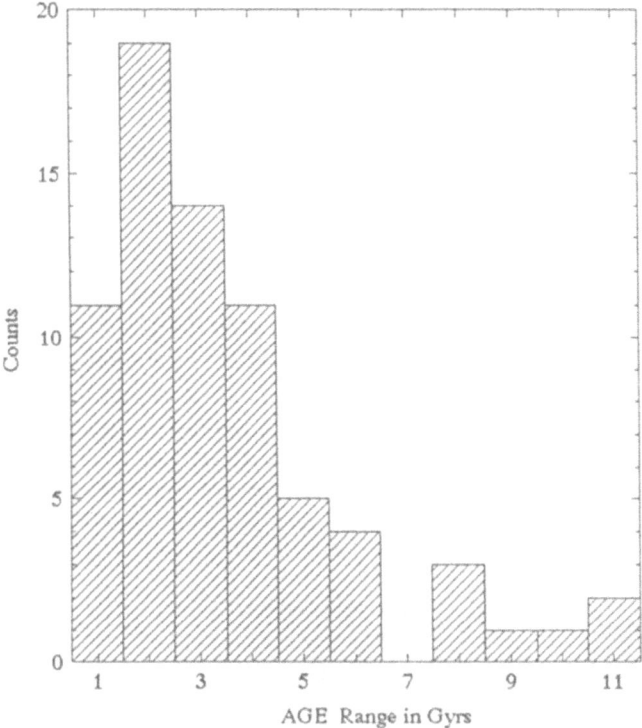

Figure 3. Age – histogram of a subset of the same sample of stars as in Figure 2, for which an age determination was possible.

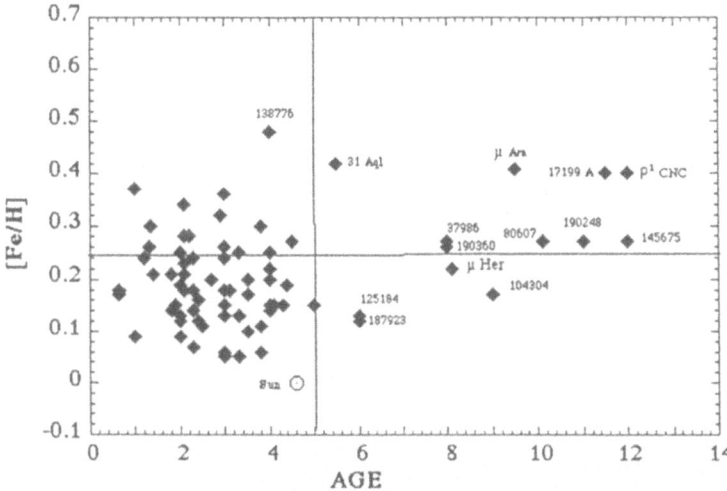

Figure 4. Metallicity versus age for the same subset of SMR stars as in Figure 3. Note that two subsets seem to appear: a cluster of stars younger than 5 Gyr and a less populous tail extending to 12 Gyr.

5. Conclusions

The number of detailed spectral analyses of SMR stars has tripled in 13 years: from 30 to more than 90. All the analysed stars have Hipparcos parallaxes and well determined radial velocities. The metallicity parameter of these stars [Fe/H] range between 0.12–0.5. The very high photometric metallicities, ($\neq$ 1.0 dex), found in these stars in the 70 ies, are no more accepted.

In comparing the observational (log T_{eff}, M_{bol}) diagram of the sample to a theoretical diagram, we get informations on the state of evolution of each star, including the age of the slightly evolved stars in the sample. It is confirmed that the SMR phenomenon has always existed from the very beginning of the thick Disk, but it is only marginally confirmed that it was more active in the past than it is now. These informations have shown that it is highly worthwhile to continue the effort in: getting accurate data for a more important sample of SMR stars in order to increase the number of actual subgiants, the only stars for which reliable ages can be obtained.

Acknowledgements

The authors want to thank Arlette Noels, Michel Mayor, Pierre Magain, Sofia Feltzing, Corinne Neuforge-Verheecke and Stephane Ury for exciting discussions and for supplying data in advance of publication.

References

Castro, S., Rich, R.M., Grenon, M., Barbuy, B. and McCarty, J.K.: 1997, *Astron. J.* **114**, 376.
Cayrel de Strobel, G.: 1987, *J. Astrophys. Astron.* **8**, 141.
Cayrel de Strobel, G.: 1991, Conference proceedings, in: P. Giannone, F. Melchiorri and F. Occhionero (eds.), *Evolutionary Phenomena in the Universe* **32**, 27, SIF, Bologna.
Cayrel de Strobel, G., Soubiran, C., *et al.*: 1997, *Astron. Astrophys. Suppl.* **124**, 299.
Chmielewski, Y., Friel, E., Cayrel de Strobel, G. and Bentolila, C.: 1992, *Astron. Astrophys.* **263**, 219.
Feltzing, S.: 1997, *IAU Highlights 1997*, Kyoto.
Feltzing, S. and Gustafsson, B.: 1997, *Astron. Astrophys. Suppl.* **129**, 237.
Flynn, C. and Morell, O.: 1997, *Mon. Not. R. Astron. Soc.* **286**, 617.
Fuhrmann, K.: 1998, *Astron. Astrophys.* **338**, 161.
Gustafsson, B., Karlsson, E., Olsson, B., Edwardsson, B. and Ryde, N.: 1998, *Uppsala Astronomical Observatory*, Preprint No. 133.
Lebreton, Y.: 1998, Proceedings of the STScT Symposium, in: M. Livio (ed.), *Unsolved Problems in Stellar Evolution*, Cambridge University Press.
Mayor, M. and Queloz, D.: 1995, *Nature* **378**, 355.
Neuforge-Verheecke, C.: 1995, Thèse Université de Liège.
Neuforge-Verheecke, C. and Magain, P.: 1997, *Astron. Astrophys.* **328**, 261.
Porto de Mello, G.F.: 1996, *Thesis: Observatorio Nacional*, Rio de Janeiro.
Whitford, A.E. and Rich, R.M.: 1983, *Astrophys. J.*

OLD METAL-RICH STARS: KINEMATICS AND ORBITS

C. SOUBIRAN

Observatoire de Bordeaux, F-33270 Floirac, France

This poster presents the kinematics of the metal-rich stars which were discussed by G. Cayrel in this conference. The 121 stars were selected to have reliable metallicities from detailed analyses of high resolution, high S/N spectra. They have also precise (U, V, W) velocities from Hipparcos astrometry and radial velocities from several sources. Hot stars and evolved stars were eliminated from the sample to avoid metal excess due to internal physical processes. The metallicities of this sample are supposed to reflect the metal abundance of the gas which formed the stars. In her review G. Cayrel showed that the SMR phenomenon has always existed from the very beginning of the thick disk, and that old metal rich stars are precious tracers of the chemical evolution. Here the origin of metal-rich stars is investigated by their kinematical properties. From the velocity distribution, several populations can be identified in the sample: a moving group, thin disc stars, old disc stars, and 5 stars with a large rotational lag typical of the thick disc, as shown in Figure 1. Admitting the usual correlation between age and kinematical behavior, this suggests that a large range of ages is present in this sample of metal-rich stars. The moving group (squares), with a spatial extension of 100 pc, has a common velocity with the Hyades, but only one of the stars is already known as a Hyades member. This could indicate that the Hyades stream extends over a larger scale than previously estimated (Eggen, 1996). The standard deviation of the V velocity is only 3 km s^{-1}. Despite a slightly larger U velocity than most of the thin disc stars, this group has a circular orbit typical of the thin disc population, and a mean metallicity of 0.20. In both figures, it can also be seen that one third of the sample has old disc or thick disc kinematics (triangles), with perigalactic distances reaching 3 kpc. Stars with the highest eccentricities show a large dispersion in [Fe/H] and the most metal rich stars have the kinematics of old stars. Assuming these stars are old, there was some metal rich gas to form them several Gyrs ago. This is not in agreement with a smooth regular chemical enrichment of the Galaxy, but corresponds more to a tumultuous evolution where fragments could have evolved independently before their mixing in the Galaxy. Grenon, in this volume, explains these high velocities by perturbations due to a galactic bar.

Astrophysics and Space Science is the original source of publication of this article. It is recommended that this article is cited as: *Astrophysics and Space Science* **265**: 353–354, 1999.
© 1999 *Kluwer Academic Publishers.*

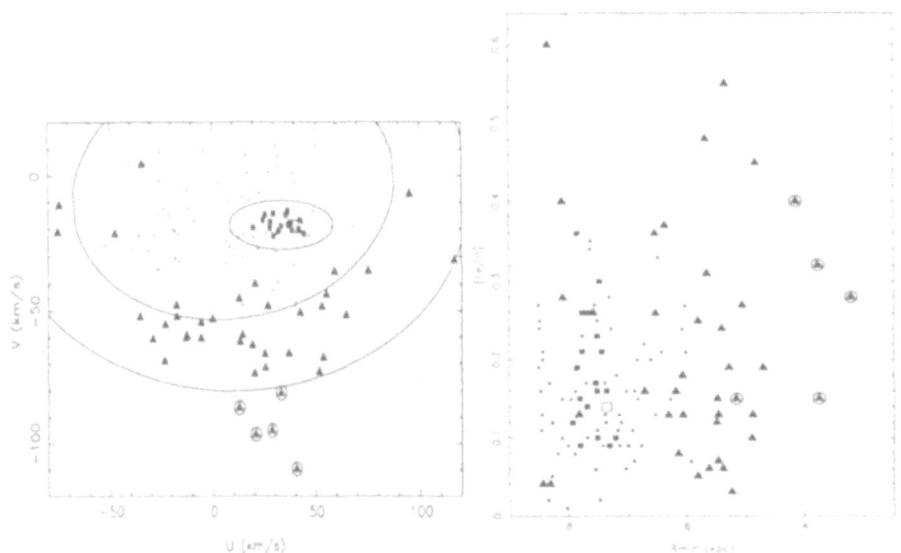

Figure 1. Left: (U,V) distribution of the sample. The ellipses correspond to the 3σ contour of the moving group, and of the young disk and old disk according to the solar motion and dispersions estimated by Dehnen and Binney (1998). Squares indicate stars within the 2σ contour of the moving group in the 3D velocity space, the open square shows the position of the Hyades, triangles corres-pond to stars outside the 3σ contour of the young disk, encirled triangles correspond to stars outside the 3σ contour of the old disk. Right: [Fe/H] versus perigalactic distance. Symbols as in left.

Acknowledgements

Special thanks to R. and G. Cayrel for letting me enjoy their huge knowledge and experience in the physics of stars over the last years. Many thanks to M. Oden-kirchen for computing and commenting the orbits and to M. Mayor for giving his radial velocities before publication. Finally I would like to thank J. Braine for correcting my English.

References

Cayrel de Strobel, G.: 1999, this volume.
Dehnen, W. and Binney, J.: 1998, *Mon. Not. R. Astron. Soc.* **298**, 387.
Eggen, O.: 1996, *Astron. J.* **111**, 1615.
Grenon M., 1999, this volume.

GLOBULAR CLUSTERS AND FIELD STARS IN THE BULGE

S. ORTOLANI

Department of Astronomy, University of Padova, Vicolo dell'Osservatorio 5, Padova I-35122, Italy

Abstract. In this paper we discuss the characteristics of the stellar content of the galactic bulge excluding the stars within a few parsec from the galactic center. The bulge clusters and the field stars are compared to the disk population. A scenario with a flattened bulge extending to about 3–4 Kpc from the galactic center is presented. There is evidence for an old bulge stellar population, decoupled from the disk.

1. Introduction

There is a growing evidence in the recent years that of galactic bulge is character-ized by a nearly solar metallicity, old stellar population, with an age comparable to the halo (Larson, 1990; Lee, 1992; Ortolani *et al.*, 1995), but there are still some claims for a younger, disk-like component (Raha *et al.*, 1992; Kiraga *et al.*, 1997; Rucinski, 1997; Ng *et al.*, 1996).

The structure of the bulge is another open issue because there is still a debate whether its shape is spherical or flattened (Wyse *et al.*, 1997; Minniti, 1995) or a bar (Spergel *et al.*, 1996), or a disk-like system (Zinn, 1985; Armandroff, 1989) and where is the transition between the disk and the bulge itself.

One of the major problems in the interpretation of the results from optical-near IR observations of the bulge stellar fields is coming from the reddening and from the contamination of the disk population projected on the bulge itself.

The results coming from radio emissions are less ambiguous.

Observations of typical disk tracers as neutral hydrogen (Kerr *et al.*, 1976) molecular lines (Burton, 1976) HII regions, pulsars (Georgelin and Georgelin, 1976; Taylor and Cordes, 1993), supernovae remnants (Clark and Caswell, 1976), clearly show a sharp cutoff of all these disk population I tracers at a distance of about 4–4.5 Kpc from the Sun, in the direction of the galctic bulge, just inside the molecular ring. The 3–4 kpc inner region is free from population I indicators (with the exception of a very limited central region with a radius of about 100 pc which is populated by the 'galactic center population', King, 1989). This excludes any inner extention of the thin disk into the galactic bulge.

These radio sources, however, are tracers of very young components only. A conservative interpretation is that the interstellar ambient in the bulge is gas and dust depleted and that the actual star formation is inhibited, but these observations

 Astrophysics and Space Science is the original source of publication of this article. It is recom-mended that this article is cited as: *Astrophysics and Space Science* **265:** 355–359, 1999.
© 1999 *Kluwer Academic Publishers.*

are not able to put constraints on the low limit of the age of star formation episodes older than about 100 million years.

The galactic bulge appears decoupled from the disk components also from the analysis of kinematic data: in spite of its relatively fast rotation, the angular momentum per mass unit of the bulge is much lower than that of the disk, with a distribution resembling more that of the halo (Wyse *et al.* (1997).

2. The Bulge Fields

Many efforts have been recently dedicated to the analysis of the stellar field bulge population, mainly in the direction of the Baade Window, with the aim to derive the metallicity and the age of its stellar population. Low resolution spectra have been obtained by Rich (1988) and Sadler *et al.* (1996), and high resolution by McWilliam and Rich (1994), while the metallicity gradient within the bulge is presented by Frogel in these proceedings. We will not discuss here these results because they are the subject of other reviews at this meeting (see Barbuy for recent data and Matteucci *et al.* for theoretical models). We just remind that there is now a quite firm evidence of a slightly less than solar abundance for the iron, with some relevant differences in individual elements, in particular α-elements which appear enhanced.

The current interpretation is an early enrichment by SNs II.

Age estimates are more controversial.

The detection of many RR Lyrae stars in the bulge with a relatively high metallicity has been taken by Lee (1992) as supporting evidence for a very old bulge, possibly older than the halo.

On the other hand, the presence of bright long period variables (LPV) and high luminosity OH/IR stars has been interpreted in the past as evidence for a relatively young stellar component (1–2 Gyr). However from the recent IR results of Guarnieri *et al.* (1997) obtained in the bulge globular cluster NGC 6553, one can clearly see that a metal rich population can produce very bright LPVs still with an old age.

Terndrup (1988), from the analysis of several field color-magnitude diagrams (CMD), concluded that the prominent, young main sequence seen in the diagrams is due to disk contamination, while the bulge population appears to have a rather old age. Early HST color-magnitude diagrams from the Baade Window by Holtzman *et al.* (1993) have been interpreted by the authors as indication of an intermediate age bulge population.

Many sources of uncertainty, however, affect the results from field CMDs because the horizontal branches (HB) are not well defined and the ages, deduced from direct comparison with the isochrones are sensitive to the adopted distance, reddening and chemical composition. Unfortunately the value of the reddening and the reddening law itself, in the direction of the galactic bulge, are not very well known. The reddening value is typically taken from independent studies on nearby globular clusters or from the color of the HB clump or from the colors and luminosities of giants distributed in larger fields. The selective to absolute absorption ratio $R = Av/E(B-V)$ is often assumed around 3.5, but large discrepancies can be found in the literature (Cardelli *et al.*, 1989; Walton and Barlow, 1993; Grebel and Roberts, 1995).

Furthermore the reddening is variable in very small fields: Armandroff (1989) proposed to explain the tilted HBs in low latitude globular clusters with a differential reddening in a few arcminute field of view, and Ortolani *et al.* (1990) showed the smearing effects in the color-magnitude diagram of the globular cluster NGC 6553, along the reddening line, due to differential reddening, in a two arcminute field. Further tests on HST images of the same cluster reveal reddening variations of 0.1–0.2 magnitudes in a 5″ scale (corresponding to about 0.1 pc at the distance of the cluster). A clumpy structure of the molecular clouds in an even smaller scale has been found in recent radio studies. AU-scale variations, corresponding to an angular size of a small fraction of an arcsecond at a distance of 1 Kpc, have been claimed by several groups (see e.g. Marscher *et al.*, 1993; Falgarone and Phillips, 1996). Appreciable molecular absorption variations have been also detected in a a year time-scale (Wiklind and Combes, 1997). If the dust is linked with the molecular clouds these results give a clear limitation to any direct method of dating the field population which is reddening sensitive.

For this reason we concentrated our studies on globular clusters projected in the direction of the galactic bulge.

3. Globular Clusters in the Bulge: The Data

Until recently no reliable data were available for most clusters in the direction of the galactic bulge (Racine and Harris, 1989; Armandroff, 1989). We now have collected CCD CMDs for 16 out of 17 known globular clusters within 5° (Barbuy *et al.*, 1998).

It is now possible for the first time to study the properties of this inner system on a statistical base. 13 more known cluster candidates at a galactic latitude below 5° have been observed at galactic longitudes extending up to 36 degrees. From the 29 studied objects 23 resulted to be genuine star clusters, while 6 are different objects (one supernova remnant, one galaxy, two planetary nebulae, one open nearby cluster and another just non-existing, see Bica *et al.*, 1993; Bica *et al.*, 1995; Ortolani *et al.*, 1996).

Distances, metallicities and reddening have been derived from the CMDs by comparison with templates.

4. Results and Conclusions

The 23 genuine globular clusters are sharply divided into two groups according to their HB morphology: red HB and blue HB. No intermediate HB have been found. The red HB ones are about twice the blue HB ones.

The discovery of a blue HB is more difficult than a red HB because, in the color-magnitude diagrams the blue tail is blended with the main sequence of the foreground disk stars. This can explain why only very few blue HB clusters have been so far detected.

Red HB clusters have also tilted red giant branches, recalling metal rich morphologies close to solar abundances. On the contrary blue HB clusters have very steep, almost vertical red giant branches similar to those of the most metal poor clusters in the halo.

The two samples, however, are well spatially mixed and also their kinematics do not seem to be sharply different.

The spatial distribution of these clusters is quite different from that discussed by Armandroff (1989) because our sample is restricted to a low galactic latitude with a scale height a factor 3 smaller.

The distribution of our clusters results much more concentrated toward the galactic center, compatible with Kent's (1991) bulge exponential distribution.

The metallicity distribution is quite flat and broad extending from about twice solar to less than one tenth solar abundance. The sample is small but resembles that of bulge stars (McWilliam and Rich, 1994).

Taking into account that also the kinematical properties of these clusters are very similar to the field stars (Minniti, 1995), the picture which comes out is one of a bulge with common parameters for the stellar population of both globular clusters and field stars, making the studied globular clusters good tracers for the bulge population.

As concerning the age, we detected the main sequence turnoff for 5 clusters. Following the reddening (and distance) free method originally proposed by Iben and Renzini (1984) based on the HB – main sequence turnoff luminosity difference, we found that four of them, NGC 6553, 6528 (Ortolani et al., 1995), NGC 6652 (Ortolani et al., 1994), and Terzan 1 (Ortolani et al., in preparation) are old, comparable to the age of 47 Tuc. Two of them (NGC 6528 and 6553), in particular, show luminosity functions identical to those of the Baade Window (see Figure 3 in Ortolani et al., 1995). The main conclusion is that the age of the bulge population is comparable to that of the bulge clusters and in turn it should not very different from the age of the halo.

References

Armandroff, T.: 1989, *Astron. J.* **97**, 375.
Barbuy, B., Bica, E. and Ortolani, S.: 1998, *Astron. Astrophys.* **333**, 117.

Bica, E., Ortolani, S. and Barbuy, B.: 1993, *Astron. Astrophys.* **270**, 117.

Bica, E., Clariá, J.J., Bonatto, C., Piatti, A.E., Ortolani, S. and Barbuy, B.: 1995, *Astron. Astrophys.* **303**, 747.

Burton, J.B.: 1976, *Annu. Rev. Astron. Astrophys.* **14**, 275.

Clark, D.H. and Caswell, J.L.: 1976, *Mon. Not. R. Astron. Soc.* **174**, 267.

Cardelli, J.A., Clayton, G.C. and Mathis, J.S.: 1989, *Astrophys. J.* **345**, 245.

Falgarone, E. and Phillips, T.G.: 1997, *Astrophys. J.* **472**, 191.

Geogelin, Y.M. and Georgelin, Y.P.: 1976, *Astron. Astroophys.* **49**, 57.

Grebel, E.K. and Roberts, W.J.: 1995, *Astron. Astrophys. Suppl.* **109**, 293.

Guarnieri, M.D., Renzini, A. and Ortolani, S.: 1997, *Astrophys. J.* **477**, L21.

Holtzman, J.A., *et al.*: 1993, *Astron. J.* **106**, 1826.

Iben, I.Jr. and Renzini, A.: 1984, *Phys. Rep.* **105**, 329.

Kent, S.M.: 1992, *Astrophys. J.* **387**, 181.

Kerr, F.J., Harten, R.H. and Ball, D.L.: 1976, *Astron. Astrophys. Suppl.* **25**, 391.

King, I.: 1989, in: R. Buser (ed.), *The Milky Way as a Galaxy*, Sauverny-Versoix, Switzerland, by Gilmore, G., King, I., van der Kruit, P., p. 41.

Kiraga, M., Paczynski, B. and Stanek, K.: 1997, *Astrophys. J.* **485**, 611.

Larson, R.B.: 1990, *Publ. Astron. Soc. Pacific* **102**, 709.

Lee, Y.W.: 1992, *Astron. J.* **104**, 1780.

Marscher, A.P., Moore E.M. and Bania, T.M.: 1993, *Astrophys. J.* **419**, L101.

McWilliam, A. and Rich, M.R.: 1994, *Astrophys. J. Suppl.* **91**, 749.

Minniti, D.: 1995, *Astron. J.* **109**, 1663.

NG, Y.K., Bertelli, G., Chiosi, C. and Bressan, A.: 1996, *Astron. Astrophys.* **310**, 771.

Ortolani, S., Barbuy, B. and Bica, E.: 1990, *Astron. Astrophys.* **236**, 362.

Ortolani, S., Bica, E. and Barbuy, B.: 1994, *Astron. Astrophys.* **286**, 444.

Ortolani, S., Renzini, A., Gilmozzi, R., Marconi, G., Barbuy, E., Bica, E. and Rich, R.M.: 1995, *Nature* **377**, 701.

Ortolani, S., Bica, E. and Barbuy, B.: 1996, *Astron. Astrophys.* **306**, 134.

Raha, N., Sellwood, J.A., James, R.A. and Kahn, F.D.: 1991, *Nature* **352**, 411.

Rich, R.M.: 1988, *Astron. J.* **95**, 828.

Rucinski, S.M.: 1997, *Astron. J.* **113**, 407.

Sadler, E., Rich, R.M. and Terndrup, D.: 1996, *Astron. J.* **112**, 171.

Spergel, D.N., Malhotra, S. and Blitz, L.: 1996, in: D. Minniti and H.W. Rix (eds.), *Spiral Galaxies in the Near-IR*, Springer-Verlag, p. 128.

Taylor, J.H. and Cordes, J.M.: 1993, *Astrophys. J.* **411**, 674.

Terndrup, D.M.: 1988, *Astrophys. J.* **96**, 884.

Walton, N.A. and Barlow, M.J.: 1993, in: H. Dejonghe and H.J. Habing (eds.), *Galactic Bulges*, Kluwer Academic Publishers, p. 337.

Wiklind, T. and Combes, F.: 1997, *Astron. Astrophys.* **328**, 48.

Wyse, R.F.G., Gilmore, G. and Franx, M.: 1997, *Annu. Rev. Astron. Astrophys.* **35**, 637.

Zinn, R.: 1985, *Astrophys. J.* **293**, 424.

HUBBLE SPACE TELESCOPE WFPC2 COLOR-MAGNITUDE DIAGRAM OF NGC 6287

L. KELLAR FULLTON

Observatoire de Genève, CH-1290 Sauverny, Switzerland

P.B. STETSON

Dominion Astrophysical Observatory, 5071 West Saanich Road, Victoria BC V8X 4M6, Canada

B.W. CARNEY

University of North Carolina at Chapel Hill, Department of Physics & Astronomy, Chapel Hill, NC 27599-3255, USA

The globular clusters in the central regions of the Galaxy hold important clues to the formation of the bulge and inner halo. The most metal-poor of these clusters may set a lower limit to the age of the inner halo if the enrichment rate of gas at the center increased monotonically with time and if the central metal-poor clusters belong to the bulge or inner halo *proper*.

In particular, NGC 6287, a central cluster with [Fe/H] ≈ -2.1 dex, was put forth as a candidate for the oldest cluster in the Galaxy by van den Bergh (1993). Recent color-magnitude diagrams (CMDs) of it have been constructed by Stetson and West (1994) in the optical and Davidge (1998) in the infrared, but due to the cluster's distance and location, neither CMD reached stars fainter than the main-

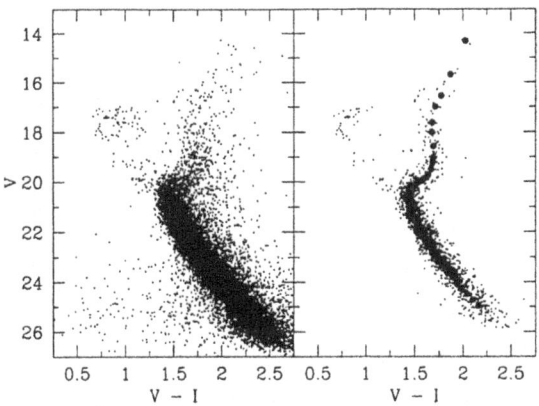

Figure 1. Color-magnitude diagram of NGC 6287.

 Astrophysics and Space Science is the original source of publication of this article. It is recommended that this article is cited as: *Astrophysics and Space Science* **265:** 361–362, 1999.
© 1999 *Kluwer Academic Publishers.*

sequence turnoff. Here, for the first time, the cluster's lower main-sequence has been clearly revealed in two orbits of *Hubble Space Telescope** time (Figure 1).

Figure 1 shows, at left, an ALLFRAME V,V−I CMD for all stars in the *HST* field of view. The large scatter is caused primarily by differential reddening. Shown at right are the cleaned CMD and fiducial sequence produced by selecting only well-measured stars in the least differentially-reddened part of the field of view. The sequence is tightened considerably, although up to 0.2 mag of differential extinction in E(V−I) remain.

The brightest portions of the CMD confirm all the features discovered by Stetson and West. Additionally, 45 blue straggler candidates are revealed, as well as a small group of five stars at the level of the horizontal branch just to the blue of the red giant branch. These stars may represent an extremely depopulated red HB, the base of the asymptotic giant branch, or something else entirely. Detailed analysis of the diagram, along with additional data at our disposal will reveal the age of NGC 6287 compared to other metal-poor globular clusters in the halo. These data include detailed stellar abundances, which will be determined by BWC from high resolution echelle spectra, and deep IR photometry, from *NICMOS* images obtained by us earlier this year.

References

Davidge, T.J.: 1998, *Astron. J.* **116**, 1744.
Stetson, P.B. and West, M.J.: 1994, *Publ. Astron. Soc. Pacific* **106**, 726.
van den Bergh, S.: 1993, *Astrophys. J.* **411**, 178.

* Based on observations with the NASA/ESA *Hubble Space Telescope*, obtained at the Space Telescope Science Institute, which is operated by AURA, under NASA contract NAS 5-26555.

DARK MATTER IN THE GALACTIC BULGE: A MICROLENSING POINT OF VIEW

C. ALARD

DASGAL, Observatoire de Paris, 77 avenue Denfert Rochereau, 75014 Paris, France

Abstract. Microlensing is one of the most promising technique to probe the density of dark matter in the Galactic Bulge. We review briefly the history of microlensing and comment on the discovery of high optical depth in the direction of the Bulge. This optical depth is several times larger than the first theoretical predictions. We will show that some of the discrepancy can be resolved by taking into account the effect of self amplification of stars into the Galactic Bar. We will also explain that the optical depth is contaminated with the contribution of faint unresolved stars. However, we emphasize that a category of sources, the bulge giants are bright enough to escape the bias due to unresolved sources. Finally we show that even if self amplification in the bar is taken into account, the optical depth to giants is hard to reproduce. We conclude by saying that in the near future this excess in the bulge optical depth should be clarified and measured with good accuracy. In particular good progress should be made when the analyze of the last observations of Bulge giants will be completed by the MACHO group. The future implementation of the image subtraction technique in the data pipelines should also help to overcome the bias in the measurement of the optical depth to turnoff stars.

1. Introduction

Measurement of the microlensing optical depth and lensing rates is a new and interesting way to probe the Bulge mass function. This is a powerful method to determine the content of dark matter in the bulge of our Galaxy that could be in the form of brown dwarf or low mass stars. The microlensing technique has developed quickly in previous years and is becoming one the important tools in astrophysics. The initial microlensing experiments were not directed at the Galactic Bulge itself. The idea was rather to probe the dark mass in the Galactic Halo by monitoring stars in the Large Magellanic Cloud. Paczyński (1986) proposed that monitoring a few millions stars in the LMC would allow to find amplification of these stars by low mass (dark) objects passing near the line of sight. Light amplification of a few stars was predicted in case of a massive Halo of low mass stars or brown dwarfs. Objects with very small masses (typically the mass of the earth or even the mass of mercury) would be visible in such experiment, provided good enough sampling of the light curves. Two collaborations EROS (Aubourg *et al.*, 1993) and MACHO (Bennett *et al.*, 1993) started almost immediately massive observational investigations with small telescopes having large fields of view. Both collaborations announced almost in the same time the detection of a microlensing event (Aubourg *et al.*, 1993; Alcock *et al.*, 1993). However, following another idea of

Astrophysics and Space Science is the original source of publication of this article. It is recommended that this article is cited as: *Astrophysics and Space Science* **265**: 363–370, 1999.
© 1999 *Kluwer Academic Publishers.*

Bohdan Paczyński (1991), the OGLE team (Udalski *et al.*, 1993) started investigations of microlensing amplification of stars located in Baade's Window, which is
located in the Bulge of our galaxy at about 4 degrees from the Galactic Center.
The basic idea was to test the microlensing technique. Due to the large number
of stars on the line of sight towards the center of the Galaxy, light amplification
by normal stars should be observed and measurable. Direct comparison of the
observed lensing rates with theoretical predictions should be possible, allowing
to check for the ability of the microlensing experiments to measure the optical
depth. However the first experimental result is telling quite a different story from
what had been predicted (Udalski *et al.*, 1994) ... the Optical depth measured by
OGLE ($\tau \simeq 3 \ 10^{-6}$) is several times larger that the first theoretical prediction for
the Galactic Bulge ($\tau \simeq 4 \ 10^{-7}$) made by Paczyński (1991). This result was later
confirmed by MACHO (Bennett *et al.*, 1994). The reason why such a discrepancy
exists is still a topic of controversy. It might indicate that a large fraction of the
Bulge mass is dark. However we will see that some of the optical depth can be
explained within a classical scheme that does not require dark matter. Although we
will see that the total optical depth that is observed is hard to reproduce. We will
discuss these question in details in the forthcoming paragraphs.

2. The Microlensing Collaborations

Before we explore the physics of microlensing towards the Bulge, we would like to
present the different collaborations that are working on Galactic Bulge fields. The
largest team is the MACHO group, they are operating a wide field CCD camera
mounted on a 1.3m telescope in Mount Stromlo observatory. This telescope is dedicated to the project. The large area of the CCD camera allows this collaboration to
cover a large area of about 10 square degrees towards low extinction Bulge fields.
The OGLE I project was also equipped with a CCD camera mounted on a 1m
telescope, although the field of view of the instrument was much smaller, and the
total area covered is less than a square degree (mostly in Baade's window). In this
first OGLE project (OGLE I), which produced most the OGLE data published to
date, the telescope was not dedicated to the project. OGLE I time allocation was
mostly composed of periods with significant moon lightning, which restricted the
observations to the I band. A new OGLE II project has been recently started. OGLE
II is now equipped with a new 1.3 m telescope, fully dedicated to the project. The
telescope is currently equipped with a single CCD, but it will be soon replaced
with a large CCD mosaic. The last project to search for microlensing towards the
Bulge is the DUO project (Alard and Guibert, 1997). DUO takes advantage of
about 200 photographic plates taken with the ESO 1 m schmidt telescope. The
lower quality of the digitized photographic images is compensated by the large
field (25 sq degrees). All these collaborations have been very successful in their

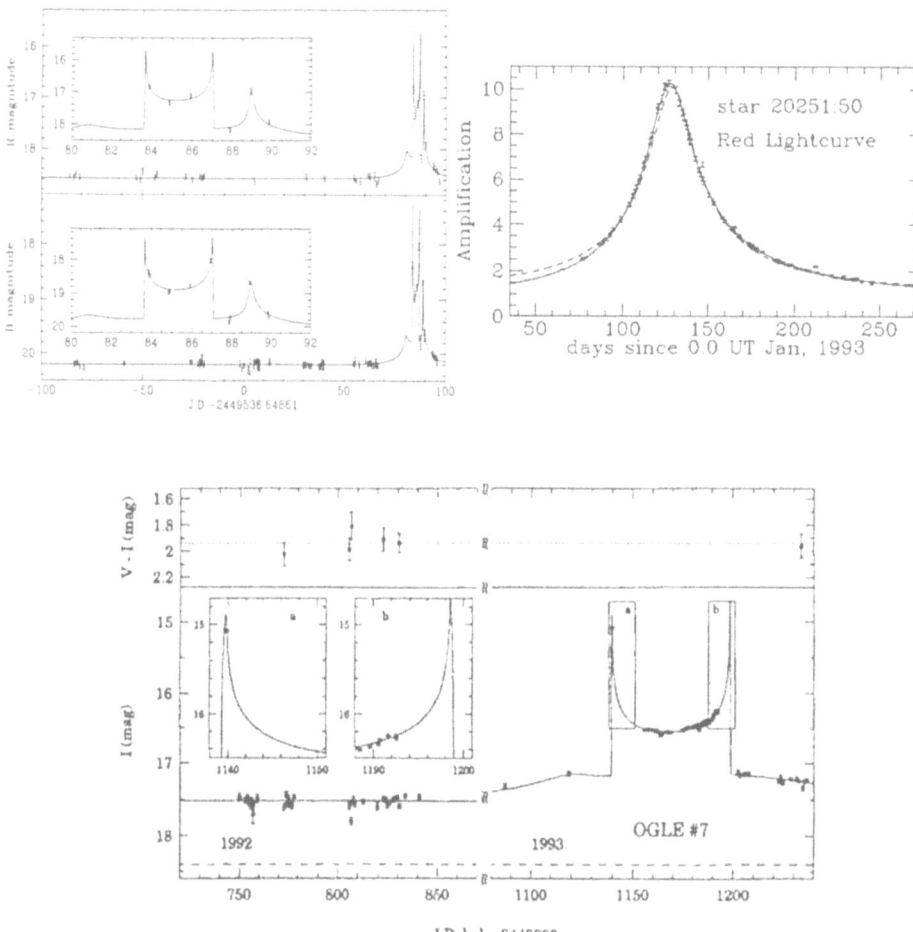

Figure 1. An illustration of the richness of the microlensing results towards the Bulge. Top left is the light curve of the DUO 2 binary lens (Alard, Mao and Guibert, 1995). Top right is the demonstration of the parallax effect by the MACHO group using an event with long duration (Alcock *et al.*, 1997). The solid line is the fit without taking into account the motion of the earth, and the dotted line shows the fit with parallax effect. The last figure present the light curve of the OGLE #7 binary lens (Udalski *et al.*, 1994).

search for microlensing, sample light curves of microlensing events are presented in Figure 1.

3. Microlensing Basics

The light deviation by a point mass forms 2 images of the source. If the lens has the mass of a star, the angular separation between the images will be very small, and we will observe a blend of these 2 images.

Using simple geometric arguments (for more details see for instance Schneider, Ehlers and Falco (1992)) one can show that the addition of the 2 images result in a light amplification of the form:

$$A = \frac{(u^2 + 2)}{u\sqrt{(u^2 + 4)}}$$

Where u is the impact parameter (or projected distance between the source and the lens).

Applying directly this formula would lead to infinite amplifications for null impact parameter. Actually this will not happen due to the finite size of the source. It can be easily understood, if we see the source as a star with some diameter. If the impact parameter of the center of the star is zero, the other points on the surface of the star will not have zero impact parameter, leading by summation to a finite amplification. Towards the Galactic Bulge this effect should be detectable only if the source is a giant (see Witt, 1995).

For an impact parameter of 1 (impact parameter equal to the Einstein radius) we get an amplification A_E which is about 1.34. It means that any star within the Einstein radius of a given lens will amplified by at least A_E. We define the optical depth as the probability that a source be amplified by at least A_E at a given time by a lens population located in a field occupying a solid angle $d\Omega$. It is easy to see that this optical depth is just the sum of the solid angles occupied by the different lenses located in the field divided by the solid angle of the field $d\Omega$. If we define the distance to the source, D_s, and the mass density of lenses $\rho(D_d)$, it is straightforward to show that the Optical depth can be expressed as:

$$\tau = \frac{4\pi G}{c^2} \int_0^{D_s} \rho(D_d) \frac{D_d(D_s - D_d)}{D_s} dD_d$$

For real calculation of the Bulge optical depth the previous formula needs to be averaged over all source distances. It requires to assume a number density law for the sources and a luminosity function. For detailed calculations, see Paczyńsky (1991) or Alard (1997).

The previous formula is used for all theoretical calculations of the optical depth. However experimental measurements of the optical depth rely on a different approach. The optical depth can be defined from the observed distribution of the microlensing events time scale, T_i (Alcock *et al.*, 1997). If we define the total duration of the observations E, and the experimental efficiency $\mathrm{eff}(T_i)$, the formula for the optical depth reads:

$$\tau_{obs} = \frac{\pi}{4E} \sum_i \frac{T_i}{\mathrm{eff}(T_i)}$$

The above formula express that the probability that a source be lensed above some threshold at a given time, is just the ratio of the averaged time spent above the threshold to the total observing time.

4. Contribution of Unresolved Sources to the Optical Depth

Due to the high density of the Galactic Bulge fields, and with the spatial resolution of the microlensing experiments, crowding effects and blending of the stellar images is important. Blending is such that within the resolution radius of a star that is detected in the image, there are a few tens of unresolved stars. This is a major problem, since one of these faint unresolved star can be amplified, and due to the blending we will attribute the lensing event to the nearby star that is visible in the image. The net effect is that we will observe a category of additional microlensing events that have nothing to do with real microlensing of the stars we are monitoring in the image. The amplitude of the effect depends on the position of the limiting magnitude (the limit beyond which star cannot be separated any more in the image) with respect to the Bulge turnoff. Unfortunately the limiting magnitude of the microlensing experiments is close to the Bulge turnoff, it is located in a region where the number of star rise very steeply with increasing magnitudes. It means that for a star near the limiting magnitude in the image (most of the stars that are monitored), there are a few other stars that we do not resolve but which are only 1 or 2 magnitudes fainter. The resulting effect is large, typically as large as the normal optical depth to the resolved stars. The effect is the largest for the DUO experiment and the smallest for OGLE which goes deeper. Even for OGLE the additional optical depth due to unresolved stars is about 60% of the normal optical depth (see Alard, 1997, and also Han, 1997), it is 190% for DUO. The effect for the MACHO experiment is somewhere between DUO and OGLE. We see that the effect of unresolved stars is a major problem, and that it might explain some of the excess observed in the optical depth. To conclude on this effect it is important to note that due to the large blending with the resolved source, the light curves of amplified unresolved objects will differ from the normal (unblended) microlensing light curves. By using the high photometric resolution provided by the image subtraction technique, Alard (1999) could demonstrate that 3 of the OGLE microlensing events show very good evidence for high blending, and consequently must result from the amplification of faint unresolved objects. It clearly demonstrates the existence of the effect.

5. Self Lensing of Stars in the Galactic Bulge

Another important clue to understand the high optical depth towards the Galactic Bulge is to consider that in the first theoretical models of microlensing a major contribution has been omitted. The basic assumption of the model was to consider amplification of a bulge star by a disk star situated close to the line of sight. However the possibility that a bulge star be amplified by another bulge star has not been taken into consideration. This is of particular importance if the Bulge has a 3 dimensional bar structure. In a line of sight towards a bar with low inclination, there is a strong geometric effect that increases the number of lenses (see Figure 2).

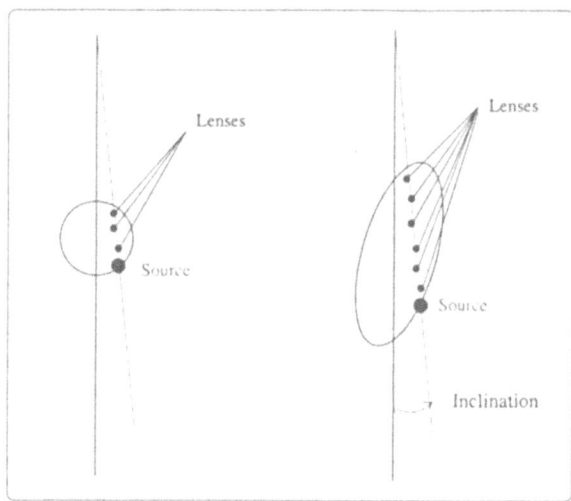

Figure 2. Demonstration of the geometrical effect due to the bar at the center of the Galaxy. Note the high density of lenses along the line of sight for the barred model, in comparison to the axysymmetric model. The effect is particularly large if the bar has low inclination with respect to the line of sight.

Due also to the elongation of the bar, the mean distance between the source and the deflector is larger than in an axysymmetric model. It is well known from gas dynamics (Binney, 1991) and photometric study of the giants (Stanek *et al.*, 1997) that the structure of our Galaxy near its center presents large assymmetries that are consistent with a bar structure. However there is no perfect agreement concerning the inclination and dimensions of the bar. The microlensing optical depth is extremely sensitive to the inclination of the bar with respect to the line of sight. The geometry of the Bar can be approached by photometric methods (Dwek *et al.*, 1995; Binney, 1997; Bissantz *et al.*, 1997), or by dynamical methods (Zhao and Mao, 1996; Fux, 1997). The optical depth calculated with these different bar models (around $1.5 \ 10^{-6}$) is always somewhat smaller than the observed optical depth (close to $3 \ 10^{-6}$). Of course it can be said that most of the additional optical depth is due to the unresolved stars. This is certainly possible for most of the stars (turnoff stars), but not for the Bulge giants. Bulge giants are much brighter than the limiting magnitude, and consequently are not much affected with blending effects. Thus the optical depth to giants is almost not contaminated with the bias due to unresolved stars. The problem is that, even if the optical depth for giants as measured by MACHO (alcock *et al.*, 1995) is somewhat smaller than for turnoff stars, it is still too large to be explained only by the effect of the Bar. Unless we assume rather large mass to light ratio for the Bulge, it is very hard to reconstruct all the optical depth we observe, even from the more optimistic model for the central Galaxy.

6. Concluding Remarks

We have seen that the large value of the optical depth towards the Galactic Bulge is difficult to explain, even when additional effects are taken into account. We also explained that the bias due to unresolved stars is an annoying effect, which amplitude is rather difficult to estimate. But we have also seen that a possible solution was to use the optical depth calculated for Bulge giants only. The recent results accumulated by the MACHO collaboration will soon provide us with large sample of microlensed Bulge giants, and should allow to measure with good accuracy the excess of optical depth towards the Bulge. Another promising approach is the developpement of the image subtraction technique (Alard and Lupton, 1998). Optimal image subtraction (very close to the photon noise) can be performed in a very reasonable computing time for the microlensing data. The image subtraction is an excellent way to eliminate all blending effects, and consequently is a natural solution to the problem of microlensing of faint unresolved stars. Unbiased quantities can be calculated even for very faint highly blended stars by using the image subtracted light curve. Thus to conclude we can say that the progress of the microlensing technique has been very quick. even if problems have been encountered in the early developpement of the technique, we see that solutions have been found and that the near future will provide us with definitive, and important results concerning the amount of dark matter in the Galactic Bulge. In addition to the measurement of the Bulge mass function, microlensing can also provide informations in other fields of astrophysics, like for instance the detection of planetary objects around stars (Albrow *et al.*, 1998), or even measurements of the size of giants stars (Sasselov, 1998). Microlensing is becoming mature, and will certainly become one of the major fields of astrophysics.

Acknowledgements

I would like to thank the organizers of this conference for the invitation to present this review.

References

Albrow, M., *et al.*: 1998, *Astrophys. J.* **509**, 687.
Alard, C.: 1999, *Astron. Astrophys.*, in press (astro-ph/9808092).
Alard, C. and Lupton, R.: 1998, *Astrophys. J.* **503**, 325.
Alard, C.: 1997, *Astron. Astrophys.* **321**, 424.
Alard, C. and Guibert, J.: 1997, *Astron. Astrophys.* **326**, 1.
Alard, C. and Guibert, J.: 1997, *Astron. Astrophys.* **300**, L17.
Alcock, C. *et al.*: 1997, *Astrophys. J.* **486**, 697.

Alcock, C. *et al.*: 1993, *Nature* **365**, 621.

Aubourg, E. *et al.*: 1993, *Nature* **365**, 623.

Bennett, D. *et al.*: 1994, *Astron. Astrophys. Suppl., Meeting* **185**, 17.03.

Bennett, D. *et al.*: 1993, *Astron. Astrophys. Suppl., Meeting* **183**, 72.06.

Binney, J.: 1997, *Mon. Not. R. Astron. Soc.* **288**, 365.

Binney, J.: 1991, *Mon. Not. R. Astron. Soc.* **270**, 703.

Bissantz, *et al.*: 1997, *Mon. Not. R. Astron. Soc.* **289**, 651.

Dwek, *et al.*: 1995, *Astrophys. J.* **445**, 716.

Fux, R.: 1997, *Astron. Astrophys.* **327**, 983.

Han, C.: 1997, *Astrophys. J.* **507**, 102.

Paczyński, B.: 1991, *Astrophys. J.* **371**, L63.

Paczyński, B.: 1986, *Astrophys. J.* **304**, 1.

Sasselov: 1998, *ASP Conf. Ser.* **154**, 383.

Schneider, P., Ehlers, J. and Falco E.: 1992, *Gravitational Lenses*, Springer-Verlag.

Stanek, C., *et al.*: 1997, *Astrophys. J.* **477**, 163.

Udalski, A., *et al.*: 1994, *Acta Astron.* **436**, 103.

Udalski, A., *et al.*: 1994, *Acta Astron.* **44**, 165.

Udalski, A., *et al.*: 1993, *Acta Astron.* **43**, 289.

Witt, H.: 1995, *Astrophys. J.* **449**, 42.

Zhao, H. and Mao, S.: 1996, *Mon. Not. R. Astron. Soc.* **283**, 1197.

CHEMO–DYNAMICAL EFFECTS OF BARS IN GALACTIC EVOLUTION

L. MARTINET and D. RABOUD

Geneva Observatory, CH–1290 Sauverny, Switzerland

Abstract. Some effects of a bar on stellar orbits and on chemical gradients are presented for spiral galaxies in general and for the Solar Neighborhood in particular.

1. Introduction

Infrared surface photometry data reveal bars which were not previously detected in bluer bands. The percentage of discovered barred galaxies is more and more increasing. Then the bars seem to be ubiquitous, at least for some important fraction of galaxy life and they certainly play an essential role in evolution of discs.

The Milky Way is a barred galaxy with the near side of the bar pointing the first galactic quadrant. Near-IR surface photometry as well as discrete source counts or gas/stellar kinematics bring evidences for that.

Effects of bars on the morphological evolution of galaxies such as peanut or box shape of the bulge or gas inflow towards the center have been extensively discussed in the past (for a recent review see Pfenniger, 1998). In the context of the present conference, since G. and R. Cayrel have been interested in kinematical and chemical evolution of our Galaxy (amongst other topics!), we specifically emphasize the effects of a bar on stellar orbits and the connection between the bar strength and the chemical gradient. In fact, in the existing modelisation of chemical evolution, the non-stationarity and the non-axisymmetry of the galactic potential have been generally neglected. The resulting simplifications could lead to erroneous conclusions.

2. Stellar Orbits

The presence of bar in our Galaxy implies characteristic orbital behaviors of stars. Let us distinguish here two categories of orbits of interest: 1) elongated orbits confined to the bar, trapped about the long–axis x_1 family of periodic orbits (bar particles), 2) 'hot' orbits which essentially display a typical chaotic behaviors, erratically wandering between regions inside the bar and outside corotation. Whereas orbits of kind 1) have an Hamiltonian $H < H(L_{1,2})$, the 'hot' orbits have $H >$

Astrophysics and Space Science is the original source of publication of this article. It is recommended that this article is cited as: *Astrophysics and Space Science* **265**: 371–374, 1999.
© 1999 *Kluwer Academic Publishers.*

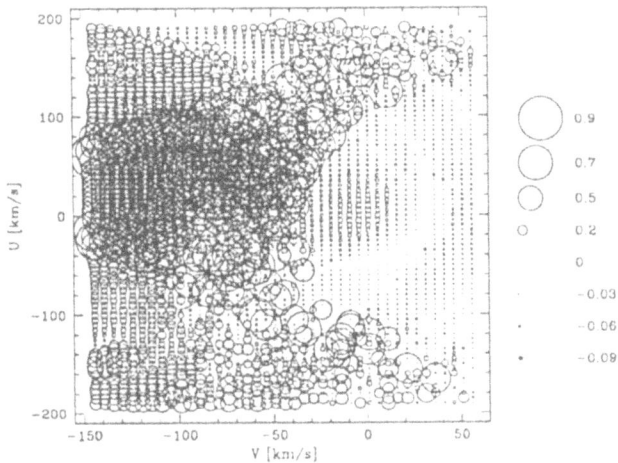

Figure 1. (u, v) plane of residual velocities in which we plot circles, the radius of which being the difference between pericentric distances in axisymmetric and perturbed potentials of the Galaxy measured in units of axisymmetric pericentric distance (see text for details).

$H(L_{1,2})$, $H(L_{1,2})$ being the Hamiltonian at the Lagrange point L_1 and L_2 for a zero velocity in the rotating frame of the bar. Examples of these kinds of orbits can be seen for ex. in Figure 15 of Sparke and Sellwood (1987). In the solar neighborhood, we are not able to observe the first category of orbits as the Sun is outside corotation. But at the present time we observe near the Sun some number of hot orbit stars which are able to visit our neighborhood after spending more or less time in the inner region of the bar. For these stars, orbital parameters such as pericentric (R_p) or apocentric (R_a) distances are more difficult to estimate than in the axisymmetric case, because they can be strongly time dependent.

In order to give an idea of the difference of R_p in a barred field (B) with respect to the axisymmetric case (A), we calculate orbits in one of Fux's models (Fux, 1997) for various initial conditions of local velocities relative to the Sun. We estimate Δ the relative difference between $R_p(A)$ and $R_p(B)$ in units of $R_p(A)$.

We symbolically represent Δ by circles of radius Δ in the (u, v) grid (Figure 1). Whereas stars on small epicyclic energy have small $\Delta (\lesssim 0.2)$, it can be seen that, for old stars of larger velocities, differences between $R_p(A)$ and $R_p(B)$ can be very sensitive (of the order of 4 kpc for ex. for $R_p(A) \approx 5$ kpc). People interested in origin places of these stars must pay attention to this fact.

3. Local Kinematics of Old Stars

A feature of local stellar motions which practically passed unnoticed in the literature is a global asymmetry of the star distribution in the (u, v) plane of velocities relative to the Sun, in addition to the vertex deviation which essentially concerns

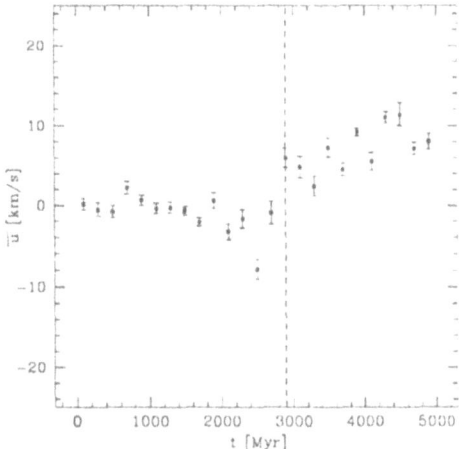

Figure 2. Time behavior of $\bar{u}(\phi = 30°)$ over 5 Gyr, averaged in bins of 200 Myr, for stars located within a torus between $\tilde{R} = 7.5$ and 8.5 (in the initial units of the simulation m08, Fux, 1997). In this simulation, the bar is completely settled at the right of the vertical dashed line.

stars of small epicyclic energies: the mean u–velocity ($\bar{u}$) of the old disc stars, as inferred from various samples, appears significantly different from zero even if we correct it for the solar motion, contrarily to the expectation for a stationary axisymmetric galaxy. This anomaly was already apparent in various papers (references in Raboud *et al.*, 1998). In particular Mayor (1972) emphasized a significant excess $\bar{u}(h) > 0$ (h = angular momentum) for stars having a mean asymmetrical drift $\langle S \rangle \approx 20 - 30$ km s^{-1}.

In Raboud *et al.* (1998), we have suggested that the presence of a bar in our Galaxy (axis ratio $b/a = 0.5$, bar pattern $\Omega_P = 50$ km s^{-1}, corotation radius $= 4.3$ kpc) would be responsible for this phenomenon. In the m08 Fux simulation we followed the mean radial motion $\bar{u}$ of particles with respect to the Galactic Center as a function of time, across a toroidal region of 1 kpc diameter at a distance $R = R_0 = 8$ kpc from the Galactic Center. Before 2 Gyrs, the global deviation from axisymmetry is negligible: $\bar{u}$ is never very different from zero during this early period. However, as soon as the bar is stabilized ($t \gtrsim 2.9$ Gyr) $\bar{u}$ significantly differs from 0 and depends on ϕ, the angle between the bar and the line Sun–Galactic Center.

In Figure 2, where the values of $\bar{u}$ have been averaged within bins of 200 Myr width, a significantly positive value is obtained when the bar is stabilized. As shown in Grenon's talk (this conference), metal-rich old disc stars are typical contributors to the kinematical anomaly mentioned above. They populate 'hot' orbits defined in the previous section. The deviation Δ for these stars can be non negligible.

4. Bars and Chemical Gradient

Martin and Roy (1994) established a correlation between the radial O/H abundance gradient in the discs of SBs and the bar axis ratio (b/a) in the sense that stronger bars have a rather flat gradient, whereas steeper gradients are observed in galaxies with weak or no bars. Locally the dependence of the chemical gradient on bar axis ratio could be weighted by the star formation efficiency. Moreover, in some galaxies, two different slopes for the O/H abundance gradient have recently clearly been inferred (references in Martinet and Friedli, 1997).

The numerical simulations reported in Martinet and Friedli (1997) show this feature and indicate that the age of the bar is another factor influencing the chemical gradient. In the bar region, during the early phase of its existence, the gradient is maintained since the gas dilution (following the significant gas inflow) is compensating the heavy-element production in the furious star formation then going on in the nuclear vicinity. After $t = 1200$ Myr, the disc abundance gradient remains essentially flat. This results of course in the presence of *two different radial abundance gradients in young strongly barred galaxies.* So, there is a 'steep-shallow' break in the slope profile obtained in the simulations, as already observed in some galaxies.

Thus, the slope profile of radial abundance gradient in SB galaxies is monitored by the strength and the age of a bar.

5. Conclusion

Two major implications of the previous considerations are:
1. In the modelisation of chemical evolution, the non-stationary effects coming from non-axisymmetric structures must be taken into account.
2. The studies of relation between kinematics and chemical composition as well as the places of star formation in the Milky Way must pay attention to the effect of a bar on orbital parameters of old stars.

References

Fux, R.: 1997, *Astron. Astrophys.* **327**, 983.
Martin, P. and Roy, J.-R.: 1994, *Astrophys. J.* **424**, 599.
Martinet, L. and Friedli, D.: 1997, *Astron. Astrophys.* **323**, 363.
Mayor, M.: 1972, *Astron. Astrophys.* **18**, 97.
Pfenniger, D.: 1998, Abundance profiles: diagnostic tools for galaxy history, *Astron. Astrophys.* **335**, L61.
Raboud, D., Grenon, M., Martinet, L., Fux, R. and Udry, S.: 1998, *Astron. Astrophys.* **335**, L61.
Sparke, L. and Sellwood, J.: 1987, *Mon. Not. R. Astron. Soc.* **225**, 653.

MODELISATION OF THE BARRED MILKY WAY

R. FUX and L. MARTINET
Geneva Observatory, Ch. des Maillettes 51, CH-1290 Sauverny, Switzerland

We have built many self–consistent 3D stellar dynamical barred models of our Galaxy extending beyond the Solar circle. These models are obtained by N-body evolution of various bar unstable axisymmetric initial conditions and include three distinct mass components: a nucleus-spheroid (NS) standing for the inner bulge and the stellar halo, a stellar disc and a dark halo (DH). The initial velocity distributions are based on radially anisotropic solutions of the hydrodynamical Jeans equations and are bounded by the escape velocity. The evolved models, sampled from the simulations with a frequency of 200 Myr, span a large variety of different barred models.

The location of the observer relative to the bar has been constrained in each model using the COBE/DIRBE K-band map corrected for foreground extinction by dust in the region $|\ell| < 30°$ and $3° < |b| < 15°$, assuming a homogeneous mass-to-light ratio and by minimising the mean quadratic relative residual between model and observed fluxes corrected for statistical noise. The best matching models, with residuals down to 0.3%, suggest a bar inclination angle $\varphi_o = 28° \pm 7°$ and a corotation radius of 4.3 ± 0.5 kpc.

After rescaling the models to a galactocentric distance of the observer $R_o = 8$ kpc and to a projected radial velocity dispersion in Baade's Window of 113 km s^{-1} (Sharples, 1990), some of the models also reproduce many other observational constraints throughout the Galaxy, like the bulge and disc stellar kinematics (see Figure 1). More details can be found in Fux (1997).

References

Fux, R.: 1997, *Astron. Astrophys.* **327**, 983.
Rich, R.M.: 1996, in: H. Morrison and A. Sarajedini (eds.), *ASP Conf. Ser.* **92**, 24.
Rodgers, A.W.: 1977, *Astrophys. J.* **212**, 117.
Sevenster, M.N., *et al.*: 1997a, *Astron. Astrophys. Suppl.* **122**, 79.
Sevenster, M.N., *et al.*: 1997b, *Astron. Astrophys. Suppl.* **124**, 509.
Sharples, R., Walker, A. and Cropper, M.: 1990, *Mon. Not. R. Astron. Soc.* **246**, 54.
te Lintel Hekkert, P., *et al.*: 1991, *Astron. Astrophys. Suppl.* **90**, 327.
Tyson, N.D. and Rich, R.M.: 1991, *Astrophys. J.* **367**, 547.

Astrophysics and Space Science is the original source of publication of this article. It is recommended that this article is cited as: *Astrophysics and Space Science* **265**: 375–376, 1999.
© 1999 *Kluwer Academic Publishers*.

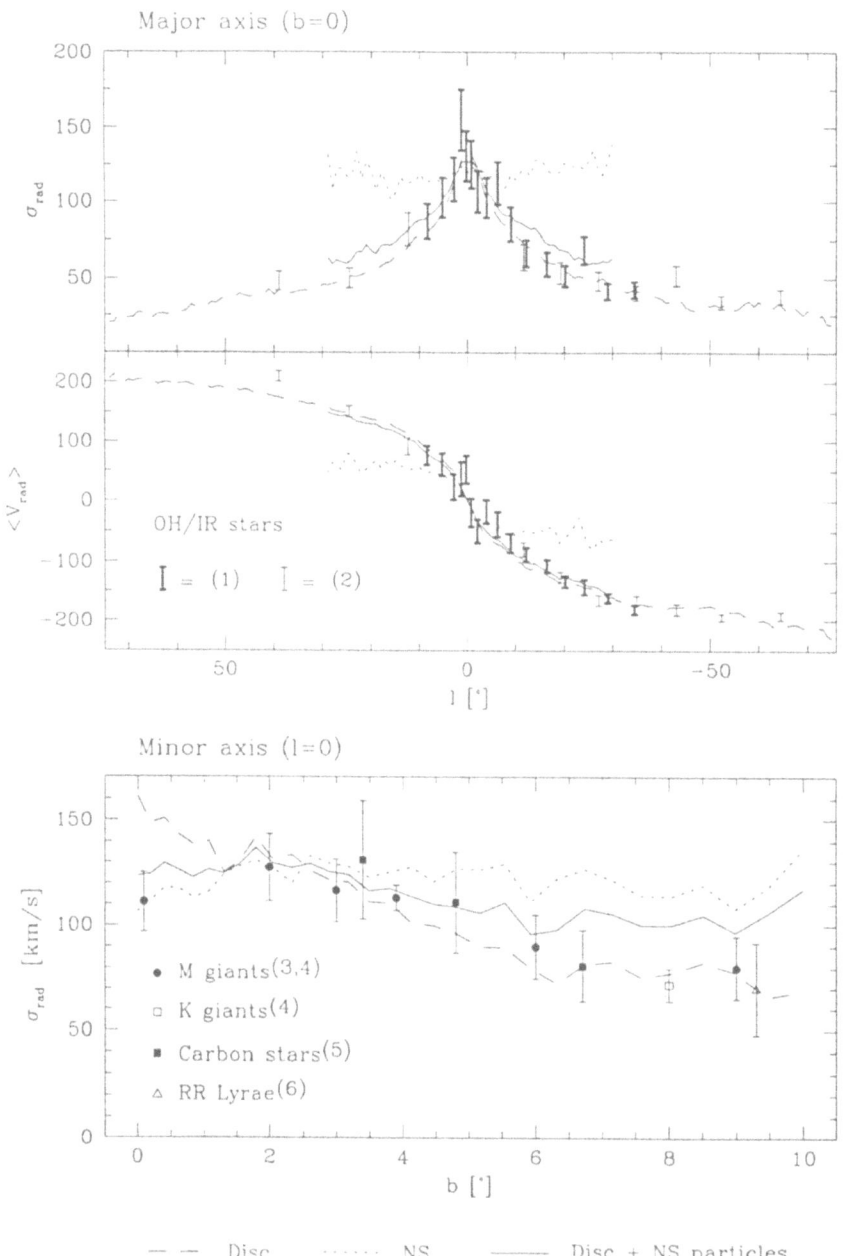

Figure 1. Projected radial velocity dispersion and galactocentric mean velocity in one of the favorite model (m08t3200). The data are from: (1) Sevenster *et al.*, 1997a, b; (2) te Lintel Hekkert *et al.*, 1991; (3) Sharples *et al.*, 1990; (4) Terndrup published by Rich, 1996; (5) Tyson and Rich, 1991 and (6) Rodgers, 1977.

SPECULATIONS ON BAR–DRIVEN EVOLUTION OF THE GALAXY

M.N. SEVENSTER

Mt. Stromlo and Siding Spring Observatories, Private Bag Weston Creek PO., 2611 ACT Weston, Australia

1. Introduction

Barred galaxies may never settle in 'true' equilibrium; observations as well as simulations show continuous transfer of gas from intermediate radii to the centre of the galaxy with subsequent star bursts and destruction of the bar itself[1,2]. Resonances with the natural freqencies of the stellar orbits continue to influence the dynamical distributions of gas and stars. It is not unlikely that altogether bars are a common and transient phenomenon in spiral galaxies, leaving more massive and thickened bulges behind. As such, bars may be the driving force behind a galaxy's evolution along the Hubble sequence (from Sd to S0). We will discuss some evidence for such evolution in the Galaxy.

Part of this evidence comes from a sample of OH/IR stars. These are oxygen–rich, intermediate–mass stars at the tip of the asymptotic giant branch. The outflow velocity of their circum–stellar envelopes is a useful statistical indicator of initial mass and thus age. This paper is a very brief summary of material presented in Sevenster (1997).

2. Scenario

(1) The Bar formed $\sim 6 - 7$ Gyr ago via a disk instability and the large–scale star formation increased considerably in the Bulge region.

(2) Subsequently, gas was redistributed. Over several Gyrs, an inner ring (3.5 kpc) formed and gas was transported from inside this ring to the Centre, increasing the central metallicity.

(3) Around 1 Gyr ago, enough mass accumulated in the Centre to ignite (several) star bursts[3,4] and to form an inner–Lindblad resonance (ILR, 1.5 kpc).

(4) Presently, gas inflow and star formation take place at low level[4,5], possibly hampered by the ILR.

(5) It is most likely that the Bar will slowly (continue to) dissolve, leaving behind a more massive (and thicker?) Bulge.

 Astrophysics and Space Science is the original source of publication of this article. It is recommended that this article is cited as: *Astrophysics and Space Science* **265:** 377–378, 1999.
© 1999 *Kluwer Academic Publishers.*

3. Discussion

This scenario gives an explanation for the absence of gas between 1.5 kpc and 3 kpc (ad 2), the nature of the 3–kpc arm (ad 2) and the CO motions in the Centre[6] as resulting from x2–orbits (ad 3). Also the observed ring of ultra–compact HII regions[7] (ad 3) and distribution of ages and metallicities of OH/IR stars in the Bulge and the Centre[3,8] find a place in this scenario (ad 1,2,3). The dynamics of the central Galaxy are not very different from those at larger radii[8] (ad 1). The structure of the galactic spiral arms may be explained as bifurcations at corotation.

We have not discussed here the influence that the Bar has had on the metallicity gradient in the disk via radial mixing and on the disks scaleheight and possible thick disk via enhanced vertical heating. The scaleheight of the Bar seems to be small, similar to that of the Disk[8]. This suggests that the usual buckling instability has not been effective in the Galaxy, if indeed the Bar has existed for several gigayears. Also, it is unlikely that strong interaction is playing a role, but that is not to say that weak (continued) triggering by for instance the Magellanic Clouds, possibly via halo wakes[9], is excluded.

If an ILR ring indeed exists than a determination of its time of formation could pin down the above scenario. Alternatively, an estimate of the ratio of the length of the Bar to the corotation radius would give a handle on the Bar's age, as this ratio may gradually decrease to 1 as bars evolve. Unfortunately, both ILR age and Bar length are far from straightforward to determine.

References

Bally, J., Stark, A., Wilson, R. and Henkel, C.: 1988, *Astrophys. J.* **324**, 223.

Comeron, F. and Torra, J.: 1996, *Astron. Astrophys.* **314**, 776.

Friedli, D. and Benz, W.: 1993, *Astron. Astrophys.* **268**, 65.

Hasan, H., Pfenniger, D. and Norman, C.: 1993, *Astrophys. J.* **409**, 91.

Krabbe, A., *et al.*: 1995, *Astrophys. J. Lett.* **447**, 95.

Sevenster, M.: 1997, dissertation Sterrewacht Leiden.

Sjouwerman, L.: 1997, dissertation Onnsala Space Observatory.

von Linden, S., Duschl, W. and Biermann, P.: 1993, *Astron. Astrophys.* **269**, 169.

Weinberg, M.: In: Buta, Crocker and Elmegreen (eds.), *Publ. Astron. Soc. Pacific C.* **91**, 517.

FORMATION OF BULGES

JOSEPH SILK and RYCHARD BOUWENS
*Departments of Astronomy and Physics, and Center for Particle Astrophysics, University of
California, Berkeley, CA 94720, USA*

Abstract. Bulges, often identified with the spheroidal component of a galaxy, have a complex
pedigree. Massive bulges are generally red and old, but lower mass bulges have broader dispersions
in color that may be correlated with disk colors. This suggests different formation scenarios. I will
review possible formation sequences for bulges, describe the various signatures that distinguish these
scenarios, and discuss implications for the high redshift universe.

1. Three Scenarios for Bulge Formation

We consider the following possibilities for bulge formation: bulges form before,
contemporaneously with, or after disks.

Bulges are old: Monolithic collapse (Eggen, Lynden-Bell and Sandage, 1962)
described the formation of population II prior to disk formation. Evidence on
the age of the inner Milky Way bulge stars generally supports an old population
that formed before the disk. Galaxies with massive bulges would have necessarily
formed by primordial collapse, major mergers at high redshifts, or infall of satellite
galaxies (Pfenniger, 1992).

The chemical evidence is less clear: the predictions of a simple monolithic
formation model cannot be easily reconciled with the observed abundance spread
in bulge stars. Indeed theory has largely supplanted a monolithic collapse picture
with a clumpy collapse model. The theory of galaxy formation has had considerable success in predicting various properties of large-scale structure. Hierarchical
formation is closer in spirit to the Searle-Zinn view of halo and bulge formation
in which many globular cluster mass clumps merge together. We regard bulge and
luminous halo formation as closely related phenomena, the bulge simply being
the core of the field star halo. Hierarchical galaxy formation involves a sequence
of successive mergers of larger and larger subgalactic scale clumps. At any given
stage, gas dissipation and settling produces disks that are destroyed in subsequent
mergers. Not all disk structure is erased: dwarf satellites and even globular clusters
may be substructure relics. The disk only forms after the last massive merger via
gas infall. In environments such as rich clusters disk infall is largely suppressed
and the cluster cores are dominated by spheroidal galaxies. Bulges are inevitably
older than disks, and formed on a dynamical time scale. Their formation is characterized by a series of intense formation episodes or starbursts, produced by each

 Astrophysics and Space Science is the original source of publication of this article. It is recommended that this article is cited as: *Astrophysics and Space Science* **265**: 379–387, 1999.
© 1999 *Kluwer Academic Publishers.*

merger. Most stars that are now in the bulge formed during the process of bulge accumulation.

Bulges are of intermediate age: One can also envisage the following prescription for bulge formation. Merging of dwarf irregular galaxies with a massive disk galaxy will result in the dwarf being stripped of gas. The angular momentum of the gas guarantees that it will eventually dissipate to provide infall into the disk. The stellar component, however, as it interacts with the disk is partially tidally disrupted, to form the thick disk, but the dense cores undergo dynamical friction, spiraling into the center to form the bulge. While this may not be appropriate to the inner Milky Way bulge, such a picture may be relevant to the outer spheroid. There is evidence for tidal streams that are continuously generated by disruption of satellites such as the LMC. The age spread of the globular star clusters is consistent with a model in which bulges and disks would be of similar age.

Bulges are young: Bulges may also form slowly by dynamical instabilities of disks. Secular evolution of disks has occurred in at least some galaxies, particularly in late-type galaxies (Kormendy, 1992; Courteau, 1996). The secular evolution of a cold disk inevitably results in gravitational instability. On galaxy scales this is dominated by the non-axisymmetric modes that induce formation of a bar. Tidal torques are expected by the bar on the disk gas, which consequently suffers angular momentum transfer and forms a massive central concentration. This in turn eventually tidally disrupts the bar, as well as undergoes a central starburst and forms the bulge. This process can repeat as infall of gas continues and the disk becomes sufficiently massive to again be gravitationally unstable. Bar disruption takes up to 100 bar dynamical timescales.

2. Signatures of Bulge Formation

The observed properties of bulges provide a fossil testament to their formation. The various signatures do not lead to any unique conclusion, however. Consider first the Milky way as a prototype for bulge formation, bearing in mind that our local neighborhood is necessarily limited in scope. The following remarks are largely summarized from an excellent review of the Milky Way bulge by Wyse, Gilmore and Franx (1997).

Ages: This should be the cleanest signature. The Milky Way bulge appears to be indistinguishable in age from the inner globular clusters that form a flattened subsystem and appear to be ~ 2 Gyr younger than the oldest globular cluster systems. However the colors of other bulges, especially in late-type disks, show a broad dispersion which may reflect age differences.

Abundances: The observed abundances of Milky Way bulge stars show a broad dispersion, suggestive of inhomogeneous enrichment, and the mean metallicity is lower than that of the disk, but higher than that of the halo. This supports the inference of an old bulge from observed ages in localized regions. In particular the outer bulge appears to be more metal-poor than the old disk. If age traces metallicity, one infers that the bulge precedes the disk. However dynamical processes could delay bulge star formation without inducing chemical evolution.

Angular momentum: The angular momentum distribution of the inner bulge resembles that of the halo (or outer bulge), rather than that of the disk. This is suggestive of a sequential formation process.

Other bulges provide a broader basis with which to search for clues to bulge formation.

Profiles: Core radii of bulges and disks are well correlated. This suggests that the formation of the two components is closely coupled.

Colors: Bulge colors show a broad dispersion, but generally track disk colors. Both metallicity and age must therefore be correlated.

Dynamics: Low luminosity bulges are rotationally supported, as are disks, but luminous bulges generally are supported by anisotropic velocity dispersion. Bulges overlap with, but generally have lower anisotropic velocity dispersion support (i.e. σ/v_{circ}) than do ellipticals. This suggests that massive bulges are distinct from disks and closer to ellipticals in dynamical origin, whereas low luminosity bulges are more closely associated with disk star formation.

Fundamental plane: Bulges lie in the identical fundamental plane as ellipticals, although there is a slight offset of the zero point. This is suggestive of a similar early formation phase for bulges to that for ellipticals.

The signatures provide mixed signals on the epoch of bulge formation. It is probably true that many bulges, especially if massive, form early, while some, especially if associated with later-type spirals, form late. The age differences provide an interesting environment with which to probe bulge formation models.

3. Bulges at High Redshift

The high redshift universe potentially provides a unique discriminant. The differences between the bulge models described above are magnified at high redshift.

We may broadly classify these bulge formation scenarios into three types: secular evolution in which bulges form relatively late by a series of bar-induced starbursts, one in which bulges form simultaneously with disks, and an early bulge formation model in which bulges form earlier than disks. Adjusting the three models to produce optimal agreement with $z = 0$ observations, we compare their high-redshift predictions with present-day observations, in particular, with data compiled in various studies based on the CFRS (Schade *et al.*, 1996; Lilly *et al.*, 1998) and the HDF (Abraham *et al.*, 1998). We have developed a simple scheme for simulating the rival models and comparing bulge colors and sizes with observations. Hubble Space Telescope photometry and color information is available for galaxy samples that extend up to redshift unity and beyond. At this epoch one may hope to detect the difference between old and young bulge models.

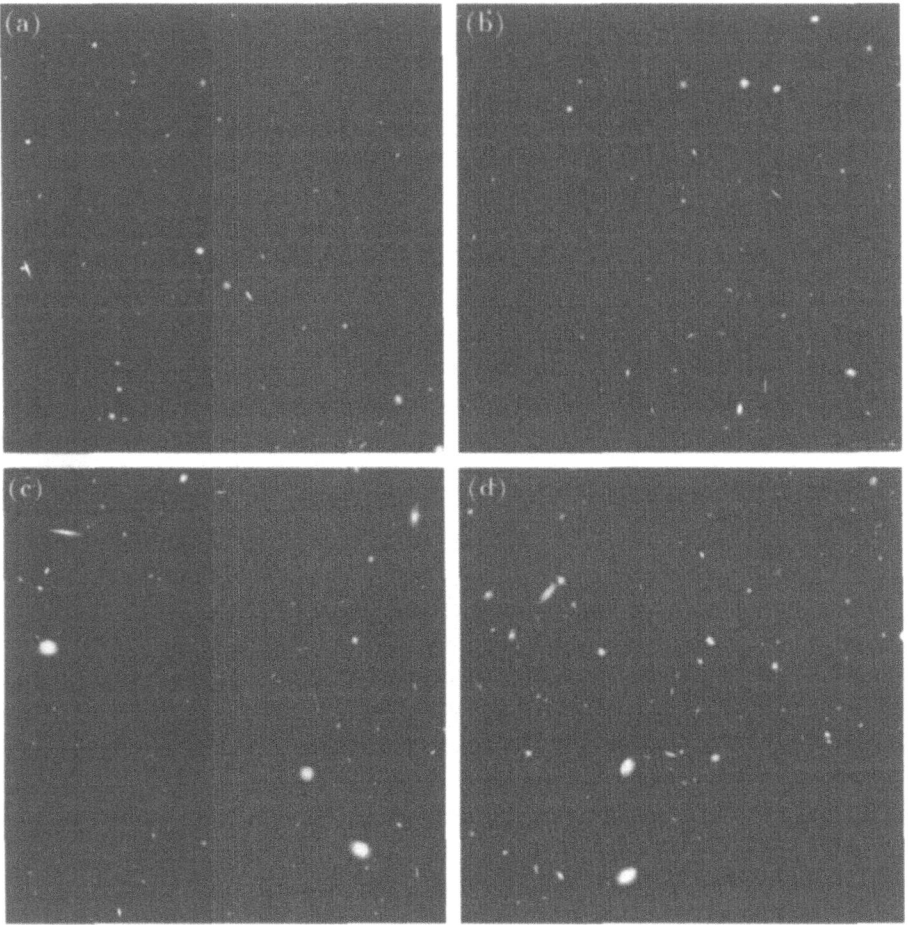

Figure 1. Comparison of simulated BVI images of a $2'' \times 2''$ patch of the HDF with the observed images (panel d). Panel (a) illustrates our secular evolution model for bulges, panel (b) illustrates our simultaneous formation model, and panel (c) illustrates our early bulge formation model. Calculations are performed using a galaxy-evolution software package written by one of the authors.

For the purposes of normalizing our models, we examine two local $z = 0$ samples (de Jong and van der Kruit, 1994; hereinafter, DJ; Peletier and Balcells, 1996, hereinafter, PB). Both are diameter-limited samples but differ in orientation selection, so that edge-on disks in the PB sample are simply redder and less prominent relative to the bulges.

Starting with the local properties of disks and a reasonable distribution of formation times, we construct a fiducial disk evolution model, to which we add three different models for bulge formation, the principle difference being simply the time the bulges form relative to that of their associated disks. We do not attempt to model the internal dynamics or structure of spirals (e.g., Friedli and Benz, 1995).

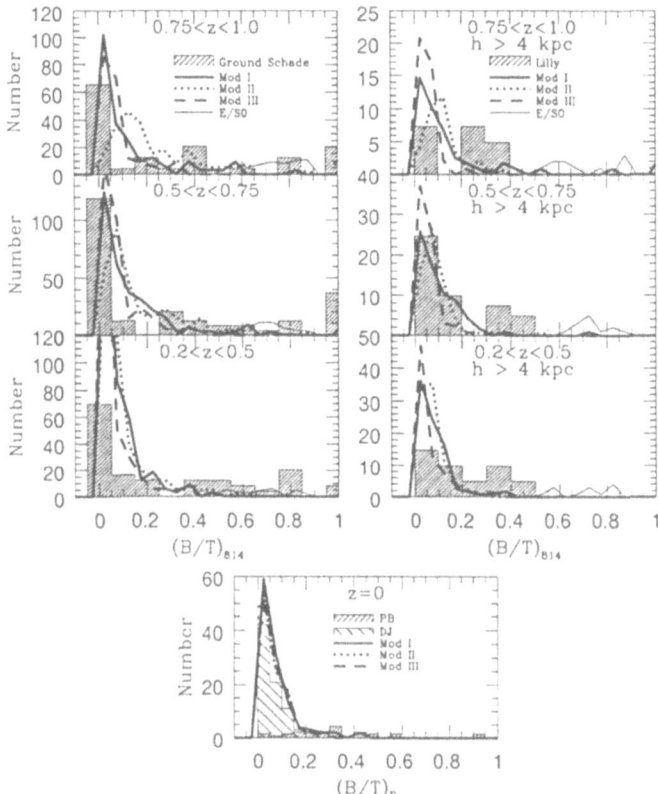

Figure 2. Comparison of the observed bulge-to-total ratios (histograms) with three bulge formation models: a secular evolution model (Mod I, solid line), a simultaneous formation model (Mod II, dotted line), and an early bulge formation model (Mod III, short dashed line) (from Bouwens, Cayon and Silk, 1999). High redshift comparisons are performed in the upper left panels against the Schade *et al.* (1996) data using the CFRS selection criteria and in the upper right panels against the Lilly *et al.* (1996) data using the CFRS selection criteria plus a size cut ($h > 4$ kpc). $E/S0$ predictions are also included in the high redshift figures (long dashed line). Models are renormalized to match the data. The de Jong and van der Kruit (1994) and Peletier and Balcells (1996) samples are used for the low-redshift comparisons.

We adopt the usual Sabc and Sdm luminosity functions (LFs) for disk galaxies (Binggeli, Sandage and Tammann, 1988). We evolve these galaxies backwards in time and in luminosity according to their individual star formation histories without number evolution. We take halo formation time to equal the time over which 0.25 of the final halo mass is assembled. We normalize by assuming a constant mass-to-light ratio where $M_{b_J} = -21.1$ corresponds to 4×10^{12} $M_{\odot}$ and we adopt the usual CDM matter power spectrum. We take star formation in the disk to commence at the halo formation time with an e-folding time that depends on the $z = 0$ galaxy luminosity, to roughly fit the $z = 0$ colour-magnitude relationship. We assume

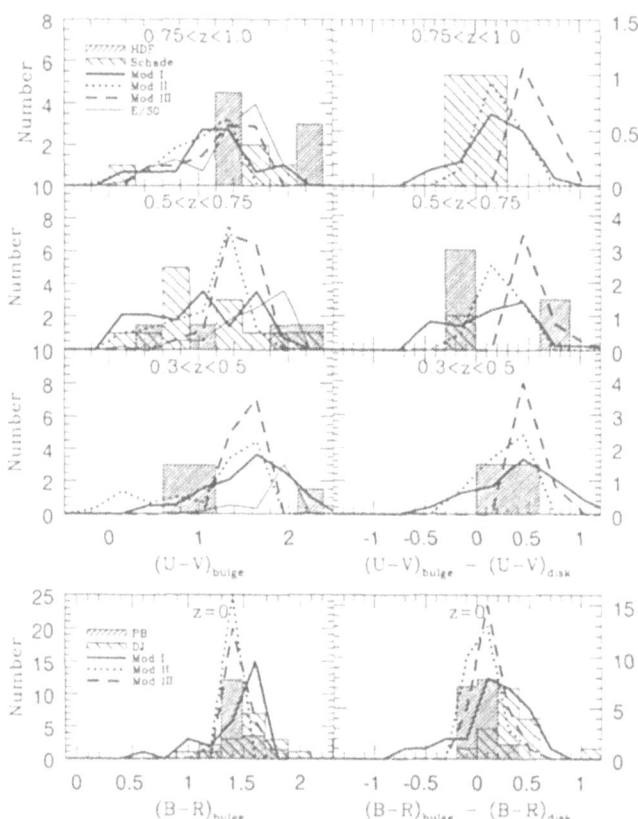

Figure 3. Comparison of the observed bulge and relative bulge-to-disk colours (histograms) with those of the models, at both high and low redshift (from Bouwens, Cayon and Silk, 1999). Model curves (renormalized to match observations and multiplied by 1.6 to increase their prominence) and low redshift data are represented as in Figure 2. The high redshift comparison includes data from the HDF for the Bouwens *et al.* (1998) sample (shaded histogram) and HST data from Schade *et al.* (1995) (open histogram).

exponential profiles for the disks with a b_J central surface brightness modified to account for the observed correlation between surface brightness and luminosity (e.g., de Jong, 1996; McGaugh and de Blok, 1997). We compute bulge spectra for the purposes of determining colours and magnitudes using the Bruzual and Charlot instantaneous-burst metallicity-dependent spectral synthesis tables (Leitherer *et al.*, 1996).

To calibrate our fiducial disk evolution models, we compare the model predictions to both the colour-magnitude relationship of disks in spirals and the cosmic history of luminosity density. There is good agreement with the colour-magnitude relationship. All models, for which bulge, disk, and E/S0 contributions have been considered, produce fair agreement with the luminosity density of the universe at all redshifts for which observable constraints are available (Lilly *et al.*, 1996; Madau *et al.*, 1996; Connolly *et al.*, 1997).

If bulges form through the merging of disk galaxies, the formation of the stars found in bulges is expected to precede the formation of stars in the disks which form out of gas that accretes around the spheroid (e.g., Kauffmann and White, 1993; Frenk *et al.*, 1996). For simplicity, we commence star formation in the bulge 4 Gyr prior to the formation of disks in our fiducial model and suppose that it continues for $\tau_{burst} = 0.1$ Gyr.

Suppose next that star formation in the bulge commences at the formation time of disks, for example because high angular momentum gas forms the disk while low angular momentum gas simultaneously forms the bulge which undergoes a mild starburst. One can imagine gas-rich satellite infall. The gas is tidally stripped and accreted onto the disk at large radii, whereas the dense cores lose angular momentum by dynamical friction and are incorporated into the bulge. A refinement of this model would allow for a sequence of early mergers that formed the bulge. However the final merger dominates the luminosity and therefore the spectral evolution, since the star formation efficiency is greatest for the most massive systems.

Finally, in secular evolution, bulges form after disks, with gas accretion onto the disk triggering the formation of a bar that drives gas inflow into the center followed by star formation (Friedli and Benz, 1995). The build-up of a central mass destroys the bar and inhibits gas inflow (Norman, Sellwood and Hasan, 1996), consequently stopping star formation in the bulge until enough gas accretes onto the galaxy to trigger the formation of a second bar, followed by a second central starburst. Somewhat arbitrarily, we suppose that the first central starbursts occur some 2 Gyr after disk formation in our fiducial model, that central starbursts last $\tau_{burst} = 0.1$ Gyr, as in the simulations by Friedli and Benz (1995), and that 2.4 Gyr separate central starbursts, in order to illustrate the general effect of a late secular evolution model for the bulge. We assume that the star formation rate follows an envelope with an e-folding time equivalent to the history of disk star formation. We thereby force star formation in the disk and the bulge to follow very similar time scales, given the extent to which they are both driven by gas infall processes.

We add a simple model for E/S0 galaxies to aid with the interpretation of observed high redshift, high B/T systems, somewhat arbitrarily assuming that the distribution of formation redshifts for the $E/S0$ population is scaled to be at exactly twice the distribution of formation redshifts.

We perform all our calculations using a galaxy evolution software package written by one of the authors for calculating how the gas, metallicity, star formation, luminosity, and colours vary as a function of time for a wide variety of morphological types, formation times, and star formation histories. With this software package, we present representative HDF simulations for our three bulge formation models in Figure 1 for comparison with the observations.

Clearly the secular evolution model, with late bulge formation, has a paucity of large B/T objects relative to the other models (Figure 2). The simultaneous bulge formation model has a large number of such galaxies simply because a large number of bulges were forming at this time, while the early bulge formation model has a slightly lower value due to the fact that bulges in this model had long been in place within their spiral hosts.

As expected, in all redshift bins, bulges are slightly bluer in the late bulge formation models than are the disks (Figure 3). A blue tail may be marginally detectable in the Schade *et al.* data in the highest redshift bins. Unfortunately, given the extremely limited amount of data and uncertainties therein, little can be said about the comparison of the models in all three redshift bins, except that the range of bulge and relative bulge-to-disk colours found in the data appears to be consistent with that found in the models.

While consistent with currently available data, our models for bulge formation are schematic and are intended to illustrate the observable predictions that will eventually be made when improved data sets are available in the near future. Our models are still quite crude, assuming among other things that the effects of number evolution on the present population of disks can be ignored to $z \sim 1$ as suggested, for example, in Lilly *et al.* (1998). In contrast, one recent analysis (Mao, Mo and White, 1998) has argued that observations favor the interpretation that a non-negligible amount of merging has taken place in the disk population from $z = 0$ to $z = 1$. For this particular interpretation, it remains to be seen how all the present stellar mass in disks could have built up if disks were continually destroyed by merging to low z given the constraints on the cosmic star formation history. Infall of metal-poor gas provides a non-destructive alternative that is supported by chemical evolution modeling of the old disk and even by observations of a reservoir of high velocity outer halo clouds.

Acknowledgements

We thank our collaborator Laura Cayon for many inspiring discussions of bulge issues. This research has been supported in part by NSF.

References

Abraham, R.G., Ellis, R.S., Fabian, A.C., Tanvir, N.R. and Glazebrook, K.: 1998, *Mon. Not. R. Astron. Soc.*, submitted.

Binggeli, B., Sandage, A. and Tammann, G.A.: 1988, *Annu. Rev. Astron. Astrophys.* **26**, 509.

Bouwens, R., Cayon, L. and Silk, J.: 1999, *Astrophys. J.*, in press.

Connolly, A.J., Szalay, A.S., Dickinson, M., Subbarao, M.U. and Brunner, R.J.: 1997, *Astrophys. J.* **486**, L11.

Courteau, S.: 1996, to appear in: D. Block and M. Greenberg (eds.), *Morphology and Dust Content in Spiral Galaxies*, Kluwer, Dordrecht.

de Jong, R.S. and van der Kruit, P.C.: 1994, *Astron. Astrophys. Suppl.* **106**, 451.

de Jong, R.S.: 1996, *Astron. Astrophys.* **313**, 45.

Eggen, O., Lynden-Bell, D. and Sandage, A.: 1962, *Astrophys. J.* **136**, 748.

Friedli, D. and Benz, W.: 1995, *Astron. Astrophys.* **301**, 649.

Kauffmann, G. and White, S.D.M.: 1993, *Mon. Not. R. Astron. Soc.* **261**, 921.

Kormendy, J.: 1992, Galactic Bulges, in: H. Dejonghe and H. Habing (eds.), *Proc IAU Symp.* **153**, 209, Kluwer, Dordrecht.

Leitherer, *et al.*: 1996, *Publ. Astron. Soc. Pacific* **108**, 996.

Lilly, S.J., LeFevre, O., Hammer, F. and Crampton, D.: 1996, *Astrophys. J.* **460**, L1.

Lilly, S.J., *et al.*: 1998, *Astrophys. J.*, submitted.

Madau, P., Ferguson, H.C., Dickinson, M.E., Giavalisco, M., Steidel, C.C. and Fruchter, A.: 1996, *Mon. Not. R. Astron. Soc.* **283**, 1388.

Mao, S., Mo, H.J. and White, S.D.M.: 1998, *Astrophys. J.*, submitted, astro-ph/9712167.

McGaugh, S.S. and de Blok, W.J.G.: 1997, *Astrophys. J.* **481**, 689.

Norman, C.A., Sellwood, J.A. and Hasan, H.: 1996, *Astrophys. J.* **462**, 114.

Peletier, R.F. and Balcells, M.: 1996, *Astron. J.* **111**, 2238–2242.

Pfenniger, D.: 1992, Galactic Bulges, in: H. Dejonghe and H. Habing (eds.), *Proc IAU Symp.* **153**, 387, Kluwer, Dordrecht.

Schade, D., Lilly, S.J., Le Fevre, O., Hammer, F. and Crampton, D.: 1996, *Astrophys. J.* **464**, 79.

MAGNESIUM-TO-IRON RATIO IN NUCLEI AND BULGES OF DISK GALAXIES

O.K. SIL'CHENKO

Sternberg Astronomical Institute and Isaac Newton Institute, Chile, Moscow Branch,
University av. 13, Moscow 119899, Russia

Abstract. The results of magnesium-to-iron ratio estimates are presented for the nuclei and central bulges of disk galaxies. A great variety of behaviours is found: the nuclei have solar Mg/Fe ratio or are Mg-overabundant, the bulges can be more or less Mg-overabundant than the nuclei. But the most bulges have nearly solar Mg/Fe ratios, irrespective of their luminosity.

Element ratios are important indicators of chemical evolution details. Particularly, the magnesium-to-iron ratio is very informative because these elements are produced by different types of SNe having progenitors of different masses. We know that giant E-galaxies are Mg overabundant (Worthey *et al.*, 1992). For disk galaxies the results remain to be controversial. Our early analysis (Sil'chenko, 1993) has shown that centers of disk galaxies of all luminosities have the solar Mg/Fe. But later Fisher *et al.* (1996) have reported Mg overabundance in the nuclei of some bright lenticular galaxies. I think that 30%-50% of early-type disk galaxies have chemically distinct metal-rich nuclei (Sil'chenko, 1994); the properties of these nuclei may not match those of the bulges because of different chemical evolution. Meanwhile up to now the majority of published works on bulge abundances do not take into account a possible presence of chemically distinct nuclei.

Using the facilities of the Special Astrophysical Observatory (SAO RAS), we search for chemically distinct nuclei since 1989. The data presented here have been obtained during 1994–1998, mostly with the Multi-Pupil Field Spectrograph of the 6 m telescope (its description in Afanasiev *et al.*, 1990). Reduction of the data has been made with the software developed in the SAO RAS (Vlasyuk, 1993). The data on individual galaxies are published in (Sil'chenko, 1996, 1998, 1999a, b, c; Sil'chenko *et al.*, 1997a, b, 1998, 1999).

The results are presented in Table I (the bulge luminosities are taken: M_B from Simien and Vaucouleurs, 1986, M_V from Baggett *et al.*, 1998, M_r from Kent, 1985). Unlike Es, the disk galaxies show quite various radial behaviours of their Mg/Fe ratio. 4 galaxies have stable [Mg/Fe] = 0 and 2 galaxies keep their [Mg/Fe] = +0.2 ÷ +0.3; in 2 cases [Mg/Fe] rises and in 4 cases drops when passes from the nuclei to the bulges. So, any conclusions on the bulges made without excluding the nuclei would be wrong.

 Astrophysics and Space Science is the original source of publication of this article. It is recommended that this article is cited as: *Astrophysics and Space Science* **265**: 389–390, 1999.
© 1999 *Kluwer Academic Publishers.*

TABLE I

Magnesium-to-iron ratios and other parameters.

Galaxy	Type	[Mg/Fe](nuc)	[Mg/Fe](bul)	M_B(bul)	M_V(bul)	M_r(bul)
N0224	SA(s)b	+0.3	+0.3	−19.47	−20.24	
N0488	SA(r)b	+0.3	+0.5	−18.86		−20.88
N1023	SB(rs)0−	+0.3	+0.1	−18.90	−19.68	
N2685	(R)SB0+pec	0	0		−19.46	
N2841	SA(r)b	+0.3	+0.2	−19.42	−22.58	
N4216	SAB(s)b	0	+0.1	−19.10	−21.77	
N4501	SA(rs)b	0	0	−18.73	−20.01	
N4826	(R)SA(rs)ab	+0.3	0		−18.87	
N5533	SA(rs)ab	+0.3	0		−22.53	−21.13
N7331	SA(s)b	+0.3	+0.1	−19.37	−21.83	
N7332	S0 pec	0	0		−19.38	
I1689	S0 pec	0	0	−18.67		

References

Afanasiev, V.L., Vlasyuk, V.V., Dodonov, S.N. and Sil'chenko, O.K.: 1990, preprint SAO, N54.
Baggett, W.E., Baggett, S.M. and Anderson, K.S.J.: 1998, *Astron. J.* **116**, 1626.
Fisher, D., Franx, M. and Illingworth, G.: 1996, *Astrophys. J.* **459**, 110.
Kent, S.M.: 1985, *Astrophys. J. Suppl.* **59**, 115.
Reshetnikov, V.P., Hagen-Thorn, V.A. and Yakovleva, V.A.: 1995, *Astron. Astrophys.* **303**, 398.
Sil'chenko, O.K.: 1993, *Pis'ma v Astron. Zh.* **19**, 701.
Sil'chenko, O.K.: 1994, *Astron. Zh.* **71**, 706.
Sil'chenko, O.K.: 1996, *Pis'ma v Astron. Zh.* **22**, 124.
Sil'chenko, O.K., Vlasyuk, V.V. and Burenkov, A.N.: 1997a, *Astron. Astrophys.* **326**, 941.
Sil'chenko, O.K., Burenkov, A.N. and Vlasyuk, V.V.: 1997b, *New Astronomy* **3**, 15.
Sil'chenko, O.K.: 1998, *Astron. Astrophys.* **330**, 412.
Sil'chenko, O.K., Burenkov, A.N. and Vlasyuk, V.V.: 1998, *Astron. Astrophys.* **337**, 349.
Sil'chenko, O.K.: 1999a, *Pis'ma v Astron. Zh.* **25**, **176**, 826.
Sil'chenko, O.K., Burenkov, A.N. and Vlasyuk, V.V.: 1999, *Astron. J.* **117**.
Sil'chenko, O.K.: 1999b, *Astron. J.* **117**, 2725.
Sil'chenko, O.K.: 1999c, *Astron. J.* **118**, 186.
Simien, F. and de Vaucouleurs, G.: 1986, *Astrophys. J.* **302**, 564.
Vlasyuk, V.V.: 1993, *Astrofiz. issled. (Izv. SAO RAS)* **36**, 107.
Worthey, G., Faber, S.M. and Gonzalez, J.J.: 1992, *Astrophys. J.* **398**, 69.

CHEMICAL EVOLUTION OF BULGES WITH NUCLEAR ACTIVITY

T.P. IDIART, R.D.D. COSTA and J.A. DE FREITAS PACHECO
Instituto Astronômico e Geofísico – USP, C.P. 3386, 01060-970 São Paulo, SP, Brazil

1. Introduction

The nature of the observed activity in many galactic nuclei is not well established yet. Different models have been proposed, like a more gradual process of star formation than those in 'normal' bulges or a bulge 'rejuvenated' by accretion of gas clouds. One way to tackle this problem is to study the stellar populations in these objects, by performing the analysis of their integrated spectra. In this work we report spectroscopical observations of a sample of 17 active bulges of galaxies classified as LINERs or HII. Mean abundances as well as mean population ages were derived from the procedure developed by Idiart *et al.* (1996a, b), and some clues on the chemical enrichment processes are given.

2. Observations and Data Reduction

All the objects were observed (at least three times each) at the Pico dos Dias Observatory (LNA/CNPq) in Brazil, using a Cassegrain spectrograph attached to the 1.60 m telescope, with the slit centered in the brightest region. Data reduction followed the standard procedure of bias, dark and falt-field corrections, extraction of the spectrum, wavelength calibration, and flux calibration through spectrophotometric standard stars observed each night. Radial velocity were corrected using the emission lines and spectra were corrected for galactic extinction.

3. Results and Discussion

The figure below shows that active bulges are on the average less metallics than non-active bulges, with metallicities comparable to galactic globular clusters. The diagram suggests a continuity between enrichment properties of active bulges, non-active bulges and ellipticals.

A multipopulational synthesis was made for the objects of our sample, using a closed box model with continuum star formation and integrated indices models

 Astrophysics and Space Science is the original source of publication of this article. It is recommended that this article is cited as: *Astrophysics and Space Science* **265**: 391–392, 1999.
© 1999 *Kluwer Academic Publishers.*

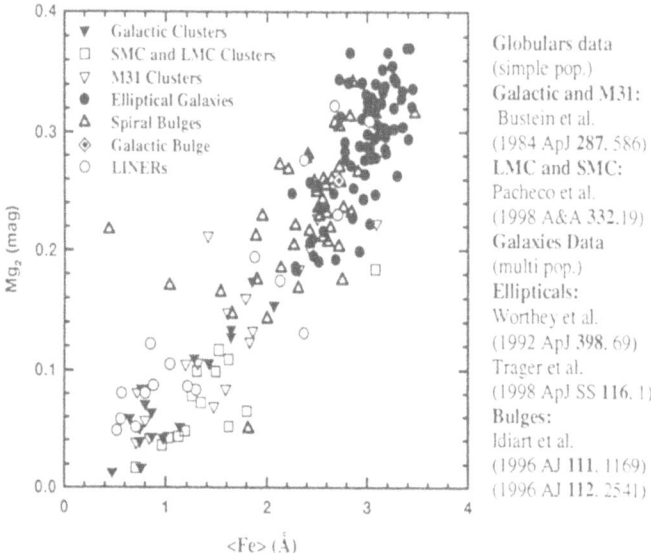

Figure 1. Absorption indices Mg$_2$ vs. ⟨Fe⟩, comparing distinct spheroidal objects. Note that inclusion of simple population data is only to compare metallicity values, since we cannot compare enrichment properties once the objects have different origins.

given by Borges *et al.* (1995). Results show that most of active bulges seem to be metal poor, with ⟨[Fe/H]⟩ ∼ −2.0, −1.4. Moreover, [Mg/Fe] is non-solar, typically ⟨[Mg/Fe]⟩ ∼ 0.45 − 0.65, indicating that chemical enrichment was essentially due to type II supernovae, and ages are comparable to non-active galaxies: 8–18 Gyrs. But, before to reach any final conclusion, it should be verified if dilution of absorption lines by a non-stellar continuum affects the measured indices for some bulges. If this is true, mean ages will be dramatically changed.

Acknowledgements

This work was supported by FAPESP.

References

Borges, A.C.A, Idiart, T.P., Freitas Pacheco, J.A. and Thévenin, F.: 1995, *Astron. J.* **110**, 2408.
Idiart, T.P., Freitas Pacheco, J.A. and Costa, R.D.D.: 1996, *Astron. J.* **111**, 1169.
Idiart, T.P., Freitas Pacheco, J.A. and Costa, R.D.D.: 1996, *Astron. J.* **112**, 2541.

BOX- AND PEANUT-SHAPED BULGES

R. LÜTTICKE and R.-J. DETTMAR

*Astronomisches Institut der Ruhr-Universität Bochum, Universitätsstr. 150, D-44780 Bochum,
Germany*

1. Introduction

Box- and peanut-shaped (b/p) bulges are not really that peculiar as it seemed in
the past and very common processes are required to explain the high frequency.
At present several mechanisms for their origin are discussed. These structures may
result from material accreted from infalling satellite companions (soft merging)
(Binney and Petrou, 1985; Whitmore and Bell, 1988). An alternative mechanism
for forming b/p bulges are instabilities or resonances animated by bars (Combes *et
al.*, 1990; Raha *et al.*, 1991). N-body simulations for stars in barred potentials have
demonstrated that with regard to the shape of bulges this theory and observational
evidence are consistent (Kuijken and Merrifield, 1995). Within this framework,
however, the question what may cause the bar becomes even more important.

2. Statistics of b/p Bulges

In a complete sample of edge-on disk galaxies selected from the RC3 (de Vaucouleurs *et al.*, 1991) with $D_{25} > 2$ arcmin and inclination greater than 65^0 ($\sim$
1350 galaxies) we characterized bulges by their degree of b/p-shape as non-b/p, or
unclassifiable using the 'Digitized Sky Survey'. The main result is that 46% of all
classifiable galaxies have a b/p bulge.

3. Box/Peanut Bulges and the Influence of Environment

The mean density parameters from the LV-Catalogue and the 'Nearby Galaxies
Catalog' are for galaxies with (only prominent b/p bulges) and without b/p bulge
with regard to the errors nearly the same. The analysis of galaxies in the Virgo-
and Ursa Major-Cluster shows that there the fraction of b/p bulges is the same as
for the complete galaxy sample. The result of an investigation of the environment
of galaxies with very prominent b/p bulges is that the mean number of companions

Astrophysics and Space Science is the original source of publication of this article. It is recommended that this article is cited as: *Astrophysics and Space Science* **265:** 393–394, 1999.
© 1999 *Kluwer Academic Publishers.*

($\Delta v < 1000$ km s^{-1}; NASA/IPAC Extragalactic Database) in a projected radius of 5 resp. 2.5 $\times$ D_{25} (RC3) around the center of a host galaxy is twice resp. more than twice the number of a sample of galaxies without b/p bulges. We get from a visual counting of all galaxies inside of the 2.5fold D_{25}-radius using the DSS for the galaxies with b/p bulge a increased number of satellites. An very surprising result comes from statistics of b/p bulges in isolated spiral galaxies ($N = 24$) (Zaritsky *et al.*, 1993, 1997). We find that 90 % of the classifiable galaxies with satellites have b/p bulges. In contrast, there is no galaxy from the control sample without satellites showing a b/p-bulge.

The normalized interaction index I (van den Bergh *et al.*, 1996) shows a definitely higher value for the sample of very prominent b/p bulges as the control sample. The first one has a value relatively near to the value of the Hubble Deep Field and the last one a value which is comparable to the Medium Deep Survey and the Shapley-Ames Catalogue.

4. Conclusion

We detect that on large scales there is no hint for a connection between the environment of galaxies with b/p bulge and this internal structure. However, box/peanut bulges preferentially occur in disk galaxies with companions and satellites and show frequently interactions. The favorable model scenarios for the development of b/p-bulges therefore are resonances at a bar triggered by galaxy interaction or by an infalling satellite in an otherwise stable disk.

However, currently it can not totally be excluded that some b/p-structures result from bars produced by dynamically cold disks through a global instability or directly from material accreted from infalling satellite companions.

References

Binney, J. and Petrou, M.: 1985, *Mon. Not. R. Astron. Soc.* **214**, 449.

Combes, F., Debbasch, F., Friedli, D. and Pfenniger, D.: 1990, *Astron. Astrophys.* **233**, 82.

de Vaucouleurs, G., de Vaucouleurs, A., Corwin, H.G. Jr., Buta, R.J. and Fouqué, P.: 1991, *Third Reference Catalogue of Bright Galaxies (RC3)*, Springer.

Kuijken, K. and Merrifield, M.R.: 1995, *Astrophys. J.* **433**, L13.

Lauberts, A. and Valentijn, E.A.: 1989, *The Surface Photometry Catalogue of the ESO-Uppsala Galaxies (LV-Catalogue)*.

Raha, N., Sellwood, J.A., James, R.A. and Kahn F.D.: 1991, *Nature* **352**, 411.

Tully, R.B.: 1988, *Nearby Galaxies Catalog*.

van den Bergh, S., Abraham, R., Ellis, R.S., Tanvir, N.R., Santiago, B.X. and Glazebrook, K.G.: 1996, *Astron. J.* **112**, 359.

Whitmore, B.C. and Bell, M.: 1988, *Astrophys. J.* **324**, 741.

Zaritsky, D., Smith, R., Frenk, C.S. and White, S.D.M.: 1993, *Astrophys. J.* **405**, 464.

Zaritsky, D., Smith, R., Frenk, C.S. and White, S.D.M.: 1997, *Astrophys. J.* **478**, 39.

VI – EVOLUTION OF THE MILKY WAY

THE EVOLUTION OF THE MILKY WAY

A self-consistent chemodynamical model

GERHARD HENSLER

*Institut für Theoretische Physik und Astrophysik, Universität Kiel, Olshausenstr. 40, D-24098 Kiel,
Germany*

Abstract. A two-dimensional chemodynamical model of the Milky Way Galaxy is presented that
can account for the structural, kinematical, and chemical pecularities of the galactic components in
a self-consistent way. The dynamics of three stellar components and the multi-phase interstellar medium consisting of clouds and intercloud gas are followed in detail. Mass interchange and energetic
interaction processes between the stars and the gas phases are treated simultaneously according to
the astrophysical experience including star formation, supernovae type I and II, planetary nebulae,
stellar winds, evaporation and condensation, drag, cloud collisions, heating and cooling, and stellar
nucleosynthesis. These processes are coupling large ranges on temporal and spatial scales, and allow
for feedback and self-regulation mechanisms, which play a significant role in galactic evolution.
In comparison with observations the capability of the chemodynamical treatment is convincingly
proved by the excellent agreement with various observations. In addition, also well-known problems
(G-dwarf problem, the discrepancy between local effective yields, etc.), which so far could be only
explained by artificial constraints, are solved in the global scenario. Here we wish also to focus on
temporal behaviours of the radial abundance gradient and abundance ratios in order to stimulate
further more specific observations and to make particular predictions which can test the validity of
used model ingredients like stellar yields.

1. Introduction

Disk galaxies consist of complex structures with at least three main components:
bulge, halo and disk. These differ in their fractions of ionized, atomic and molecular gas, their dust content and their stellar populations. Since the Milky Way Galaxy
(MWG) is plausibly the best studied disk galaxy, the observations of its stars and
its interstellar medium (ISM) can be used to test galactic evolutionary models and
to disentangle the evolutionary scenario in order to answer the addressed fundamental questions concerning initial conditions and the formation and evolution of
the MWG and of disk galaxies in general, e.g.: When, how and on what timescales did the Galactic components form? Was there any connection between them?
Which external influences have affected the structure? Is the solar neighbourhood
representative for an 'average disk'? Is the sun still at its birth-place environment
and at the same galactocentric distance?

From the evolutionary signatures of stars, i.e. their kinematics, chemical composition and ages, one derives information about the Galaxy at early evolutionary
stages. The number of stellar remnants (e.g. white dwarfs) and late dwarfs provides

Astrophysics and Space Science is the original source of publication of this article. It is recommended that this article is cited as: *Astrophysics and Space Science* **265**: 397–407, 1999.
© 1999 *Kluwer Academic Publishers.*

a partial insight into the early star-formation history. In addition, the total amount of heavy elements produced in our MWG can be estimated from element abundances in the present ISM. It is, however, complicated to trace their temporal production from abundances of solar-like low-mass stars of different ages (Edvardsson *et al.*, 1993) and their reliability is not yet complete. In general, several other observational indicators, like star-gas mass fractions, supernova (SN) rates or colour measurements place further constraints on galactic models.

At present, two major and basically different strategies for modelling galaxy evolution are followed: firstly, dynamical investigations which include hydrodynamical simulations of isolated galaxy evolution and of protogalactic interactions reaching from cosmological perturbation scales to direct mergers; and, on the other hand, studies which neglect any dynamical effects but consider either the whole galaxy or particular regions and describe the temporal evolution of mass fractions and element abundances in detail. Simulating different galactic regions e.g. by a closed-box model, however, presupposes that these are neither energetically nor dynamically coupled. Thus, in this picture a galaxy is only the sum of different isolated subsystems without any connection, despite the initial conditions. Even multi-zone models which include gas exchanges between different galactic regions (e.g. Ferrini *et al.*, 1994) are devoid of self-consistent gas dynamics. Actually, more sophisticated numerical investigations separate gas and stars dynamically and combine the global dynamical evolution of gas packages of a single gas phase with their internal, i.e. isolated chemical evolution (see e.g. Berczik, 1999 and this conference). Although those models can e.g. successfully reproduce the radial abundance gradients, they have to fail where the gas phases which differ by characteristic elements disconnect dynamically and large-scale streaming motions combine with local mixing effects (see below).

Firstly, in this paper we wish to refer to a few particular observations within the MWG that rise problems for interpretations if simple models are used. Consequently, we will proceed to the self-consistent chemodynamical (*cd*) treatment as the most appropriate one for galaxy evolution. Because of already published details of the *cd* formulation (Theis, Burkert, Hensler, 1993, hereafter: TBH; Samland, Hensler and Theis, 1997, hereafter: SHT) and results of a proper *cd* model for the MWG in the comprehensive paper by SHT, in this paper we intend to highlight only the most important issues of the *cd* model. However, we wish to stress correspondingly some essentials of the *cd* prescription which are necessary to reproduce and to understand the observed details. Finally, because *cd* models follow the full evolutionary history, one can formulate prospects for new observations and for further tests of the models by tracing back the age dependence of particular variables.

2. Evolutionary Signatures

For the case of a closed box a linear relation between the time-dependent metallicity $Z(t)$ and the initial-to-temporal gas ratio $ln[M_{g,0}/M_g(t)]$ follows analytically, where the slope is determined by the yield y, i.e. the metallicity release per stellar population. Deviations from this simple relation are explained by lower 'effective' yields y_{eff} due to outflow of metal-rich gas from an open volume or to infall of low-metallicity (presumably primordial) gas. Such dynamical effects can only be properly treated if simulations allow for the energetics, the composition, and the dynamical state of the galactic gas, as well as the relevant interaction processes, in a self-consistent manner. This includes the pollution of the different gas phases with characteristic elements by means of various stellar mass-loss mechanisms, the gas-phase transitions, and the self-consistent large-scale dynamics of hot intercloud gas (ICM) and cool gas (CM); the latter falls mainly in from a reservoir of protogalactic gas. It follows that galactic regions and components experience mutual dynamical influences and mass exchanges so that their evolutions are not decoupled.

Let us e.g. consider various chemical peculiarities of different kinds of disk galaxies which require the use of sophisticated multi-phase descriptions of the ISM in studies of the global scenario of galaxy evolution, because they stem from the dynamics and phase transitions:

1) In the solar vicinity at least three severe problems arise by considering the metallicity distribution of F or G dwarfs, stars that live sufficiently long to trace the evolution of a galaxy: The age-metallicity relation, y of the metallicity distribution and, finally, the lack of metal-poor G dwarfs (the well-known G-dwarf problem). Various influences on the evolution of the solar neighbourhood have therefore been invoked by different authors and are applied by artificial parametrisations ranging from time-dependent accretion of pristine halo gas to temporal variations of the stellar initial mass function. Although they lack self-consistency, the results can often provide very helpful basic insights into influences of distinct effects.

2) It is also fundamentally relevant to explain the observed differences of y in bulge, disk and halo of the MWG (Pagel, 1987).

3) That the solar oxygen abundance (and most α elements) is significantly higher than of both solar-like neighbourhood stars (Edvardsson et al., 1993) with the same Fe/H and even of young stars, respectively, could obtrude the conclusion that the sun is born at a smaller galactocentric distance and moved outward since its birth (Wielen et al., 1996). This involves the inherent assumption, however, that the radial abundance gradient was temporally unchanged. This as well as any cause of its development are still unsolved.

4) There is much evidence that metal-absorption line systems in QSO spectra arise from the gaseous halos of forming galaxies or of forming disks. As a working hypothesis it is assumed that the early hot halo gas is produced by SNeII, because the observed abundance ratios agree well with SNII yields (Reimers et al., 1992). The observed existence of N both at those early stages of galaxy evolution and in

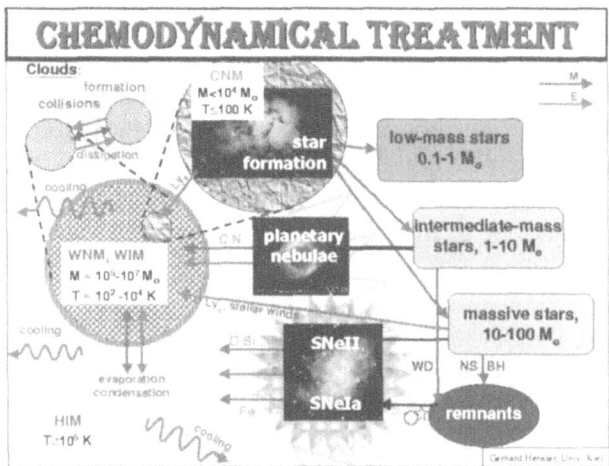

Figure 1. The different gaseous and stellar components taken into account in the *cd* treatment and their mutual interactions by mass, momentum and energy exchange. In addition, their dynamics are treated simultaneously. (This figure is taken from a transparency of the conference talk.)

the hot ISM phase requires its primary production on short timescales, i.e. already by massive stars, and/or the mixture from CM to hot gas when it is released from intermediate-mass giants.

5) The N/O ratio in HII regions of the MWG and other spirals reach higher values at larger oxygen abundances than dwarf galaxies (DGs) and is distributed along the line of secondary N production, but is scattered at low N/O over a wide range in O like in DGs (van Zee *et al.*, 1998).

3. The Chemodynamical Treatment

For systems and sites of low gravitational energy we know from empirical studies and theoretical investigations that the ISM is on average held in balance by counteracting processes like heating and cooling, turbulence and dissipation (Burkert and Hensler, 1989; Hensler *et al.*, 1999b). Since these processes are non-linearly coupled, the effect of neglecting one of them will alter the evolution completely. As an example this becomes obvious for the self-regulation of star formation (SF). A number of studies exists with particular attention to the influences of stellar radiation, supernova heating, and the evaporation/condensation balance between the two chemically and dynamically distinct gas phases, the cloudy medium (CM) and the hot ICM (Franco and Cox, 1983; Ikeuchi *et al.*, 1984; McKee, 1989; Bertoldi and McKee, 1995; Köppen *et al.*, 1995, 1998). Accordingly, the evolution of DGs is locally self-regulated and determined by large-scale outflows (Hensler *et al.*, 1993, 1999a). But this picture changes in cases when superbubble shells fragment and trigger SF (Ehlerova *et al.*, 1997) and when external processes like extended

DM halos, tidal fields, intergalactic gas pressure, and gas infall affect the evolution and cause the morphological differences of DGs.

Self-regulation with SNII energy deposition also characterizes the structure and evolution of galactic disks which reach a lower effective gravitational potential in rotational equilibrium (Burkert *et al.*, 1992; Firmani and Tutukov, 1992; Rosen and Bregman, 1995).

To approach global models of galaxy evolution which achieve the structural differences and details, an appropriate treatment of the dynamics of stellar and gaseous components is essential. In addition, at least the following processes have to be taken into account: SNe, SF, heating, cooling, stellar mass loss, condensation and evaporation. This includes the treatment of the multi-phase character of the ISM as well as the star-gas interactions and phase transitions. Since gas and stars evolve dynamically, and because several processes both depend on their metallicities but also influence the element abundances in each component, these models are called **chemodynamical**. The network of *cd* processes is sketched in Figure 1.

As can be shown and must be emphasized, however, the number of free parameters in the *cd* scheme is small, because they are either theoretically evaluated (like e.g. evaporation and condensation) or empirically determined (like e.g. stellar winds), or because they force self-regulation in a way that is independent of the parameterisation. The only free parameters which cannot approach particular 'equilibrium values' due to the absence of feedback are the stellar initial mass function, the momentum transfer by drag and the initial conditions, although the initial density distribution and gas-phase fractions are basically not affecting the model evolution. Because of limited space we refer the interested reader to more comprehensive descriptions of the *cd* treatment in TBH and SHT and to different applications (non-rotating galaxies: TBH, Hensler *et al.*, 1993, 1999a; vertical settling of the galactic disk: Burkert *et al.*, 1992; disk galaxies: Samland and Hensler, 1996; the MWG: SHT; DGs: Hensler and Rieschick, 1999).

4. The Milky Way's Chemodynamical Evolution

Here we will at first briefly refer to some of the striking results from the published *cd* model (SHT) in comparison to the observational features of the MWG which demonstrate the convincing success of the *cd* treatment. The particular emphasis is laid on the self-consistency of one single model with the temporally and locally equal treatment of the mentioned processes and without their artificial variations or parametrisations.

The model starts from an isolated spheroidal, rotating but purely gaseous cloud with a mass of $3.7 \cdot 10^{11}$ M$_\odot$, a radius of 50 kpc, and an angular momentum of about $2 \cdot 10^7$ M$_\odot$ pc^2 Myrs^{-1}, corresponding to a spin parameter $\lambda = 0.05$. We assume that the protogalaxy consists initially of CM and ICM with a density distribution of Plummer-Kuzmin-type (Satoh 1980) with 10 kpc scalelength. The initial CM/ICM

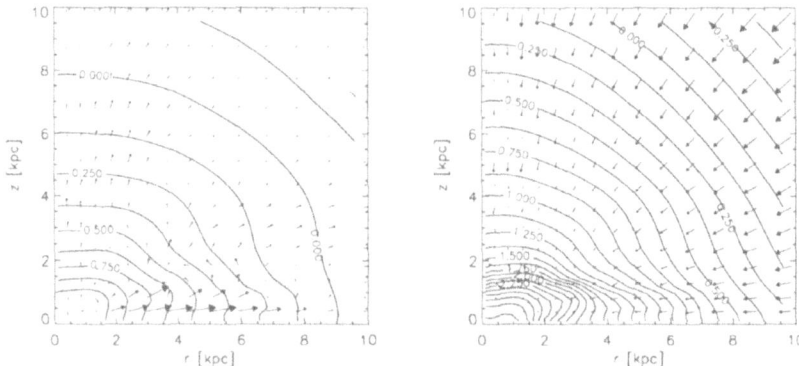

Figure 2. Surface densities and velocities of hot intercloud medium (ICM, left) and cloudy medium (CM, right) 7 Gyrs after the onset of the galactic collapse. The ICM in the galactic plane has velocities up to 230 km s^{-1}, while the velocities of the infalling CM are less than 60 km s^{-1}. (Figure 4 in SHT).

mass division (99%/1%) does not affect the later collapse, because the onset of SF determines the physical state within less than 10^7 years.

Since almost excellent agreement of the *cd* model after 15 Gyrs is found with the presently observed structure of the MWG, it may be safe to assume that also the evolutionary behaviour of different properties under considerations can be reliably deduced. The 15 Gyrs old model is e.g. able to reproduce the different metallicity distributions and y_{serif} of the halo, the bulge, and the solar vicinity (see Figure 7 in SHT), respectively, and also to solve the G-dwarf problem as an effect of large-scale dynamics of the metal-enriched ICM and its condensation and metal pollution of the CM (Figure 2). Furthermore, the radial abundance gradients (see Figure 6 in SHT), abundance ratios like O/Fe versus Fe/H (see Figure 9 in SHT), mass fractions of the components, SF rate in the disk, etc. fit the observations strikingly.

Also interestingly, one can trace the temporal run e.g. of SN and PN rates (see Figure 8 in SHT) and of the SF rates in different MWG regions (see Figure 3 in SHT) what allows the dating of their formation, showing a delay of the disk formation with respect to the halo extinction by almost 4 Gyrs and a bulge evolution extended over 5 Gyrs with at least two SF episodes. While the disk age is in agreement with other determinations (e.g. from white dwarfs and open clusters), the latter is in clear disagreement to the single-burst model for the bulge by Matteucci (this conference).

The temporal evolution of the oxygen abundance within the equatorial plane (Figure 3) stresses the fact that the large-range oxygen contamination by the ICM outflow leads to a steep increase of the oxygen abundance to about 10% solar and an inward moving condensation front already during the halo formation epoch. More important is the constancy of the radial oxygen distribution, i.e. the non-existence of an abundance gradient before the last 4–5 Gyrs. Nevertheless, from this result it cannot be excluded that the sun was born at a smaller galactocentric

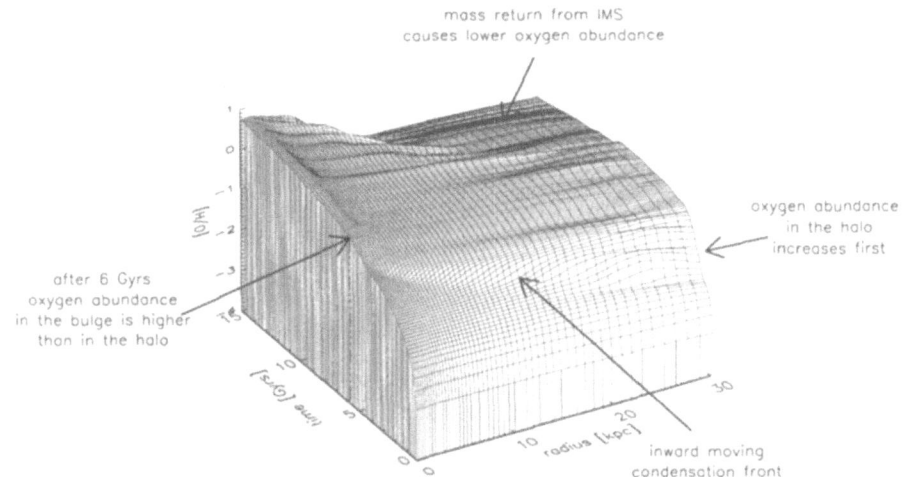

Figure 3. Temporal evolution of the radial oxygen abundance of the CM in the MWG galactic plane (Figure 5 in SHT; from S94).

distance but it contradicts to the assumption that the radial abundance gradient has remained temporally constant. Strikingly, the radial oxygen abundance of the *cd* model after 15 Gyrs reveals a flattening at galactocentric distances below 5 kpc and beyond 14 kpc in agreement with recent indications (Vilchez and Esteban, 1996; Rudolph *et al.*, 1997).

For a better understanding of the oxygen abundance in the CM, an element that is exclusively released to the ICM by SNeII, the gas mixing between ICM and CM is considered temporally in detail in Figure 4 (Samland 1994, hereafter S94) for four regions in the galactic plane. The sun is located between $R = 5$ kpc and $R = 10$ kpc, while the $R = 20$ kpc cut represents the halo region. One perceives the higher oxygen abundance of the ICM at all times. Clearly discernible is the larger O/H overabundance of the ICM at early epochs because of a dominance of cloud evaporation due to low densities. The faster enrichment of the CM in the outermost part is then caused by a higher specific condensation rate. Since the disk forms not before 7 Gyrs after the onset of the galaxy collapse (SHT), the O/H of the ICM exceeds that of the CM (and by this the HII regions) by 0.3 to 0.7 dex. The abundance variations of ICM and CM can be plausibly interpreted as follows: [O/H]$_{ICM}$ increases if SNII explosions add fresh oxygen and decreases when evaporation of less O-rich clouds reduces the mean abundance;
[O/H]$_{CM}$, on the other hand, rises due to condensation of hot metal-rich ICM and decreases on average by infalling (probably primordial) clouds of generally lower oxygen content.

Since one of the essential processes which are contained within the *cd* treatment is the transition of mass and energy from the one to the other gas phase by means of

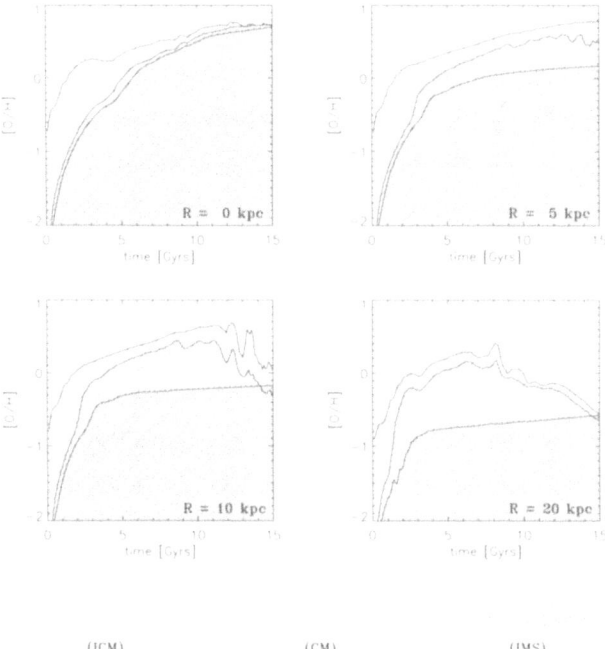

Figure 4. Temporal changes of the mean oxygen abundance of ICM, CM, and intermediate-mass stars within the equatorial plane at 4 different galactocentric radii (from S94). The reasons of variations are explained in the text.

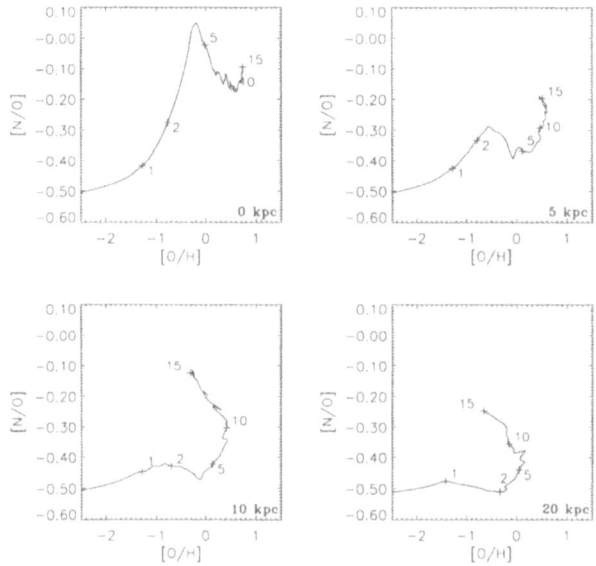

Figure 5. Time dependent relation of N/O vs. O/H of the CM within the equatorial plane at 4 different galactocentric radii (from S94). The tickmarks denote the galaxy age in Gyrs after the onset of the collapse.

condensation or evaporation, what depends on the physical state of both gases, the temporal behaviour of abundance ratios can also be derived from the *cd* models. The here refered *cd* model of the MWG (SHT) has used 'old' stellar yields that had not taken primary and secondary N production by massive stars into account like the most recent stellar evolution models by Woosley and Weaver (1995) have achieved. Here the separate production of oxygen by massive stars and of nitrogen purely by intermediate-mass stars are applied. Therefore, one can consider the enrichment of two distinct elements, one that is released into the CM while another one is expelled to the ICM. This provides information about the gas mixing of the ISM over the evolution of the galaxy.

In Figure 5 (from S94) the N/O ratio vs. O/H are presented again at different radii within the equatorial plane. In contrast to the non-selfconsistent analytical approachs, the validity of *cd* model could be proven by their prediction of [N/O] vs. [O/H] with the ages of stars as indicated in Figure 5. A few N and O abundance determinations of halo stars exist already. Tomkin and Lambert (1984) e.g. presented at first the [N/O] vs. [O/H] discrepancy between halo and disk stars. While the disk stars lie along the secondary N production, in the Fuhrmann and Reetz sample (this conference and private communication) also the halo stars follow this line so that differences to other authors must be studied in detail. Recent *cd* models of DGs (Hensler and Rieschick, 1999; Rieschick and Hensler, 1999) have indeed demonstrated the relevance of accounting for the new nitrogen yields from massive stars in order to reproduce the regime of their observed [N/O] vs. [O/H] (Garnett, 1990). This absence of additional nitrogen production in the presented model might explain why the evolutionary tracks in Figure 5 do not find agreement with solar values and can only be interpreted qualitatively.

5. Discussion

Here we have refered to a published *cd* model of the MWG and presented additional results which demonstrate both the agreement of one single global model with numerous detailed and regional observations as well as with the global structure. The necessity of the full but complex *cd* treatment of the galaxy components is justified. Several observational features like abundance distributions and ratios can only be understood in a self-consistent global evolutionary scenario if the *cd* treatment is taking the dynamics of the components and their relevant interaction processes into account. It has been shown that a large-scale coupling of different galactic regions by dynamical effects as well as the small-scale mixing between the gas phases due to condensation/evaporation affect the observational signatures. As long as the long-range streaming with inherent small-scale interactions under the inclusion of a two-phase ISM with their small-scale spatial resolution cannot be treated in other numerical simulations, e.g. in present SPH models, they miss self-consistency and cannot reproduce the observations in their global extent.

Furthermore, recent publications which investigated the effect of stellar motions in order to explain chemical peculiarities like solar SiC anomalies (Clayton, 1997) and the radial abundance gradient in the MWG (Allen *et al.*, 1998) are misleadingly denoted as chemodynamical but are far from its proper definition. Obviously, these models lack the self-consistent treatment of at least two essential attributes of the *cd* prescription, that are the dynamical calculation of each component separately in their combined self-consistent gravitational potential as well as the empirical application of mutual energy, mass, and, by the latter, inherent element transitions. They only deal with one limited aspect of *cd* models, namely to consider the element distribution by means of stellar dynamics, but start their studies already with assumed metal abundances. Nevertheless, they provide an important insight into special dynamical effects and contribute to our understanding of chemical details. One has to emphasize that a reliable *cd* model has to reproduce the complete set of observational features on different galactic scales, i.e. for all existing components and observed variables as the whole. It is not sufficient for an understanding of the global galactic evolution to fit single particular observed properties.

The time dependence of element abundances and their ratios can serve as reliable diagnostic tools of galaxy evolution and the physical state of the ISM. Their combination provides a new chance for detailed deconvolution and, furthermore, to constrain stellar yield models (Samland, 1998). While the *cd* models also provide element abundances in the hot ICM, their observational validation is still a problem. Although *cd* models can only deal with a limited numerical volume because of computational restrictions, the initial gas reservoir, i.e. the initial radius of the protogalaxy, has to be set sufficiently large in order to account for a halo formation by mass accumulation of intergalactic clouds and for later gas infall into the disk. Although, the here applied grid code is still limited to two dimensions because of the numerically expensive treatment of complex processes and spatial resolution, numerical methods are always progressively developing. Nevertheless, one has to keep in mind, that *cd* investigations present a giant leap with respect to non-dynamical chemical or purely gasdynamical studies.

Acknowledgements

Cooperations and discussions with Markus Samland, Joachim Köppen, Andreas Rieschick, Christian Theis, and Andi Burkert and their contributions to the field of chemodynamics are gratefully achnowledged. I thank Markus Samland for providing me with original figures from his thesis which was supported by the *Deutsche Forschungsgemeinschaft* (DFG) under grant No. He 1487/5. The numerical calculations are partly performed at the computer centers RZ Kiel, ZIB Berlin, and HLRZ Jülich.

References

Allen, C., Carigi, L. and Peimbert, M.: 1998, *Astrophys. J.* **494**, 247.

Bertoldi, F. and McKee, C.F.: 1995, in: R.Y. Chiao (ed.), *Amazing Light*, Springer, New York.

Berczik, P.: 1999, *Astron. Astrophys.*, submitted.

Burkert, A. and Hensler, G.: 1989, in: J.E. Beckmann and B.E.J. Pagel (eds.), *Evolutionary Phenomena in Galaxies*, Cambridge University Press, p. 230.

Burkert, A., Truran, J.S.W. and Hensler, G.: 1992, *Astrophys. J.* **391**, 651.

Clayton, D.D.: 1997, *Astrophys. J.* **484**, L67.

Edvardsson, B., Andersen, J., Gustafsson, B., *et al.*: 1993, *Astron. Astrophys.* **275**, 101.

Ehlerova, S., Palous, J., Theis, C. and Hensler, G.: 1997, *Astron. Astrophys.* **328**, 111.

Ferrini, F., Molla, M., Pardi, M.C. and Diaz, A.I.: 1994, *Astrophys. J.* **427**, 745.

Firmani, C. and Tutukov, A.V.: 1992, *Astron. Astrophys.* **264**, 37.

Franco, J. and Cox, D.P.: 1983, *Astrophys. J.* **273**, 243.

Garnett, D.R.: 1990, *Astrophys. J.* **360**, 142.

Hensler, G., Theis, Ch. and Burkert, A.: 1993, in: D. Alloin and G. Stasinska (eds.), *Proc. 3rd DAEC Meeting, The Feedback of Chemical Evolution on the Stellar Content of Galaxies*, Observatoire de Paris, p. 239.

Hensler, G., Gallagher, J.S. and Theis, Ch.: 1999a, *Astrophys. J.*, submitted.

Hensler, G., Köppen, J., Samland, M. and Theis, Ch.: 1999b, in: N. Arimoto and W. Duschl (eds.), *Proc. of the German-Japanese Workshop, Galaxy Evolution*, Springer, in press.

Hensler, G. and Rieschick, A.: 1999, in: T.X. Thuan *et al.* (eds.), *Proc. XVIIIth Rencontre de Moriond*, in press.

Ikeuchi, S., Habe, A. and Tanaka, Y.D.: 1984, *Mon. Not. R. Astron. Soc.* **207**, 909.

Köppen, J., Theis, Ch. and Hensler, G.: 1995, *Astron. Astrophys.* **296**, 99.

Köppen, J., Theis, Ch. and Hensler, G.: 1998, *Astron. Astrophys.* **328**, 121.

McKee, C.F.: 1989, *Astrophys. J.* **345**, 782.

Pagel, B.E.J.: 1987, in: G. Gilmore and B. Carswell (eds.), *The Galaxy*, Reidel, pp. 341.

Reimers, D., Vogel, S., Hagen, H.-J., *et al.*: 1992, *Nature* **360**, 561.

Rieschick, A. and Hensler, G.: 1999, in preparation.

Rosen, A. and Bregman, J.N.: 1995, *Astrophys. J.* **440**, 634.

Rudolph, A.L., Simpson, J.P., Haas, M.R., *et al.*: 1997, *Astrophys. J.* **489**, 94.

Samland, M.: 1994, *PhD. thesis*, University of Kiel, Shaker Verlag (S94).

Samland, M.: 1998, *Astrophys. J.* **496**, 155.

Samland, M. and Hensler, G.: 1996, *Rev. Mod. Astron.* **9**, 277.

Samland, M., Hensler, G. and Theis, Ch.: 1997, *Astrophys. J.* **476**, 544 (SHT).

Satoh, C.: 1980, *Publ. Astron. Soc. Jpn.* **32**, 41.

Theis, C., Burkert, A. and Hensler, G.: 1992, *Astron. Astrophys.* **265**, 465 (TBH).

Tomkin, J. and Lambert, D.L.: 1984, *Astrophys. J.* **279**, 220.

van Zee, L., Salzer, J.J. and Haynes, M.P.: 1998, *Astrophys. J.* **497**, L1.

Vilchez, J.M. and Esteban, C.: 1996, *Mon. Not. R. Astron. Soc.* **280**, 720.

Wielen, R., Fuchs, B. and Dettbarn, C.: 1996, *Astron. Astrophys.* **314**, 438.

Woosley, S.E. and Weaver, T.A.: 1995, *Astrophys. J. Suppl.* **101**, 181.

CHEMO-SPECTRAL EVOLUTION OF THE MILKY WAY AND OF SPIRAL DISKS

S. BOISSIER and N. PRANTZOS
Institut d'Astrophysique de Paris, 78 bis, boulevard Arago, 75014 Paris, France

Our model and its application to the Milky Way are discussed in detail in Boissier and Prantzos (1998). The stellar yields of Woosley and Weaver (1995) are used to compute the chemical evolution of the Galaxy, while the spectral one is coherently followed with the stellar evolution tracks of the Geneva group (Schaller *et al.*, 1992; Charbonnel *et al.*, 1996) covering all the evolutionary phases (MS to AGB) for several metallicities and the synthetic stellar spectra library of Lejeune *et al.* (1997) providing spectra as a function of $(T_{eff}, \log(g), Z)$. Extinction from dust is calculated as in Guiderdoni and Rocca-Volmerange (1987) with the geometry of stars and dust of Xu *et al.* (1997). The Star Formation Rate is assumed to be locally a Schmidt Law ($\alpha \Sigma_g^{1.5}$), and to depend on galactic radius ($\alpha \ 1/R$) according to the theory of star formation induced by the passage of density waves Prantzos and Aubert, 1995 and references therein). The Initial Mass Function is the one of Kroupa *et al.* (1993). Finally, an exponential infall with time-scale $\tau = 7$ Gyr in the Solar neighborhood is assumed, to account for the local G-Dwarf distribution. All of the major observational features in the Solar neighborhood (surface densities, luminosities, colours, SFR) are reproduced by the model presented above.

The disk is simulated by independent zones (no radial flows) featuring the mean history at a given radius. Each zone evolves at its own rate, fixed by the efficiency of star formation (with the 1/R factor) and the time-scale for infall varying from 2.7 Gyr at 3 kpc to 7 Gyr at the solar radius: 8 kpc (inside-out disk formation). The results of that simple model compare well with the current gas, stellar, and SFR profiles and they predict the observed abundance gradient of $d[O/H]/dR \sim -0.07$ dex kpc^{-1} as well as the integrated quantities: luminosities in various wavelengths, gas fraction, SN rates ... . Some of the obtained profiles are shown in Figure 1. We also find a scale-length dependence on wavelength (the redder, the shorter), that results mainly from the evolution of stellar populations and to a less extent from extinction.

This simple model can be used to investigate the average properties of large samples of spiral galaxies on the one hand, and to reproduce the profiles of various quantities in individual galactic disks on the other. Those aspects will be presented in future papers.

Astrophysics and Space Science is the original source of publication of this article. It is recommended that this article is cited as: *Astrophysics and Space Science* **265**: 409–410, 1999.
© 1999 *Kluwer Academic Publishers.*

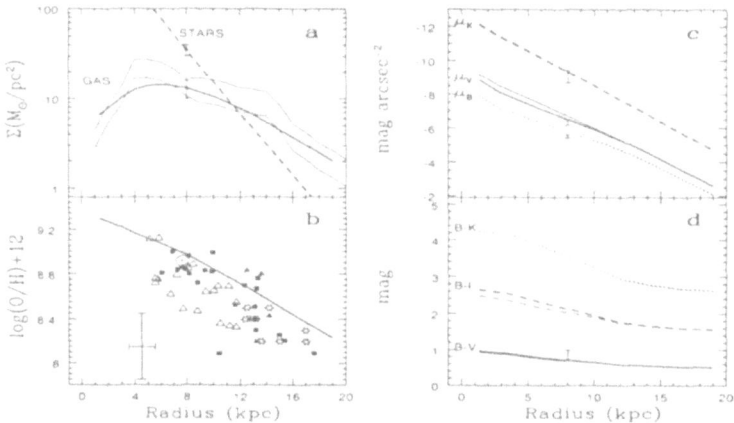

Figure 1. Present day profiles in the Milky Way and comparison to observations. a) Surface densities of gas (thick solid curve) and stars (dashed). Data at R = 8 kpc are from Kulkarni and Heiles, 1987 (gas) and Gilmore *et al.*, 1990 (stars). The gaseous profile is from Dame, 1993. b) Oxygen abundance gradient compared with the data: Fich and Silkey, 1991 (filled triangles); Shaver *et al.*, 1983 (open triangles); Vilchez and Esteban, 1996 (stars); Smartt and Rolleston, 1997 (squares). c) B, V, K-Surface brightness profiles whithout extinction (faint line) and with extinction (thick line). Local values are from Pagel, 1997 (V); Van der Kruit, 1986 (B); and Kent *et al.*, 1991 (K). The scalelengths spread from ~ 2.6 kpc in the K-band to ~ 3.9 kpc in the B-band. d) Colors whithout extinction (faint line) and with extinction (thick line). The local B–V is from Van der Kruit, 1986.

References

Boissier S. and Prantzos N.: 1998, *Mon. Not. R. Astron. Soc.*, **307**, 857.

Charbonnel, C., Meynet, G., Maeder, A. and Schaerer, D.: 1996, *Astron. Astrophys. Suppl.* **115**, 339.

Dame, T.: 1993, in: S. Holt and F. Verter (eds.), *Back to the Galaxy*, AIP, p. 267.

Fich, M. and Silkey, M.: 1991, *Astrophys. J.* **366**, 107.

Kent, S., Dame, T. and Fazio, G.: 1991, *Astrophys. J.* **378**, 131.

Kroupa, P., Tout, C. and Gilmore, G.: 1993, *Mon. Not. R. Astron. Soc.* **262**, 545.

Kulkarni, S. and Heiles, C.: 1987, in: D. Hollenbach and H. Thronson (eds.), *Interstellar Processes*, Kluwer, p. 87.

Lejeune, T., Cuisinier, F. and Busser, R.: 1997, *Astron. Astrophys. Suppl.* **125**, 229.

Pagel, B.: 1997, in: *Nucleosynthesis and Chemical Evolution of Galaxies*, Cambridge University Press, p. 211.

Prantzos, N. and Aubert, O.: 1995, *Astron. Astrophys.* **302**, 69.

Gilmore, G., Wyse, R. and Kuijken, K.: 1989, in: J. Beckman and B. Pagel (eds.), *Evolutionary Phenomena in Galaxies*, Cambridge University Press, p. 172.

Guiderdoni, B. and Rocca-Volmerange, B.: 1987, *Astron. Astrophys.* **186**, 1.

Schaller, G., Schaerer, D., Maeder, A. and Meynet, G.: 1992, *Astron. Astrophys. Suppl.* **96**, 269.

Shaver, P., McGee, R., Newton, L., Danks, A. and Pottasch, S.: 1983, *Mon. Not. R. Astron. Soc.* **204**, 53.

Smartt, S. and Rolleston, W.: 1997, *Astrophys. J. Lett.* **481**, L47.

van der Kruit, P.: 1986, *Astron. Astrophys.* **157**, 230.

Vilchez, J. and Esteban, C.: 1996, *Mon. Not. R. Astron. Soc.* **280**, 720.

Woosley, S. and Weaver, T.: 1995, *Astrophys. J. Suppl.* **101**, 181.

Xu, C., Buat, V., Boselli, A. and Gavazzi, G.: 1997, *Astron. Astrophys.* **324**, 32.

GALACTIC CHEMICAL EVOLUTION: FROM THE LOCAL DISK TO THE DISTANT UNIVERSE

N. PRANTZOS

Institut d'Astrophysique de Paris, 98 bis boulevard Arago, 75104 Paris, France

Abstract. A brief review is presented of our current understanding of the chemical evolution of the Milky Way and, in particular, of the solar neighbourhood and the disc of our Galaxy, in the light of recent theoretical and observational results. We explore the implications of this understanding for studies of 'cosmic chemical evolution'.

1. Introduction

Recent observations of star formation rate and gas-phase abundances of various elements in the high red-shift Universe, prompted studies of 'cosmic chemical evolution'. In that respect, it should be noted that our Galaxy is, by far, the best known system, as far as its chemical evolution is concerned. Indeed, the wealth of available data, especially in the solar neighbourhood, constrain seriously the parameters of simple models of chemical evolution and point to a rather well defined history for the local disc. Understanding the Milky Way evolution is then crucial in any attempt of studying 'cosmic chemical evolution'.

In Section 2 we present a brief review of the observational data for the solar neighborhood and the Milky Way disk and a phenomenological model that accounts weel for these data. Assuming that our Galaxy is a typical spiral, one can calculate the properties of disc galaxies as a function of redshift (in the framework of a given cosmological model) and compare to the observed properties of the extragalactic universe: global star formation rate, gas content and metal abundances in gas clouds. The conclusions of such a comparison appear in Section 3.

2. The Milky Way

In the solar neighbourhood (defined as a cylinder of $\sim$ 1 kpc radius at a distance $R_\odot = 8.5$ kpc from the galactic center), the main observables relevant to chemical evolution are (Prantzos and Silk, 1998): (i) the current surface densities of gas, stars and total amount of matter, and the current star formation rate (SFR); (ii) the metallicity at solar birth ($Z_\odot$) and today (Z_O); (iii) the oxygen vs. Fe (O-Fe) relationship; (iv) the age-metallicity relationship; (v) the metallicity distribution of long-lived G-type stars.

Astrophysics and Space Science is the original source of publication of this article. It is recommended that this article is cited as: *Astrophysics and Space Science* **265**: 411–415, 1999.
© 1999 *Kluwer Academic Publishers.*

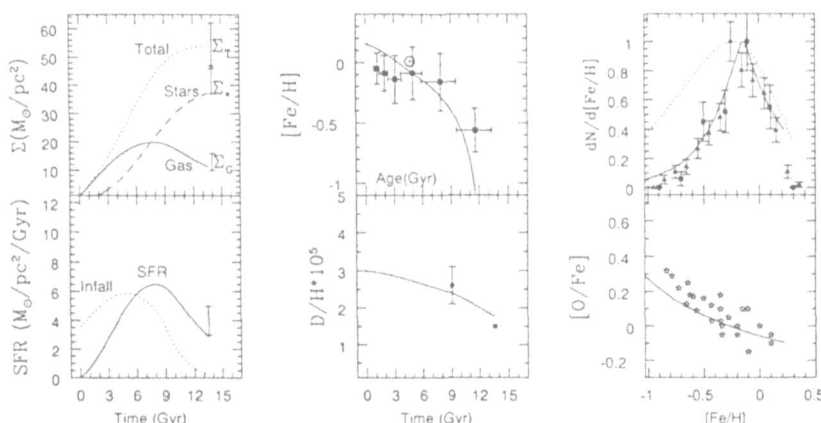

Figure 1. Results of a model for the chemical evolution of the local disc. The adopted SFR is proportional to $\Sigma_G^{1.5}$, while the infall rate is a broad gaussian with $\Delta\tau = 6$ Gyr, i.e. the local disc is formed very slowly. *Left-Up*: Evolution of the gaseous, stellar and total mass and comparison of the final results to observations of current local Σ_G, Σ_* and Σ_T, respectively. *Left-Down:* Evolution of Infall rate and SFR and comparison of the latter to its observed current value. *Middle-Up:* Age-metallicity relationship. *Middle-Down:* Deuterium evolution. *Right-Up:* Metallicity distribution of G-dwarfs *solid line* and comparison to observations; *dotted line*: a closed box model. *Right-Down*: O vs. Fe evolution: the observed decline is attributed to the late injection of Fe from SNIa.

These data, interpreted in the framework of a simple model of chemical evolution, imply that the local disc was formed on timescales of many Gyr, a conclusion also supported by recent chemodynamical models for the evolution of the Milky Way (Samland *et al.*, 1997). Models with slow infall of primordial composition reproduce reasonably well the above constraints (Figure 1).

The data for the rest of the disk concern only its current state, not its past history. The results of a simple-minded model for the rest of the disk appear on Figure 2. The adopted SFR $\propto \Sigma_G^{1.5}/R$ is based on the idea that star formation in spiral galaxies is triggered by density waves (e.g. Wyse and Silk, 1989). The radial variation in the SFR efficiency and of the infall timescale are the only new parameter introduced in the model, all the others been already fixed by the modelisation of the solar neighbourhood. This simple model can reproduce the various observed gradients in the disk (Prantzos and Silk, 1998) and predicts an important D gradient (Prantzos, 1996).

3. Implications for 'Cosmic Chemical Evolution'

In recent years, observations of high redshift systems started revealing some views of the early phases of chemical evolution in the Universe. These observations concern the 'global' SFR and HI content of the Universe (i.e. both averaged over suf-

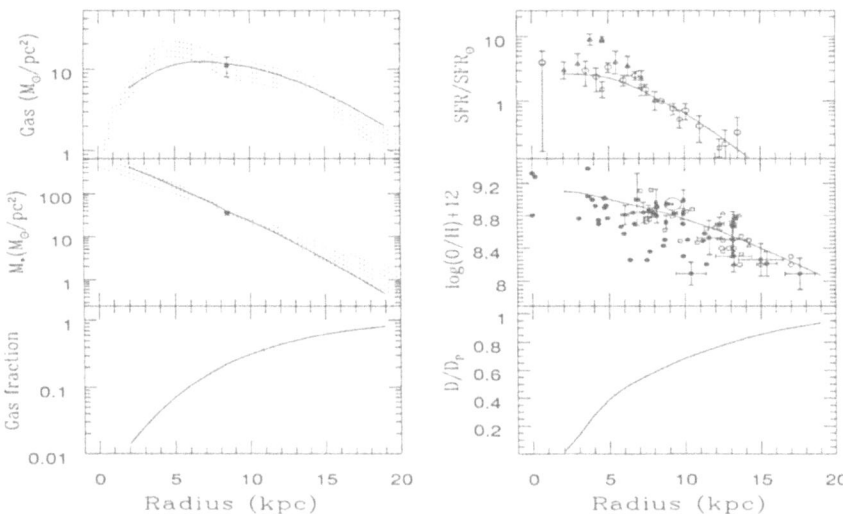

Figure 2. Results of a simple (independent ring) model for the chemical evolution of the Milky Way disc at a galactic age $T = 13$ Gyr, and comparison to observations. The adopted SFR is SFR $\propto \Sigma_G^{1.5}/R$ and the adopted infall rate is gaussian in time with $\Delta\tau = 6$ Gyr in all the zones. *Left*, from top to bottom: final profiles of the surface density of gas, stars and of the gas fraction, respectively. *Solid lines*: model results; *shaded regions* correspond to observations for the disc, and data points at $R = 8.5$ kpc to solar system values. *Right*, from top to bottom: current SFR, O and D profiles, respectively. SFR is normalised to its local value and D to its primordial one D_P. Data for O are from HII regions (*open symbols* and B-stars (*filled symbols*).

ficiently large volumes), as well as the abundances of various elements in gaseous systems, in the line of sight of quasars.

Since the only system with data sufficient to constrain its history is the Milky Way, it may be interesting to see how its evolution, modelled in Section 2 and translated in a cosmological framework, compares to the observables of 'cosmic evolution'. A favourable comparison would imply that the Milky Way is a really 'average' galaxy, but this would be a rather improbable result. The aim of that comparison is rather to see to what extent the Milky Way evolution differs from the 'average' one (Prantzos and Silk, 1998).

The results are shown in Figure 3. The model SFR is normalised to the current SFR of the local Universe (at redshift $z = 0$), since most of the current SFR is taking place in spirals; all other quantities are normalised accordingly. It can be seen that the Milky Way SFR does not increase between $z = 0$ and $z \sim 1$ as steeply as indicated by observations, although it peaks at about the right redshift. Milky Way type spirals can then account only partially for the observed evolution of the cosmic SFR; this is a reasonable conclusion, since other galaxy types (ellipticals? mergers? earlier type spirals? dwarfs?) are also expected to have a large contribution to the cosmic SFR at $z > 1$. Also, the amount of gas used by spirals acording to our model accounts fairly well for the neutral gas detected as a function

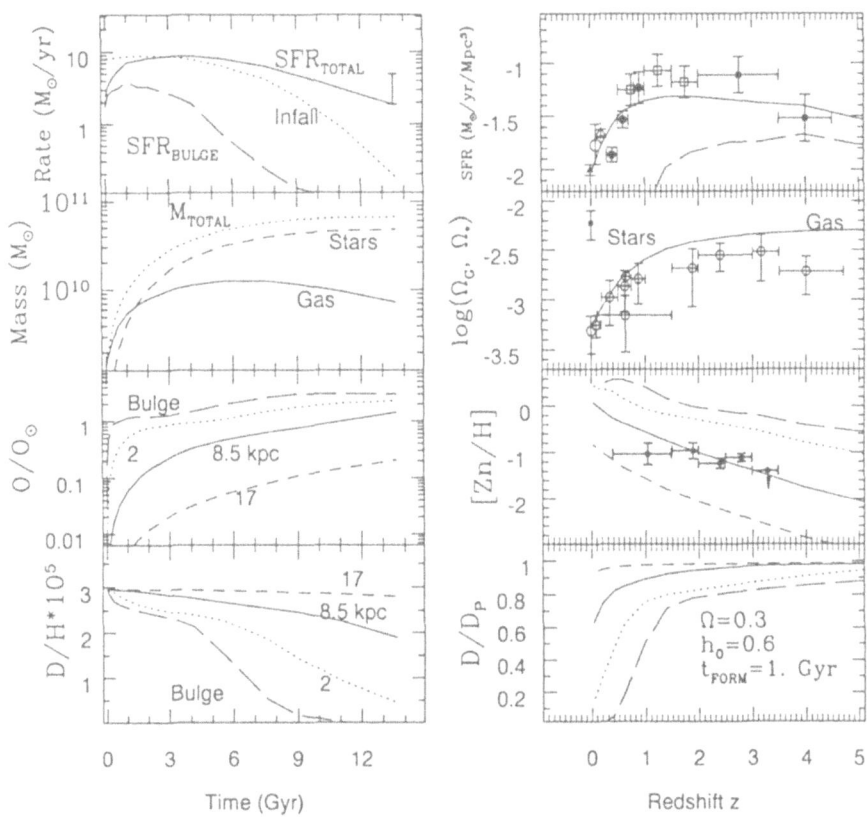

Figure 3. History of the Milky Way disc as a function of time (*left*) and of redshift (*right*), assuming that our galaxy is a typical spiral; a cosmological model with $\Omega = 0.3$, $h_0 = 0.6$ and galaxy formation starting 1 Gyr after the Big Bang is adopted. *Left*, from top to bottom: History of a) SFR and infall rate; b) gaseous, stellar and total mass; c) overall metallicity in four different zones, at distances of 2, 8.5 and 17 kpc from the galactic center, and in the bulge, respectively; d) D abundance in the same regions. *Right*, from top to bottom: a) the 'cosmic' SFR of disc galaxies, when normalised to the current local value ($z = 0$), does not show the steep observed increase back to $z \sim 1$; data from Madau (1998); b) Evolution of gas and stars; data for HI gas density from Storri-Lombardi *et al.* (1996); c) The evolution of Zn/H in various regions of spiral galaxies (only four are shown here, corresponding to those on the left); d) The corresponding evolution of D.

of the redshift. However, still larger reservoirs of gas, undetected up to now, should obviously exist in order to account for the fomation of other types of galaxies.

The interesting result of that exercice (i.e. placing the Milky Way in a cosmological setting) is that the evolution of Zn in the various disc zones brackets well the observations as a function of the redshift. It may well be indeed, that most of the lines of sight to quasars intercept various parts of (proto)galactic discs. This conclusion is opposite to the one reached in Lu *et al.* (1996), which was based on a comparison of the data to the local age-metallicity relationship. Notice that for geometrical reasons, the probability of detecting outer disc regions is higher, favouring systematically lower abundances than those spanned during the LD history.

Finally, in the framework of the same model, it turns out that large D depletion may indeed take place in inner galactic discs, but only at low redshifts. D abundances observed at high redshifts in disc galaxies should always be close to the primordial value. If the observed values differ indeed by the large factors reported in the literature, then at least one of the corresponding systems has evolved in a radically different way from a typical spiral.

References

Madau, P.: 1998, preprint astr-ph/9801005.
Lu, L., Sargent, W. and Barlow, T.: 1996, *Astrophys. J. Suppl.* **107**, 475.
Pei, Y. and Fall, M.: 1995, *Astrophys. J.* **454**, 69.
Pettini, M., Smith, L., King, D. and Hunstead, R.: 1997, *Astrophys. J.* **486**, 665.
Prantzos, N. and Aubert, O.: 1995, *Astron. Astrophys.* **302**, 69.
Prantzos, N.: 1996, *Astron. Astrophys.* **310**, 106.
Prantzos, N. and Silk, J.: 1998, *Astrophys. J.* **507**, 229.
Samland, M., Hensler, G. and Theis, C.: 1997, *Astrophys. J.* **476**, 544.
Storri-Lombardi, L., McMahon, R. and Irwin, M.: 1996, *Mon. Not. R. Astron. Soc.* **283**, L79.
Wyse, R. and Silk, J.: 1989, *Astrophys. J.* **379**, 700.

TIME-SCALE FOR ACCRETION OF MATTER

F. COMBES

DEMIRM, Observatoire de Paris, 61 Av. de l'Observatoire, F–75014 Paris, France

Abstract. Mass accretion is the key factor for evolution of galaxies. It can occur through secular evolution, when gas in the outer parts is driven inwards by dynamical instabilities, such as spirals or bars. This secular evolution proceeds very slowly when spontaneous, and can be accelerated when triggered by companions. Accretion can also occur directly through merging of small companions, or more violent interaction and coalescence. We discuss the relative importance of both processes, their time-scale and frequency along a Hubble time. Signatures of both processes can be found in the Milky Way. It is however likely that our Galaxy had already gathered the bulk of its mass about 8–10 Gyr ago, as is expected in hierarchical galaxy formation scenarios.

1. Introduction

There are two essential processus for a galaxy to accrete mass: either it accretes gas regularly, through internal dynamics, producing radial flows, from gas in the outer parts (the reservoir could be in the Local Group); or the accretion occurs in more violent events, galaxy interactions or mergers. The first process, that will be called secular accretion, can also be triggered and enhanced by the passage of companions, so that the two processus are in fact inter-related. Let us try to estimate the corresponding time-scales involved.

2. Secular Evolution

The galaxy can be considered as a giant accretion disk: to minimise its energy, it has the tendency to concentrate, and accrete gas from the outer parts towards the center. But the angular momentum is a barrier: only through tangential forces, creating torques, can the angular momentum be exchanged and transfered outwards in order that the mass flows inwards. One can apply an analog of the theory of viscous disks (Lin and Pringle, 1989). In that frame, viscous torques are the way to re-distribute angular momentum. If the time-scale for this re-distribution, τ_{vis}, is of the same order of magnitude as the time-scale to form stars τ_*, then an exponential stellar disk is created, and this generates also an exponential distribution of metallicity (Tsujimoto *et al.* 1995).

Astrophysics and Space Science is the original source of publication of this article. It is recommended that this article is cited as: *Astrophysics and Space Science* **265**: 417–424, 1999.
© 1999 *Kluwer Academic Publishers.*

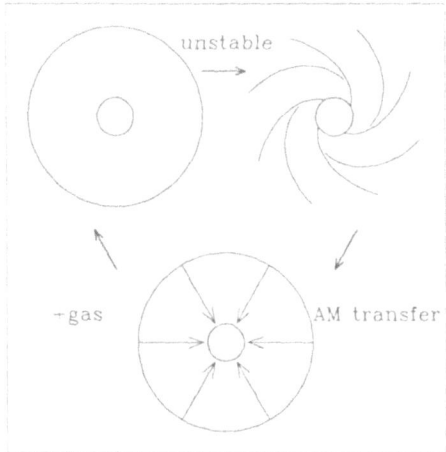

Figure 1. Gravitational instabilities regulate the transfer of angular momentum, through a feedback process, so that we can talk of 'gravitational viscosity': when fresh gas has been accreted, or radially transported inwards, the disk becomes cold ($Q < 1$); the disk then develops waves, that create non-axisymmetry and gravity torques; they transfer the angular momentum outwards (trailing waves), and matter is driven inwards. The waves heat the disk, until $Q \sim 1$. The disk needs the presence of gas to cool down again, and close the loop.

2.1. GRAVITY TORQUES

What is the nature of the viscosity, effective in galactic disks? Normal viscosity is not efficient, due to the very low density of the gas. One could think of macroturbulent viscosity, but the time-scales are longer than the Hubble time at large radii, and could be effective only inside the central 1 kpc. Instead, if the galaxy disks develop non-axisymmetric density waves such as spirals or bars, gravity torques are then very effective at transfering the angular momentum outward. This led Lin and Pringle (1987) to propose a prescription for an effective kinematic viscosity for self-gravitating disk undergoing gravitational instabilities.

The basis of this prescription is described in Figure 1. Gravitational instabilities are suppressed at small scales through the local velocity dispersion c, and at large scale by rotation. The corresponding limiting scales are the Jeans scale for a 2D disk $\lambda_J \sim c^2/(G\mu)$, and $\lambda_c \sim G\mu/\kappa^2$, where μ is the disk surface density, and κ the epicyclic frequency. Scales between λ_J and λ_c are unstable, unless c is larger than $\pi G\mu/\kappa$, or the Toomre $Q = \frac{c\kappa}{\pi G\mu}$ is larger than 1.

If the disk is cold at the beginning (general case for the gas), instabilities set in, which heat the disk until $Q \sim 1$, and those instabilities provide the necessary angular momentum tranfer, or viscosity, to concentrate the mass. Since the size of the region over which angular momentum is tranferred is $\sim \lambda_c$, and the time-scale is a rotation period, $2\pi/\Omega$, the effective kinematic viscosity is $\nu \sim \lambda_c^2\Omega$, and the typical viscous time $\tau_\nu \sim R^2\Omega^3/(G^2\mu^2)$.

Why should there be approximate agreement between the two time-scales, viscosity and star-formation? This comes from the fact that the two processes depend exactly on the same physical mechanism, i.e. gravitational instabilities. As shown empirically by Kennicutt (1989), the Toomre parameter Q appears to control star-formation in spiral disks. Therefore, if the regulating instabilities have time to develop, one can expect that $\tau_\nu \sim \tau_*$, as required for exponential light and metallicity distribution.

In summary, if the gravitational viscous time-scale is governing the mass accretion, the time-scale for accretion by secular evolution is of the order of a few dynamical times. But there must exist a reservoir of gas in the outer parts. Evidence that such accretion exists can be found in present day galaxies: the strong frequency of large-scale asymmetries of gas observed in the outer parts, or lopsidedness, cannot be explained without continuous or repeated accretions (e.g. Richter and Sancisi, 1994). The presence of gas warps in almost every disk galaxies observed in HI cannot be explained by a persistant mode, but is likely to be the manifestation of gas accretion with a different angular momentum orientation than the normal to the galaxy disk (Binney, 1992).

3. Galaxy Interactions

There is plenty of evidence that the Milky Way has accreted mass or has even encountered merging events in the past (e.g. Searle and Zinn, 1978; Zinn, 1993). Moving groups are detected in the halo (Majewski, 1993, and this volume), and the thick disk is best explained by an interaction event (Bienaymé, this volume). This is in line with hierachical formation.

In hierarchical cosmological scenarios, galaxies have formed by successive mergers of larger and larger entities. Primordial fluctuations of dark matter exist at all scales, but small-scale perturbations become non-linear and collapse first, and are then incorporated in larger structures collapsing later on. Small halos then merge into larger ones; the baryonic structures that had condensed in the halos potential well, can also merge, with some delay. While the merging scenario of dark halos is quite well understood and reproduced in simulations, or quantified analytically by Press-Schechter formalism, the scenario concerning the baryonic component, i.e. galaxies, is less known, because complex physics intervenes (cooling, star formation, feedback, IMF, metal abundance) in supplement to gravity. As a general statement, it is thought that galaxy merging directly follows dark halos merging, as soon as the virial velocities inside the merged dark structure is below a certain threshold (roughly equal to the escape velocity for individual galaxies, from dynamical friction efficiency). Since structures forming now have quite large virial velocities, galaxy merging becomes less and less efficient.

Galaxy interactions were undoubtedly more frequent in the past, and many groups have tried to quantify the effect. Already Toomre in 1977 has estimated the number of mergers from their observed frequency at $z = 0$ just taking into account the probability of excentricities of binary orbits. Statistics of close galaxy pairs from faint-galaxy redshift surveys have shown that the merging rate increases as a power law with redshift, as $(1+z)^m$ with $m = 4 \pm 1.5$ (e.g. Yee and Ellingson, 1995). Lavery *et al.* (1996) claim that ring galaxies are also rapidly evolving, with $m = 4 - 5$, although statistics are still insufficient. Many other surveys, including IRAS faint sources, or quasars, have also revealed a high power-law. Governato *et al.* (1998) from numerical simulations of standard ($\Omega = 1$) and open ($\Omega_0 = 0.3$) CDM models find that the number density of interacting binaries is proportional to $(1 + z)^{4.2}$ and $(1 + z)^{2.5}$ respectively.

The number of mergings for a given galaxy is still quite uncertain observationally; the merger frequency, and the peak merging epoch are very sensitive to the values of universe density (Ω) and the cosmological constant (Λ, see e.g. Carlberg, 1991). The observation of relatively thin stellar disks has been advanced as a constraining argument.

3.1. THICKENING OF DISKS

Galaxy interactions can easily thicken or even destroy a stellar disk (e.g. Gunn, 1987). The fragility of stellar disks with respect to thickening has been used by Toth and Ostriker (1992) to constrain the frequency of merging and consequently the value of the cosmological parameter Ω. They claim for instance that the Milky Way disk have accreted less than 4% of its mass within the last $5\ 10^9$ yrs. Numerical simulations have tried to quantify the thickening effect (Quinn *et al.*, 1993; Walker *et al.*, 1996). They show that the stellar disk thickening can be large and sudden, but it is strongly moderated by gas hydrodynamics and star-formation processes, since the thin disk can be reformed continuously through gas infall. Velazquez and White (1998) through N-body simulations find that analytical derivations overstimate by factors 2–3 the disk heating and thickening; while prograde satellites do heat the disk, retrograde ones produce essentially a coherent tilt of the disk. Companion accretions therefore cause stellar warps and asymmetric disks.

Galaxies presently interacting have their ratio h/z_0 of the radial disk scalelength h to the scaleheight z_0 1.5 to 2 times lower than normal (Bottema, 1993; Reshetnikov and Combes, 1997). However, since galaxies have experienced many interactions in the past, including the presently isolated galaxies, all these perturbations, thickening of the planes and radial stripping, must be transient, and disappear after an interaction time-scale, i.e. one Gyr. Present galaxies are thought to be the result of merging of smaller units, according to theories of bottom-up galaxy formation; a typical galaxy has accreted most of its mass, and the existence of shells and ripples attests of the frequency of interactions (Schweizer and Seitzer, 1992). This implies that the global thickness of galaxy planes can recover their small values after galaxy interactions. Or in other words, the disk of present day spirals has been essentially assembled at low redshift (Mo *et al.*, 1998).

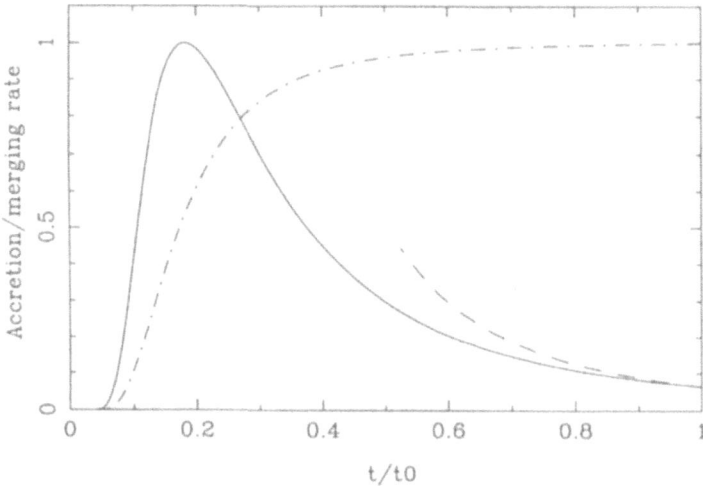

Figure 2. Indicative accretion rate as a function of time, for a structure of the Milky Way mass (10^{12} $M_\odot$ halo). Full line: normalised accretion rate, as expected from the Press-Schechter formalism (standard CDM model) for the dark halos, and modified for baryonic systems merging; Dashed line: power law $m = 4.5$ as a function of $(1 + z)$ for comparison; Dash-Dotted line: corresponding fraction of the present mass assembled as a function of time.

3.2. MERGING HISTORY OF THE MILKY WAY

Since the number of galaxy interactions and mergers were much larger in the past, it is likely that most accretion events occured long ago in the Milky Way. This is compatible with the observation of globular clusters for example: Unavane *et al.* (1996) conclude that there has not been more than 10% of the mass accreted over the last 10 Gyr. The same is true for many galaxies of the Local Group, except maybe for the SMC (Sarajedini *et al.*, 1998).

Although we are still witnessing the on-going mergers of some dwarf galaxies bound to the Milky Way (Sagittarius dwarf, Ibata *et al.*, 1994; Magellanic Clouds, Lin and Lynden-Bell, 1977; Putman *et al.*, 1998), the bulk of the mass must have been accreted some Gyrs ago. Many debris stay coherent as tidal streams for the Milky Way life-time: Johnston (1998) estimates that up to 10% of the sky is covered by those debris, and this could have consequences for microlensing experiments (Zhao, 1998). If the time-scale for mixing after stellar accretion can be long, more than a Hubble time, the time-scale for gas accretion and processing is quite short (less than a Gyr). High velocity clouds (HVCs) could represent a huge reservoir of gas around the Milky Way, if we interprete correctly their distance. If the latter is around 40 kpc from the center, they correspond to a total of 10^{11} $M_\odot$ of HI gas (Blitz *et al.*, 1998). Simulations of the clouds infalling towards the Local Group gravity center are compatible with the observations. HVCs will then be the analog of the Lyman-limit absorbing clouds in absorption in front of quasars.

Embedded in dark mini-halos, they could be the building blocks of the Galaxy (Blitz et al., 1998).

To have a statistical estimate of these accretion times, we can use the empirical law of interactions increase with redshift, in $(1 + z)^m$ as described earlier. An other approach is to consider the probability of formation of halos of mass $M = 10^{12} M_\odot$, such as the Milky Way. This is obtained through the well-known Press-Schechter (1974) formalism, or 'excursion set' derivation (Bond et al., 1991). The model is based on the existence of a Gaussian random field for initial density fluctuations, on the follow up of their linear growth, and on spherical collapse theory. This leads to the differential mass function:

$$f(\sigma(M), t) = \frac{\delta_c}{\sqrt{(2\pi)}D(t)\sigma^3} \exp[-\frac{\delta_c^2}{2D(t)^2\sigma^2}]$$

where $\sigma(M)$ is the mass variance, $D(t)$ the linear growth factor and δ_c the critical overdensity $\frac{\delta\rho}{\rho}$ for collapsed structures (extrapolated in linear theory $= 1.69$). The derivative of this formula with respect to time (or redshift) gives the rate of change of density of halos of a given mass M. But this is the difference between their increase as the result of mergers of smaller halos, and their decrease due to combining into halos of larger masses (and can become negative). To have the rate of merging, Carlberg (1990a) assumes that the halos of mass larger than M come essentially from merging masses between $M/2$ and M; a more exact calculation is derived by Lacey and Cole (1993), and leads to an analytical expression for the merging or accretion rate as a function of time, to form the Milky Way halo, for example. This analytical expression is in very good agreement with N-body simulations (Lacey and Cole, 1994).

Once the rate of merging of halos is estimated, the probability of merging the baryonic systems must be taken into account, to compare with observations, and this introduces large uncertainties. Carlberg (1990b) proposes a simple criterion for baryonic systems to merge: either they merge in a dynamical time-scale if their relative velocity is lower than the escape speed from the systems, or they never merge in a Hubble time. He then derives the baryonic merging rate by considering only the fraction of the Maxwellian distribution for velocities less than the threshold for merging. Lacey and Cole (1993) show that the merging efficiency depends strongly on the excentricity of the initial orbits, through dynamical friction. The results between circular orbits and highly elongated ones (pericenter $= 5\%$ of apocenter) can vary by factors 3 or more. An average estimate of the accretion rate as a function of time for baryonic matter is plotted in Figure 2. It is very close at low redshift to a power-law of power $m = 4.5$, the value expected for this model universe, through numerical simulations. In this scenario, it can be seen that most of the Galaxy (90% of the mass) was built 9 Gyr ago (if $t_0 = 15$ Gyr).

4. Conclusion

Mass accretion can be considered to occur through two processus: a secular accretion from gas in the outer parts of galaxies, or in the near halo (may be HVCs), that occurs on a dynamical time-scale; and through galaxy interactions and mergings, that has progressively built the galaxy in the past. Both processus can occur on widely different time-scales, according to the environment. For instance, Low Surface Brightness (LSB) galaxies appear unevolved systems, with large gas fraction, low mass concentration, and their long evolution time-scale is attributed to their poor environment (Bothun *et al.*, 1997).

The Milky Way on the contrary, as an HSB, had a very rapid evolution time-scale, and is likely to have accreted the bulk of its mass 8–10 Gyrs ago. Its evolution is compatible with the statistical mean obtained in hierachical scenarios, from the Press-Schechter formalism for the dark halos merging, implemented with simple recipes for the merging of baryonic systems.

References

Binney, J.: 1992, *Annu. Rev. Astron. Astrophys.* **30**, 51.
Bond, J.R., Kaiser, N., Cole, S. and Efstathiou, G.: 1991, *Astrophys. J.* **379**, 440.
Bothun, G., Impey, C. and McGaugh, S.: 1997, *Publ. Astron. Soc. Pacific* **109**, 745.
Bottema, R.: 1993, *Astron. Astrophys.* **275**, 16.
Blitz, L., Spergel, D.N., Teuben, P.J., *et al.*: 1998, *Astrophys. J.*, in press (astro-ph/9803251).
Carlberg, R.G.: 1990a, *Astrophys. J.* **350**, 505.
Carlberg, R.G.: 1990b, *Astrophys. J.* **359**, L1.
Carlberg, R.G.: 1991, *Astrophys. J.* **375**, 429.
Governato, F., Gardner, J.P., Stadel, J., *et al.*: 1998, astro-ph/9710140.
Gunn, J.E.: 1987, in: S.M. Faber (ed.), *Nearly Normal Galaxies*, Springer, New York, p. 459.
Ibata, R.A., Gilmore, G. and Irwin, M.J.: 1994, *Nature* **370**, 194.
Johnston, K.V.: 1998, *Astrophys. J.* **495**, 297.
Kennicutt, R.C.: 1989, *Astrophys. J.* **344**, 685.
Lacey, C. and Cole, S.: 1993, *Mon. Not. R. Astron. Soc.* **262**, 627.
Lacey, C. and Cole, S.: 1994, *Mon. Not. R. Astron. Soc.* **271**, 676.
Lavery, R., Seitzer, P., Walker, A.R., *et al.*: 1996, *Astrophys. J.* **467**, L1.
Lin, D.N.C. and Pringle, J.E.: 1987, *Mon. Not. R. Astron. Soc.* **225**, 607.
Lin, D.N.C. and Lynden-Bell, D.: 1977, *Mon. Not. R. Astron. Soc.* **181**, 59.
Majewski, S.R.: 1993, *Annu. Rev. Astron. Astrophys.* **31**, 575.
Mo, H.J., Mao, S. and White, S.D.M.: 1998, *Mon. Not. R. Astron. Soc.* **295**, 319.
Richter, O-G. and Sancisi, R.: 1994, *Astron. Astrophys.* **290**, L9.
Schweizer, F. and Seitzer, P.: 1992, *Astron. J.* **104**, 1039.
Searle, L. and Zinn, R.: 1978, *Astrophys. J.* **225**, 357.
Toomre, A.: 1977, in: B.M. Tinsley and R.B. Larson (eds.), *The Evolution of Galaxies and Stellar Populations*, Yale Univ. Obs., New Haven, p. 401.
Toth, G. and Ostriker, J.P.: 1992, *Astrophys. J.* **389**, 5.
Tsujimoto, T., Yoshii, Y., Nomoto, K. and Shigeyama, T.: 1995, *Astron. Astrophys.* **302**, 704.
Press, W.H. and Schechter, P.: 1974, *Astrophys. J.* **187**, 425.
Putman, M.E., *et al.*: 1998, *Nature* **394**, 752.

Quinn, P.J., Hernquist, L. and Fullagar, D.P.: 1993, *Astrophys. J.* **403**, 74.

Reshetnikov, V. and Combes, F.: 1997, *Astron. Astrophys.* **324**, 80.

Sarajedini, A., Geisler, D., Harding, P. and Schommer, R.: 1998, *Astrophys. J.*, in press (astro-ph/9809275).

Unavane, M., Wyse, R.F.G. and Gilmore, G.: 1996, *Mon. Not. R. Astron. Soc.* **278**, 727.

Velazquez, H. and White, S.D.M.: 1998, *Mon. Not. R. Astron. Soc.*, preprint (astro-ph/9809412).

Yee, H.K.C. and Ellingson, E.: 1995, *Astrophys. J.* **445**, 37.

Zhao, H.: 1998, *Mon. Not. R. Astron. Soc.* **294**, 139.

Zhao, H.: 1998, *Astrophys. J.* **500**, L149.

Zinn, R.: 1993, in: H. Graeme, H. Smith and Jean P. Brodie (eds.), *The Globular Cluster-Galaxy Connection*, ASP Series, p. 38.

A NEW PICTURE FOR THE GALAXY FORMATION

C. CHIAPPINI

Depto. de Astronomia, Observatório Nacional, Brazil

F. MATTEUCCI

Dipartimento di Astronomia, Università di Trieste, Italy

Abstract. The observational constraints are of fundamental importance to build a realistic chemical evolution model. With respect to these constraints the last years have been of crucial importance and, in the case of the Milky Way, the new observational data required a revision of the previous chemical evolution models (see Chiappini *et al.*, 1997 – CMG97) and Pagel and Tautvaisiene, 1995 – PT95). The results obtained by CMG97 from a careful comparison between model predictions and observational constraints strongly suggest that the previously adopted picture for the Galaxy formation in which the gas shed from the halo was the main contributor to the thin disk formation, is not valid anymore. With our detailed chemical evolution model we are able to put some constraints on the IMF variation and on the Deuterium primordial value.

1. Results

In this model (CMG97) we included the calculation of the evolution of the halo and disk as a result of two main infall episodes. Our best model reproduces quite well the most recent data on the distribution of the G-dwarf stars in the solar-vicinity, one of the most important constraints for the evolution of the Galaxy, as well as all the main observational data concerning the solar vicinity and the whole disk. In this work a new result is that the only model that is in good agreement with recent observational data is the one that assumes that the galactic thin disk was not formed from gas shed from the halo but was formed mainly from extragalactic gas. This is the fundamental prediction of this new chemical evolution model for the Milky Way and in this new picture the disk was formed slowly (with a timescale of 8 Gyrs at the solar vicinity). This means that at high redshift we should expect to see smaller disks in size. This new scenario for the formation of the Galaxy is also in agreement with recent results by PT95 and with the observational results by Gratton (1998) and Beers and Sommer-Larsen (1995). Furthermore, this model predicts that the abundance gradients steepen with time and that they are slightly flatter in the outer parts of the galactic disk. This work also suggests that the Galaxy formed from inside out, in agreement with results from dynamical models.

We explored the effects of adopting an initial mass function (IMF) variable in time on the chemical evolution of the Galaxy (Chiappini *et al.*, 2000). In order to do

Astrophysics and Space Science is the original source of publication of this article. It is recommended that this article is cited as: *Astrophysics and Space Science* **265**: 425–426, 1999.
© 1999 *Kluwer Academic Publishers.*

that we adopted the two-infall model discussed above, and a new IMF derived by Padoan *et al.* (1997). We show that this model is in agreement with the available observational constraints of the Galaxy and with the present day mass function. This model predicts that the IMF depends on time but such time variation is important only in the early phases of the Galactic evolution, when the IMF is biased towards massive stars. The abundance gradients along the galactic disk are flatter when adopting the above IMF. It is clear from these calculations that an IMF which is a strong function of time will not lead to a good agreement with the observational constraints suggesting that if the IMF varied this variation should have been small.

We also investigated (Tosi *et al.*, 1998) alternative galactic evolution scenarios able to predict a high astration factor for Deuterium in order to reconcile the current abundance of D in the interstellar medium with the high primordial value suggested for instance by Rugers and Hogan (1996). We have explored a variety of chemical evolution models including infall of processed material and early supernovae-driven winds with the aim of identifying models with large D-destruction which are still consistent with the observations of stellar-produced heavy elements. When such models are compared with data they confirm that only a modest destruction of deuterium (less than a factor of 3) is permitted implying a 2σ upper bound for the deuterium primordial value of $(D/H)_P \leq 5.0 \times 10^{-5}$.

References

Beers, T.C. and Sommer-Larsen, J.: 1995, *Astrophys. J. Suppl.* **96**, 175.

Chiappini, C., Matteucci, F. and Gratton, R.: 1997, *Astrophys. J.* **477**, 765 – CMG97.

Chiappini, C., Matteucci, F. and Padoan, P.: 2000, *ApJ* **52**: Issue Jan. 10 Part I, in press.

Gratton, R.: 1998, this volume.

Padoan, P., Norlund, A.P. and Jones, B.J.T.: 1997, *Mon. Not. R. Astron. Soc.* **288**, 43.

Pagel, B.E.J. and Tautvaisiene, G.: 1995, *Mon. Not. R. Astron. Soc.* **276**, 505 – PT95.

Rugers, M. and Hogan, C.J.: 1996, *Astrophys. J.* **459**, L1.

Tosi, M., Steigman, G., Matteucci, F. and Chiappini, C.: 1998, *Astrophys. J.* **498**, 226.

DISENTANGLING METALLICITY AND AGE FOR TURNOFF STARS

R.C. PETERSON

Astrophysical Advances, Palo Alto, CA 94301, USA

Abstract. This is a progress report on calculations of near-ultraviolet spectra with Ben Dorman at NASA/Goddard, for the ultimate purpose of extracting age and metallicity from extragalactic spectra. We are calculating from first principles a grid of spectra covering 2200–3400 Å using the Kurucz program SYNTHE, beginning with stars of metallicity less than one-fifth solar ([Fe/H] < −0.7). For these stars, LTE calculations using known opacities and line lists including only transitions measured in the laboratory, coupled with standard line-blanketed LTE models, provide a satisfactory match to the spectra of turnoff stars of temperatures T_{eff} = 5750 K – 6250 K. For more metal-rich stars, two problems arise: lines without a laboratory identification become increasingly influential, and the cores of all strong lines become too strong. These problems must be addressed to match near-UV spectra of turnoff stars of solar metallicity or higher.

1. Introduction

Characterizing the age and metallicity of galaxies more than a few billion years old has puzzled astronomers for decades. Roughly speaking, the age of a galaxy is indicated by the temperature of the hottest 'turnoff' stars, those just evolving from the main sequence (Dorman and O'Connell, 1995), while the metallicity is deduced from the strength of atomic absorption features. Unfortunately, spectral features are also sensitive to temperature. In the optical wavelength region, where both giants and turnoff stars contribute flux, Worthey (1994) concluded that two stellar populations which differ in age and metallicity by d log age $\sim 3/2d$ log Z will be indistinguishable in all colors and most optical spectral indices. In the near UV, 2200–3400 Å, red-giant stellar flux is negligible and the turnoff population dominates. Once its temperature is known, both age and abundances are forthcoming. The contribution of very hot stars, if any, can be estimated from the far-UV and subtracted, being virtually free of near-UV lines.

Currently, however, interpretation of near-UV spectra rests on empirical relationships between spectral line indices among solar-neighborhood stars (e.g. Fanelli *et al.*, 1990; Ponder *et al.*, 1998). Unfortunately, these are sometimes at odds with physical expectations. For example, Fanelli *et al.* note in F and G stars that the strength of the MgII 2800 Å resonance doublet increases as metallicity decreases from solar to one-tenth solar abundance. In principle, such behavior could be explained by theoretical calculations of UV synthetic stellar spectra. In practice, spectral calculations succeed well at low metallicities, but cannot reproduce major

Astrophysics and Space Science is the original source of publication of this article. It is recommended that this article is cited as: *Astrophysics and Space Science* **265**: 427–430, 1999.

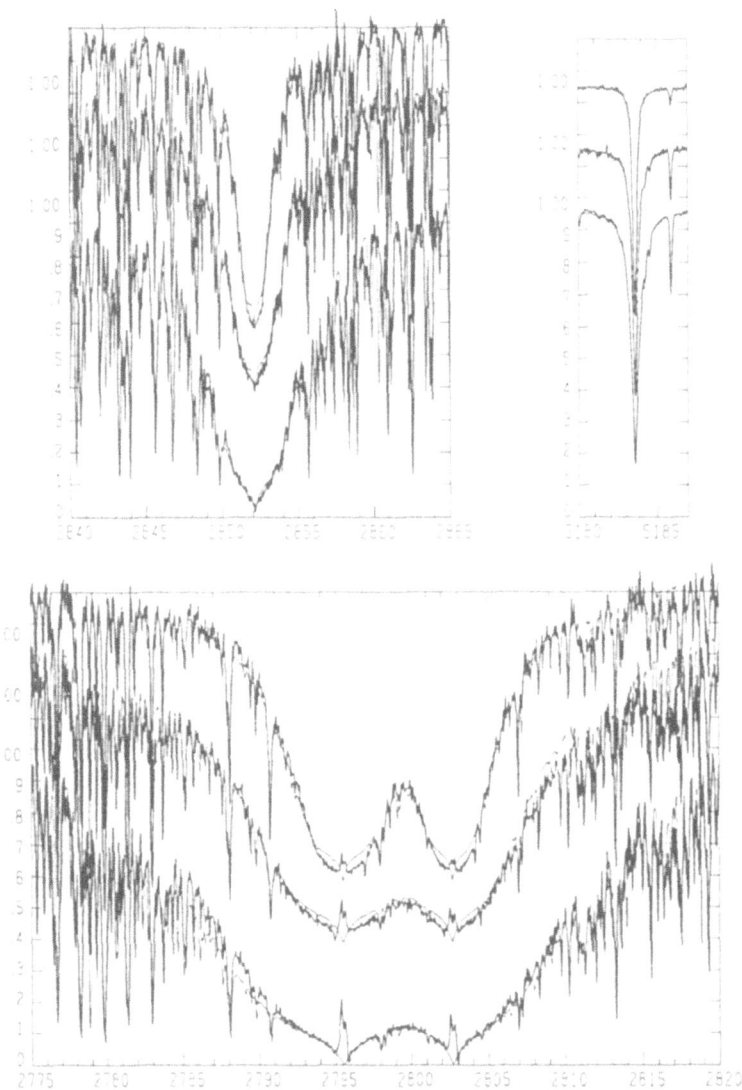

Figure 1. Fits of regions with Mg lines.

absorption features in turnoff stars of solar abundance. In this work we illustrate that the line list is missing many weak transitions which become strong in strong-lined stars, and also note that the calculations of strong lines are too strong in the core.

2. Calculations of UV Spectra

Ben Dorman and I, with NASA support, are calculating near-UV spectra using updated Kurucz versions of his program SYNTHE, his model atmospheres, and his line lists. The plane-parallel, static models assume radiative and local thermodynamic equilibrium (LTE), and are line-blanketed with a solar elemental abundance distribution. We modify their species number densities to reflect the light-element abundance enhancement common among halo metal-poor stars of the Galaxy (Wheeler, Sneden and Truran, 1989). The line list incorporates atomic and molecular hydride transitions identified in the laboratory. All line data, models, and programs can be downloaded from Kurucz's Web site at http://cfaku5.harvard.edu.

We have compared such calculations against spectra for several well-studied metal-poor stars of nearly solar temperature and gravity in plots of normalized residual flux versus wavelength in Å. The figure shows our calculations (light line) compared to observed spectra (heavy line) for three stars in three wavelength regions each containing one or more strong magnesium lines. The stars (with model T_{eff}, log g, [Fe/H], and [Mg/Fe]) are HD 19445 (6000 K, 4.5, −2.0, +0.4), HD 94028 (6000 K, 4.4, −1.4, +0.3), HD 201891 (5900 K, 4.2, −1.0, +0.3). The UV observations are from the Hubble Space Telescope, and the optical are from Lick Observatory; they have FWHM resolutions of 25 000 and 38 000 respectively. The calculations for the Mgb region near 5170 Å employ the gf-values of Peterson, Dalle Ore and Kurucz (1993) based on fits to the solar spectrum.

The MgII doublet at 2800 Å fits well with standard line parameters, but for the MgI resonance line at 2852 Å we were forced to reduce the gf-value by 0.3 dex. The calculation is clearly missing lines (e.g. the 2815–2820 Å region). We have calculated additional trial spectra with other diatomic molecules, and with the predicted lines of the hydrides, but these do not account for the features. We plan further investigations of cool, strong-lined stars such as Hyades dwarfs to attempt to fill these gaps empirically.

Similar plots of the entire near-UV region for these and more metal-rich stars show that fluxes are well reproduced for the stars illustrated, but increasingly less so for stronger-lined stars. Since the discrepancies are worst in weak-lined regions, we conclude that the integrated contribution of missing lines apparently increases to the point where the spectral flux is significantly overestimated in relatively line-free regions. Moreover, stronger-lined stars show a tendency for the cores of strong lines to be too strong in the cores (Peterson, Carney and Smith, 1999). Those authors propose a change to the surface temperature gradient as a remedy.

Together these discrepancies currently make it very difficult to derive the temperature of a near-solar-metallicity turnoff star from a near-UV spectrum of moderate resolution. However, the incorporation of missing lines and an increased surface temperature should go a long way to address this difficulty.

References

Dorman, B. and O'Connell, R.W.: 1995, in: C. Leitherer, U. Fritze-von Alvensleben and J.P. Huchra (eds.), *From Stars to Galaxies: The Impact of Stellar Physics on Galaxy Evolution*, ASP, San Francisco, p. 105.

Dorman, B. and O'Connell, R.W.: 1997, in: W.H. Walter and M.N. Fanelli (eds.), *The Ultraviolet Universe at Low and High Redshift: Probing the Progress of Galaxy Evolution*, AIP, New York.

Fanelli, M.N., O'Connell, R.W., Burstein, D. and Wu, C.-C.: 1990, *Astrophys. J.* **364**, 272.

Peterson, R.C., Dalle Ore, C.M. and Kurucz, R.L.: 1993, *Astrophys. J.* **404**, 333.

Peterson, R.C., Carney, B.W. and Smith, H.A.: 1999, *Astrophys. J.*, submitted.

Ponder, J.M., *et al.*: 1998, *Astron. J.* **116**, 2297.

Wheeler, J.C., Sneden, C. and Truran, J.W., Jr.: 1989, *Annu. Rev. Astron. Astrophys.* **27**, 279.

Worthey, G.: 1994, *Astrophys. J. Suppl.* **95**, 107.

METALLICITIES OF NEARBY K-DWARFS

EIRA KOTONEVA

Tuorla Observatory, Piikkiö, Finland

CHRIS FLYNN

Tuorla Observatory, Piikkiö, Finland

One of the main issues in the chemical evolution of galaxies is the so called 'G-dwarf problem' (Pagel and Patchett, 1975) in which the observed stellar metal distribution differs from the predicted in the sense that there are too many metal deficient stars. The same problem was now been seen among K-dwarfs, indicating the problem is a general one (Flynn and Morell, 1997).

We have used the Hipparcos catalogue to choose all K-dwarfs with absolute magnitude M_V between $5.5 < M_V < 7.3$, within a radius of 54 parsecs with an apparent visual magnitude $V < 8.2 + 1.1 \sin |b|$, where b is the galactic latitude. This sample consists of 642 stars and is complete. 408 of these stars were found from the Geneva photometric catalogue (Rufener, 1989) and 364 from the Strömgren catalogue of Hauck and Mermilliod (1998).

We have measured metallicities for 248 of these stars using $R - I$ photometry from Bessell (1990) and Geneva and Strömgren based metallicity indicators. For stars with Geneva colours metallicities were found using existing relations for K-dwrafs (Flynn and Morell, 1997). We have developed a Strömgren photometric metallicity indicator for the stars with Strömgren colours. We used 34 G and K dwarfs from Flynn and Morell (1997) to define a relation between the Strömgren colours, the abundance [Fe/H] and effective temperature T_{eff}. For these stars accurate, spectroscopically determined metallicity abundances were known and the effective temperatures were estimated using Cousins $R - I$ photometry from the Gliese catalogue (Bessell, 1990).

In Figure 1a the relation between spectroscopically determined metallicity and the Strömgren based metallicity is shown. The scatter around the one-to-one relation is ≈ 0.2 dex.

Iron abundances [Fe/H] have been converted to oxygen abundances [O/H] as in Flynn and Morell (1997) using their Equation 6. When comparing data and models oxygen abundances are more appropriate because oxygen is produced in short lived stars and can be treated using the convenient approximation of instantaneous recycling (Pagel, 1989).

Astrophysics and Space Science is the original source of publication of this article. It is recommended that this article is cited as: *Astrophysics and Space Science* **265**: 431–432, 1999.
© 1999 *Kluwer Academic Publishers. Printed in the Netherlands.*

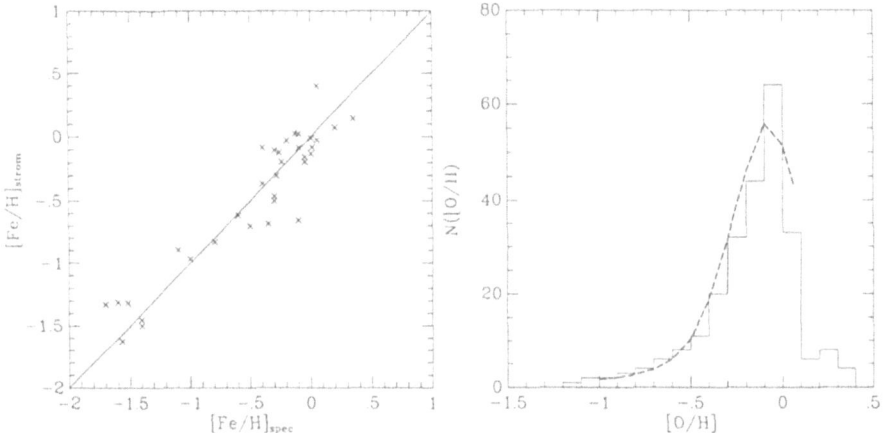

Figure 1. Left panel shows the relation between the abundances using Strömgren b_1 colour, Cousins $R - I$ and spectroscopically determined abundances. The right panel shows the histogram of oxygen abundances of 248 stars. This histogram shows clearly that there is a 'K-dwarf problem'. The Pagel (1989) inflow model is shown as a dashed line and fits the data quite well.

The metallicity histogram for all 248 stars is shown in Figure 1b. In this figure the K-dwarf problem is clearly seen. The histogram is peaked near solar metallicity and the number of metal deficient stars is small.

Note that the sample has not been corrected for two effects. Firstly, there is a kinematic bias against higher velocity stars in nearby samples (Sommer-Larsen, 1991). Secondly, the sample is slightly biased toward too many metal weak stars since these are more likely to be in the photometric catalogues we used. In the future we plan to observe the complete sample and thus remove these biases.

References

Bessell, M.: 1990, *Astron. Astrophys. Suppl.* **83**, 357.

Flynn, C. and Morell, O.: 1997, *Mon. Not. R. Astron. Soc.* **286**, 617.

Hauck, B. and Mermilliod, M.: 1998, *Astron. Astrophys. Suppl.* **129**, 431.

Sommer-Larsen, J.: 1991, *Mon. Not. R. Astron. Soc.* **249**, 368.

Pagel, B.E.J.: 1989, *Rev. Mex. Astr. Astrofis.* **18**, 153.

Pagel, B.E.J. and Patchett, B.E.: 1975, *Mon. Not. R. Astron. Soc.* **172**, 13.

Rufener, F.: *Astron. Astrophys. Suppl.* **78**, 469.

STUDY OF EARLY-TYPE STARS POPULATIONS AND OF THEIR ORIGIN

F. ROYER

DASGAL, Observatoire de Paris, Meudon, France

1. Introduction

In order to provide an insight into the different populations of early-type stars, a sample of high velocity stars has been gathered from the Hipparcos catalogue. The hypothesis of blue-stragglers as high-velocity early-type stars is tested in term of kinematics to explain their possible origin.

2. Sample

The observed sample is composed of Hipparcos stars whose absolute magnitudes where re-calibrated using the LM method (Luri *et al.*, 1995). The new distances computed with this method allow us to obtain the total velocity $\mathcal{V}$ (with respect to the Sun) and the galactic spatial components U, V and W (Meillon *et al.*, 1998). The stars bluer than $B - V = 0.4$ and with total velocities larger than $\mathcal{V} = 65$ km s^{-1} were selected.

Among more than 9 000 B-A-F stars with the 5 astrometric parameters and radial velocity, 121 high-velocity early-type stars were selected.

3. Kinematics

In order to isolate the stars with disk-like kinematical behaviour, we have used the SEMMul algorithm (Celeux and Diebolt, 1986), on the V and W-distributions of our sample.

56% of the sample belong to the disk component:

$$\langle V \rangle = 182 \pm 4 \text{ km s}^{-1} \quad \sigma_V = 26 \pm 2 \text{ km s}^{-1}$$
$$\langle W \rangle = -8 \pm 4 \text{ km s}^{-1} \quad \sigma_W = 24 \pm 2 \text{ km s}^{-1}$$

The rest of the sample is mainly composed of known field horizontal-branch stars (Philip, 1987) as well as more luminous stars far above the location of the

Astrophysics and Space Science is the original source of publication of this article. It is recommended that this article is cited as: *Astrophysics and Space Science* **265**: 433–434, 1999.
© 1999 *Kluwer Academic Publishers.*

blue horizontal branch in the H-R diagram, some of these massive stars are known to be *runaways*.

4. Simulations

A blue-straggler Main-sequence has been built using the Catalogue of blue stragglers in open clusters (Ahumada and Lapasset, 1995). $B - V$ color indexes and the M_V absolute magnitudes are both function of age τ in the range $7.5 < \log \tau < 10.0$. Using the Besançon population synthesis model (Robin and Crézé, 1986; Haywood, 1998), blue-stragglers have been simulated in the 1 kpc-radius volume around the Sun using the computed main-sequence and the kinematical status of coeval stars in the red giant branch. The proportion of $\frac{1}{4}$ of the red giants (Wheeler, 1979) is taken as a rule of thumb to account for the blue stragglers rate.

The high-velocity stars are selected in the same way as the observed sample. The resulting sample of 'disk' stars is composed half of simulated blue-stragglers, half of distribution-tail stars. There is no stars more luminous than $M_V = -1$. The distribution tail stars are mainly located in the color range $0.3 < B - V < 0.4$ whereas the simulated blue-stragglers cover a wider range. The velocity distribution is $\langle V \rangle = 212$ km s^{-1} $\sigma_V = 25$ km s^{-1} $\langle W \rangle = -11$ km s^{-1} $\sigma_W = 17$ km s^{-1} to be compared with the previous ones.

5. Discussion

The high-velocity early-type stars are a mixture of different populations: blue-horizontal thick disk or halo stars passing in our neighbourhood and high-velocity disk stars. The contribution of blue-stragglers as long lived stars on the main-sequence provide an adequate explanation for A-type main sequence high-velocity stars as found by Shields and Twarog (1988), whereas other mechanisms have to be invoked for more massive stars.

References

Ahumada, J. and Lapasset, E.: 1995, *Astron. Astrophys. Suppl. Ser.* **109**, 375.

Celeux, G. and Diebolt, J.: 1986, *R.S.A.* **34**, No. 2, 35.

Haywood, M.: 1998, private communication.

Meillon, L., Crifo, F., Gómez, A. E., Udry, S. and Mayor, M.: 1997, *Hipparcos – Venice'97*, ESA SP-402, 591.

Philip, A.G.D.: 1987, *IAU Colloq. 95: Conference on Faint Blue Stars*, 67.

Robin, A. and Crézé, M.: 1986, *Astron. Astrophys.* **157**, 71.

Shields, J.C. and Twarog, B.A.: 1988, *Astrophys. J.* **324**, 859.

Wheeler, J.C.: 1979, *Astrophys. J* **234**, 569.

SPECTROSCOPY OF SUPERGIANTS WITH IR–EXCESSES: RV TAU TYPE STARS

V.G. KLOCHKOVA and V.E. PANCHUK

Special Astrophysical Observatory, Nizhnij Arkhyz, 357147 Russia

1. Chemical Composition of Pulsating Supergiants

Using high resolution CCD spectra obtained with the echelle-spectrometers of the 6 m telescope, we determined by the model atmospheres method the parameters T_e, log g and the detailed chemical composition for 4 RV Tau type pulsating stars in the galactic field: AC Her, U Mon, RV Tau and AI CMi. For more details see: Klochkova, Panchuk (1996, 1998). For the first time from high resolution spectra the chemical composition of the star RV Tau, the prototype of this class of pulsating stars, has been derived.

We concluded that both the metallicity and the chemical abundance pattern for pulsating supergiants are not uniform. Only in the case of AC Her they are consistent with the expected ones for a post–AGB halo star: underabundance of iron group elements [Fe/H] = −0.82, overabundance of CNO (C/O > 1) and s–process elements La, Ce, Pr, Nd, Sm (the average value [s/Fe] = +0.32). At the same time in the case of similar metal–poor ([Fe/H] = −0.68) star U Mon having also the large carbon excess [C/Fe] = +0.77 and C/O > 1 we have found no excess of s–process elements: the average value [s/Fe] = −0.53. It is important that the abundances of Zn, S, the elements hard-to-condense, are essentially enhanced relative to the Fe content in the atmospheres of both stars. This suggests that the initial metallicity of these stars, being close to the normal (solar) value, has been modified by selective separation in their gaseous-dusty envelope.

For RV Tau itself the solar metallicity (average content of iron group elements is $[X/H]_\odot = −0.04$ dex) and the altered proportion of CNO elements permits us to suspect the evolution status of a young supergiant. Another unexpected result for this object is that sulphur has been revealed to be largely in excess with respect to iron: [S/Fe] = +0.82. We do not know how it could be explained for the object with the solar metallicity! This may be either real sulphur excess or a manifestation of possible non–LTE effects for the atoms of sulphur. Any traces of selective depletion of chemical elements are absent in the atmosphere of RV Tau. The chemical composition of AI CMi is typical of metal–deficient post–AGB supergiants with overdeficiency of s–process elements. The atmospheres of all stars studied are

Astrophysics and Space Science is the original source of publication of this article. It is recommended that this article is cited as: *Astrophysics and Space Science* **265**: 435–436, 1999.
© 1999 *Kluwer Academic Publishers.*

TABLE I

Main parameters of stars studied

Name	IRAS	T_e, K	$\log g$	$[Fe/H]_\odot$
RV Tau	04440+2605	5600	1.0	+0.07
U Mon	07284–0940	4950	0.0	–0.69
AI CMi	07331+0021	4500	0.0	–1.13
AC Her	18281+2149	6100	1.5	–0.82

enriched in sodium which could be explained both by nucleosynthesis during the preceding evolution and by systematic errors due to non–LTE effects for this atom.

2. The Phenomenon of Variability of RV Tau Type Stars

Pulsating stars of this type have unusual photometric and spectroscopic properties that distinguish this class of objects from related W Vir type stars and semi–regular variable supergiants. The main distinguishing characteristic of the stable enough periodical pulsations of stars of RV Tau type is the presence of two minima on the phase light curve. We proposed (Klochkova and Panchuk, 1998) that in interpretation of peculiar photometric and spectroscopic properties of RV Tau type stars the influence of their envelopes on the photometric and spectroscopic behaviour have to be taken into account. Then the double–peaked photometric curve at the one–peaked behaviour of equivalent widths of lines during the period of a RV Tau type star could be caused by asynchronous variability of physical conditions inside its atmosphere and envelope. The response of an extended envelope to a change of physical state in the star atmosphere is delayed.

References

Klochkova, V.G. and Panchuk, V.E.: 1996, Spectroscopy of the PPN–candidate AI CMi = IRAS 07331+0021, *Bull. Spec. Astrophys. Observ.* **41**, 5–27.

Klochkova, V.G. and Panchuk, V.E., 1998, Spectroscopy of supergiants with IR–excesses: RV Tau type stars, *Pis'ma Astron. Zhurn.* **24**, 754–767.

CHEMICAL ABUNDANCES OF FAST ROTATORS: APPLICATION OF THE SPECTRUM SYNTHESIS METHOD TO A AND F TYPE STARS OF THE HYADES OPEN CLUSTER

O. VARENNE

Observatoire astronomique de Strasbourg, 11, rue de l'Université, F-67000 Strasbourg, France

1. Introduction

Much work has been done on the chemical abundances of the G and F dwarf stars of the Hyades open cluster because of their often low apparent rotational velocities ($6 \leq v \sin i \leq 20$ km s^{-1} for F5-F9 dwarf stars) that facilitate the abundance determination. Actually, the abundances of only a few chemical species (mainly iron, lithium and oxygen) have been derived for a large sample of F stars. For 'normal' A stars, the situation is much more critical since 71% (of all 'normal' A stars) have a $v \sin i$ value greater than 100 km s^{-1}. In consequence, chemical adundances studies have been carried out mainly on the chemically peculiar (CP) A stars (Am and Ap). Though 'normal' A stars represent 80% of A-type stars (including 'normal' and CP A stars), we actually do not know what could be the abundances of superficially normal A stars and their relationship with those of peculiar types.

TABLE I

Hyades abundances derived from the 29 F-type stars.

Element	$\langle [\frac{N_H}{H}] \rangle$	σ	$\log[\frac{N}{N_H}]_\odot$
O	+0.05	0.13	−3.13
Na	+0.09	0.18	−5.67
Si	+0.05	0.09	−4.45
Ca	+0.00	0.14	−5.64
Fe	−0.06	0.15	−4.50
Ni	+0.00	0.13	−5.75
Ba	−0.06	0.32	−9.87

Astrophysics and Space Science is the original source of publication of this article. It is recommended that this article is cited as: *Astrophysics and Space Science* **265**: 437–438, 1999.
© 1999 *Kluwer Academic Publishers.*

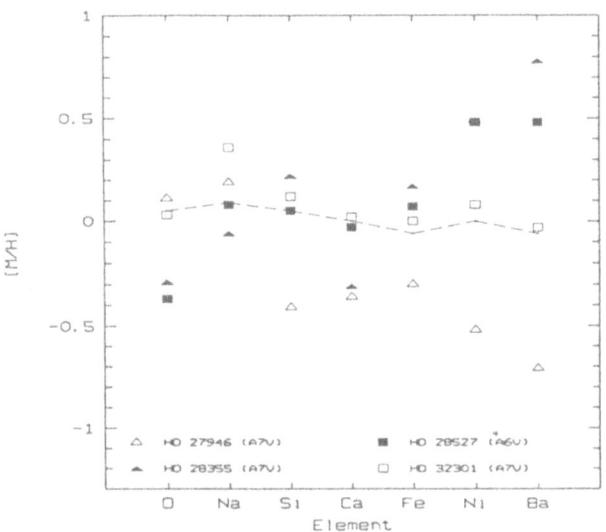

Figure 1. Abundances for the 4 A-type stars ($70 \leq v \sin i \leq 183$ km s^{-1}). The Hyades mean abundances (using the 29 F-type stars) are depicted by the dash line.

2. Chemical Abundances Analysis

2.1. OBSERVATIONS AND METHOD

High resolution spectra ($R = \lambda/\Delta\lambda \sim 35000 - 60000$) with high signal over noise ratio (S/B ~ 200) centered on $\lambda6160$Å have been obtained with the AURE-LIE spectrometer (Observatoire de Haute-Provence). The effective temperatures and gravities have been determined using the Strömgren photometry indices. The model atmospheres have been calculated using Kurucz's ATLAS9 code and the abundances of O, Na, Si, Ca, Fe, Ni, Ba have been derived using the spectrum synthesis method (Takeda's, 1995, code).

2.2. RESULTS

The Hyades mean abundances calculated from the 29 F-type stars ($15 \leq v \sin i \leq 132$ km s^{-1}) are gathered in Table I. Individual abundances for the 4 A stars are displayed in Figure 1.

Reference

Takeda, Y.: 1995, *Publ. Astron. Soc. Jpn.* **47**, 287.

LITHIUM LINES IN SPECTRA OF THE COOLEST M-DWARFS AND BROWN DWARFS – THEORETICAL ASPECTS

YA.V. PAVLENKO

Main Astronomical Observatory, Golosiiv 252650, Kyiv-22, Ukraine

1. Classical Lithium Test

'Lithium test' was proposed by R. Rebolo *et al.* (1992) to tell substellar objects from low-luminosity stars. Pavlenko *et al.* (1995) showed that the resonance doublet of Li I λ 670.8 nm may be observed in the spectra of brown dwarfs. Recently Li lines were observed in spectra of several young dwarfs in open clusters as well free-floating brown dwarfs (Rebolo, 1998).

2. Li in PMS Stars

Lithium lines in spectra of low mass dwarfs are well known tracers of stellar evolution. Cool pre-main sequence (PMS) stars with powerful convective envelopes preserve their initial lithium only during their first few millions years (see Magazzu *et al.*, 1992; Pavlenko and Oppenheimer, 1998).

Strong lithium lines may be formed in the outermost layers of M-dwarf atmospheres with chromospheric-like features. Still the impact of the structures on Li I lines (N.B!: in NLTE!) is rather weak (Pavlenko, 1998a).

3. Modern Lithium Test

Due to the low temperatures and high pressures regime the 'gas-dust' phase transition occurs for many species in atmospheres of the coolest (M and L-)dwarfs ($T_{eff} < 2700$ K, see Tsuji *et al.* (1996) for details). As result, a) temperatures rise in the outermost layers, b) molecular bands of VO and TiO become weak (or even disappear!) in spectra of the late L-dwarfs (see Rebolo, 1998). It means new chances for the lithium test (Pavlenko, 1998b).

The overall shape of L-dwarf spectra at λ > 600 nm is governed by the huge K I and Na I resonance doublets (Pavlenko, 1998b). To fit the observed L-dwarf spectra we suggest the presence of the 'additional (dusty) opacity' (Pavlenko *et al.*,

 Astrophysics and Space Science is the original source of publication of this article. It is recommended that this article is cited as: *Astrophysics and Space Science* **265**: 439–441, 1999.
© 1999 *Kluwer Academic Publishers.*

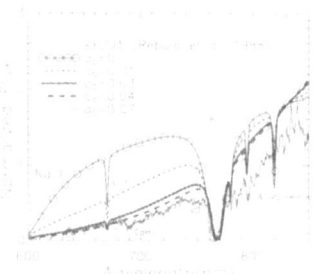

Figure 1. Fit of observed spectrum of L-dwarf Kelu1 (Rebolo *et al.*, 1998). Additional opacity $a_\lambda = a_o * (\lambda/770)^4$ (Pavlenko *et al.*, 1998) for the 'dusty' C-model atmosphere 2000/5.0/0 of Tsuji (1998) was implemented.

1998) in L-dwarf atmospheres which may affect Li I lines also (Figure 1). It means, that we should study the 'dusty properties' of L-dwarf atmospheres before the use of Li I lines for the 'lithium test'.

4. Conclusion

Lithium test survives even in the case of the L-dwarfs.
Physics of the processes of formation of spectra of the L-dwarfs differs drastically from the case of M-dwarfs. Still the internal sense of the lithium test has not been changed.

Acknowledgements

The author thanks Profs. T. Tsuji for providing 'dusty' models, Prof. R. Rebolo and Dr M.R. Zapatero Osorio for fruitful collaboration. I am indebted to LOC, SOC, and Ministre de l'Education Nationale (programme ACCES) for the financial support of my participation in the Meeting.

References

Magazzu, A., Rebolo, R. and Pavlenko, Ya.V.: 1992, *Astroph. J.* **392**, 159–171.
Pavlenko, Ya.V. and Oppenheimer, B.: 1998, in: K. Dupree (ed.), *Proceedings of 10th Cambridge Workshop 'Cool stars, stellar systems and the Sun'*, Cambridge, in press.
Pavlenko, Ya.V.: 1998a, *Astron. Rept.* **42**, 501–507.
Pavlenko, Ya.V.: 1998b, *Astron. Rept.* **42**, 787–792.
Pavlenko, Ya.V., Zapatero Osorio, M.R. and Rebolo, R.: 1998, in: *Euroconference on VLMSBD and Brown Dwarfs in Stellar Clusters and Associations*, La Palma, May 11–15th, in press.

Pavlenko, Ya.V., Rebolo, R. and Martín, E.L. and García López, R.J.: 1995, *Astron. Astrophys.* **308**, 807–818.

Rebolo, R., Martín, E.L. and Magazzu, A.: 1992, *Astrophys. J.* **389**, L83–L86.

Rebolo, R.: 1998, in: K. Dupree (ed.), *Proceedings of 10th Cambridge Workshop 'Cool Stars, Stellar Systems and the Sun'*, Cambrige, in press.

Tsuji, T., Ohnaka, K. and Aoki, W.: 1996, *Astron. Astrophys.* **305**, L1–L5.

Tsuji, T.: 1998, private communications.

LITHIUM IN X-RAY SELECTED ACTIVE COOL STARS

L. PASTORI and G. TAGLIAFERRI
Brera Astronomical Observatory, Via E. Bianchi 46, 22055 Merate, Italy

L.E. PASINETTI FRACASSINI
Department of Physics, University of Milano, Via G. Celoria 16, 20133 Milano, Italy

As part of a larger program to study in the optical region the physical characteristics of X-ray selected cool active stars, we have analysed in the LiI 6708 Å line region 56 stars of spectral types F5-K8 detected by EXOSAT and ROSAT surveys. High resolution spectra were taken with the McMath telescope at Kitt Peak. Lithium abundances have been computed with n-LTE models and rotation velocities derived by using the fxcor IRAF package with a sample of standard stars; colour indices $B - V$ have been taken from the Hipparcos and Ticho Catalogues or obtained from FeI equivalent widths; X-ray luminosities have been computed using parallaxes from Hipparcos Catalogue. The ages have been estimated comparing Li EW with those of stars belonging to the Hyades, Pleiades and IC 2602. Figure 1 compares our values with the typical Li values in the Hyades, the Li upper envelope of the rapid rotating Pleiades stars, the Li upper envelope of IC 2602 stars ($\sim$ 30 Myr old), and the Li upper envelope for T Tauri stars (see also Figure 2 by Pastori *et al.*, 1998). A large fraction of stars is found in the interval 70–300 Myr. We find good correlations between Li abundances and B-V indices for single stars (excluding BY Dra stars) and components of visual binaries, and for different ages. Of course, the correlation is less satisfactory for the whole sample, as expected. For the single stars we find also a weak correlation between Li abundances and L_x (Figure 2). No clear trends are shown by the diagrams log $N(Li)$, L_x vs. Vsini for the single stars (Figures 3 and 4), with a general increase of these quantities for $V \sin i > 10$ km sec^{-1}. The increase of L_x shows a more regular trend for stars younger than the Pleiades. Our sample includes also EK Dra which is considered an ideal and very active analog of the young Sun at a time when it had just arrived on the ZAMS and was forming its planetary system. We find for this star: $\log(N_{Li}/N_H) = 3.105$, $V \sin i = 15.9$ km sec^{-1}, $L_x = 1.03 \times 10^{30}$ erg sec^{-1}. The age is estimated to be between that of the Pleiades and IC 2602.

Astrophysics and Space Science is the original source of publication of this article. It is recommended that this article is cited as: *Astrophysics and Space Science* **265**: 443, 1999.
© 1999 *Kluwer Academic Publishers.*

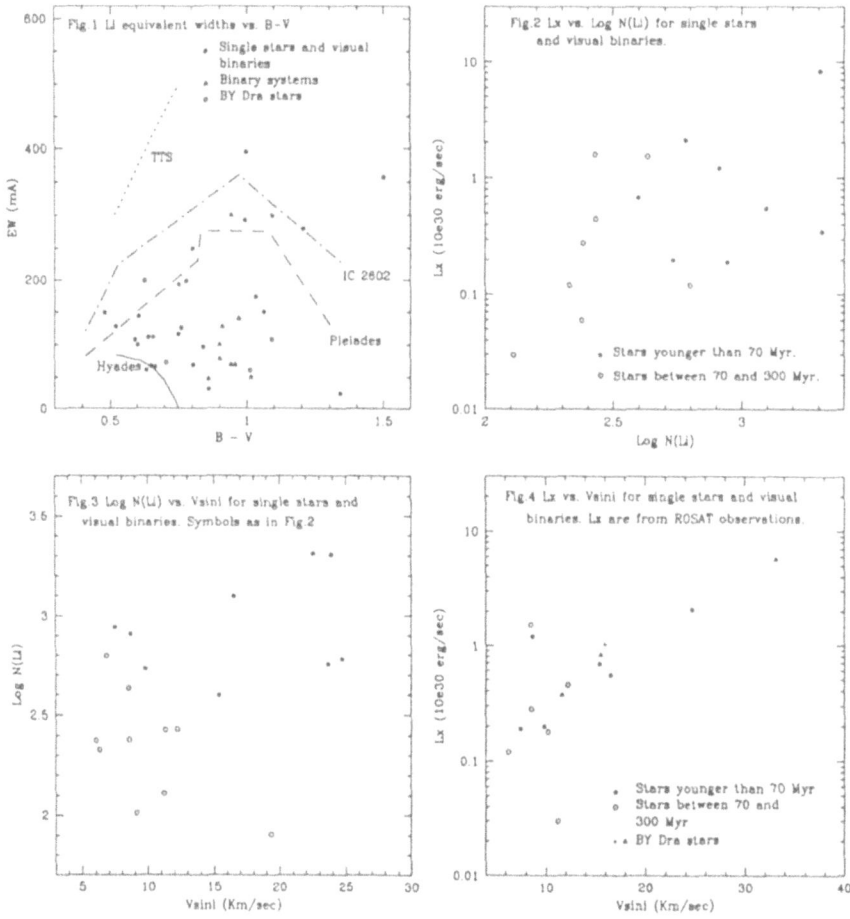

Figure 1.

Reference

Pastori, L., Tagliaferri, G., Cutispoto, G. and Pallavicini, R.: 1998, in: A. Dupree and G.A. Book-
    binder (eds.), *Proc. 10th Cambridge Workshop on Cool Stars, Stellar Systems and the Sun*, in
    press.

SUPERLITHIUM HD9746 – A BRILLIANT CASE OF NLTE IN STELLAR SPECTRA

YA. PAVLENKO and L. YAKOVINA
Main Astronomical Observatory, Kyiv-22, 252650, Ukraine

I.S. SAVANOV
Crimea Astrophysical Observatory, Nauchnyi, 334313, Ukraine

1. Algorithm

We obtained LTE and NLTE lithium abundances in the atmosphere of lithium-rich chromospheric-active K-giant HD9746 using both resonance doublet (670.8 nm) and subordinate triplet (610.3 nm) Li I:

√ Spectra of the Sun and β Gem (K0 III) were used for the verifying of line lists, model atmospheres, etc.

√ LTE lithium abundances were obtained using the synthetical spectra analysis.

√ NLTE lithium abundance corrections were derived from the comparison of LTE and NLTE curves of growth.

2. Procedure

CCD-spectra of HD9746 and β Gem (spectral resolution $\approx$ 0.013 nm, S/N > 100) were obtained in Crimean astrophysical observatory in 1993–1994.

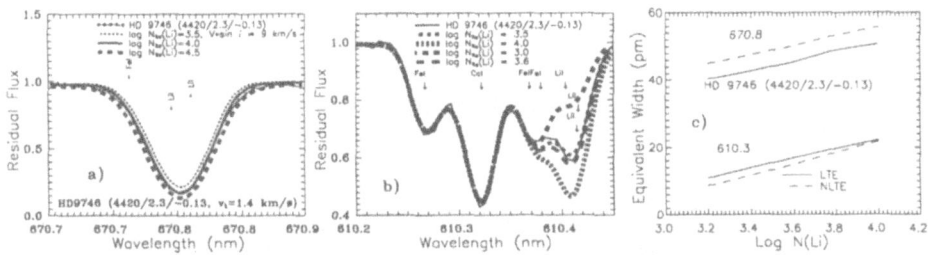

Figure 1. Fit of the regions with Li I λ 670.8 nm (a) and λ 610.3 nm (b) lines in the HD9746 spectrum (solid lines) by the LTE synthetic spectra (dashed lines). c) LTE and NLTE curves of growth of $\lambda\lambda$ 670.8 nm and 610.3 nm Li I lines.

 Astrophysics and Space Science is the original source of publication of this article. It is recommended that this article is cited as: *Astrophysics and Space Science* **265:** 445–447, 1999.
© 1999 *Kluwer Academic Publishers.*

Model atmospheres of β Gem (4865/2.75/–0.04, Drake and Smith, 1991), and HD9746 (4420/2.3/–0.13, Brown *et al.*, 1979; C,N,O abundances from Berdiugina and Savanov, 1994) were computed by the SAM941 program (Pavlenko, 1999). Line opacity was taken into account using the opacity sampling approach. For the Sun we used model atmosphere of Kurucz (1993).

Atomic and molecular lines lists were taken from databases VALD (Piscunov *et al.*, 1995) and (Kurucz, 1993). Some atomic lines were omitted and oscillator strengths of FeI 670.743, FeI 610.217, CaI 610.272 nm lines were changed using the spectra of the Sun and β Gem.

NLTE analysis of the lithium lines formation was carried out for 20-level model of the Li I atom by the complete linearization method (see Pavlenko and Magazzu, 1996 and Pavlenko, 1997 for details).

3. Results

LTE analysis of the spectrum of HD9746 gives $\log N_{\text{LTE}}(Li) = 4.0$ and 3.55 for 670.8 and 610.3 nm lines, respectively. We selected the lithium abundances with accuracy 0.05 dex but our final accuracy cann't be better 0.3 dex due to unsertainty in stellar parameters from Brown *et al.* (1989).

A comparison of LTE and NLTE curves of growths provides abundance corrections due to NLTE effects $\Delta \log N(Li) = \log N_{\text{NLTE}}(Li) - \log N_{\text{LTE}}(Li)$ as -0.4 dex and 0.1 dex for 670.8 and 610.3 nm, respectively. It gives NLTE lithium abundances in atmosphere of HD9746: 3.6 for 670.8 nm and 3.65 for 610.3 nm lines. So, we found:
NLTE lithium abundances obtained for resonance 670.8 and subordinate 610.3 nm Li lines are practically the same.

Log $N_{\text{NLTE}}(Li) = 3.6$ is higher than 'cosmic' lithium abundance (3.3 dex). It may be the evidence of the short-time lithium production in HD9746.

Acknowledgements

YP is indebded SOC and LOC for kindly invitation to participate in the Meeting. He thanks Ministre de l'Education Nationale (programme ACCES) for the financial support of his participation in the Meeting.

References

Berdiugina, S.V. and Savanov, I.S.: 1994, CNO abundances and $^{12}C/^{13}C$ ratios in the atmospheres of lithium-rich giants, *Astron. Lett.* **20**, 639–643.

Brown, J.A., Sneden, C., Lambert, D.L. and Dutchover, E. Jr.: 1989, A search for lithium-rich giant stars, *Astrophys. J. Suppl.* **71**, 293–322.

Drake, J.J. and Smith, G.: 1991, A fine analysis of calcium and iron lines in the spectrum of the K0 giant Pollux – accurate fundamental parameters, *Mon. Not. R. Astron. Soc.* **251**, 369–378.

Kurucz, R.L.: 1993, *CD-ROM's*, NN 1–22.

Pavlenko, Ya.V. and Magazzu, A.: 1996, Theoretical LTE and non-LTE curves of growth for LiI lines in G-M dwarfs and subgiants, *Astron. Astrophys.* **311**, 961–967.

Pavlenko, Ya.V.: 1997, Statistical equilibriim of lithium in the atmospheres of late type stars: lithium rich G-K giants, *Astron. Lett.* **23**, 720–726.

Pavlenko, Ya.V.: 1999, Model atmospheres of R CrB stars, *Astron. Rept.* **2**, 94–103.

Piskunov, N.E., Kupka, F., Ryabchikova, T.A., Weiss, V.V. and Jeffery, C.S.: 1996, VALD: The Vienna Atomic Line Data Base, *Astron. Astrophys. Suppl.* **112**, 525–535.

THE FAST EVOLUTION OF SAKURAI'S OBJECT

T. KIPPER

Tartu Observatory, Tõravere, EE2444, Estonia

V.G. KLOCHKOVA

Special Astrophysical Observatory RAS, Nizhnij Arkhyz 357147, Russia

The fast brightening of Sakurai's object (V4334 Sgr) has been attributed to the final helium–shell flash of a PNN (Duerbeck and Benetti, 1996). Such a flash returns the star into the domain of AGB stars in the Hertzsprung–Russell diagram. Few descendants of helium–shell flash are known (planetary nebulae A30, A78 and variable star FG Sge). Now the brightening of Sakurai's object offers a unique opportunity to monitor the sequence of events after the helium flash.

During the first year after the discovery quite a few papers were published on the object (Benetti *et al.*, 1996; Duerbeck and Benetti, 1996; Asplund *et al.*, 1997; Kipper and Klochkova, 1997). All those contributions show that Sakurai's object is hydrogen–deficient, but rich in carbon and in the light *s*–process elements. The abundance changes found so far are in agreement with the assumption that the star is a very rapidly evolving final helium–shell flash object.

In this note we report of the analysis of the spectra obtained in 1997.

The appearance of the spectra is very much different from the ones in 1996. In spring 1997 the spectrum of Sakurai's object resembles that of a quite late–type carbon star (around C5) with prominent C_2 Swan bands.

We found the following atmospheric parameters: effective temperature, $T_{eff} = 6200$ K, logarithmic gravity, $\log g = 0.6$, and microturbulent velocity, $\xi_t = 7$ km s^{-1}.

The strong molecular spectrum hinders the analysis of metallic line spectrum and the derived abundances are of low accuracy. From the present analysis we found that the abundances of most elements have not changed within uncertainties during 1996–1997. The abundance of Li has slightly decreased. Note, that in 1996 Li was amongst the elements with the largest abundance increase. This could indicate that the chemical composition of the object's atmosphere is not yet homogeneous.

The H_α line is weaker in 1997 than in 1996. This indicates that the hydrogen abundance could have further decreased comparing with 1996 but the H_α wings are heavily blended and could not be used for quantitative abundance determination.

The carbon isotopic ratio is $^{12}C/^{13}C = 5 \pm 3$ for all observing dates.

 Astrophysics and Space Science is the original source of publication of this article. It is recommended that this article is cited as: *Astrophysics and Space Science* **265**: 449–450, 1999.
© 1999 *Kluwer Academic Publishers*.

The sodium abundance is derived only from 2 Na I lines around λ6160. For the NaI D lines region the continuum is not well defined. The lines are much weaker than they should be at the derived temperature. Also, the relative intensities in the doublet are reversed. One reason for that could be the emission components in this line. No other emission lines, however, are visible in our spectra. On the low resolution spectra ($R \approx 7000$) obtained on March, 08 1998, when the star has faded almost by 2 magnitudes, the Na I D emission is clearly visible.

Sakurai's object evolution during 1996–1997 has been remarkably fast. Effective temperature has fallen more than 1000 K during a year. This has lead to considerable changes in appearance of its spectrum. Due to low continuous absorption coefficient in its hydrogen–deficient atmosphere and high carbon abundance the late–type carbon star spectrum is formed at still comparatively high temperature.

The low carbon isotopic ratio, $^{12}C/^{13}C = 5 \pm 3$, has not been observed in the spectra of other R CrB stars. The carbon isotopic bandheads were also not found in the other possible last He–flash star FG Sge. The low carbon isotopic abundance ratio means that ^{12}C produced by He–burning has been afterwards exposed to hot protons to make ^{13}C through the first two reactions of $^{12}C(p, \gamma)^{13}N(e^+, \nu)^{13}C(p,\gamma)^{14}N$ and the further burning out of ^{13}C to ^{14}N by α–capture reactions in He–burning shell has not (yet) occured after the final He–flash. Some of ^{13}C must have undergone α–captures as the abundances of light s–process elements are enhanced. Those could also originate from previous He–flashes.

References

Asplund, M., Gustafsson, B., Lambert, D.L. and Rao, N.K.: 1997, A stellar endgame – the born-again Sakurai's object, *Astron. Astrophys.* **321**, L17–L20.

Benetti, S., Duerbeck, H.W., Seitter, W.C. and Harrison, T.: 1966, Novalike variable in Sagittarius, *IAU Circ.*, 6325.

Duerbeck, H.W. and Benetti, S.: 1996, Sakurai's object – a possible final helium flash in a planetary nebula nucleus, *Astrophys. J.* **468**, L111–L114.

Kipper, T. and Klochkova, V.G.: 1997, The chemical composition of Sakurai's Object, *Astron. Astrophys.* **324**, L65–L68.

halo of M31 is metal rich. If the dwarf spheroidal galaxy population we observe presently is characteristic of the 'building blocks' thought to have been disrupted in the construction of the halo, then it would be impossible for the field population to originate from disrupted dwarf spheroidal galaxies, the so-called 'building block' hypothesis loosely based on Searle and Zinn (1978). The metal rich halo of M31 has a broad abundance distribution consistent with the one-zone model with wind outflow; there is almost no overlap between the abundance distribution of the halo field stars and those of the M31 dwarf spheroidals (or the Milky Way dwarf spheroidals). There may be some galaxies, perhaps even the Milky Way, where disrupted dwarf spheroidal galaxies such as the Sgr dwarf have contributed much mass to the halo. However, this is not the case in M31. Use of HST and eventually NGST to measure the abundance distributions in stellar halos in the Local Group and beyond will give some idea of how the stellar halos are formed. I would speculate that there is a connection between the formation of the bulges and of the halos. No doubt, this must be the case in M31: the high metallicity of the otherwise extremely low surface brightness halo is difficult to understand, otherwise.

3.1. DETAILED CHEMISTRY AS A CONSTRAINT ON FORMATION HISTORY

The bulk of metals are produced in supernovae. Massive star supernovae explode very early on, within a few million years following the formation of the most massive stars. Heavy elements are present in the highest redshift objects known; claims have been made for super metal rich enhancements in quasars (Ferland *et al.*, 1996). Massive SNe mostly yield the lighter 'alpha' elements such as O, Mg, and Si because the iron-peak elements are locked into the neutron star left behind. While the precise origin of Type I SNe is not known, they produce mostly iron-peak elements. Type I supernovae almost certainly require the presence of a white dwarf or the merger of a binary WD, and so therefore will appear no earlier than 10^8 yr after the initial star formation event, and their contribution to heavy element enrichment might lag by 1 Gyr or more.

The oldest stars preserve samples of the gas enriched by these primordial SN events and may be the best link between events in the early Universe and the present day fossil record. Indeed, they are true fossils of these early times. The ratio of $[\alpha/\text{Fe}]$ relative to Solar abundance is considered to be a kind of clock; stars formed when only type II SNe contributed to the nucleosynthesis will be rich in alpha elements and also deficient in the s-process (which can only be produced in the envelope of an AGB star). The ratio of Ba/Eu (s- to r-process) if found to be low in the most metal poor halo stars and this is consistent with them being among the first stellar generations (cf. Spite, 1994; McWilliam, 1997).

In the case of the Galactic bulge, McWilliam and Rich (1994) find that some of the alpha elements, Mg and Ti, are enhanced. A recent analysis of one high resolution spectrum of BW I-194 from the Keck telescope finds enhancement of Mg, Ti, Ca, Si, and O at the +0.3 dex level, consistent with a formation timescale

Turnoff photometry can date the relative ages of these oldest stars to within 1 Gyr, and can establish whether the oldest stars in all galaxies are the same age.

The notion of a simultaneous ignition for the oldest stars is difficult to reconcile with the idea that hierarchical mergers are the dominant process in galaxy building, unless the merging took place very early. The two dominant ideas in building the halo are the Searle and Zinn (1978) accretion model (SZ), and the Eggen, Lynden-Bell and Sandage (1962) rapid collapse model (ELS). In the SZ approach, disrupted dwarf spheroidal such as the Sgr dwarf spheroidal contribute their stars and globular clusters to the Pop II field halo, which grows over many Gyr. The idea that the halo continues to accrete globular clusters offers an attractive explanation of the second parameter problem, in that the candidate young clusters must just have been accreted from dwarf spheroidals. Lee, Demarque and Zinn (1994) use theoretical fits to the plot of horizontal branch type as a function of metallicity, which predicts that most of the outer halo globular clusters must be 1–2 gyr younger than main bulk of old halo clusters. However, these 'young' clusters tend to have horizontal branches which are red for their metallicity. The fornax dwarf spheroidal galaxy and the galaxies thought to be associated with the Sgr dwarf have by and large *blue* horizontal branches (Smith, Rich and Neill, 1998). If the possibly young clusters in the Galactic halo originated in dwarf spheroidal galaxies, then they had cluster populations entirely different from those found in the Sgr and Fornax dwarf spheroidal galaxies.

3. Bulge Formation and Halo Formation

If Population III stars did not make the first metals, perhaps other violent star-forming activity produced both metals and the high velocity winds required in order to distribute the metals widely. Renzini (1998) emphasizes that the high velocity outflows seen in some of the blue-dropout galaxies could also be capable of distributing metals over a wide volume.

If metal enriched winds were important in the early formation of galaxies, there is a clear signature left in the abundance distribution of long-lived stars. The classic one-zone model of chemical evolution in a closed volume produces the characteristic wide abundance distribution (cf. Rich, 1990) with a width of 1 dex in [Fe/H]. If there is a steady outflow of material which is lost to the system, the abundance distribution continues to have the same shape, but the mean abundance (yield) declines.

In the M31 halo, V, I photometry with HST/WFPC2 reveals a wide range of abundance and a red giant branch remarkably similar to that in the Milky Way bulge just 500 pc from the nucleus (Rich, Mighell and Neill, 1996). And M31 is not alone in having a metal rich halo. NGC 3115 (Elson *et al.*, 1997) and 5128 (Harris *et al.*, 1999) also have halo field populations with very high metallicity. Despite the presence of 5 metal poor ([Fe/H] = -1.4) dwarf spheroidals near M31, the stellar

VII – EVOLUTION OF THE EXTERNAL GALAXIES

CONNECTING THE DISTANT UNIVERSE WITH THE LOCAL FOSSIL RECORD

R. MICHAEL RICH

Department of Physics and Astronomy, University of California at Los Angeles, Math Sciences 8979, LA, CA 90095-1562, USA

Abstract. This review considers implications regarding galaxy formation and evolution that can be drawn from study of the ages, abundances, and kinematics of stellar populations in the Local Universe. The wide abundance range in the Galactic bulge and in the halo of M31 is consistent with chemical evolution in a starburst with wind outflow. We question the notion that the Galactic halo population is assembled from disrupted dwarf spheroidal galaxies, based on the presence of metal rich stars in Local Group halos. The alpha-element enhancements of bulge giants are those expected from a system forming on a timescale of less than 1 Gyr. Direct measurement of the star formation history from turnoff photometry in Local Group galaxies is not in complete agreement with the universal star formation rate inferred from high redshift studies. We argue that the properties of the Local Universe constrain galaxy formation theory just as strongly as the findings gleaned from high redshift galaxies.

1. Introduction

Thanks to the identification of candidate young galaxies at high redshift (the UV-dropout galaxies) we have a sample of distant objects whose properties may be plausibly connected to the progeny of the present-day galaxy population. The number density and luminosities (Steidel *et al.*, 1996), and clustering properties (Steidel *et al.*, 1998) are all consistent with this conclusion. The discovery and definition of this high redshift galaxy population ranks as one of the major achievements in the area of galaxy evolution. However, we cannot securely identify any individual galaxy in this population as the progenitor (or one of the progenitors of) of a present-day galaxy such as our own. Adopting the parlance of behavioral science, we are unable to do a 'longitudinal study' of any individual galaxy, that is, to observe its history from birth to the present day. Among the extremes of possible alternative histories would be the slow assembly of a galaxy by mergers versus an early starburst that forms a large fraction of the stars of a galaxy. The identification of the UV-drop galaxies does not answer the question of when the disks and bulges formed, or if these galaxies have anything to do with the present day galaxy population.

In this article, I will consider important clues from the fossil record of the local galaxy population that may be studied with HST and eventually with NGST. At the

Astrophysics and Space Science is the original source of publication of this article. It is recommended that this article is cited as: *Astrophysics and Space Science* **265**: 453–460, 1999.
© 1999 *Kluwer Academic Publishers.*

present time, the most exciting data are being produced from the high redshift studies. But what, for example, would happen if we could produce a color-magnitude diagram for the M31 disk that surpassed in quality that produced by Hipparcos for the Solar vicinity? If we knew the age distribution of the stars in Local Group galaxies, we would have a an independent (and more complete) handle on the star formation history than is possible from the Madau *et al.* (1996) plot. The two independent approaches must give consistent answers.

I remind the reader that we are in the era of high redshift studies, and that the data in this area are indeed exciting. However, choices can be made that may either assist or discourage the equally important exploration of the fossil record in our local Universe. For example, whether or not to have sensitivity in the V band for NGST, which could greatly advance the cause of stellar populations studies with this new facility.

2. The First Stars

When did the first stellar objects form? Ostriker and Gnedin (1996) revive the notion of a Population III of stars formed at $z \sim 10 - 20$ off of the cooling of H_2. Such a notion is attractive, since it might explain why metals are so widely distributed at high redshift. On the other hand, such an event must leave a fossil record: if stars form in the primordial dark matter potential wells with a normal mass function (or at least forming a few long-lived low mass stars) their progeny must be observable today. Perhaps they already are known: remarkable abundance trends appear in extreme Galactic halo stars with [Fe/H] < -3 (McWilliam *et al.*, 1995). At the present time, the sample is too small to tell if there are distinct kinematic signatures (velocity dispersion and spatial distribution characteristic of the dark matter) as might be expected if the stars formed well before the assembly of the Milky Way. McWilliam is extending surveys for such stars to the dwarf spheroidals, using narrow-band calcium filters to select candidate ultra-metal poor stars.

At $z \sim 4$, the Tolman dimming reaches 5 mag: it may not be possible to observe the formation of the low surface brightness stellar populations of galaxies, such as the field halo stars. While actively star forming globular clusters may fall within reach of NGST, it may make the most sense to age date the halo field population by relying on the local fossil record. There is growing evidence that the oldest stars in galaxies have about the same age. The extreme metal poor globular cluster Hodge 11 in the LMC is an identical twin of M92 (Mighell *et al.*, 1996). Harris *et al.* (1997) find NGC 2419, 100 kpc distant yet a Milky Way satellite, also has the identical color-magnitude diagram to M92. Finally, Olsen *et al.* (1999) find that the oldest star clusters in the LMC are identical to the oldest halo clusters in the Milky Way. NGST, and perhaps the HST Advanced Camera, could extend these investigations to the halo of M31, and to many dwarf galaxies in the Local Group.

of 0.5–1.0 Gyr (Matteucci, 1998). This conclusion largely rests on the analysis of a single spectrum. If we are to utilize the fossil record properly, it would be desirable to analyse hundreds of spectra.

Even so, there is some growing consistency between these results and what is observed at high redshift. The abundance distribution of giants in the bulge is > 1 dex wide and follows the 'simple' closed volume model distribution function (Rich, 1990). Recalling the broad abundance range in the M31 halo, we may be seeing the clear imprint of a violent starburst with wind outflow. Franx and Illingworth (1990) point out that the local color of elliptical galaxies is in general correlated with the local escape velocity. In a metal enriched wind, the mean abundance declines as a function of position, but the shape of the abundance distribution does not change (Rich, 1990). Although the UV-dropout galaxies at high redshift show clear signs of wind outflow, it is not clear what relationship those objects have to bulges of the present day. However, if the signs of alpha enhancement in the bulge turn out to be strong enough there may be a basis to connect the two. One obstacle standing in the way is the CDM hierarchical models of galaxy formation that favor the assembly of galaxies through a series of merger events (Kaufmann, 1996). The crucial step in connecting the distant Universe with present day spheroids will be measurements of the mass of these distant galaxies, which should be completed in the next few years. However, given the well established correlations between color, linestrength, and luminosity for early-type galaxies, it is difficult to find a clear place for late mergers in general. Finally, although the presently observed disruption of the Sagittarius dwarf spheroidal galaxy (Ibata et al., 1996) appears as strong evidence in the fossil record favoring the building block hypothesis, the abundance distribution and stellar content (e.g. carbon stars) is drastically different from the bulge.

4. Bimodal Populations of Globular Clusters

About a third of the Galaxy's globular clusters lie in the direction of the Galactic Center. The abundance distribution of Galactic globular clusters is bimodal, with metal rich clusters concentrated toward the Galactic Center (Morgan, 1959; Harris, 1976). Extragalactic globular cluster systems also frequently have a bimodal abundance distribution, with the metal rich clusters following the light of the spheroid while the metal poor clusters are spatially extended (cf. Geisler et al., 1996). The centrally concentrated metal rich clusters have been ascribed to formation in later merger events (Ashman and Zepf, 1992), and to a 'disk' system (Armandroff and Zinn, 1988) with scaleheight too large to comfortably be ascribed to the Galactic bulge (Rich, 1993). Improved photometry of the clusters toward the Galactic center now reveals a centrally concentrated system with same vertical scale height and abundance distribution as the Galactic bulge (Barbuy et al., 1998). HST turnoff photometry for two of these clusters, NGC 6528 and 6553 shows that they are indistinguishable in age from the prototypical 47 Tuc (e.g. old Galactic Globular Clusters). The bulge field population has the same age as these clusters, to within 10% (Ortolani *et al.*, 1995). There is growing evidence that links these metal rich clusters closely to the Galactic bulge, rather than to the thick disk, even in their kinematics (Cotè, 1999). Elliptical galaxies have red globular clusters that follow the spheroid light, and more spatially extended systems of metal poor clusters (cf. Geisler *et al.*, 1996). The first quantitative age estimates from spectroscopy argue that the red clusters in M87 are old (Cohen *et al.*, 1998). We believe that the red globular clusters in these bimodal populations are almost certainly old and not due to recent merger events.

The first high dispersion spectra have been obtained for members of NGC 6553, one of the strongest lined clusters in the Galaxy which is even mentioned in Morgan's paper as being similar to the bulge of M31. In fact, the coolest giants in NGC 6553 have such strong TiO absorption that the giant branch droops in V, with the tip being as faint as the clump (Figure 1). The same phenomenon is seen in the field population of the Galactic bulge (Rich *et al.*, 1998) and in the field of the M31 bulge (Jablonka *et al.*, 1999). *The color-magnitude diagram of NGC 6553 is representative of old metal rich populations.* There is currently a strong debate over the metallicity of NGC 6553. Barbuy *et al.* (1998) analyze 2 giants and find [Fe/H] = −0.55 a mere 0.15 dex more iron rich than 47 Tuc. However, they find that [α/Fe] = +0.6 making Z, the heavy element abundance, Solar. They argue that enhancement of Ti and O, not the iron peak elements, is responsible for the morphology of the color-magnitude diagram. Cohen *et al.* (1999) disagree and find that [Fe/H] = −0.2 with modest α enhancements. Cohen *et al.* draw their conclusion from HIRES spectra of red clump stars at R = 34 000. In NGC 6553 there is strong differential reddening on very small scales, and we believe that the effective temperatures derived by Cohen from photometry (rather than from the spectra) may be off. A small difference in the temperature scale may be enough to explain the difference in abundance. If the iron abundance of NGC 6553 is as low as Barbuy finds, the implication is that elliptical galaxies (which have possibly enhanced magnesium) might also be deficient in iron. This abundance pattern would be the signature of an early, rapid formation history.

5. The Star Formation History, and the Potential of NGST

In its original form, the Madau plot suggests that most of the stars in the Universe were formed ≈ 10 Gyr ago, specifically, that the peak of metal production occurred then. At this meeting, Tolstoy showed a plot with a much different history, based on Local Group galaxies. In fact, if the inner disks and spheroids are included, the bulk of stars appear to have formed more than 10 Gyr ago. Our current sample of detailed star formation histories are derived from HST images of low surface brightness fields. One cannot reach the old turnoff at M_V = +4 in crowded, high surface brightness fields. Unfortunately, most of the mass is in high surface brightness regions – inner disks and bulges. If we are interested in the universal star formation history, we need to measure the distribution of stellar ages in the high surface brightness regions in addition to the dwarf spheroidal galaxies that are easily attainable today. If a Next Generation Space Telescope of 8 m diameter were diffraction limited at 1 micron, it could reach the old turnoff within 1 kpc of the nucleus of M31. Further, such a telescope could image the giant branch in the halo of every galaxy in the Virgo cluster, as well as the field, and it could do so with the same ease and precision attainable in the M31 halo. If galaxy halos are primarily formed from the accumulated debris of dwarf spheroidals, we might expect little variation in the metallicities of halos from galaxy to galaxy. But if the formation of the halo is connected with the formation of spheroid, we might find strong correlations between halo metallicity and total galaxy mass or Hubble type. Because the capability to sort stars by metallicity is crucial, a diffraction-limited NGST capable of imaging at least to 6000 A would be crucial for such an effort to be successful. Once the high redshift universe is tested agains the cold reality of the fossil record, there is a strong possibility of arriving at truly robust descriptions of the formation and evolution of galaxies.

References

Armandroff, T.E.: 1993, in: G. Smith and J. Brodie (eds.), *The Globular Cluster-Galaxy Connection, ASP* **48**, 48.

Ashman, K.A. and Zepf, S.: 1992, *Astrophys. J.* **384**, 50.

Barbuy, B., Bica, E. and Ortolani: 1998, *Astron. Astrophys.* **333**, 117.

Barbuy, B., Renzini, A., Ortolani, S., Bica, E. and Guarnieri, M.D.: 1999, *Astron. Astrophys.* **341**, 539.

Cohen, J.G., Blakelee, J.P. and Rypken, A.: 1998, *Astron. J.* **115**, 377.

Coté, P.: 1999, in preparation.

Eggen, O.J., Lynden-Bell, D. and Sandage, A.R.: 1962, *Astrophys. J.* **136**, 748.

Ferland, G.J., Baldwin, J.A., *et al.*: 1996, *Astrophys. J.* **461**, 683.

Franx, M. and Illingworth, G.: 1990, *Astrophys. J.* **359**, L41.

Geisler, D., Lee, M.G. and Kim, E.: 1996, *Astron. J.* **111**, 1529.

Harris, W.E.: 1976, *Astron. J.* **81**, 1095.

Harris, G.L.H., Harris, W.E. and Poole, G.B.: 1999, *Astron. J.* **117**, 855.

Harris, W.E., *et al.*: 1997, *Astron. J.* **114**, 1030.

Kauffmann, G.: 1996, *Mon. Not. R. Astron. Soc.* **281**, 487.

Lee, Y-W, Demarque, P. and Zinn, R.: 1994, *Astrophys. J.* **423**, 248.

Ibata, R.A., Wyse, R.F.G., Gilmore, G., Irwin, M.J. and Suntzeff, N.B.: 1997, *Astron. J.* **113**, 634.

Jablonka, P., *et al.*: 1999, *Astrophys. J.*, in press.

Madau, P., Ferguson, H.C., Dickinson, M.E., Giavalisco, M., Steidel, C.C. and Fruchter, A.: 1996, *Mon. Not. R. Astron. Soc.* **283**, 1388.

McWilliam, A. and Rich, R.M.: 1994, *Astrophys. J. Suppl.* **91**, 749.

McWilliam, A., Preston, G.W., Sneden, C. and Searle, L.: 1995, *Astron. J.* **109**, 2736.

Mighell, K.J., Rich, R.M., Shara, M. and Fall, S.M.: 1996, *Astron. J.* **111**, 2314.

Morgan, W.: 1959, *Astron. J.* **64**, 432.

Olsen, K.A.G., Hodge, P.W., Mateo, M., Olszewski, E.W., Schommer, R.A., Suntzeff, N.B. and Walker, A.R.: 1998, *Mon. Not. R. Astron. Soc.* **300**, 665.

Ortolani, S., *et al.*: 1995, *Nature* **377**, 701.

Ostriker, J. and Gnedin, N.Y.: 1996, *Astrophys. J.* **472**, L63.

Renzini, A.: 1998, in: *ASP Conf.* **146**, 298.

Rich, R.M.: 1990, *Astrophys. J.* **362**, 604.

Rich, R.M., Mighell, K.J. and Neill, J.D.: 1996, Formation of the Galactic Halo ... Inside and Out, in: H. Morrison and A. Sarajedini (eds.), *ASP Conf. Series* **92**, 544.

Searle, L. and Zinn, R.: 1978, *Astrophys. J.* **225**, 357.

Smith, E.O., Rich, R.M. and Neill, J.D.: 1998, *Astron. J.* **115**, 2369.

Spite, M.: 1992, in: B. Barbuy and A. Renzini (eds.), *IAU Symp. 149, Stellar Populations of Galaxies*, 123.

Steidel, C.C., Giavalisco, M., Pettini, M., Dickinson, M. and Adelberger, K.L.: 1996, *Astrophys. J.* **462**, 17.

Steidel, C.C., Adelberger, K., Dickinson, M., Giavalisco, M. Pettini, M. and Kellogg, M.: 1998, *Astrophys. J.* **492**, 428.

CHEMICAL EVOLUTION OF THE MAGELLANIC CLOUDS

B.E.J. PAGEL

Astronomy Centre, CPES, Sussex University, Brighton BN1 9QJ, UK

G. TAUTVAIŠIENĖ

Institute of Theoretical Physics and Astronomy, Goštauto 12, Vilnius 2600, Lithuania

Abstract. Previous models for the chemical evolution of the Magellanic Clouds have assumed either a steepened IMF compared to the solar neighbourhood or preferential expulsion of oxygen and α-particle elements by selective galactic winds. These assumptions were largely motivated by a belief that the O/Fe ratio in the Clouds is substantially lower than in the Galaxy, but the difference appears to have been exaggerated: Galactic supergiants have a similar O/Fe ratio as Cloud supergiants, there is no corresponding effect in Mg and other α-elements and a combination of data from planetary nebulae, H II regions and supernova remnants indicates an O/Fe ratio more or less equal to solar.

Consequently new analytical models for the chemical evolution of the Magellanic Clouds have been developed, assuming chemical yields and time delays identical to those we previously assumed for the solar neighbourhood, but assuming (in addition to infall) non-selective galactic winds and burst-like modes of star formation represented by discontinuous variations in the star formation rate per unit gas mass. We find adequate agreement with age-metallicity relations and element:element ratios within their substantial uncertainties, whereas our LMC model turns out to give an excellent fit to the anomalous Galactic halo stars discovered by Nissen and Schuster (1997). It also gives an enhanced SNIa/SNII ratio compared to the solar neighbourhood, due to the assumption that the SFR has declined in the past 1 to 2 Gyr.

1. Introduction

The Magellanic Clouds are of particular interest as Irregular (and in the case of the SMC) dwarf glaxies that can be studied in much more detail than other examples, and the evolution of dwarf galaxies raises a number of issues, notably the origin of the metallicity-luminosity correlation (Skillman, Kennicutt and Hodge, 1989). Does this arise from homogeneous or selective loss of gas from shallow potential wells, or are there systematic variations in the true yield resulting from different stellar initial mass functions (cf. Carigi *et al.*, 1995; Carigi, Colín and Peimbert, 1998)? Furthermore, the Clouds, in common with dwarf Irregular and blue compact galaxies, show evidence for 'bursting' or 'gasping' modes of star formation which can have significant effects on chemical evolution, notably the ratio of time-delayed elements like iron from SNIa to rapidly produced elements like oxygen and magnesium coming from massive core-collapse SNII, Ib and Ic (cf. Gilmore and Wyse, 1991).

Astrophysics and Space Science is the original source of publication of this article. It is recommended that this article is cited as: *Astrophysics and Space Science* **265**: 461–468, 1999.
© 1999 *Kluwer Academic Publishers.*

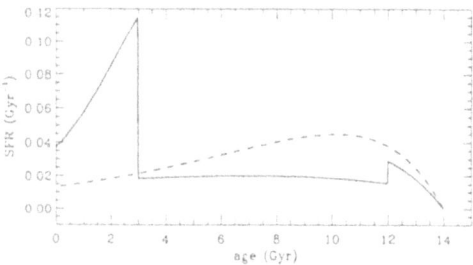

Figure 1. SFR history of the LMC according to our model. The full curve shows the bursting model and the broken-line curve the corresponding smooth model.

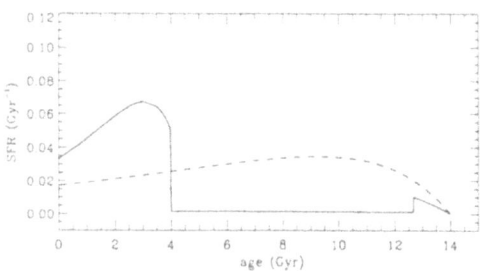

Figure 2. SFR history of the SMC according to our model. The full curve shows the bursting model while the broken curve shows the corresponding smooth model.

Some notable features of the Clouds that need to be considered in chemical evolution models are:

– The age distribution of clusters in the LMC shows a distinct gap between about 4 and 9 Gyr (Geisler *et al.*, 1997); field stars exist in this gap, but HR diagrams obtained with HST indicate that the star formation rate (SFR) was higher at times before the gap and higher still in the last 3 Gyr or so (Geha *et al.*, 1998). In the SMC, the cluster age distribution is more uniform (van den Bergh, 1991; Olszewski, Suntzeff and Mateo, 1996), but the age-metallicity relation is distinctly flat between ages of about 4 and 9 Gyr (Da Costa and Hadzidimitriou, 1998; Mighell, Sarajedini and French, 1998).

– The Clouds are losing gas in the Magellanic Stream and possibly also in superbubbles. This tends to lower the effective yield.

– The gas fraction in the LMC is about the same as in the solar neighbourhood (∼ 15%), but the overall metallicity is lower. The metallicity of the SMC is lower still, but in this case the gas fraction is higher.

– Certain element:element abundance ratios, notably O/Fe, are lower in the Clouds than in Galactic stars with the same Fe/H. This could result from a different IMF, bursting star formation or metal-enhanced winds. However, in our view the lowness of the O/Fe ratio has been exaggerated.

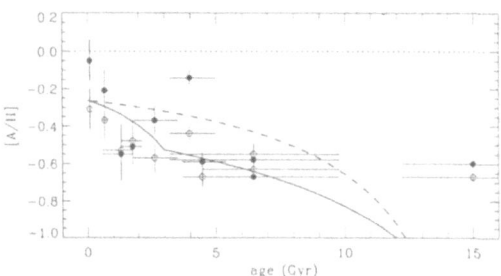

Figure 3. Age-abundance relation for α-elements in the LMC. Curves as in Figures 1, 2, data for planetary nebulae from Dopita *et al.* (1997). Open circles show oxygen; filled circles give the average of Ne, S and Ar.

2. Critique of Previous Chemical Evolution Models

Three significant chemical evolution models have been previously put forward for the Magellanic Clouds:
- Russell and Dopita (1992) considered models with inflow but no galactic wind and smooth star formation rates. They achieved low metallicity and reduced O/Fe by assuming steeper IMFs than for the solar neighbourhood; this both reduces all the true yields and diminishes the role of SNII relative to SNIa.
- Tsujimoto *et al.* (1995) used similar ideas to those of Russell and Dopita, but introduced bursting models of star formation as well as smooth ones and made detailed comparisons with their own supernova models.
- Pilyugin (1996), modelling the LMC only, argued against a modified IMF, since there is no direct evidence, and appealed instead to galactic winds, both homogeneous to reduce the overall yields and selective to reduce O/Fe. In his model, star formation is represented by a series of bursts initiating the selective winds and distributed in time like the ages of the clusters.

Our critique of these models is that the evidence for a reduced role of SNII relative to SNIa over the history of the Clouds is very weak, apart from the fact that the *current* ratio of SNIa/SNII events in the LMC is higher than in our Galaxy (Hughes *et al.*, 1995), which is a different matter (see Section 5 below). The ratios of α-particle elements (Ne, Mg, Si, S, Ca, Ti) to iron are pretty close to solar (see Figure 6), whereas the deficit in O/Fe relative to solar, [O/Fe] ≃ −0.2, determined from iron in supergiants and oxygen in various sources, is shared by Galactic cepheids and other supergiants (Hill, Barbuy and Spite, 1997; Luck *et al.*, 1998). Thus there may be problems with supergiant abundances, or the supergiants may be more representative of the local interstellar medium than is the Sun itself; but the Magellanic Clouds probably do not differ significantly from the local ISM in their O/Fe ratio and certainly not in the ratio of other α-particle elements to iron. This conclusion has recently received support from ASCA X-ray abundance measurements in LMC supernova remnants (Hughes *et al.*, 1998). We have dealt

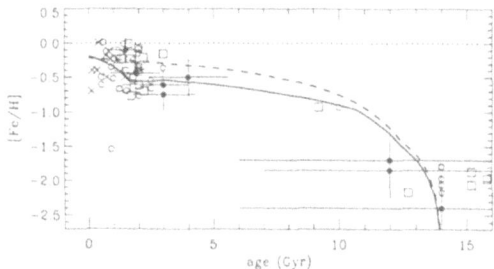

Figure 4. Age-metallicity relation for the LMC. Curves as in previous figures. Data sources: Olszewski *et al.* (1991), open circles; Girardi *et al.* (1995), crosses; Geisler *et al.* (1997), open squares; de Freitas Pacheco *et al.* (1998), filled circles.

with the oxygen problem in a somewhat arbitrary manner by simply adding 0.2 dex to [O/Fe] deduced from Magellanic Cloud supergiant spectra.

3. Our Models

Our analytical models (Pagel and Tautvaišienė, 1998) attempt to account for the chemical evolution of the Clouds using the same yields and time delays as we previously assumed *ad hoc* for the solar neighbourhood, thus doing away with the idea of differing IMFs; we also assume an absence of selective outflow of SNII products, although we do assume homogeneous outflow so as to get the effective yields down. We also assume inflow of unprocessed material (which helps to bring down the number of old stars compared to young ones as in the classical 'G-dwarf problem') and consider both smooth and bursting models for the star formation rates.

Table I summarises the main features of our models. The net star formation rate is assumed to be ω times the mass of gas, where $\omega = $ const. $= \langle \omega \rangle$ in smooth models and undergoes two discontinuous changes in bursting models. The inflow rate is assumed to vary as $\omega \exp(-\langle \omega \rangle t)$ and the outflow rate is assumed to be η times the net SFR, leading to the supposed star formation rate histories shown in Figures 1 and 2. Table II shows the yields (in units of the solar abundance of the corresponding element) and associated time delays, the delay of 1.33 Gyr referring to SNIa and the others to various contributions to the r and s-processes discussed in Pagel and Tautvaišienė (1997). These parameters are assumed to be the same in the Clouds as in the solar neighbourhood and we use the instantaneous recycling and delayed production approximations as before. Details of our formalism are given by Pagel and Tautvaišienė (1998).

TABLE I

Model parameters

	LMC	SMC	Local Gal
$\langle \omega \rangle$ (Gyr^{-1})	0.18	0.12	0.3
Outflow parameter η	1.0	2.0	0
Final gas fraction	0.15	0.29	0.14
Final abundance in units of yield	0.78	0.43	1.4
Burst models ω (0 to 2 Gyr)	0.15		
ω (2 to 11 Gyr)	0.08		
ω (11 to 14 Gyr)	0.5		
ω (0 to 1.33 Gyr)		0.1	
ω (1.33 to 10 Gyr)		0.01	
ω (10 to 14 Gyr)		0.35	

TABLE II

Yields and time delays (after Pagel and Tautvaišienė, 1995, 1997)

Δ (Gyr)	0	0.023	0.037	1.33	2.67
O	0.70				
Mg	0.88				
Si, Ti	0.70			0.12	
Ca	0.56			0.18	
Fe	0.28			0.42	
Sr, La	0.013	0.11	0.29		0.29
Y, Ba	0.010	0.08	0.30		0.30
Zr	0.022	0.17	0.28		0.28
Ce	0.020	0.16	0.26		0.26
Pr, Sm	0.055	0.46	0.11		0.11
Nd	0.040	0.33	0.17		0.17
Eu, Dy	0.080	0.64	0.01		0.01

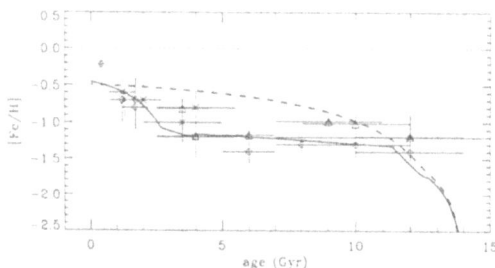

Figure 5. Age-metallicity relation for the SMC. Curves as in previous figures. Data sources: open circles, Da Costa (1991); triangles, Da Costa and Hadzidimitriou (1998); crosses, de Freitas Pacheco *et al.* (1998).

4. Comparison with Observations

Figure 3 shows the age-abundance relation for oxygen and α-particle elements in the LMC, and the data appear to favour the bursting model. Bursting models also seem to be favoured by the age-metallicity relations shown in Figures 4 and 5 for the two Clouds (cf. Mighell *et al.*, 1998). Figure 6 shows O and α-element to iron ratios against [Fe/H]. Bearing in mind that the supergiant data show mainly scatter, rather than evolution in time as envisaged in models, and that the supergiant oxygen abundances have been 'fudged' by +0.2 dex to compensate for [O/Fe] in Galactic supergiants, the agreement is about as good as can be expected. On the other hand, our model gives serendipitously superb agreement (without any fudging) with the 'anomalous' Galactic halo stars with little or no O and α enhancement found by Nissen and Schuster (1997); this agreement also applies to the s-process elements Y and Ba. Other plots of elements relative to iron in the two Clouds are given in Pagel and Tautvaišienė (1998). The chief failure of our models (in common with others) is their inability to fit the enhanced La, Nd and Eu abundances found relative to iron in the SMC by Hill (1997).

5. Relative Supernova Rates

In models with a steepened IMF, the total SNIa/SNII ratio is enhanced with respect to the solar neighbourhood which may be representative of Sbc galaxies. This effect is absent from our models, but the *current* ratio is enhanced due the assumed declining star formation rate in the last few Gyr shown in Figures 1 and 2. Making the crude assumption that all SNIa lifetimes are close to the 'typical' value $\Delta = 1.3$ Gyr, we have at time T

$$\frac{R_{\text{SNIA}}(T)}{R_{\text{SNII}}(T)} \propto \frac{\dot{s}(T - \Delta)}{\dot{s}(T)}. \tag{1}$$

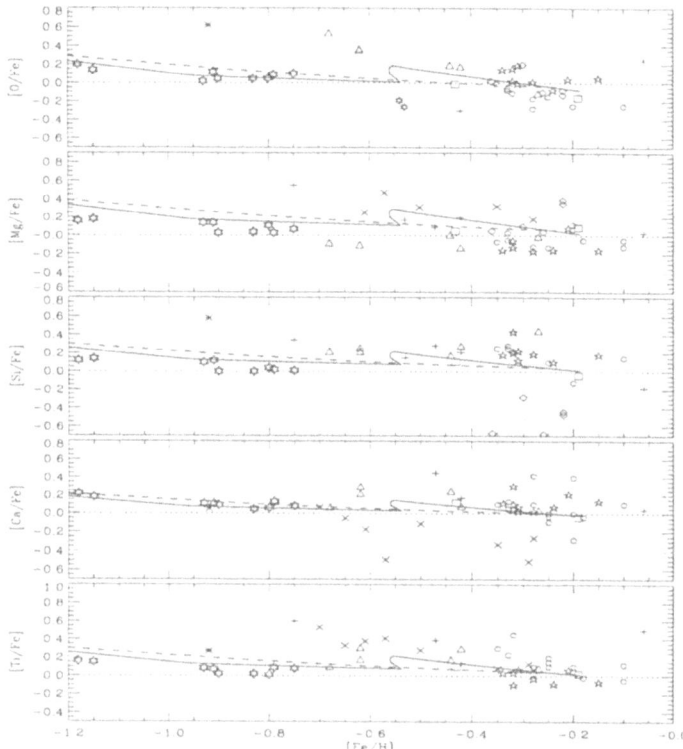

Figure 6. Element: iron ratios for oxygen, α-particle elements and Ti in the LMC and in the 'anomalous' Galactic halo stars of Nissen and Schuster (1997) shown by large six-pointed stars. Other data are from supergiants and B-type stars with [O/Fe] increased by 0.2 dex and [Si/Fe] reduced by 0.1 dex.

This ratio in our models is 1.2 for the Milky Way, 1.8 for the LMC and 1.5 for the SMC, giving an enhancement factor of 1.5 in the current SNIa/SNII ratio for the LMC compared to the Galaxy, where that ratio has been estimated to be between 0.11 and 0.25 (van den Bergh and McClure, 1994). In Sdm-Im galaxies, there is an estimate for this ratio of 0.3 (Cappellaro *et al.*, 1993), while in the LMC itself a ratio between 0.25 and 1 has been found from X-ray observations of SNR (Hughes *et al.*, 1995). Our LMC model is qualitatively in agreement with these estimates, but the prediction is sensitive to the star formation history assumed; we suggest that high SNIa/SNII ratios generically indicate a star formation rate that has declined over the past 1 to 2 Gyr.

References

Cappellaro, E., *et al.*: 1993, *Astron. Astrophys.* **273**, 383.
Carigi, L., Colín, P. and Peimbert, M.: 1998, *Astrophys. J.*, in press, *Astron. J.* **115**, 1934.

Carigi, L., Colín, P., Peimbert, M. and Sarmiento, A.: 1995, *Astrophys. J.* **445**, 98.

Da Costa, G.S.: 1991, in: R. Haynes and D. Milne (eds.), *The Magellanic Clouds, IAU Symp.* **148**, Kluwer, pp. 183.

Da Costa, G.S. and Hadzidimitriou, D.: 1998, *Astron J.* **115**, 1934.

de Freitas Pacheco, J.A., Barbuy, B. and Idiart, T.P.: 1998, *Astron. Astrophys.* **332**, 19.

Dopita, M.A., Vassiliadis, E., Wood, P.R., *et al.*: 1997, *Astrophys. J.* **474**, 188.

Geha, M.C., Holtzmann, J.A., Mould, J.R., Gallagher, J.S. III, *et al.*: 1998, *Astron. J.* **115**, 1045.

Geisler, D., Bica, E., Dottori, H., Claría, J.H., *et al.*: 1997, *Astron. J.* **114**, 1920.

Gilmore, G. and Wyse, R.F.G.: 1991, *Astrophys. J. Lett.* **367**, L55.

Girardi, L., Chiosi, C., Bertelli, G. and Bressan, A.: 1995, *Astron. Astrophys.* **298**, 87.

Hill, V.: 1997, *Astron. Astrophys.* **324**, 435.

Hill, V., Barbuy, B. and Spite, M.: 1997, *Astron. Astrophys.* **323**, 461.

Hughes, J.P., Hayashi, I., Helfand, D., *et al.*: 1995, *Astrophys. J. Lett.* **444**, L81.

Hughes, J.P., Hayashi, I. and Koyama, K.: 1998, *Astrophys. J.* **505**, 732.

Luck, R.E., *et al.*: 1998, *Astron. J.* **115**, 605.

Mighell, K.J., Sarajedini, A. and French, R.S.: 1998, *Astron. J.* **116**, 2395.

Nissen, P.E. and Schuster, W.J.: 1997, *Astron. Astrophys.* **326**, 751.

Olszewski, E.W., Suntzeff, N.B. and Mateo, M.: 1996, *Annu. Rev. Astron. Astrophys.* **34**, 511.

Pagel, B.E.J. and Tautvaišienė, G.: 1995, *Mon. Not. R. Astron. Soc.* **276**, 505.

Pagel, B.E.J. and Tautvaišienė, G.: 1997, *Mon. Not. R. Astron. Soc.* **288**, 108.

Pagel, B.E.J. and Tautvaišienė, G.: 1998, *Mon. Not. R. Astron. Soc.* **299**, 535.

Pilyugin, L.: 1996, *Astron. Astrophys.* **310**, 751.

Russell, S.C. and Dopita, M.A.: 1992, *Astrophys. J.* **384**, 508.

Skillman, E.D., Kennicutt, R.C. and Hodge, P.W.: 1989, *Astrophys. J.*, **347**, 875.

Tsujimoto, T., Nomoto, K., Yoshii, Y., Hashimoto, M., Yanagida, S. and Thielemann, F.-K.: 1995, *Mon. Not. R. Astron. Soc.* **277**, 945.

van den Bergh, S.: 1991, *Astrophys. J.* **369**, 1.

van den Bergh, S. and McClure, R.D.: 1994, *Astrophys. J.* **425**, 205.

ABUNDANCES IN MAGELLANIC CLOUDS YOUNG CLUSTERS:

and their implication on Magellanic evolution

V. HILL

ESO K. Schwarschild Str. 2, D-85748 Garching bei München

M. SPITE

DASGAL, Observatoire de Meudon, F-92195 Meudon CEDEX, France

Abstract. Detailed abundances are measured in cool (K type) supergiants in LMC young populous clusters (NGC 2004, NGC 2100 and NGC 1818), and compared to those, recently observed in
samples of field supergiants in both Magellanic Clouds. This completes the work already done on the
SMC cluster NGC 330 (Hill *et al.*, 1997b). The analysis of the field supergiants showed no evidence
of any abundance dispersion among the stars, compatible with a well mixed gas in these galaxies.
However, previous determinations of the metallicity of NGC 330 and NGC 1818 gave values well
below the respective field values. This difference is seeked, using a homogeneous method for the
field and cluster samples, and the abundance ratios of various elements are examined.

1. Introduction

The Magellanic Clouds young stellar population includes a number of *populous*
clusters, which have no counterpart in our own galaxy. Very young (10^6 to 10^7 yrs)
and yet very rich, these clusters have long been targets both for cluster formation
and stellar evolution studies. Among them, particular attention was drawn onto
NGC 330 (SMC) and NGC 1818 (LMC), since the early attempts to measure their
metallicities indicated a metal content lower than that of the surrounding coeval
stars, by factors of 5 and 4 respectively (Carney *et al.*, 1985, and Richtler *et al.*,
1989). How could a cluster form with a metallicity *lower* than that of its surroundings? All the more since the young population in the field of the MCs seems to
be fairly well mixed: no abundance gradients or large scatter are found among the
supergiants (Hill *et al.*, 1995; Hill, 1997a; Luck *et al.*, 1998), nor among the H II
regions (Russell and Dopita, 1990). Further spectroscopic studies confirmed that
NGCN 330 was indeed metal poor ([Fe/H] = -1 dex, Spite *et al.*, 1991). However, the reddening and hence the metallicity for this cluster has been questioned
(Bessell, 1991). The real metal deficiency of the MC young clusters is thus still not
understood.

Taking advantage of our recent study of six type K supergiants in the SMC field
(Hill, 1997a) providing an ideal reference sample for this type of stars, we undertook the study of similar stars in four MC Young Populous Clusters. The models,
methods and physical data being the same for all stars, most of the systematic

Astrophysics and Space Science is the original source of publication of this article. It is recommended that this article is cited as: *Astrophysics and Space Science* **265**: 469–472, 1999.

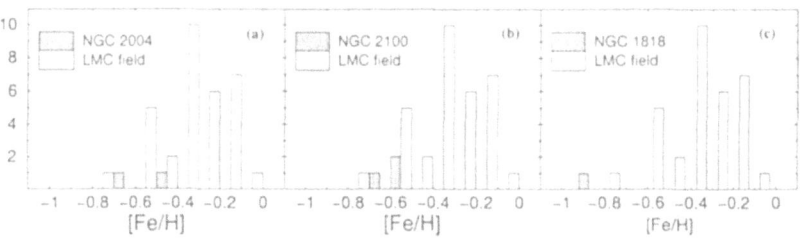

Figure 1. Histograms of the [Fe/H] distribution of cluster and field stars in the LMC: (a–c): NGC 2004, NGC 2100 and NGC 1818 (this work) and the LMC field (compilation by Luck *et al.* (1998).

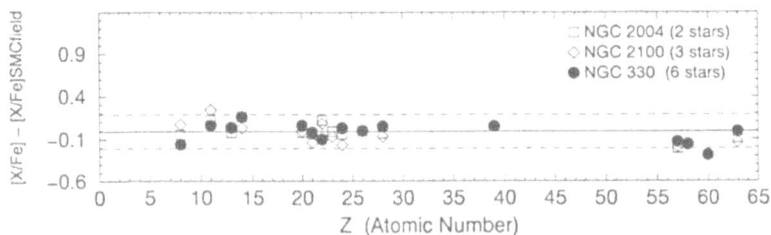

Figure 2. Mean [X/Fe] abundances of each of the four clusters, normalized to the SMC mean of the field K stars: $[X/Fe]_{cluster} - [X/Fe]_{field}$, versus atomic number of the element.

uncertainties cancel out when differential abundances (clusters versus field) are considered. The programs were conducted at the NTT (ESO) to obtain spectra of cool supergiants in: NGC 330 ($\sim$ 20 Myrs) in the SMC, NGC 1818 ($\sim$ 17 Myrs), NGC 2004 ($\sim$ 8 Myrs) in the LMC, and NGC 2100 ($\sim$ 12 Myrs) in the 30 Dor region of the LMC. The results for six stars in NGC 330 (SMC) were reported in Hill (1997b), showing that the difference between the cluster and the field stars was small, if any, and compatible with the small scatter found among the field stars. The LMC clusters are still under analysis, so the results reported here come from roughly half of the available data.

2. Clusters Versus Field Stars

The mean iron abundances found for the LMC clusters from the program stars are respectively [Fe/H]= −0.81 dex (NGC 1818, 1 star), [Fe/H] = −0.57 dex (NGC 2004, 2 stars) and [Fe/H] = −0.57 dex ±0.06 dex (σ_{rms}) (NGC 2100, 3 stars).

The histograms Figure 1 show that only in NGC 1818 are the field and cluster stars strikingly different. There still seems to be a systematical tendency of the four clusters (the three LMC clusters and NGC 330 in the SMC following Hill 1997b) to be more metal deficient than their surroundings. However, the difference is small (< 0.2 dex, hardly significant) in all cases but NGC 1818. However, since only one

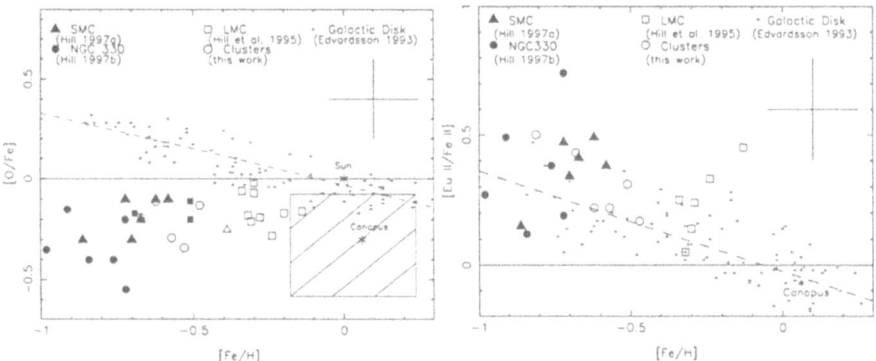

Figure 3. [O/Fe] and [Eu/Fe] ratios as a function of [Fe/H] for supergiants in both clouds (field and clusters) and in the Galaxy.

star in NGC 1818 has been analyzed yet, no firm conclusion can be drawn about this apparently peculiar cluster before other stars are analyzed.

Figure 2 displays for each cluster, the mean [X/Fe] abundance ratios versus atomic number, normalised to the mean of the SMC K type supergiants sample ($[X/Fe]_{cluster} - [X/Fe]_{field\ SMC}$). This plot shows that the difference in the detailed chemical composition of the clusters and the field of the MCs is extremely small, seldom exceeding the intrinsic expected uncertainty (dotted lines). This striking similarity of the abundance pattern favors a formation of young clusters from matter which evolved similarly to that of the bulk of the MCs.

3. α Elements and Heavy Neutron Capture Elements

Oxygen and α elements are of particular interest for the study of the chemical evolution of a galaxy, because they are thought to be produced in large quantities by the short-lived type II supernovae (SNIIe), in contrast with iron which is only marginally produced by SNIIe and more efficiently produced in the longer lived type Ia supernoave. It is now widely accepted that the α elements are not overabundant in the young matter of the Magellanic Clouds (in contrast with the Galactic dwarfs of similar metallicity). This can clearly be seen in Figure 3 (left), where both the field and clusters K supergiants [O/Fe] values are plotted in both MCs, together with Galactic data (the dashed box represent the location of the solar neighbourhood supergiants). Striking also is the *similarity of the Magellanic and Galactic supergiants oxygen over iron ratios.* This feature may be simply explained by a star formation history occuring more slowly in the Clouds than in the Galaxy; a bursting star formation history could also produce such [O/Fe] ratios in the young population. In either cases there is *no need to ask for a IMF different in the Clouds and the Solar neighbourhood.*

On the other hand, the heavy r- process neutron capture elements (Europium is the best example) are also thought to be produced mainly by the SNIIe. In the disk of our Galaxy, the behaviour of [Eu/Fe] indeed resembles that of the [O/Fe] (Figure 3). However, in the MCs, the heavier s- and r- neutron capture process elements are found to be significantly overabundant (by respectively $\sim$ 0.3 and $\sim$ 0.4 dex both in field and cluster stars). In Figure 3 (right) are plotted the [Eu/Fe] values for field and clusters in both MCs. The small symbols are the data for the Galactic disk (Woolf *et al.*, 1995).

This difference of behaviour between α and r- process elements in the MCs is very difficult to understand, since both types of elements are thought to be produced in SNIIe and more work deserves to be done on this topic to unravel the true history of chemical enrichment in the Clouds.

References

Bessell, M.: 1991, in: R. Haynes and D. Milne (eds.), *IAU symp.* **147**, Kluwer, p. 273.

Carney, B.W., Janes, K.A. and Flowers, P.J.: 1985, *Astron. J.* **90**, 1196.

Edvardsson, B., Andersen, J., Gustafsson, B., Lambert, *et al.*: 1993, *Astron. Astrophys.* **275**, 101.

Hill, V.: 1997b, in: *IAU symp.* **186**, in press.

Hill, V.: 1997a, *Astron. Astrophys.* **324**, 435.

Hill, V., Andrievsky, S. and Spite, M.: 1995, *Astron. Astrophys.* **293**, 347.

Luck, E., Moffett, T., Barnes, T. and Gieren, W.: 1998, *Astrophys. J.* **115**, 605.

Richtler, T., Spite, M. and Spite, F.: 1989, *Astron. Astrophys.* **225**, 351.

Russell, S. and Dopita, M.: 1990, *Astrophys. J. Suppl.* **74**, 93.

Spite F., Spite M. and Richtler, T.: 1991, *Astron. Astrophys.* **252**, 557.

Woolf, V., Tomkin, J. and Lambert, D.L.: 1995, *Astrophys. J.* **453**, 660.

CHEMO-DYNAMICAL EVOLUTION OF DISK GALAXIES, SMOOTHED PARTICLES HYDRODYNAMICS APPROACH

Chemo-Dynamical SPH code

PETER BERCZIK

Main Astronomical Observatory of Ukrainian National Academy of Sciences, 252650, Golosiiv, Kiev-022, Ukraine

Abstract. A new Chemo-Dynamical Smoothed Particle Hydrodynamic (CD-SPH) code is presented. The disk galaxy is described as a multi-fragmented gas and star system, embedded into the cold dark matter halo. The star formation (SF) process, SNII, SNIa and PN events as well as chemical enrichment of gas have been considered within the framework of standard SPH model. Using this model we try to describe the dynamical and chemical evolution of triaxial disk-like galaxies. It is found that such approach provides a realistic description of the process of formation, chemical and dynamical evolution of disk galaxies over the cosmological timescale.

1. The CD-SPH Code and Results

The simplicity and numerical efficiency of the SPH method were the main reasons why we chose similar technique for the modelling of the evolution of complex, multi-fragmented triaxial protogalactic systems. We used our own modification of the hybrid N-body/SPH method (Berczik and Kravchuk, 1996a; Berczik and Kravchuk, 1996b; Berczik and Kravchuk, 1997; Berczik, 1998), which we call as chemo-dynamical SPH (CD-SPH). The 'dark matter' and 'stars' were included into the standard SPH algorithm as the N-body collisionless system of particles, which can interact with the gas component only through the gravitation (Katz, 1992). The star formation process and supernova explosions were included into the scheme in the manner proposed by (Raiteri *et al.*, 1996; Carraro *et al.*, 1997) but with our own modifications.

For description of the process of the gas material convertion into stars we modify the standard SPH star formation algorithm (Katz, 1992; Navarro and White, 1993) taking into account the presence of chaotic motions in the gaseous environment and the time lag between initial development of suitable conditions for star formation and star formation itself (Berczik, 1998). After the formation, these 'star' particles return the chemically enriched gas to surrounding 'gas' particles due to SNII, SNIa and PN events. For the description of this process we use the approximation proposed by (Raiteri *et al.*, 1996). We consider only the production of ^{16}O and ^{56}Fe, and try to describe the full galactic time evolution of these elements, from the beginning to present time (i.e. $t_{evol} \approx 13.0$ Gyr).

 Astrophysics and Space Science is the original source of publication of this article. It is recommended that this article is cited as: *Astrophysics and Space Science* **265**: 473–477, 1999.
© 1999 *Kluwer Academic Publishers.*

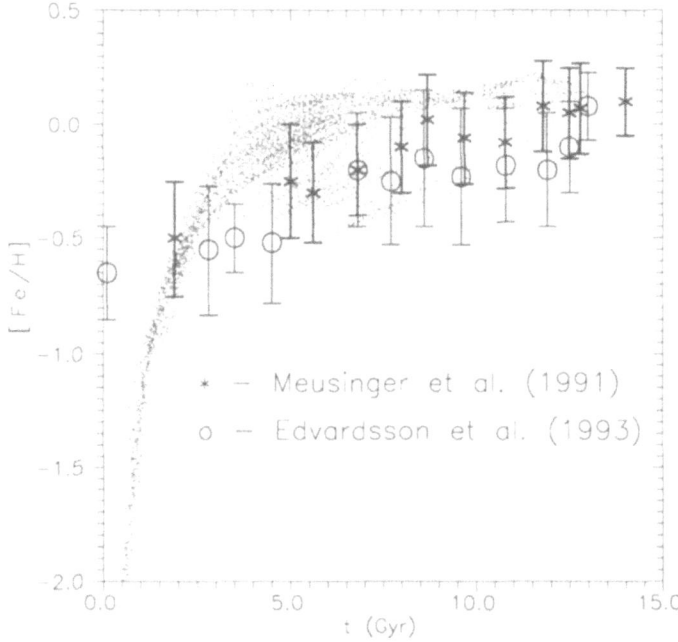

Figure 1. [Fe/H](t). The age metalicity relation of the 'star' particles in the 'solar' cylinder (8 kpc < r < 10 kpc).

In our calculations, as a first approximation, it is assumed that the model galaxy halo contains the CDMH component with Plummer-type density profiles (Dauphole and Colin, 1995).

The SPH calculations were carried out for $N_{gas} = 2109$ 'gas' particles. According to (Navarro and White, 1993; Raiteri *et al.*, 1996), such number seems to be quite enough to provide qualitatively correct description of the system behaviour. Even such small number of 'gas' particles produces a $N_{star} = 31631$ 'star' particles at the end of calculation.

As initial model (relevant for CDM-scenario) we took constant-density homogeneous gaseous triaxial configuration ($M_{gas} = 10^{11}\ M_{\odot}$) within the dark matter halo ($M_{halo} = 10^{12}\ M_{\odot}$). We set $A = 100$ kpc, $B = 75$ kpc and $C = 50$ kpc for semiaxes of system. Such triaxial configurations are reported in cosmological simulations of the dark matter haloes formation (Eisenstein and Loeb, 1995; Frenk *et al.*, 1988; Warren *et al.*, 1992). We set the smoothing parameter of CDMH: $b_{halo} = 25$ kpc.

To check the SF and chemical enrichment algorithm in our SPH code, we use the chemical characteristics of the disk in 'solar' cylinder (8 kpc < r < 10 kpc). The age metalicity relation of the 'star' particles in the 'solar' cylinder [Fe/H] (t) we present in Figure 1. We presented in Figure 2 the metallicity distribution N_*([Fe/H]). The [O/Fe] vs. [Fe/H] distribution we presented in Figure 3. In figures

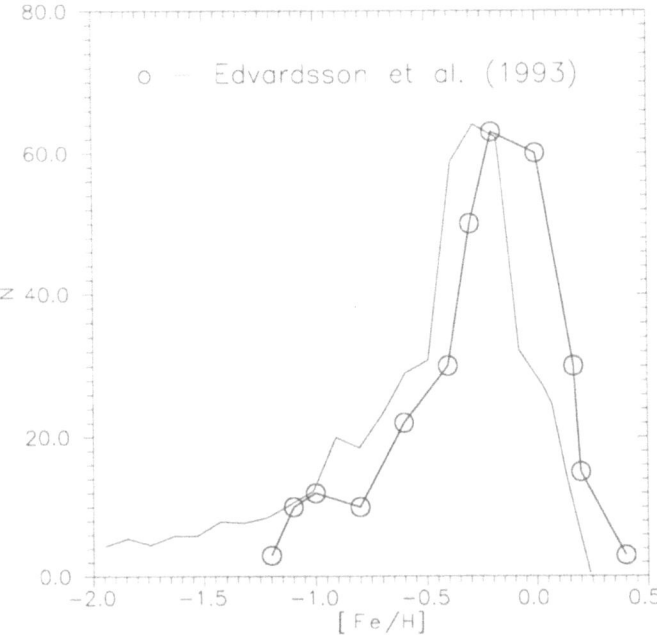

Figure 2. N_*([Fe/H]). The metallicity distribution of the 'star' particles in the 'solar' cylinder (8 kpc $< r <$ 10 kpc).

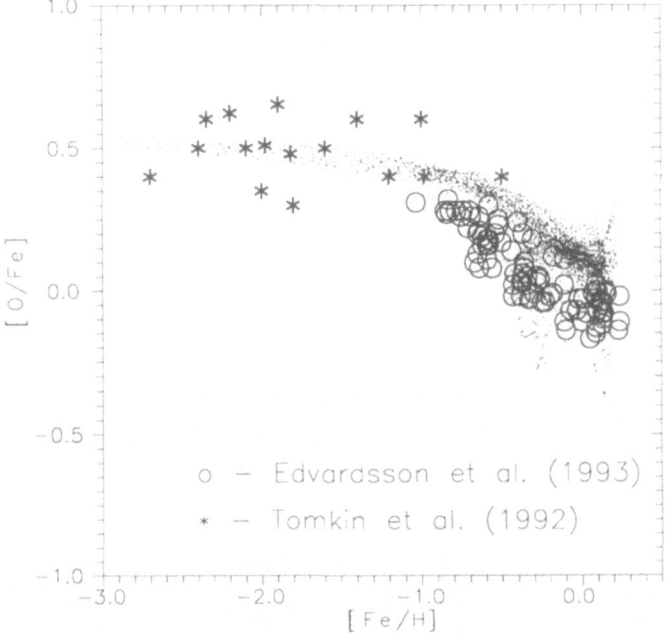

Figure 3. The [O/Fe] vs. [Fe/H] distribution of the 'star' particles in the 'solar' cylinder (8 kpc $< r <$ 10 kpc).

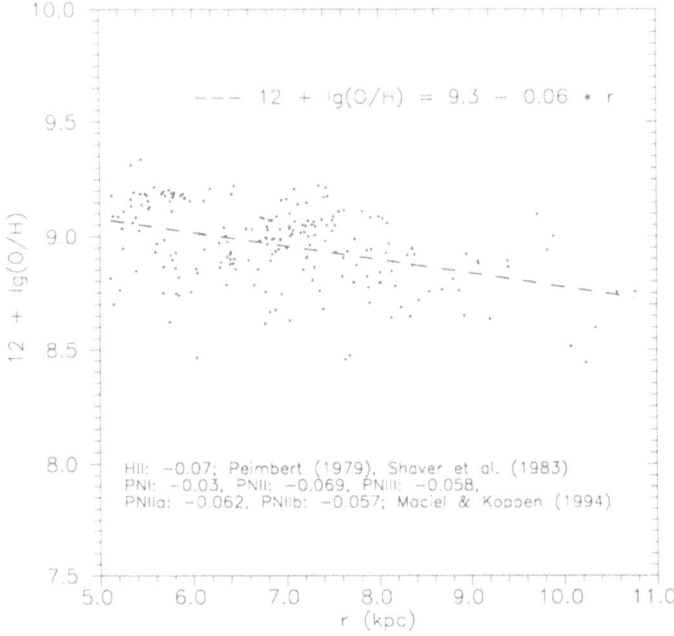

Figure 4. [O/H](*r*). The [O/H] radial distribution.

we also present the observational data from (Meusinger *et al.*, 1991; Edvardsson *et al.*, 1993; Tomkin *et al.*, 1992). The [O/H] radial distribution [O/H](*r*) we presented in Figure 4. At the distances 5 kpc $< r <$ 11 kpc the model radial abundance gradient is -0.06 dex kpc^{-1}. Our model has a very good agreement with oxygen radial gradient in the our Galaxy (Peimbert, 1979; Shaver *et al.*, 1983).

1.1. CONCLUSION

This simple model provides a good, self-consistent picture of the process of galaxy formation and star formation in the galaxy. The dynamical and chemical evolution of modelled disk-like galaxy is coincident with the results of observations for our own Galaxy. The results of our modelling give a good base for a wide use of proposed SF and chemical enrichment algorithm in other SPH simulations.

Acknowledgements

The author is grateful to S.G. Kravchuk, L.S. Pilyugin and Yu.I. Izotov for fruitful discussions during the preparation of this work. This research was supported by a grant from the American Astronomical Society.

References

Berczik, P. and Kravchuk, S.G.: 1996a, *Astrophys. Space. Sci.* **245**, 27.

Berczik, P. and Kravchuk, S.G.: 1996b, in: *Proceedings of the Workshop and Spring School: 'The Interaction of Stars with Their Environment'*, 23–25 May 1996, Visegrad, Hungary. Communic. from the Konkoly Obs. of the Hun. Acad. of Sci. No. 100 (Vol. 12, Part 2), p. 313.

Berczik, P. and Kravchuk, S.G.: 1997, in: *Proceedings 2nd INTEGRAL workshop*, ESA Preprint Series No. SP-382 (March 1997), p. 525.

Berczik, P.: 1998, Presented as a poster in the Isaac Newton Institute Workshop: 'Astrophysical Discs', June 22–27, 1998, Cambridge, U.K. (astro-ph/9807059).

Carraro, G., Lia, C. and Chiosi, C.: 1997, *Mon. Not. R. Astron. Soc.* in press (astro-ph/9712307).

Dauphole, B. and Colin, J.: 1995, *Astron. Astrophys.* **300**, 117.

Edvardsson, B., Andersen, J., Gustaffson, B., *et al.*: 1993, *Astron. Astrophys.* **275**, 101.

Eisenstein, D.J. and Loeb, A.: 1995, *Astrophys. J.* **439**, 520.

Frenk, C.S., White, S.D.M., Davis, M. and Efstathiou, G.: 1988, *Astrophys. J.* **327**, 507.

Katz, N.: 1992, *Astrophys. J.* **391**, 502.

Maciel, W.J. and Koppen, J.: 1994, *Astron. Astrophys.* **282**, 436.

Meusinger, H., Reimann, H.G. and Stecklum, B.: 1991, *Astron. Astrophys.* **245**, 57.

Navarro, J.F. and White, S.D.M.: 1993, *Mon. Not. R. Astron. Soc.* **265**, 271.

Peimbert, M.: 1979, in: W.B. Burton (ed.), *The Large Scale Characteristic of the Galaxy*, Reidel, Dordrecht, p. 307.

Raiteri, C.M., Villata, M. and Navarro, J.F.: 1996, *Astron. Astrophys.* **315**, 105.

Shaver, P.A., McGee, R.X., Newton, L.M., Danks, A.C. and Pottasch, S.R.: 1983, *Mon. Not. R. Astron. Soc.* **204**, 53.

Tomkin, J., Lemke, M., Lambert, D.L. and Sneden, C.: 1992, *Astron. J.* **104**, 1568.

Warren, M.S., Quinn, P.J., Salmon, J.K. and Zurek, W.H.: 1992, *Astrophys. J.* **399**, 405.

THE INFLUENCE OF MERGING EVENTS ON THE DISK COMPONENT OF SPIRAL GALAXIES

U. SCHWARZKOPF and R.-J. DETTMAR

Astronomisches Institut, Ruhr-Universität, Universitätsstraße 150, D-44780 Bochum, Germany

Abstract. We report on first results of a comprehensive study of interacting and merging processes between spiral galaxies and small satellites to investigate the effects of such events on the disk component of spirals. Analysis of our newly obtained photometric data of about 125 edge-on galaxies in optical and in NIR shows that there are considerable differences between interacting and non-interacting galaxies concerning their absolute disk scale parameters as well as their ratios (h/z_0). In comparison with normal spirals, the average heating factor perpendicular to the disk plane of mergers is about 1.5. The most striking feature of the distributions of (h/z_0) for both normal galaxies and mergers, is the total lack of typical flat disk axis ratios (i.e. that of late type ones) of (h/z_0) > 6 for mergers.

1. Observations

As a consequence of the large galaxy sample ($n \approx 125$) needed for this study, observations were obtained with different telescopes during the last 2 years: 2.2-m ESO/MPI, 1.54-m Danish, 24-inch Bochum Telescope (La Silla, Chile), 2.2-m, 1.23-m telescope (Calar Alto, Spain), 42-inch telescope (Lowell Obs., Arizona), and 1-m telescope (Hoher List, Eifel). For most of the R-band images we obtained supplementary observations in Near Infrared (bands H, K') to avoid extreme dust extinction near the galactic plane and to get information about colors and different aging disk populations.

2. Galactic Disk Models

A detailed description of applied disk models and about this contribution's results can be found in Schwarzkopf (1998), and in the WWW:
http://aibn91.astro.uni-bonn.de/webgk/ws98/
http://www.astro.ruhr-uni-bochum.de/schwarz/galaxies/index.html.

Astrophysics and Space Science is the original source of publication of this article. It is recommended that this article is cited as: *Astrophysics and Space Science* **265**: 479–480, 1999.
© 1999 *Kluwer Academic Publishers.*

3. Statistics of Disk Scale Parameters h/z_0

3.1. ... NON INTERACTING EDGE-ON'S

We found that the ratio of horizontal and vertical scale parameters (h/z_0) for normal (i.e. non interacting galaxies) correlates well with the morphological type of galaxies, meaning that (h/z_0) increases from an average $(h/z_0) = 1$ for early types like S0 to $(h/z_0) = 7.0$ for late type spirals Sc/Sd, with an error of about ± 2; there is a smooth transition between these two extremes.

3.2. ... INTERACTING/MERGING EDGE-ON'S

The correlation between h and z_0 is destroyed when all mergers are taken into account, i.e. mergers possess disturbed disks and do not follow the (h/z_0)-trend (the ratio (h/z_0) is about half of that of normal disks!).

4. Distribution of Disk Thickening

The distribution of ratio (h/z_0), most clearly an indicator for vertical disk thickening, is completely different for normal and for merging galaxies, resp.! The center of both distributions is at $(h/z_0) = 3.9$ and $(h/z_0) = 2.7$, resp., leading to a ratio of vertical disk thickening of about 1:1.5.

In comparison to normal galaxies, the distribution for mergers shows a clear drop-off for ratios $(h/z_0) > 4$ and no tail at extreme axis ratios > 7.

References

Hernquist, L. and Mihos, J.C.: 1995, *Astrophys. J.* **448**, 41.
Reshetnikow, V. and Combes, F.: 1997, *Astron. Astrophys.* **324**, 80.
Schwarzkopf, U. and Dettmar, R.-J.: 1997, *Astronomische Gesellschaft Abstract Series* **13**, 238.
Schwarzkopf, U. and Dettmar, R.-J.: 1998, *25. GK-Meeting Workshop on Dwarf Galaxies*, Shaker, in press.
van der Kruit, P.C. and Searle, L.: 1981a,b, *Astron. Astrophys.* **95**, 105.
Walker, I.R., Mihos, J.C. and Hernquist, L.: 1996, *Astrophys. J.* **460**, 121.

EVOLUTIONARY PROPERTIES OF HIGH REDSHIFT GALAXIES SEEN THROUGH CLUSTER-LENSES

GUSTAVO BRUZUAL

C.I.D.A., AP 264, Mérida 5101-A, Venezuela

ROSER PELLÓ

Observatoire Midi-Pyrénées, 14 Av. Edouard Belin, F-31400 Toulouse, France

1. Introduction

Clusters of galaxies acting as gravitational lenses allow us to study the stellar content and evolutionary state of galaxies at much fainter levels than the usual spectroscopic limit. By selecting background galaxies which are close to the critical lines at high-z and with a photometric redshift $z \geq 1$, one maximizes the probability of discovering galaxies with $z \approx 1 - 5$. The amplification close to the critical lines is typically $\Delta m \sim 2$ to 3 mag, but it is still ~ 1 mag at a distance of 1′ from the cluster center (Fort and Mellier, 1994). Only well studied cluster-lenses, with a mass distribution highly constrained by multiple images, are useful for this purpose, in order to keep the amplification uncertainties below $\Delta m_{lensing} \sim 0.3$ mag.

The aim of this program is to determine the redshift distribution, the luminosity function and the main characteristics of the galaxy population at $z \geq 1$. The sample of galaxies obtained through cluster-lenses is much less biased in luminosity or towards galaxies with a strong star-formation activity than the field sample. This is particularly true if the photometric redshift is computed through a large wavelength domain, including the near-IR. As an example, we concentrate here on the high-z sources found behind A2390 ($z = 0.231$), a gravitational lens which has been extensively studied during recent years (see relevant references in Kneib *et al.*, 1999 and Pelló *et al.*, 1999). Similar work is presently going on for sources behind A2218 and A370.

2. Results

We have studied two different cases in Abell 2390: the highly reddened ISO objects (A,B–C,D, Lémonon *et al.*, 1998) at $z \sim 1$, and the multiple images H3 and H5 at

 Astrophysics and Space Science is the original source of publication of this article. It is recommended that this article is cited as: *Astrophysics and Space Science* **265**: 481–482, 1999.
© 1999 *Kluwer Academic Publishers.*

$z \sim 4$. The optical and near-IR SEDs of objects B–C and D ($z = 0.913$) appear brighter in the near-IR and fainter in the blue bands compared to A ($z = 1.033$). The SEDs of these objects can be fitted by different synthetic spectra from the Bruzual and Charlot (1993, updated 1998) atlas. However, the fits are not unique and a degeneracy in the SFR-age-metallicity-extinction space is clearly seen. The best fits to B–C and D with initial-burst models at ages younger than 10^8 yr (the most likely range, taking the spectroscopic information into account) are obtained with a rest-frame $A_V \sim 3$, a stable result independent of metallicity. The extinction law is of SMC type (Prévot *et al.*, 1984). The corrected magnitudes for objects B–C and D (lensing and absorption) are very similar, $M_B = -20.8(-21.1)$ and $M_B = -20.7(-21.0)$ respectively, for $q_0 = 0.5(0.1)$. The total mass involved in the burst is $\sim 10^{10} M_\odot$ in both cases ($q_0 = 0.5$). Using constant SFR models, we obtain good fits in both cases with a rest-frame $A_V \sim 3.0 - 4.0$, and a mean corrected SFR of 40 to $50 M_\odot h_{50}^{-2}$ yr^{-1} with $q_0 = 0.5$ and a lower mass-limit for star-formation of 1 $M_\odot$. According to these results, the two lensed sources detected by ISOCAM at $z = 0.913$ are strongly reddened star-forming galaxies.

H3 and H5 are at $z = 4.05$ (Pelló *et al.*, 1999). Lyα is seen in emission in both galaxies, which are necessarily dominated by massive OB stars at the wavelengths seen in the visible bands. Two kinds of SFRs were considered: an instantaneous burst, and a continuous star-forming system (Salpeter IMF, mass-limits $0.1 M_\odot \leq m \leq 125 M_\odot$). H3 is better constrained than H5, the weight of the old stellar population being more important in H3 than in H5, leaving less room to play with the reddening value. Both H3 and H5 are well fitted by instantaneous burst models of 0.2 solar or even higher metallicities, but H5 is also in fair agreement with a reddened system undergoing a continuous star-formation activity. H3 should be 20 to 70 Myr old ($Z = Z_\odot$, $A_V = 0.9$), and is definitively older than H5 (9 to 45 Myr) for a instantaneous burst model. The two $z \sim 4$ sources are intrinsically bright ($M_B^* - 2$ to -1 mag, depending on A_V).

References

Bruzual A.G. and Charlot, S.: 1993, *Astrophys. J.* **405**, 538.
Fort, B. and Mellier, Y.: 1994, *Astron. Astrophys. Rev.* **5**, 239.
Kneib, J.P., *et al.*: 1999, in preparation.
Lémonon, *et al.*: 1998, *Astron. Astrophys.* **334**, L21.
Pelló, R., *et al.*: 1999, *Astron. Astrophys.*, submitted.
Prévot, *et al.*: 1984, *Astron. Astrophys.* **132**, 389.

DAMPED LYMAN-α SYSTEMS

P. PETITJEAN

Institut d'Astrophysique de Paris, 98bis, Boulevard Arago, 75014 Paris, France

C. LEDOUX

Observatoire de Strasbourg, 11, rue de l'Université, 67000 Strasbourg, France

Abstract. Damped Lyman-α systems observed in the spectra of high-redshift quasars are considered as the progenitors of present-day galaxies. Indeed, the large neutral hydrogen column densities observed and the presence of metals imply that the gas is somehow closely associated with regions of star formation. The nature of the absorbing object is unclear however.

The discussion of whether high-redshift damped Lyman-α systems are produced through large, fast-rotating, thick disks or through building-blocks of galaxies is important since it is related to how present-day galaxies form, either through initial formation of large disks and subsequent accretion of gas or as a result of merging of pregalactic clumps.

We argue here that damped Lyman-α systems are progenitors of *any kind* of galaxy (dwarfs and large spirals); they trace peaks in the spatial density distribution of the gas.

1. Introduction

Damped Lyα (hereafter DLA) systems are defined as systems with neutral hydrogen column density $N(\mathrm{H\,I}) > 2 \times 10^{20}$ cm^{-2}. This definition is artificial since damped wings appear for lower column densities ($N(\mathrm{H\,I}) > 10^{19}$ cm^{-2}; see e.g. Petitjean, 1998). It has been introduced assuming that these lines should be characteristic of galactic disks at high redshift (Wolfe *et al.*, 1986). This definition may introduce a systematic bias in the discussion of what is the nature of these systems and it would be most valuable to compare the properties of systems with $19 < \log N(\mathrm{H\,I}) < 20.3$ and $\log N(\mathrm{H\,I}) > 20.3$.

For $\log N(\mathrm{H\,I}) > 19$, the optical depth at the Lyman limit is large enough so that hydrogen is neutral. The gas is either cold ($T < 1000$ K) and contains molecules (e.g. Srianand and Petitjean, 1998) or warm ($T \sim 10^4$ K) for the highest or lowest column densities respectively (Petitjean *et al.*, 1992). As a consequence of the shape of the column density distribution, $\mathrm{d}^2n/\mathrm{d}N\mathrm{d}z \propto N^{-\beta}$ with $\beta \sim 1.5$, most of the mass is in the systems of highest column densities. The number density of the latter decreases with time presumably as a consequence of star-formation (Wolfe *et al.*, 1986; Lanzetta *et al.*, 1995; see however Turnshek *et al.*, 1998). Indeed, the cosmic density of neutral hydrogen in DLA absorbers at $z \sim 3$ is similar to that of stars at the present time (Wolfe, 1995; Storrie-Lombardi *et al.*, 1996).

Astrophysics and Space Science is the original source of publication of this article. It is recommended that this article is cited as: *Astrophysics and Space Science* **265**: 483–486, 1999.
© 1999 *Kluwer Academic Publishers.*

Metallicities and dust content have been derived from zinc and chromium observations (Pettini *et al.*, 1994, 1997a). The [Zn/Cr] ratio is an indicator of the presence of dust if it is assumed, that, as in our Galaxy, zinc traces the gaseous abundances whereas chromium is heavily depleted into dust grains (see Lu *et al.*, 1996 and Pettini *et al.*, 1997b for a critical discussion of this assumption). The typical dust-to-gas ratio determined this way is about 1/30 of that in the Milky-Way (Pettini *et al.*, 1997a; Vladilo, 1998). This amount of dust could bias the observed number density of DLA systems (Fall and Pei, 1993; Boissé *et al.*, 1998). Metallicities are of the order of a tenth solar and tend to decrease from $z \sim 1 - 2$ to $z > 3$ (Pettini *et al.*, 1997a; Boissé *et al.*, 1998). At any redshift however, the scatter is large and it may be hazardous to draw premature conclusions from the small samples available.

Recently, Prochaska and Wolfe (1997b) have used Keck spectra of 17 DLA absorbers to investigate the kinematics of the neutral gas from unsaturated low-excitation transitions such as Si II λ1808. They show that the absorption profiles are inconsistent with models of galactic haloes with random motions, spherically infalling gas and slowly rotating hot disks. The CDM model (Kaufman, 1996) is rejected as it produces disks with rotation velocities too small to account for the large observed velocity broadening of the absorption lines. Models of thick disks ($h \sim 0.3R$, where h is the vertical scale and R the radius) with large rotational velocity ($v \sim 225$ km s^{-1}) can reproduce the data. In a subsequent paper however, Haehnelt *et al.* (1998) use hydrodynamic simulations in the framework of a standard CDM cosmogony to demonstrate that the absorption profiles can be reproduced by a mixture of rotational and random motions in merging protogalactic clumps. The typical virial velocity of the haloes is about 100 km s^{-1}.

The issue of whether DLA systems originate in thick disks has been questioned previously. In particular, distribution of DLA systems is inconsistent with that of stars in the thick disk of our Galaxy (Pettini *et al.*, 1997b; see however Wolfe, 1998). Arguments in favor of DLA systems being associated with dwarf galaxies have been reviewed by Vladilo (1998). However, it has been shown recently that DLA systems at intermediate redshift are associated with galaxies of very different morphologies (Le Brun *et al.*, 1997). This strongly suggests that the objects associated with high-redshift DLA absorbers are progenitors of present-day galaxies of *any kind*.

2. The Metal Line Profiles of DLA Systems

In a recent paper (Ledoux *et al.*, 1998), we address the problem of the kinematics of damped systems without any a priori model in mind. We emphasize that large velocity broadenings arise most often in peculiar systems and that high and low-ionization species show some correlation in their kinematics. We have compiled a sample of 26 DLA systems from the literature. For each of them we have measured

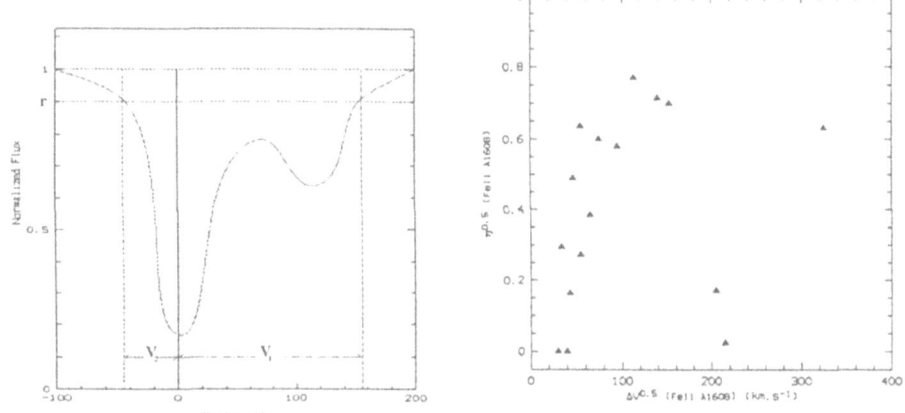

Figure 1. Left: Definition of the line parameters, the velocity broadening $\Delta V = V_1 - V_2$ and the asymmetry parameter $\eta = |V_1 + V_2|/(V_1 - V_2)$. Right: Asymmetry parameter η versus velocity broadening for the Fe II lines measured at $\tau_v \sim 0.7$.

the velocity broadening of the profiles ΔV for the low and high ionization species at different optical depths. ΔV_{low}^r is the velocity width of a low-excitation line measured as the velocity separation of the wavelengths at which the residual in the normalized spectrum is r. We consider $r = 0.9$ and 0.5.

Except for a few cases, there is a redshift at which the optical depth of unsaturated low-excitation transitions (Zn II, Fe II) reaches a maximum. We take this redshift as the origin of velocities. V_1 (> 0) and V_2 (< 0) are the largest and smallest velocities for which the residual in the normalized spectrum is r. We define an asymmetry parameter $\eta = |V_1 + V_2|/(V_1 - V_2)$ that is close to zero for a symmetric profile and close to one for a one-sided profile (see Figure 1).

Most of the profiles have $\Delta v < 100$ km s^{-1}, while a few have $\Delta v > 150$ km s^{-1}. It can be seen in Figure 1 that the asymmetry parameter η is correlated with the velocity broadening up to $\Delta V \sim 150$ km s^{-1}. For larger velocity broadenings however this correlation disappears. This approach complements the edge-leading test of Prochaska and Wolfe (1997b) Indeed in models of fast rotating thick-disk, the largest optical depth is to be found on one edge of the absorption (see Figure 1) and some correlation between asymmetry and velocity broadening is expected. The velocity broadening becomes larger because the wing becomes more prominent. As a consequence, the line becomes more asymmetric. The correlation shown in Figure 1 is suggestive of rotational motions in sub-systems on scales smaller than $\Delta V < 150$ km s^{-1}. It seems however that one has to be very careful when discussing the kinematics of systems with larger velocity broadenings. A detailed inspection of a few of them leads to the conclusion that they often show the sub-systems that are expected if the objects are in the process of merging. The claim that the CDM model should be rejected has to be considered with caution (Haehnelt *et al.*, 1998).

References

Boissé, P., Le Brun, V., Bergeron, J. and Deharveng, J.M.: 1998, *Astron. Astrophys.* **333**, 841.

Fall, S.M. and Pei, Y.C.: 1993, *Astrophys. J.* **402**, 479.

Haehnelt, M.G., Steinmetz, M. and Rauch, M.: 1998, *Astrophys. J.* **495**, 647.

Kauffmann, G.: 1996, *Mon. Not. R. Astron. Soc.* **281**, 475.

Kulkarni, V.P., Fall, S.M. and Truran, J.W.: 1997, *Astrophys. J.* **484**, L7.

Lanzetta, K.M., Wolfe, A.M. and Turnshek, D.A.: 1995, *Astrophys. J.* **440**, 435.

Le Brun, V., Bergeron, J., Boissé, P. and Deharveng, J.M.: 1997, *Astron. Astrophys.* **321**, 733.

Ledoux, C., Petitjean, P., Bergeron, J., Wampler, E.J. and Srianand, R.: 1998, *Astron. Astrophys.* **337**, 51.

Lu, L., Sargent, W.L.W., Barlow, T.A., Churchill, C.W. and Vogt, S.S.: 1996, *Astrophys. J. Suppl.* **107**, 475.

Petitjean, P.: 1998, astro-ph/9810418.

Petitjean, P., Bergeron, J. and Puget, J.L.: 1992, *Astron. Astrophys.* **265**, 375.

Pettini, M., Smith, L.J., Hunstead, R.W. and King, D.L.: 1994, *Astrophys. J.* **426**, 79.

Pettini, M., King, D.L., Smith, L.J. and Hunstead, R.W.: 1997a, *Astrophys. J.* **478**, 536.

Pettini, M., Smith, L.J., King, D.L. and Hunstead, R.W.: 1997b, *Astrophys. J.* **486**, 665.

Prochaska, J.X. and Wolfe, A.M.: 1996, *Astrophys. J.* **470**, 403.

Prochaska, J.X. and Wolfe, A.M.: 1997a, *Astrophys. J.* **474**, 140.

Prochaska, J.X. and Wolfe, A.M.: 1997b, *Astrophys. J.* **487**, 73.

Savage, B.D. and Sembach, K.R.: 1996, *Annu. Rev. Astron. Astrophys.* **34**, 279.

Srianand, R. and Petitjean, P.: 1998, *Astron. Astrophys.* **335**, 33.

Storrie-Lombardi, L.J., McMahon, R.G. and Irwin, M.J.: 1996, *Mon. Not. R. Astron. Soc.* **283**, L79.

Turnshek, D.A.: 1998, in: P. Petitjean and S. Charlot (eds.), *Evolution and Structure of the IGM*, XIII IAP Colloquium, Editions Frontières, p. 263.

Vladilo, G.: 1998, *Astrophys. J.* **493**, 583.

Wolfe, A.M., *et al.*: 1986, *Astrophys. J. Suppl.* **61**, 249.

Wolfe, A.M.: 1995, in: G. Meylan (ed.), *QSO Absorption Lines*, Springer, Berlin, p. 13.

Wolfe, A.M. and Prochaska, J.X.: 1998, *Astrophys. J.* **494**, L15.

Wolfe, A.M.: 1998, in: P. Petitjean and S. Charlot (eds.), *Evolution and Structure of the IGM*, XIII IAP colloquium, Editions Frontières, p. 243.

FORMATION EPOCH OF *FIELD* EARLY-TYPE GALAXIES

T. KODAMA

Institute of Astronomy, University of Cambridge, Madingley Road, Cambridge CB3 0HA, UK

R.G. BOWER and E.F. BELL

Department of Physics, University of Durham, South Road, Durham DH1 3LE, UK

It has been shown that the bulk of stars in *cluster* early-types form at high redshift such as $z > 2$ (see Kodama *et al.*, 1998, and the references therein). In contrast, stellar population of the *field* early-types has not been well studied yet, and remained controversial. The hierarchical galaxy formation models suggest that the formation epoch of field early-types is younger on average than the cluster counterparts by a few Gyrs (e.g., Kauffmann, 1996). Observationally, Larson, Tinsley and Caldwell (1980) claimed that the CMR scatter was larger for field early-types than for cluster early-types. However, Bernardi *et al.* (1998) recently found almost identical $Mg_2 - \sigma$ correlation of early-types in cluster, group, and the field, therefore claimed the coevality of early-types irrespective of environment.

In this paper we discussed the formation epoch of field early-types, on the basis of the CMR of early-types in the Hubble Deep Field (HDF). See Kodama, Bower and Bell (1999) for details. We chose Franceschini *et al.*'s (1998) sample which consists of 35 morphologically selected early-types in the HDF with a limiting magnitude of $K < 20.15$. Fifteen of them have spectroscopic redshifts, and for the rest we estimated photometric redshifts using Kodama, Bell and Bower (1999). Many available photometric passbands (U_{300}, B_{450}, V_{606}, I_{814}, J, H, K) allowed us to transform all the galaxies onto the single C-M plane in the rest-frame by applying K-correction to each galaxy. Because the sample is at a range of different redshifts in the field, we also applied evolution correction to make up a single coeval C-M diagram observed at the median redshift $z = 0.9$. We do not know *a priori* the amount of evolution of each galaxy since it depends on the stellar population which we try to investigate. Therefore we first applied the same amount of passive evolution correction to all the galaxies assuming the high formation redshift of $z_{form} = 4.4$, and then applied additional correction to each galaxy depending on its colour deviation from the CMR.

The final C-M diagram is shown in Figure 1. *The ridge-line of the CMR is well defined even in the field early-types at $z = 0.9$.* This is a striking result, as this fact itself indicate the oldness of these galaxies. For comparison, the CMRs of Coma cluster transformed to $z = 0.9$ and a distant cluster at $z = 0.9$ are plotted together. Both of these relations trace just along the red sequence of the HDF. The detailed

 Astrophysics and Space Science is the original source of publication of this article. It is recommended that this article is cited as: *Astrophysics and Space Science* **265**: 487–488, 1999.
© 1999 *Kluwer Academic Publishers.*

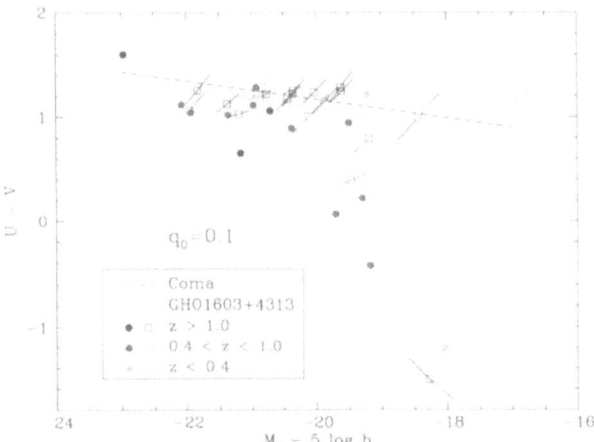

Figure 1. The colour-magnitude diagram for early-types in the HDF presented in the rest frame at $z = 0.9$. Filled symbols indicate galaxies with spectroscopic redshifts, open symbols refer to those with photometric redshifts only. The error bars correspond to the typical photometric redshift error of 0.08. The gray scale of the symbols indicates their redshift. The dashed line shows the CMR of Coma (Bower, Lucey and Ellis, 1992) transformed to $z = 0.9$ after the passive evolution ($z_{form} = 4.4$) and the aperture correction. The dot-dashed line shows the CMR of a cluster GHO 1603+4313 at $z = 0.9$ from Stanford *et al.* (1998) after K-correction.

analysis of the residual colour distribution with respect to the Coma CMR does suggest that at least nearly half of the field early-types at $z = 0.9$ should be as old as cluster early-types such as $z_{form} > 2$. Also true that there are some blue, hence young, galaxies well below the red sequence. Although most of them are faint and show peculiarities in morphologies, it might suggest some field early-types could have continuously formed down to the observed epoch.

References

Bernardi, M., *et al.*: 1998, *Astrophys. J. Lett.* **508**, 143.
Bower, R.G., Lucey, J.R. and Ellis, R.S.: 1992, *Mon. Not. R. Astron. Soc.* **254**, 601.
Franceschini, A., *et al.*: 1998, *Astrophys. J.* **506**, 600.
Kauffmann, G.: 1996, *Mon. Not. R. Astron. Soc.* **281**, 487.
Kodama, T., Arimoto, N., Barger, A.J. and Aragón-Salamanca, A.: 1998, *Astron. Astrophys.* **334**, 99.
Kodama, T., Bell, E.F. and Bower, R.G.: 1999, *MNRAS* **302**, 152.
Kodama, T., Bower, R.G. and Bell, E.F.: 1999, *MNRAS* **306**, 561.
Larson, R.B., Tinsley, B.M. and Caldwell, C.N.: 1980, *Astrophys. J.* **237**, 692.
Stanford, S.A., Eisenhardt, P.R.M. and Dickinson, M.: 1998, *Astrophys. J.* **492**, 461.

GAS-RICH LOCAL DWARF STAR-FORMING GALAXIES AND THEIR CONNECTION WITH THE DISTANT UNIVERSE

D. KUNTH

Institut d'Astrophysique de Paris, 98 bis Boulevard Arago, 75014 Paris, France

Abstract. I discuss the properties of gas-rich forming galaxies. I particularly emphasize the latest results on Lyα emission that are relevant to the search of distant young galaxies. The interdependance of the Lyα escape with the properties of the ISM in starburst galaxies is outlined. A new model from G. Tenorio-Tagle and his collaborators explains Lyα profiles in starburst galaxies from the hydrodynamics of superbubbles powered by massive stars. I stress again that since Lyα is primarely a diagnostic of the ISM, it is mandatory to understand how the ISM and Lyα are related to firmly relate Lyα to the cosmic star–formation rate.

1. Introduction

Among galaxies that pertain to our local Universe the gas-rich ones and in particular the ones with violent episodes of massive star formation attract our attention. Along the Hubble sequence the fractional gas content (i.e. the gas to total mass ratio) increases as one moves from early types to late types and reaches values as large as 0.1 to 0.5 for Irr galaxies. More extreme cases are found amongst LSBs (Low surface brightness galaxies) and the dwarfs star-forming galaxies in which the total amount of gas (mainly atomic gas as the molecular gas content in these low mass objects is still unknown) could be the dominant component. Our recent view of the distant Universe opens the possibility that small gaseous systems gradually merge into large ones. Galaxy formation appears more gradual that had been anticipated and one would expect to see some left over debris in our local Universe after a priviledged epoch of star formation at around z of 1–2. Dwarf star-forming galaxies naturally comes into discussion for at least two reasons – both linked to the distant Universe. First they appear so chemically unevolved that they represent the best site to look at for the determination of the primordial Helium abundance. Indeed, their measured helium abundances are accurately determined and are close to the primordial helium value since the He products from stellar burning is presumably minimal. Second they offer ideal local counterparts of what should be distant protogalaxies at the onset of their first massive star formation episodes. I will focus on this later point in the rest of this review.

Astrophysics and Space Science is the original source of publication of this article. It is recommended that this article is cited as: *Astrophysics and Space Science* **265**: 489–499, 1999.
© 1999 *Kluwer Academic Publishers*.

2. The Dwarf Star-Forming Galaxies

2.1. THEIR OVERALL PROPERTIES

Dwarf star-forming galaxies are sometimes called Blue Compact Dwarf (BCD) or HII galaxies because their ongoing star-formation rate large with respect to their integrated past star formation rate so that their overall spectroscopic appearance is that of a Giant HII region. Although HII galaxies also comprise a wider class of objects in term of mass and luminosity I shall refer to this terminology in the following discussion. Local HII galaxies have absolute luminosities dominated by the young stellar population and range from $M_B = -13$ to -17 (if only restricted to dwarfs). They are bluer than ordinary irregular galaxies (with typically $U - B = -0.6$ and $B - V = 0.0$) and are rich in ionized gas. The equivalent width of the Hβ line ranges from 30 to 200 Å implying a burst star formation no older than a few Myrs. The role of massive stars is to energize the ISM and enrich it with heavy metals and some helium. These galaxies are chemically unevolved and their metal distribution peaks at 1/10th the solar value, with the most deficient galaxy IZw18 at only 1/50th. Their neutral gas content is relatively high, amounting a few $10^7\ M_\odot$ in general, i.e. a large fraction of the luminous mass. On the other hand there are some indications that the dynamical mass is much larger, (Carignan and Beaulieu, 1993). UV to IR studies clearly indicate that massive star formation is a discontinuous process concentrated to short bursts intervened by longer passive periods. Their UV spectra show small amount of dust that makes them ideal targets to study locally.

2.2. THEIR LINK WITH THE PRIMEVAL GALAXIES

The major motivation in follow-up studies of unevolved and nearly dust-free HII galaxies is their possible ressemblance with distant protogalaxies in their early phase of star-formation. Assuming that protogalaxies would undergo a rapid collapse on a short dynamical timescale (Eggen et al., 1962), the expectation is that an L* galaxy could be detected from its redshifted Lyα emission line. Indeed in a dust free case and a normal IMF such a galaxy could pop up with a star formation rate of nearly 1000 $M_\odot$ yr^{-1} producing a Lyα luminosity of 10^{45} erg/sec (this amounts few percent of the total luminosity). At a redshift of 3 this translates to 10^{-15} erg cm^{-2} sec^{-1}, in principle easily detectable even from deep spectrocopic searches or narrow band imaging using panoramic CCDs. The failure to detect such a strong Lyα is an additional piece of evidence that the overall process of galaxy formation is more complex: galaxies form from smaller building blocks and/or luminous protogalaxies have to be found at much larger redshift and/or are enshrouded in large dusty cocoons.

3. The Early IUE Observations and Interpretations

While galaxies at their very early stage could be nearly dust-free hence easily detectable from their Lyα emission (Partridge and Peebles, 1967); early ultraviolet observations of nearby HII galaxies, have revealed a much weaker Lyα emission than predicted by starburst models and simple models of galaxy formation. In some other galaxies Lyα was non-existent or even appeared as a broad absorption profile (Deharveng *et al.*, 1986; Hartmann *et al.*, 1984, 1988; Kunth *et al.*, 1997, 1998; Meier and Terlevich, 1981; Terlevich *et al.*, 1993). For young star-forming galaxies without so much dust as to suppress Lyα , large equivalent widths are expected in the range of $100 - 200(1 + z)$ Å (Charlot and Fall, 1993). However it was early realized that pure extinction by dust would be unable to explain the low observed Lyα/Hβ (although Calzetti and Kinney, 1992, tentatively proposed that proper extinction laws would correctly match the predicted recombination value). Valls-Gabaud (Valls-Gabaud, 1993) on the other hand suggested that ageing starbursts could reduce Lyα equivalent widths because they are affected by strong underlying stellar atmospheric absorptions. Early IUE data have provided evidence for an anticorrelation between the Lyα/Hβ ratio and the HII galaxy metallicity. These results were attributed to the effect of resonant scattering of Lyα photons and their subsequently increased absorption by dust (Charlot and Fall, 1993; Chen and Neufeld, 1994; Giavalisco *et al.*, 1996; Meier and Terlevich, 1981, and references therein).

Finally Charlot and Fall (1993) advocated that the structure of the interstellar medium (porosity and multi-phase structure) is most probably an important factor for the visibility of the Lyα line.

4. HST Observations

New observations performed with the HST indicate that the velocity structure in the interstellar medium plays a key role in the transfer and escape of Lyα photons. Kunth *et al.* (1994) and Lequeux *et al.* (1995) have used the Goddard High Resolution Spectrograph (GHRS) onboard the Hubble Space Telescope (HST) to observe Lyα and the interstellar lines O I 1302.2 Å and Si II 1304.4 Å. The O I 1302 Å and Si II 1304 Å allow to measure with reasonable accuracy the mean velocity at which the absorbing material lies with respect to the star-forming region of a given galaxy. Surprisingly, Lyα was observed only in absorption in IZw 18, the most metal–poor starburst galaxy known (Kunth *et al.*, 1994). Meanwhile to add to the confusion, a Lyα emission line showing a clear P–Cygni component, has been detected in the star-forming galaxy Haro 2, dustier than IZw 18 and with $Z = 1/3 \, Z_\odot$ (Kunth *et al.*, 1994). The detection of such a profile in the Lyα emission line led us to postulate that the line was visible because the absorbing neutral gas was velocity–shifted with respect to the ionized gas. This was confirmed by the analysis of the

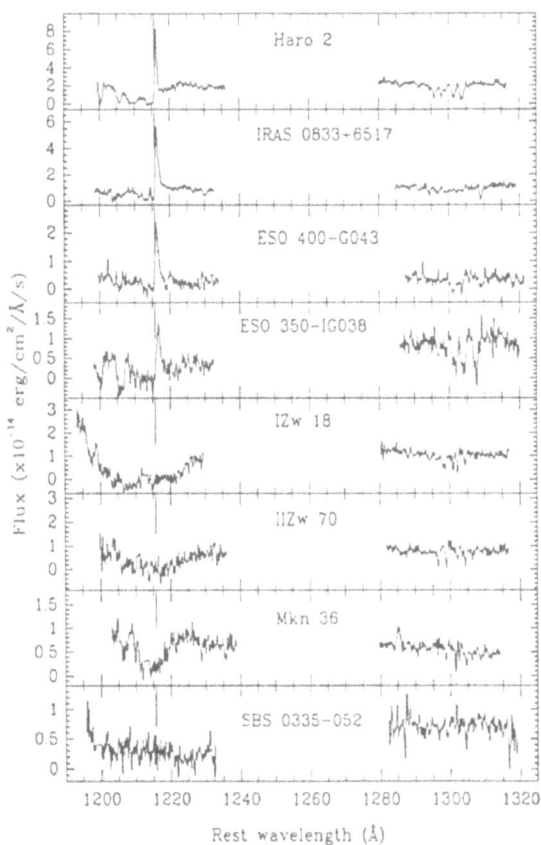

Figure 1. GHRS spectra of all the galaxies in Kunth *et al.* (1998). The spectra have been shifted to rest velocity assuming the redshift derived from optical emission lines. Vertical bars indicate the wavelength at which the Lyα emission line should be located. The geocoronal emission profile has been truncated for the sake of clarity. The spectra have been plotted after rebinning to 0.1 Å per pixel and smoothed by a 3 pixel box filter.

UV O I and Si II absorption lines (blue–shifted by 200 km s^{-1} with respect to the optical emission lines) and that of the profile of the Hα line (Legrand *et al.*, 1997). Observations were subsequently made on additional galaxies by Kunth *et al.* (1998) while Thuan and Isotov (1997) have obtained GHRS spectra of two more starburst galaxies, namely Tol65 and T1214-277. Tol65 reveals a broad damped Lyα absorption while T1214-277 shows a pure Lyα emission profile with an equivalent width of 70 Å and with no blue absorption. The individual spectra of galaxies observed in Kunth *et al.* (1998) are shown in Figure 1.

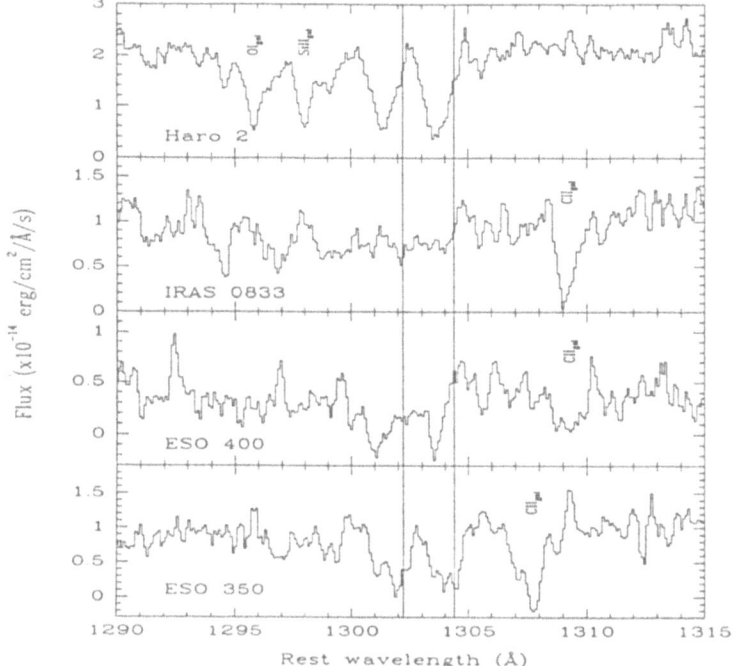

Figure 2. Detail of the O I and Si II region for the galaxies with Lyα emission. The vertical bars indicate the wavelength at which the O I and Si II absorption lines should be located, according to the redshift derived from optical emission lines. Some Galactic absorption lines have been marked. Note that the metallic lines appear systematically blueshifted in these galaxies with respect to the systemic velocity. In some cases there is no significant absorption at all at zero velocity.

5. The Role of the Velocity Structure of the ISM

Three types of observed lines have been identified so far: pure Lyα emission; broad damped Lyα absorption centered at the wavelength corresponding to the redshift of the HII emitting gas and Lyα emission with blue shifted absorption features, leading in some cases to P–Cygni profiles.

Lyα emission with deep blueward absorption troughs evidence a wide velocity field. The equivalent widths of the Lyα emission range between 10 and 37 Å hence much below the value predicted by Charlot and Fall (1993) for a dust–free starburst model. In all cases, interstellar absorption lines (OI, SiII) are significantly blueshifted with respect to the HII gas (see Figure 2). On the other hand, if the HI is static with respect to HII, the destroyed Lyα photons are those emitted by the HII region and the interstellar lines are not displaced (see Figure 3).

In galaxies with low dust content – IZw 18 has a dust-to-gas ratio at least 50 times smaller than the Galactic value (Kunth *et al.*, 1994), it remains possible to weaken observationally Lyα by simple multiple resonant scattering from a static neutral gas and produce an absorption feature. The H I cloud surrounding these

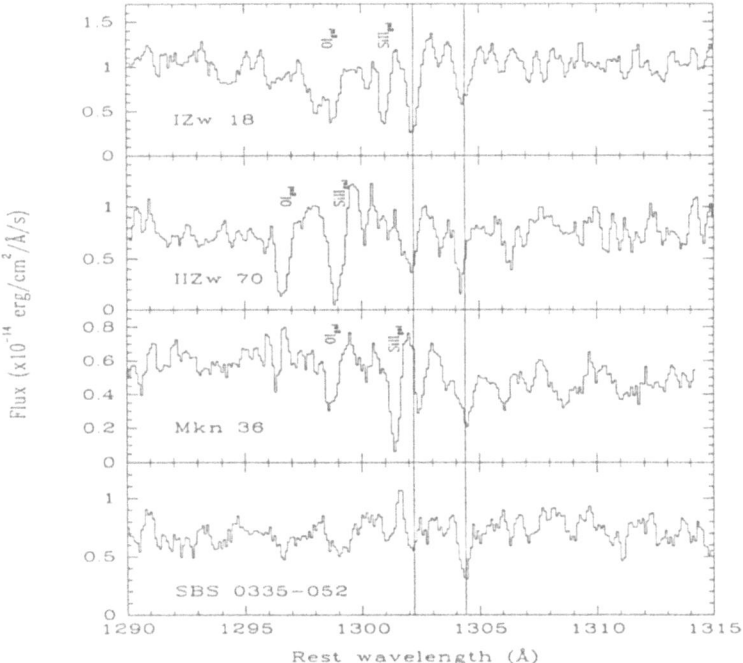

Figure 3. O I and Si II region for the galaxies showing damped Lyα absorptions. Details as in Figure 2. Note that in these galaxies the metallic lines are essentially at the same redshift as the ionized gas, indicating the presence of static clouds of neutral gas, as discussed in the text.

galaxies might be leaking Lyα photons through its external surface. The Lyα line would then become very hard to detect because of its low surface brightness. On the other hand, the situation in Haro 2 is quite different. The Lyα P Cygni profile suggests a rather modest amount of neutral gas of the order of N (H I) $= 7.7 \times 10^{19}$ atoms cm^{-2}. The crucial point here is that the neutral gas responsible for the absorption is not at the velocity at which the Lyα photons were emitted but is being pushed out by an expanding envelope around the H II region, at velocities close to 200 km s^{-1}. This interpretation is strengthened by the presence of the blueshifted absorptions of O I, Si II and Si III precisely due to outflowing gas in front of the ionizing hot stars, (see also Legrand *et al.*, 1997). Data on other H II galaxies with detected Lyα emission confirm that Haro 2 is not an isolated case. Most spectra show Lyα emission with a broad absorption on their blue side except for ESO 350-IG038 in which the emission is seen atop of a broad structure requiring several filaments. When metallic lines are detected, they are always blueshifted with respect to the ionized gas, further supporting the interpretation. In the case of ESO 350-IG038 the velocity structure seems to be more complicated and several components at different velocities are identified on the metallic lines. Note that there might be a secondary peak emission in the blue side of the main line in the spectrum of ESO-400-G043.

The main conclusion drawn from this set of data is that complex velocity structures are determining the Lyα emission line detectability, showing the strong energetic impact of the star-forming regions onto their surrounding ISM. We want to stress that this effect seems to be almost independent of the dust and metal abundance of the gas.

Thuan and Izotov (1997) have detected strong Lyα emission in T1214-277, with no clear evidence of blueshifted Lyα absorption. They argue that a significant fraction of the area covered by the slit along the line of sight is essentially free from neutral gas, suggesting a patchy or filamentary structure of the neutral clouds. Such a geometry would be unlikely in galaxies surrounded by enormous H I clouds, as in IZw 18 and similar objects.

6. The Evolution of Superbubbles in Extended HI Halos

To account naturally for the variety of Lyα line detections in star-forming galaxies Tenorio-Tagle *et al.* (1998), have proposed a scenario based on the hydrodynamics of superbubbles powered by massive starbursts. This scenario is visually depicted in Figure 4. The overpowering mechanical luminosity ($E_0 \geq 10^{41}$ erg s^{-1}) from massive starbursts ($M_{stars} = 10^5 - 10^6 \, M_\odot$) is known to lead to a rapidly evolving superbubble able to blowout the gaseous central disk configuration into their extended HI haloes.

As shown by Koo and McKee (1992) the superbubble will blowout and thus massive starbursts will lead to superbubble blowout phenomena even in massive galaxies such as the Milky Way. One can then predict the venting of the hot superbubble interior gas through the Rayleigh-Taylor fragmented shell, into the extended HI halo where it would push once again the outer shock, quite early in the evolution of the starburst ($T_{blowout} \leq 2$ Myr), allowing to build a new shell of swept up halo matter (see Figure 4b).

The blowout or the fragmentation of the shell allows the expansion of the hot interior gas into the extended low density HI halo, and the leakage of the uv photons emitted by the starburst. These photons establish an ionized conical HII region with its apex at the starburst. The density of the halo steadily decreases with radius, and one can show that the conical sector of the HII region will extend all the way to the galaxy outer edge, i.e. several kpc. Furthermore, the low halo density implies a long recombination time ($t_{rec} = 1/(\beta n_{halo}) \geq 10^7$ yr), with the implication that the ionized conical sector of the galaxy halo becomes, and remains, transparent to the ionizing flux produced during the ensuing lifetime of the ionizing massive stars (≤ 10 Myr). Then clearly, the Lyα photons produced at the central HII region will also be able to travel freely in such directions. The escape of uv photons from the galaxy would be particularly important during the early stages

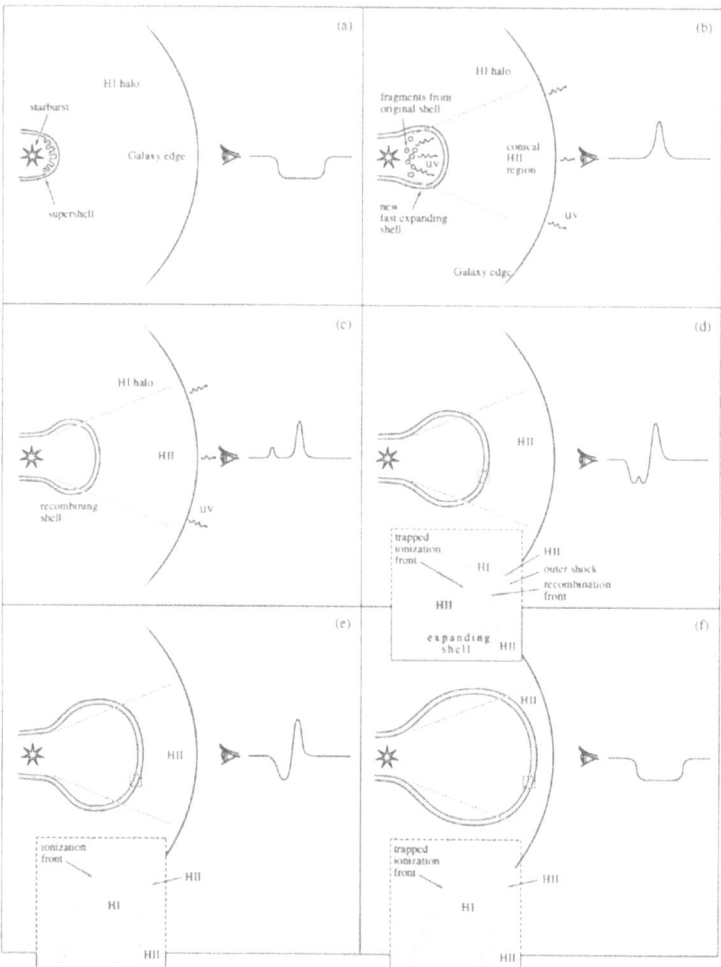

Figure 4. The evolution of a superbubble as a function of time.

after blowout and until the new shell of swept up halo matter condenses enough material as to allow its recombination. If the shock progresses with speeds of a few hundred ($\leq$ 400) km s^{-1}, (i.e. with a Mach number $M \leq 40$) it will promote the rapid cooling of the shocked gas that will cool down to the HII region temperature making the shock isothermal, and thus causing compression factors of several hundreds. Compression leads to recombination in time-scales ($t_{rec} = 1/\beta n_{shock}$) of less than 10^5 yr, and this immediately and steadily will reduce the number of stellar uv photons leaking out of the galaxy. At the same time recombination in the fast expanding shell will lead to a correspondingly blueshifted Lyα emission, as depicted in Figure 4c. Once the shell presents a large column density ($\sim 10^{19}$ atoms cm^{-2}), as it grows to dimensions of a few kpc, it will trap the ionization front. Note that from then onwards, recombinations in the shell will inhibit the further escape of ionizing photons from the galaxy (compare Figure 4b, c and d).

The trapping of the ionization front, makes the shell acquire a multiple structure with a photoionized inner edge, a steadily growing central zone of HI, and an outer ionized sector where the recently shocked ionized halo gas is steadily incorporated. The growth of the central layer eventually will cause sufficient scattering and absorption of the Lyα photons emitted by the central HII region, leading to a blueshifted Lyα absorption. As long as recombinations continue to occur at the leading edge of the shell, a blueshifted Lyα in emission will appear superposed to the blueshifted absorption feature (see Figure 4d). Recombination at the leading edge will become steadily less frequent, depleting the blueshifted Lyα in emission. This is due when the shell and its leading shock move into the outer less dense regions of the halo, and the shell recombination time, despite the compression at the shock, becomes larger than the dynamical time. At this stage, an observer looking along the conical sector of the HII region will detect a P-Cygni-like Lyα line profile as shown in Figure 4e.

Geometrical dilution of the uv flux will begin to make an impact as the super-bubble grows large. This and the drop in the uv photon production rate, caused by the death of the most massive stars after $t = t_{ms}$, will enhance the column density of neutral material in the central zone of the recombined shell to eventually cause the full saturated absorption of the Lyα line (see Figure 4f). Full saturated absorption has usually been accounted for by the large column density of the extended HI envelope of these galaxies and thus, as in all models, many different orientations will match the observations.

7. Discussion

7.1. THE RELEVANCE OF THE SUPERBUBBLE MODEL

The main implication of the evolution depicted in Figure 4 is that it is the feedback from the massive stars that leads to the large variety of Lyα emission profiles. P-Cygni Lyα profiles are predicted when observing along the angle subtended by the conical HII region but only once the ionization front is trapped by the sector of the superbubble shell evolving into the extended halo. This will produce the fast moving layer of HI at the superbubble shell, here thought to be responsible for the partial absorption observed in sources such as Haro 2, ESO 400-G043 (which probably exhibits a secondary blueshifted Lyα emission) and ESO 350-IG038 Kunth *et al.* (1998).

Damped Lyα absorption is seen in several galaxies. We note that these objects are all gas–rich dwarf galaxies whereas in most cases but Haro 2, the HII galaxies that exhibit Lyα in emission or with a P-Cygni profile, are on the higher luminosity side of the distribution ($M \leq -18$).

Pure Lyα emission is observed in C0840+1201 and T1247-232 (Terlevich *et al.*, 1993; IUE) or T1214-277 (Thuan and Izotov, 1997; HST). Such a line implies no absorption and thus no HI gas between the starburst HII region and the observer, as when observing the central HII region after the superbubble blowout, within the conical HII region carved in the extended HI halo. It is not a straigthforward issue to estimate what is the fraction of Lyman continuum radiation that leaks out from galaxies. This scenario however predicts a short but significant evolutionary phase (between blowout and the trapping of the ionization front by the fast expanding shell) during which a large amount of *uv* radiation could leak out of a galaxy into the intergalactic medium. Detailed numerical calculations of the scenario proposed here are currently underway. These results and further implications of the model will be reported in a forthcoming communication.

7.2. THE GALAXIES AT HIGH-REDSHIFT

The effect of neutral gas flows helps to understand why luminous high-redshift objects have only been found up to now with linewidths larger than 1000 km s^{-1}. High–redshift galaxies with very strong (EWs > 500 Å) extended Lyα emission are characterized by strong velocity shears and turbulence ($v > 1000$ km s^{-1}); this suggests an AGN activity, in the sense that other ISM energising mechanism than photoionization by young stars may be operating. On the other hand Steidel *et al.* (1996) have recently discovered a substantial population of star–forming galaxies at $3.0 < z < 3.5$ that were selected but from the presence of a very blue far-UV continuum and a break below 912 Å in the rest frame. These Lyman–break galaxies (LBG) are much more luminous but similar to our local starbursts in the sense that 50% show no Lyα emission whereas the rest does, but with weak EWs no larger than 20 Å at rest. The Lyα profiles of this population looks also very similar to those of our local starburst galaxies (Franx *et al.*, 1997; Pettini *et al.*, 1997; see also Dey *et al.*, 1998 for the $z = 5.34$ galaxy).

New Lyα emitters are now found at high-redshift from surveys using large telescopes with narrow-band filters (Hu *et al.*, 1998; Pettini *et al.*, 1997). Limits down to a few 10^{-18} erg cm^{-2} sec^{-1} are now reachable and give access, in principle, to galaxies with fainter continuum magnitudes than the LBG. It is not clear whether these Lyman–alpha–galaxies (LAG) represent the earliest stages of the galaxy formation process as suggested by Hu *et al.* (1998). Are these galaxies less dusty than the LBGs? Is there an intrinsically fainter population of dust-free star–forming galaxies, possibly connected to the bright-end of the star-forming dwarfs population (-18 to -20?) we see today? What are their masses and spatial distribution (see also Pascarelle *et al.*, 1998 for a recent HST study)? Near-IR spectroscopic studies with large ground based telescopes will in the future help to settle some of these questions (Pettini *et al.*, 1998).

The possible increase of the star formation rates as one moves at higher redshift (between 3 and 6) reported by Hu *et al.* (1998) must be taken with caution since from the preceeding discussion, it is clear that a significant, yet unknown fraction of the youngest galaxies may not be Lyα emitters. Indeed, since Lyα is primarily a diagnostic of the ISM, it is only when we fully understand how the ISM and Lyα are related can we hope to firmly relate Lyα to the SFR.

References

Calzetti, D. and Kinney, A.L.: 1992, *Astrophys. J.* **399**, L39.

Carignan, C. and Beaulieu, S.: 1993 *Astrophys. J.* **347**, 760.

Charlot, S. and Fall, S.M.: 1993, *Astrophys. J.* **415**, 580.

Chen, W.L. and Neufeld, N.A.: 1994, *Astrophys. J.* **432**, 567.

Deharveng, J.M., Joubert, M. and Kunth, D.: 1986, in: D. Kunth, T.X. Thuan and J. Tran Thanh Van (eds.), *The First IAP workshop: 'Star-Forming Dwarf Galaxies and related objects'*, Editions Frontières, p. 431.

Dey, A., Spinrad, H., Stern, D., Graham, J. and Chaffee, F.H.: 1998, *Astrophys. J.* **498**, L93.

Eggen, O.J., Lynden–Bell, D. and Sandage, A.: 1962, *Astrophys. J.* **136**, 748.

Franx, M., Illingworth, G.D., Kelson, D.D., van Dokkum, P.G. and Kim-Vy, T.: 1997, *Astrophys. J.* **486**, L75.

Giavalisco, M., Koratkar, A. and Calzetti, D.: 1996, *Astrophys. J.* **466**, 831.

Hartmann, L.W., Huchra, J.P. and Geller, M.J.: 1984, *Astrophys. J.* **287**, 487.

Hartmann, L.W., Huchra, J.P., Geller, M.J., O'Brien, P. and Wilson, R.: 1988, *Astrophys. J.* **326**, 101.

Hu, E.M., Cowie, L.L. and McMahon, R.G.: 1998, *Astron. J.*, in press (astro-ph/9803011).

Koo, B-C. and McKee, C.F.: 1992, *Astrophys. J.* **388**, 93.

Kunth, D., Lequeux, J., Sargent, W.L.W. and Viallefond, F.: 1994, *Astron. Astrophys.* **282**, 709.

Kunth, D., Lequeux, J., Mas-Hesse, J.M., Terlevich, E. and Terlevich, R.: 1997, *Rev. Mex. Astr. Astrofis.* **6**, 61.

Kunth, D., Mas-Hesse, J.M., Terlevich, E., Terlevich, R., Lequeux, J. and Fall, M.: 1998, *Astron. Astrophys.* **334**, 11.

Legrand, F., Kunth, D., Mas-Hesse, J.M. and Lequeux, J.: 1997, *Astron. Astrophys.* **326**, 929.

Lequeux, J., Kunth, D., Mas-Hesse, J.M. and Sargent, W.L.W.: 1995, *Astron. Astrophys.* **301**, 18.

Meier, D.L. and Terlevich, R.: 1981, *Astrophys. J.* **246**, L10.

Partridge, R. and Peebles, P.J.E.: 1967, *Astrophys. J.* **146**, 868.

Pascarelle, S.M., Windhorst, R.A. and Keel, W.C.: 1998, astro-ph9809181.

Pettini, M., Steidel, C.C, Adelberger, K.L., *et al.*: 1997, to appear in: J.M. Shull and C.E. Woodward and H. Thronson (eds.), *ORIGINS*, ASP Conference Series.

Pettini, M., *et al.*: 1998, astro-ph9806219.

Steidel, C.C., Giavalisco, M., Pettini, M., Dickinson, M. and Adelberger, K.: 1996, *Astrophys. J.* **462**, 17.

Tenorio-Tagle, G., Kunth, D., Terlevich, E., Terlevich, R. and Silich, S.A.: 1998, in preparation.

Terlevich, E., Diaz, A.I., Terlevich, R. and Garcia-Vargas, M.L.: 1993, *Mon. Not. R. Astron. Soc.* **260**, 3.

Thuan, T.X. and Izotov, Y.I.: 1997, *Astrophys. J.* **489**, 623.

Valls-Gabaud, D.: 1993, *Astrophys. J.* **419**, 7.

STAR FORMATION IN ACCRETION DISKS AROUND MASSIVE BLACK HOLES AND PREGALACTIC ENRICHMENT

SUZY COLLIN

DAEC, Observatoire de Paris, Section de Meudon, 1 Place Janssen, F92195 Meudon, France

JEAN-PAUL ZAHN

DASGAL, Observatoire de Paris, Section de Meudon, 1 Place Janssen, F92195 Meudon, France

Abstract. Broad Absorption Lines (BALs) prove the existence of a high velocity outflowing gas with metallicities larger than solar in the central few parsecs of high redshift quasars. At the same distance from the black hole, accretion disks in quasars and Active Galactic Nuclei (AGN) are locally gravitationally unstable, and clumps must form with a size of the order of the scale height of the disk. This is hardly a coincidence, and we have tried to link these two facts. We have assumed that the unstable clumps give rise to protostars, which become massive stars after a rapid stage of accretion, and explode as supernovae, producing strong outflows perpendicular to the disk and inducing outward transfer of angular momentum in the plane of the disk. As a consequence a self-regulated disk made of gas and stars where supernovae sustain the inflow mass rate required by the AGN is a viable solution in this region of the disk. This model could explain the BALs, and could also account for a pregalactic enrichment of the intergalactic medium and of the Galaxy, if massive black holes formed early in the Universe.

1. Motivations

1.1. EXISTENCE OF HIGH METALLICITIES AND OUTFLOWS IN HIGH REDSHIFT QUASARS

About 10% of high redshift quasars display broad absorption lines (BALs), on the blue side of emission lines (see for instance the review of Turnshek, 1995). They are attributed to outflowing ionized matter covering the continuum source, and in some cases also the broad line emission region (BLR). Most experts share the opinion that, although BAL QSOs are different from non BAL QSOs in several respects, all QSOs possess a BAL region, but it can be seen only when the angle of view is favorable.

Abundances in the BAL gas are derived from photoionization models, relying on ionic column densities deduced from line equivalent widths. In spite of the uncertainties due for instance to line saturation, there is strong evidence that the BAL gas has high abundances of heavy elements, at least solar, and most probably 10 times larger than solar (cf. for instance Hamann, 1997). The line profiles prove outflowing motions, with velocities of 10^4 up to $3\ 10^4$ km s^{-1}, with a rate of mass

 Astrophysics and Space Science is the original source of publication of this article. It is recommended that this article is cited as: *Astrophysics and Space Science* **265**: 501–505, 1999.
© 1999 *Kluwer Academic Publishers.*

ejection: $\dot{M}_{out} \sim 10^{-2} N_{21}(Z/Z_\odot)^{-1}(\Omega/4\pi)_{-1}V_9 R_{pc} M_\odot$ yr^{-1}, where $(Z/Z_\odot)$ is an average of CNO abundances referred to the solar ones, N_{21} is the column density expressed in 10^{21} cm^{-2}, V_9 is the velocity of the outflow in 10^9 cm s^{-1}, R_{pc} the distance in parsec, and $(\Omega/4\pi)_{-1}$ the opening angle of the flow expressed in 10^{-1}. This mass flow rate is small compared to the accretion rate, $M_{acc} \sim 2 L_{46}M_\odot$ yr^{-1} (the efficiency of mass energy conversion is set equal to 0.1), but it may represent only a small fraction of the total mass outflow, the 'emerged part of the iceberg'. Indeed the volumic filling factor of the BAL region is shown to be $f_{vol} \sim 10^{-6} N_{21}(Z/Z_\odot)^{-1}T_5L_{46}^{-1}R_{pc}^{-2}$, where T_5 is the temperature in 10^5 K. The BAL region is thus made of many clouds (or filaments, or shells), with a small filling factor. If there is a more diffuse medium (presumably hot and contributing to the strong X-ray absorption characteristic of BAL quasars, as proposed by Murray *et al.*, 1995), or if the column density is underestimated due to a bad evaluation of the – unknown – ionizing continuum, the outflow rate might be larger, and easily reach $\dot{M}_{accr}$ (for a brief review see Collin, 1998).

Finally the precise location of the BAL region is unknown, but can be estimated roughly using photoionization arguments. It is at least of the order of the BLR, which scales with the luminosity as $\sim 0.3\sqrt{L_{46}}$ pc, where L_{46} is the luminosity expressed in 10^{46} ergs s^{-1}, and it is at most equal to 100 pc.

1.2. ACCRETION DISKS AT LARGE DISTANCE FROM THE BLACK HOLES

It is widely admitted that AGN are massive black holes fueled via accretion disks. The observation of the 'UV bump' (Shields, 1978; Malkan and Sargent, 1982, and many subsequent papers) argues in favor of geometrically thin and optically thick disks, possibly embedded in a hot X-ray emitting corona. At large radii ($R \geq$ 0.1 pc) these disks have two serious problems: they are not able to transport rapidly enough the gas to the black hole, and they are gravitationally unstable (Toomre, 1964; Goldreich and Linden-Bell, 1965). Farther from the center ($R \geq 100$ pc) the supply of gas can be provided by gravitational torques or by global non axisymmetric gravitational instabilities, but this is not true in the intermediate region where the disk is locally but not globally self-gravitating. Curiously this intermediate region coincides with the BLR and the BAL region. In our opinion it is hardly a coincidence. It is why we have developed a model relating these facts.

2. Star Formation and Evolution in Accretion Disks

The intermediate region in accretion disks is generally identified with the 'molecular torus' in the Unified Scheme of AGN (cf. for instance Krolik and Begelman, 1988). It could be made of marginally unstable randomly moving clouds, where the 'viscosity' of the disk would be provided by cloud collisions (Begelman, Frank and Shlosman, 1989). In Collin and Zahn (1999) we have adopted the opposite view that if an unstable fragment begins to collapse, the collapse will continue until a protostar is formed, unless the collapse time is larger than the characteristic time for mass transport in the disk. We have examined the evolution of these stars, and whether they could supply the angular momentum transport required by the AGN.

The unstable region of the disk is molecular and dense compared to galactic molecular clouds (Collin and Huré, 1998). Unstable fragments should collapse rapidly and give rise to compact objects (planets or protostars). These objects accrete at a high rate and in less than 10^6 years they acquire a mass of a few tens of $M_\odot$, according to a mechanism proposed by Artymowicz, Lin and Wampler (1993) for stars captured by the disk. When the stars explode as supernovae, the supernova shells break out of the disk, producing strong outflows, while in the radial and in the azimuthal directions the expansion of the supernovae is limited by the shear of the keplerian motion, so finally the gaseous disk is able to support a large number of massive stars and supernovae while staying relatively homogeneous. On the other hand these supernovae induce a transfer of angular momentum towards the exterior, as shown by the numerical simulations of Rozyczka, Bodenheimer and Lin (1995). A very crude stationary disk model made of stars embedded in a quasi homogeneous gas can then be built for the intermediate region of the disk, where the transfer of angular momentum needed to sustain the mass transport to the black hole is ensured by the supernovae.

This homogeneous model works in a ring located between 0.1 and 10 pc for a black hole mass of 10^6 $M_\odot$, and between 1 and 100 pc for a black hole mass of 10^8 $M_\odot$ or larger, whatever the abundances, but for relatively low luminosities ($L_{46} < 1$). In more luminous quasars, the number of supernovae is too large, and the disk becomes entirely paved with hot cavities surrounded by shocks, so the thin disk should be replaced by a thick disk (a 'torus'). Star formation may then take place in the compressed regions of the supernova shells. Moreover the mass inflow from the periphery of the disk is most probably variable with time, as it would be the case if it is achieved by large molecular complexes comparable to those present near the Galactic center, with successively 'low states' and 'high states' where the disk is alternatively 'quiescent', and 'active', with an intense supernova activity and super-Eddington outflows.

3. Implications of the Model

Collin and Zahn (1999) have shown that a typical stationary rate of supernovae is 10^{-2} to 10^{-1} M_8 yr^{-1}, where M_8 is the mass of the black hole expressed in 10^8 $M_\odot$. Each supernova releases about 10 $M_\odot$ of metals out of the nucleus, so the mass of metals ejected from the disk during the accretion phase is of the order of the mass of the black hole itself. The mass outflow rate due to the supernovae accounts easily for the observations of BALs in low luminosity quasars. For high luminosity objects, the non stationary process should be invoked. The ejection occurs in high velocity thin shells, in agreement with the structure of the BAL region, and the location of the phenomenon ($\leq$ 100 pc) is in agreement with the observations. Finally the fact that the opening angle of the BAL region is equal to a fraction of 4π is explained, the ejection taking place mainly in a cone in the direction of the disk axis. Note that in this model the metallicity of the gas fuelling the black hole can be very small, while the observed outflow is always enriched.

A second outcome of the mechanism is to account for a pregalactic enrichment, if massive black holes are created early in the process of galaxy formation, and if galaxy formation takes place through an hierarchical scenario (Silk and Rees, 1998). Massive galaxies will retain their gas, and the supernova shells will compress the interstellar medium, trigger star formation like in the interstellar medium, and induce a starburst. A fraction of the enriched gas will escape from the galaxy, not only due to the nuclear supernovae explosions, but also to the induced starburst. Small galaxies will not retain their gas, which will escape with the enriched gas produced by the supernovae. It will pollute the intergalactic medium (IGM). In particular, if the formation of the black holes precedes the formation of galaxies, it will lead to a pregalactic enrichment of the IGM. If the universe at high redshift is dominated by a homogeneous population of compact and spheroidal galaxies (Steidel *et al.*, 1996) which are the progenitors of massive galaxies, the enrichment of IGM induced by the black holes could be quite homogeneous.

Simply taking the integrated comoving mass density of *observed* quasars, which corresponds to about 10^{-6} of the closure density (Soltan, 1982, and further studies) and a typical black hole mass of $10^8 \ M_\odot$, the mechanism will provide about 10^{-6} of the closure density in metals, i.e. after mixing with the IGM, an average metallicity of a few $10^{-3} \ \Omega(\mathrm{IGM})_{0.02} Z_\odot$, close to the metallicity observed in the Lα forest which constitutes the main fraction of the IGM.

Finally, our Galaxy is presently not active and the black hole in the center has a small accretion rate ($< 10^{-4} \ M_\odot \ \mathrm{yr}^{-1}$), so it has most probably accreted a large fraction of its mass ($2 \ 10^6 \ M_\odot$) during an early period. The previous estimation leads to an ejection of a few $\sim 10^4 \ M_\odot$ of metals. After mixing with a hydrogen halo of $10^{11} \ M_\odot$, it gives a metallicity of a few 10^{-5} solar, close to that observed in the oldest halo stars.

References

Artymowicz, P., Lin, D.N. and Wampler, E.J.: 1993, *Astrophys. J.* **409**, 592.
Begelman, M.C., Frank, J. and Shlosman, I.: 1989, in: F. Meyer *et al.* (eds.), *Theory of Accretion disks*, Kluwer Academic Publishers.
Collin, S. and Huré, J.M.: 1998, *Astron. Astrophys.* **341**, 385.
Collin, S.: 1998, to appear in the proceedings of the Conference 'From atomic nuclei to galaxies', held in Haifa, Israel, Physics Reports, edited by O. Regev.
Collin, S., Zahn, J.-P.: 1999, *Astron. Astrophys.* **344**, 433.
Goldreich, P. and Lynden-Bell, D.: 1965, *Mon. Not. R. Astron. Soc.* **130**, 97.

Hamann, F.: 1997, *Astrophys. J. Suppl.* **109**, 279.

Krolik, J.H. and Begelman, M.C.: 1988, *Astrophys. J.* **329**, 702.

Malkan, M.A. and Sargent, W.L.W.: 1982, *Astrophys. J.* **254**, 22.

Murray, N., Chiang, J., Grossman, S.A. and Voit, G.M.: 1995, *Astrophys. J.* **451**, 498.

Rozyczka, M., Bodenheimer, P., Lin, D.N.C.: 1995, *Mon. Not. R. Astron. Soc.* **276**, 597.

Shields, G.A.: 1978, *Nature* **272**, 423.

Silk, J. and Rees, M.J.: 1998, *Astron. Astrophys.* **331**, L1.

Soltan, A.: 1982, *Mon. Not. R. Astron. Soc.* **200**, 115.

Steidel, C., Giavalisco, M., Pettini, M., Dickinson, M. and Adelberger, K.L.: 1996, *Astrophys. J.* **462**, L17.

Toomre, A.: 1964, *Astrophys. J.* **139**, 1217.

Turnshek, D.: 1995, in: Meylan (ed.), *QSO Absorption Lines*, ESO Workshop, New York, Springer.

STAR FORMATION HISTORIES FROM FAINT GALAXY COUNTS

B. ROCCA-VOLMERANGE

Institut d'Astrophysique de Paris, 98bis Bd Arago F-75014, Paris, France

M. FIOC

Goddard Space Flight Center, Greenbelt, MD 20771, USA

Abstract. Multispectral faint galaxy counts, including the deepest Hubble Deep Field, are interpreted with the help of our evolution model PEGASE (Fioc and Rocca-Volmerange, 1997). The best fits correspond to galaxy formations at high redshifts, a pure luminosity evolution and classical luminosity functions. The adopted cosmology is a flat universe with the matter density parameter $\Omega_M = 0.3$ and a cosmological constant $\Omega_\Lambda = 0.7$. A solution with $\Omega_M = 0.01$ (open universe) is also acceptable. But a flat universe with $\Omega_M = 1$ is clearly excluded. The star formation histories for galaxy types are derived from scenarios of evolution. The comparison with results already published in the litterature, arises puzzling problems needing a further analysis of star formation tracers, specifically for bright galaxies.

1. Introduction

The cosmic star formation history (SFH) is recognized to bring fundamental clues for our knowledge of galaxy evolution. This tracer of galaxy activity throughout the universe as a function of space and time brings a significant link between models and observations. In fact, the various tracers of star formation are not equivalent. Chemical abundances, nebular emission lines, far-UV to NIR stellar energy distributions (SEDs) respectively trace either a past or a current star formation activity and either young or old stellar populations of various mass ranges. The samples of redshift surveys and faint galaxy number counts can be used as tracers of populations of all ages. They are interpreted with the help of evolution scenarios. SFH is derived for each galaxy type when their population number density fit at best the observed number count distributions. Multispectral galaxy counts from the deepest surveys are observed with the most recent generation of telescopes (HST, Keck and nearly VLT). An interpretation of faint galaxy counts is proposed by Fioc and Rocca-Volmerange, 1998, (FRV98) using the model PEGASE (FRV97) (see also Pozzetti *et al.*, 1996 with GISSEL (Bruzual and Charlot, 1993)). In our modelling, a peculiar attention is given to bright counts in FRV97, solving the problem of the slope of bright counts, so that the local environment is no more galaxy deficient. The calibration of bright counts is an essential constraint to analyse the faintest ones. FRV98 also include an analysis of the galaxy luminosity function and evol-

 Astrophysics and Space Science is the original source of publication of this article. It is recommended that this article is cited as: *Astrophysics and Space Science* **265**: 507–511, 1999.
© 1999 *Kluwer Academic Publishers.*

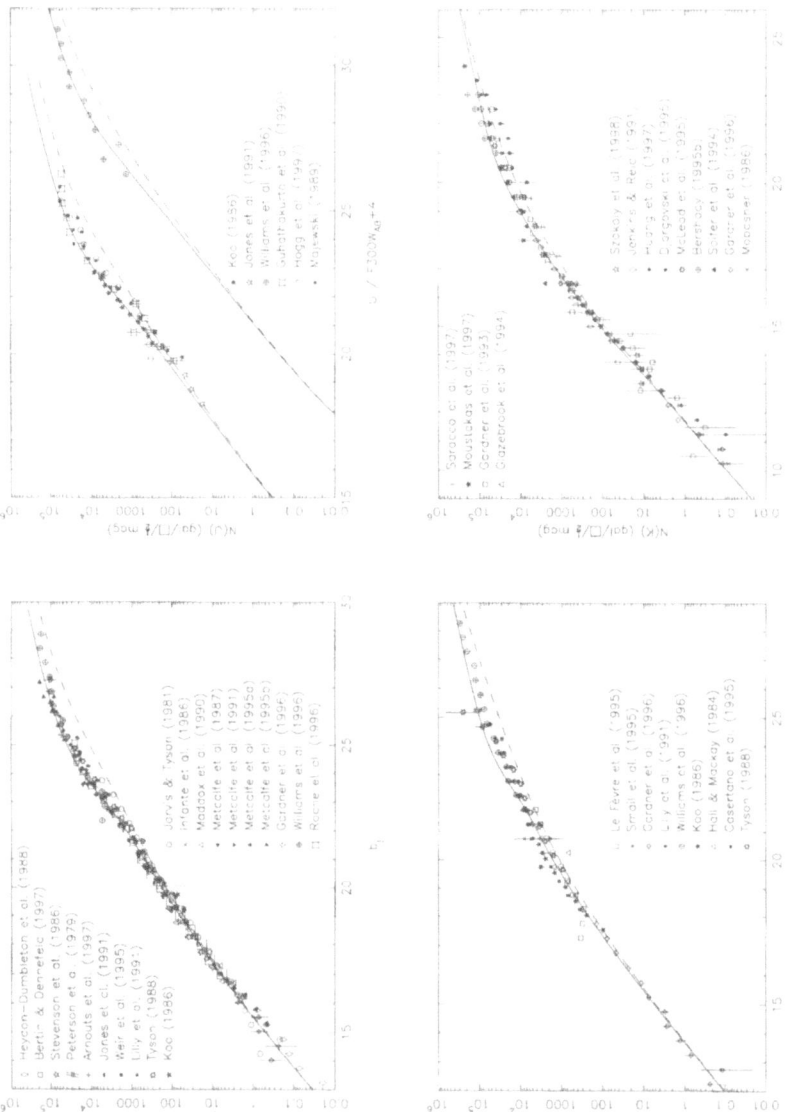

Figure 1. Faint (left) and bright (right) galaxy count models through b_J, I and K filters with a normalization at $b_J \simeq 16$ and a luminosity function from (Heyl *et al.*, 1996), compared to observations. The adopted cosmology corresponds to $H_0 = 65$ km s^{-1} Mpc^{-1}, $\Omega_0 = 0.3$ and $\Lambda_0 = 0.7$. Dashed line is the case without evolution.

ution in the far-UV (2000 Å). The authors interpret the bright counts FOCA2000 from Armand and Milliard, 1994, in coherency with the optical and near-IR galaxy counts. The debated question of the blue excess of galaxies is preferentially solved in the far-UV. The result is a specific luminosity function of very blue bursting dwarf galaxies, added to the standard luminosity function of Heyl *et al.*, 1996.

2. Modelling Faint Counts With PEGASE

The new spectrophotometric model PEGASE (FRV97), available on

ftp.iap.fr at *pub/from_users/pegase*

predicts energy distributions, including nebular emission, of eight types of galaxies. A new version with metallicity effect (Fioc and Rocca-Volmerange, 1999) is now operational and will be nearly proposed for users on the WEB. Synthetic spectra of the atlas are fitting the SEDs of the Hubble Sequence at $z = 0$. The model PEGASE is carefully built from the far-UV to the near-infrared between 0 to 20 Gyr with a time resolution of 1 million years. The most advanced phases of stellar evolution, asymptotic giant branch (AGB) with thermal pulses and post-AGB (Groenewegen, 1995; Schoenberner, 1983; Bloeker, 1995) were connected to the classical evolutionary tracks from Padova or Geneva groups describing the main sequence up to the early AGB-phase. The stellar library has a significant coverage of the Hertzsprung-Russell diagram. Eight templates of galaxies of the Hubble sequence are defined by comparison with observables at $z = 0$. Our spectrophotometric model is continuous on the largest wavelength range of stellar emission. Modelling faint counts from such a model is depending on three parameters: i) the adopted evolution scenarios of galaxies from which are computed the cosmological and evolutionary $(k + e)$ corrections applied to the $z = 0$ template spectra ii) the cosmological parameters: the Hubble constant H_0, the density parameter Ω_0 and a possible cosmological constant Λ_0 iii) the luminosity functions of galaxies, observed at $z = 0$ or higher redshifts. In the following, we adopt galaxy evolution scenarios with classical timescales of star formation, leaving open the hypothesis of number density evolution proposed by Rocca-Volmerange and Guiderdoni, 1990; Broadhurst *et al.*, 1992 for redshifts $z \leq 1$ and to be confirmed by deeper redshift surveys. On recent arguments (Turok, 1996; Ferreira *et al.*, 1998), we adopt an open universe $\Omega_0 = 0.3$ with a cosmological constant $\Lambda_0 = 0.7$ to save a flat universe. The adopted luminosity function is derived from the 2dF survey (Heyl *et al.*, 1997).

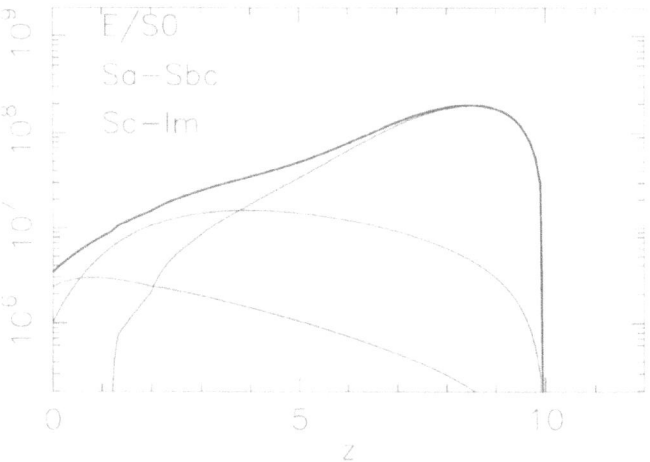

Figure 2. Star formation histories by galaxy types, derived from our model of faint galaxy counts (FRV98).

3. Star Formation Histories

The cosmic star formation histories from very deep ground-based and HST multi-band surveys gave typical results with an increase by a factor 10 from $z = 0$ to $z = 1$ and a rapid decrease for $z > 1$ (Connolly *et al.*, 1997; Madau, 1996; Treyer *et al.*, 1998; Tresse and Maddox, 1998) (the two last references give only one point). Various indicators (UV broadbands, H_α) are used by the authors. The star formation history corresponding to the fits of faint counts is shown in Figure 2. Differences with other results in the litterature are not merely explained. At low redshifts, they could be due to a population of dwarf galaxies, in relation with the poorly understood faint end of the luminosity function. But in the blue band, this population is taken into account by FRV98. An effect of fading luminosity or a drastic change in comoving space density is not needed to fit galaxy counts. For $z > 1$, the depression observed in Figure 2 is signicantly smoother than in previous results. The origin of the discrepancy is not clear and deserves a deeper comparison (Rocca-Volmerange and Fioc, in preparation).

References

Bruzual, G. and Charlot, S.: 1993, *Astrophys. J.* **405**, 538.
Broadhurst, T.J., Ellis, R. and Glazebrook, K.: 1992, *Nature* **355**, 55.
Bloecker, T.: 1995, *Astron. Astrophys.* **299**, 755.
Connolly, A.J., Szalay, A.S., Dickinson, M., Subbarao, M.U. and Brunner, R.J.: 1997, *Astrophys. J.* **486**, L11.
Fioc, M. and Rocca-Volmerange, B.: 1997, *Astron. Astrophys.* **326**, 950.
Fioc, M. and Rocca-Volmerange, B.: 1998, *Astron. Astrophys. J.*, submitted.

Groenewegen, M.A.T. and de Jong, T.: 1993, *Astron. Astrophys.* **267**, 410.

Ferreira, P.G., Juszkiewicz, R., Feldman, H.A., Davis, M. and Jaffe, A.H.: preprint.

Heyl, J., Colless, M., Ellis, R. and Broadhurst, T.: 1997, *Mon. Not. R. Astron. Soc.* **285**, 613.

Madau, P., Ferguson, H., Henry, C., Dickinson, M.E., Giavalisco, M., Steidel, C. and Fruchter, A.: 1996, *Mon. Not. R. Astron. Soc.* **283**, 1388.

Pozzetti *et al.*: 1998, *Mon. Not. R. Astron. Soc.* **298**, 1133.

Rocca-Volmerange, B. and Guiderdoni B.: 1990, *Mon. Not. R. Astron. Soc.* **247**, 166.

Schoenberner, D.: 1983, *Astrophys. J.* **272**, 708.

Tresse, L. and Maddox, S.: 1998, *Astrophys. J.* **495**, 691.

Treyer, M.A., Ellis, R.S., Milliard, B., Donas, J. and Bridges, T.: 1998, *Mon. Not. R. Astron. Soc.* **300**, 303.

Turok, N. (ed.): 1996, in: *Critical Dialogues in Cosmology*, Princeton.

Williams, R.E., *et al.*: 1996, *Astron. J.* **112**, 1335.

METALLICITY DISTRIBUTIONS OF ELLIPTICAL GALAXIES AND GLOBULAR CLUSTER SYSTEMS

Absorption Line Strength Gradients in Elliptical Galaxies

NOBUO ARIMOTO

Institute of Astronomy, School of Science, University of Tokyo, 2-21-1, Osawa, Mitaka, Tokyo 181, Japan

CHIAKI KOBAYASHI

Department of Astronomy, School of Science, University of Tokyo, 7-3-1, Hongo, Bunkyo-ku, Tokyo 113, Japan

Abstract. Gradients of absorption line indices are studied and mean stellar metallicities are estimated for 46 elliptical galaxies. The mean stellar metallicities range from $\langle[\text{Fe/H}]\rangle \simeq -0.8$ to $+0.2$ and ellipticals with smaller central velocity dispersions tend to have lower $\langle[\text{Fe/H}]\rangle$; thus the mass-metallicity relation holds not only for the galaxy center but also for the whole part of the galaxy. There is an evidence that the magnesium is enhanced systematically in all ellipticals by 0.2 dex with respect to the iron. Giant elliptical galaxies show lack of metal-poor stars (the G-dwarf problem). Metal-poor globular clusters of ellipticals formed well in advance of the formation of metal-rich ones which formed simultaneously with the bulk of stars of mother galaxies under the influence of galaxy chemical enrichment. The bimodal [Fe/H] distribution of globular clusters does not necessarily mean that elliptical galaxies formed by the mergers of disc galaxies.

1. Introduction

How elliptical galaxies formed is one of the key questions of modern astronomy. Two competing scenarios are so far proposed: Ellipticals galaxies formed monolithically by gravitational collapse of gas cloud with considerable energy dissipation (e.g., Larson, 1974; Carlberg, 1984; Arimoto and Yoshii, 1987), or ellipticals formed via mergers of relatively small star-forming galaxies (e.g., Cole *et al.*, 1994). Frustratingly, observational evidences are confusing and controversial.

In this paper we study gradients of absorption line strengths of elliptical galaxies, derive their mean stellar metallicities, and compare stellar metallicity distributions with metallicity distributions of globular cluster systems surrounding ellipticals. The line strength gradients of elliptical galaxies have been extensively studied in the last two decades (e.g., Faber, 1977; Davies, Sadler and Peletier, 1993; Gonzalez, 1993; Carollo and Danziger, 1994a,b; Fisher, Franx and Illingworth, 1995). It has been demonstrated that $Mg_{2,0}$ at the galaxy centre correlates tightly with velocity dispersion σ_0 (e.g., Davies *et al.*, 1987). Ellipticals having larger $Mg_{2,0}$ tend to have steeper Mg_2 gradients (Gorgas *et al.*, 1990; Carollo *et al.*, 1993;

Astrophysics and Space Science is the original source of publication of this article. It is recommended that this article is cited as: *Astrophysics and Space Science* **265:** 513–521, 1999.
© 1999 *Kluwer Academic Publishers.*

Gonzalez and Gorgas, 1996). Since brighter ellipticals have larger $Mg_{2,0}$ (Davies and Sadler, 1987), Gonzalez and Gorgas (1996) suggested that all ellipticals should have the same mean stellar metallicity, whatever their luminosities and masses are. If this is true, the metallicity-luminosity relation is nothing but a *local* one that holds at the galaxy centre and may not reflect major formation process of ellipticals. Therefore, it is particularly important to estimate accurately the mean stellar metallicities of ellipticals and study the relationships with other global features.

2. Line Strength Gradients

The radial gradients of Mg_2, $Fe_1(5270 \text{ Å})$, $Fe_2(5335 \text{ Å})$, and H_β are available for 105 ellipticals in the literature. However, not all the data have the same quality and some have rather poor calibrations, sky subtractions, velocity dispersion corrections, and resolution corrections. Excluding the data of poor qualities, we estimated by ourselves the gradients of Mg_2 (and other indices if available) for 46 ellipticals (Bender and Surma, 1992; Davies *et al.*, 1993; Carollo *et al.*, 1993; Hes and Peletier, 1993; Carollo and Danziger, 1994a,b). We then converted the line-strength gradients to the metallicity gradients with a help of the index-metallicity relations for 17 Gyr old single stellar populations calculated by Worthey (1994).

More than 80% of the sample show the metallicity gradients. No sign of bimodality is seen in the distribution of gradients. Mean metallicity gradient is $\Delta[Fe/H] / \Delta \log x = -0.32$, which is less steeper than $\Delta[Fe/H]/\Delta \log x = -0.5$ predicted by Carlberg (1984).

3. Mean Stellar Metallicity

We consider an elliptical galaxy of spherically symmetric shape whose surface brightness profile follows the Sersic law:

$$I(x) = I_e \exp\left[-b\left\{\left(\frac{x}{x_e}\right)^{\frac{1}{n}} - 1\right\}\right], \tag{1}$$

with

$$I_e = \frac{1}{2n\pi e^b b^{-2n} \Gamma(2n)} \frac{L_*}{x_e^2}, \tag{2}$$

where Γ is a general Gamma function and b is given as a function of n. The parameter n correlates with the galaxy luminosity; bright ellipticals have $n = 4$ profile (de Vaucouleurs' law) while fainter ellipticals tend to have smaller n with an exponential $n = 1$ profile at the faintest end (Binggeli and Jerjen, 1997).

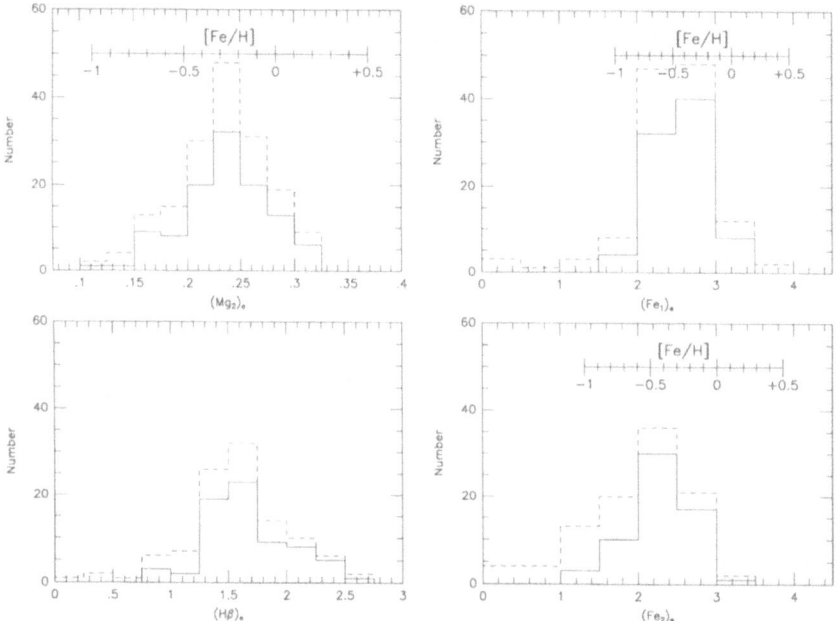

Figure 1. The distribution of the line strength indices Mg_2, Fe_1, Fe_2, and H_β at the efective radius for 105 ellipticals (dashed lines) and for 46 elliptical galaxies studied in this paper (solid lines). Metallicity scales are taken from Worthey (1994).

The observed Mg_2 gradients suggest that the projected metallicity distribution is exponential:

$$\tilde{Z}(x) = Z_e \left(\frac{x}{x_e}\right)^{-c}. \tag{3}$$

The mean stellar metallicity of the galaxy is then given by:

$$\langle[Fe/H]\rangle = \log\{b^{cn} \frac{\Gamma(2n - cn)}{\Gamma(2n)}\}[Fe/H]_e \simeq [Fe/H]_e. \tag{4}$$

For the present sample, we obtain $\langle c\rangle \simeq 0.3$. Therefore, for the typical elliptical galaxy having $n = 4$ and $c \sim 0.3$, the coefficient $b^{cn}\Gamma(2n - cn)/\Gamma(2n)$ in Equation (4) is ~ 1.1, thus the mean stellar metallicity is very close to the projected metallicity measured at one effective radius. Arimoto *et al.* (1997) already showed that this is the case for smaller sample of galaxies if de Vaucouleurs' surface brightness profile is assumed. In this study, we generalized it to the Sersic's surface brightness profile and found that so far as the slope c is smaller than 0.4, as is the case for most ellipticals, the mean metallicity is approximately given by $[Fe/H]_e$.

Figure 1 shows the four different indices measured at one effective radius x_e. The upper left panel indicates that the mean stellar metallicity of a typical elliptical galaxy is about a half solar, $\langle[Fe/H]\rangle \simeq -0.3$, if Mg_2 is used for the estimate.

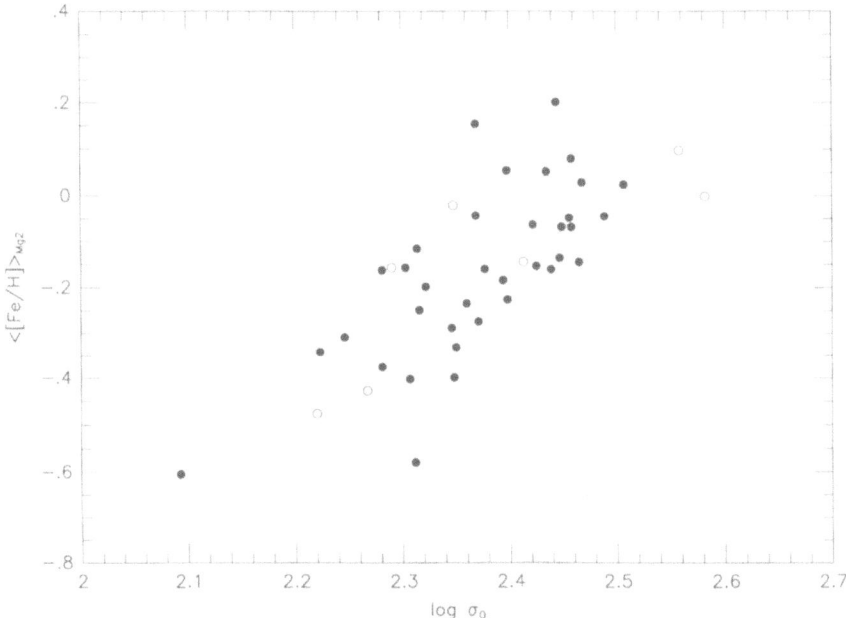

Figure 2. The mean stellar metallicity versus central velocity dispersion relation for elliptical galaxies (filled circles) and cD galaxies (open circles). ⟨[Fe/H]⟩ are calculated from Mg_2 index. The relation is tight and pararell to the $[Fe/H]_0$ versus $\log \sigma_0$ relation but is systematically lower by 0.3 dex.

On the other hand, the two right panels suggest that ⟨[Fe/H]⟩ $\simeq$ -0.5 if Fe_1 or Fe_2 is used instead. *In any case, the mean stellar metallicity of elliptical galaxy is considerablly lower than what used to be considered* (e.g., Arimoto and Yoshii, 1987).

We note that the mean metallicity of the present sample ranges from ⟨[Fe/H]⟩ $\simeq$ -0.8 to $+0.2$; thus *the mean stellar metallicity of elliptical galaxies is not universal*, contrary to what Gonzalez and Gorgas (1996) claimed. We do not find any evidence supporting that the Mg_2 gradient correlates with either $[Fe/H]_0$ or ⟨[Fe/H]⟩. Clearly a careful selection of the observational data is crucial in the study of line strength gradients.

4. Global Scaling Relations

4.1. METALLICITY-MASS RELATION

The colour-magnitude (CM) relation of ellitical galaxies and the $Mg_{2,0} - \log \sigma_0$ relation at the galaxy centre are usually interpreted as the metallicity-mass relation taking its origin at an early epoch of galaxy formation. In this section, we study if a similar relation holds for ⟨[Fe/H]⟩. Figure 2 shows a surprisingly tight correlation between ⟨[Fe/H]⟩ and σ_0. [Fe/H] is calculated from Mg_2 index. The present sample galaxies have almost an identical $Mg_{2,0} - \log \sigma_0$ relation to that of 572 elliptical galaxies observed by Davies *et al.* (1987). We should note that the $\langle[Fe/H]\rangle - \log \sigma_0$ relation has nearly the same slope to that of the $[Fe/H]_0 - \log \sigma_0$ relation but the zero point is 0.3 dex lower; i.e., *the mean stellar metallicity is about a half of the central metallicity regardless of the galaxy mass.*

4.2. MAGNESIUM ENRICHMENT

By comparing the mean metallicities given by Mg_2 and Fe_1 gradients, we studied whether the magnesium is enhanced with respect to the iron in elliptical galaxies. It has been quite often claimed that the magnesium and perphaps other α-elements are overabundant in elliptical galaxies at the galaxy centre (Worthey, Faber and Gonzalez, 1992; Davies *et al.*, 1993; Gonzalez, 1993; Fisher *et al.*, 1995). In partic-ular, Fisher *et al.* (1995) demonstrated that ellipticals with larger central velocities tend to have larger central [Mg/Fe] values. We have studied if this is also true for the mean abundances of these elements and find that the mean metallicities calculated from Mg_2 index are systematically higher than those given by Fe_1 index by ~ 0.2 dex. If (perphaps big if) Mg_2 and Fe_1 reflect the abundances of the magnesium and the iron, respectively, this implies that the magnesium is enhanced by ~ 0.2 dex with respect to the iron in whole system of the galaxy. We have checked if there is any systematic difference in the gradients of Mg_2 and Fe_1 but find no clear difference. This means that $\langle[Mg/Fe]\rangle \simeq +0.2$ at any radius in elliptical galaxies. We also note that we do not see any trend of increasing $\langle[Mg/Fe]\rangle$ towards higher $\langle[Fe/H]\rangle_{Mg_2}$; or in other words, $\langle[Mg/Fe]\rangle \simeq +0.2$ in all elliptical galaxies inde-pendent of their mass, in contrast to what was claimed for the central [Mg/Fe]. This can be most naturally explained if the star formation stopped in elliptical galaxies before the bulk of SN Ia begin to explode, as was already predicted by a galactic wind model by Arimoto and Yoshii (1987).

However, a word of caution is to be given. The magnesium enhancement might not be real. Although Fe_1 is less fragile to the velocity dispersion gradient than Fe_2, the observational errors of Fe_1 are much larger than those of Mg_2. It is not yet fully understood whether Mg_2 gives [Mg/H] and Fe_1 traces [Fe/H]. An increase of metallicity increases the opacity and changes the structure of stellar atmosphere, as a result of which the indices will be strengthened; thus Mg_2 should also be sensitive to [Fe/H]. Therefore, one should keep in mind that the ratio of Mg_2 to Fe_1 may not directly give the [Mg/Fe] ratio.

5. G-Dwarf Problem

The metallicity distributions we obtained from the line strength gradients are the projected distributions of mean stellar metallicity (Equation (3)). With a help of Abell Integral, we converted them into the three dimensional metallicity distribu-tions (3DZDs). If we assume that all stars within a spherical shell $(r, r + \Delta r)$

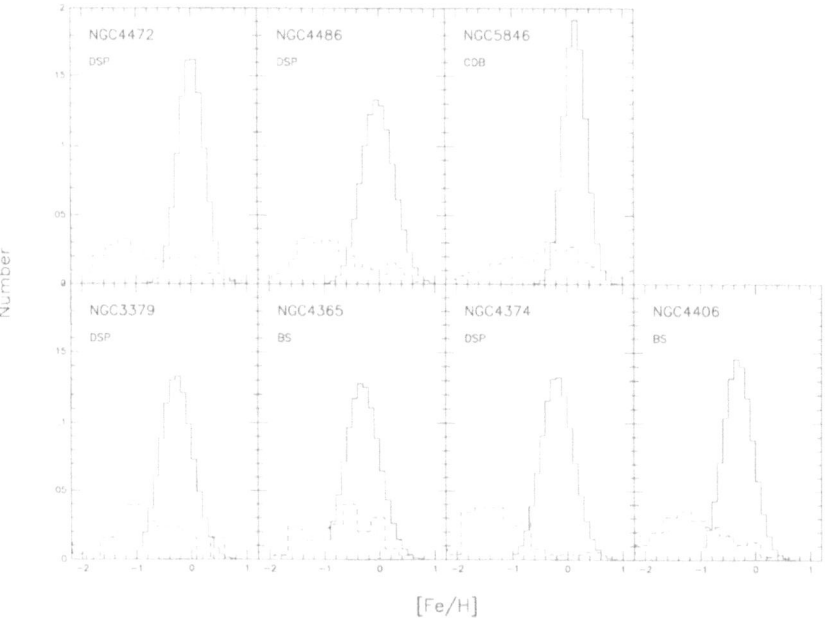

Figure 3. [Fe/H] distributions of elliptical galaxies and their globular cluster systems. The SMDFs of ellipticals are normalized to unity and the [Fe/H] distributions of GCSs are normalized to 1/10 of galaxy SMDFs.

have identical metallicity that is given by the 3DZD and that the stellar mass-to-light ratio is constant throughout the galaxy, we can calculate the number of stars in each metallicity bin (i.e., the stellar metallicity distribution function, SMDF) by using the surface brightness profile given by the Sersic law. Here we assume that all stars in ellipticals have the same mass. It may not be true that all stars in each shell have the same [Fe/H] given by the 3D-Zd. Instead, stars at each shell may have an intrinsic spread of metallicity. Since star formation would last longest at the galaxy centre, the metallicity spread is likely to become maximum at the centre. However, Greggio (1997) synthesized Mg_2 indices in elliptical galaxies and showed that the metallicity spread at the galaxy centre is negligiblly small. In the present study, we introduced the dispersion $\sigma_{[Fe/H]} = 0.2$ dex, taken from the metallicity distribution function of G-dwarfs in the solar neighbourhood disc (Pagel, 1989), for the whole part of a galaxy and convolved the SMDFs.

Figure 3 shows the resulting SMDFs for seven ellipticals NGC3379, NGC4365, NGC4374, NGC4406, NGC4472, NGC4486, and NGC5846. The SMDF of ellipticals has a single gaussian profile. The location of the peak gives directly the *effective* yield of chemical enrichment. The effective yields for giant ellipticals (e.g., NGC4472) are nearly solar, suggesting a flat IMF ($x \simeq 1.1$) fo these galaxies. The effective yields decrease towards fainter ellipticals, probablly due to a feedback from galactic winds. The mean dispersion of [Fe/H] distribution is $\sigma_{[Fe/H]} \simeq$

0.30 ± 0.05 and is almost independent of the central velocity dispersion. The stellar metallicity in NGC4472 is in the range of $-0.7 \leq$ [Fe/H] $\leq +0.7$. Stars at the galaxy centre have extremely high metallicity. Clearly, giant elliptical galaxies lack metal-poor stars, thus the G-dwarf problem of elliptical galaxies do exist.

6. Globular Cluster Systems

Ellipticals have a significant number of globular clusters. The specific number frequency of globular clusters is at least twice larger than that of disc galaxies (Harris, 1991). Globular cluster systems (GCSs) of many elliptical galaxies show bimodal metallicity distributions (e.g., Forbes, Brodie and Grillmair, 1997). Why GCSs are so numerous in elliptical galaxies and why GCSs have the bimodal [Fe/H] distributions? At which stage of galaxy formation the GCSs formed? To answer these questions, we compare the SMDFs of elliptical galaxies and the [Fe/H] distributions of their surrounding GCSs. The results are illustrated in Figure 3. The GCS data are taken from Ajhar, Blakeslee and Tonry (1994) for NGC3379, NGC4365, NGC4374, and NGC4406, Geisler, Lee and Kim (1996) for NGC4472, and Lee and Geisler (1993) for NGC4486. The number of globular clusters is normalized to 1/10 of the SMDF of the galaxy. Except for NGC4374, all ellipticals show evidences for the bimordal [Fe/H] distribution in their GCSs. So far as these 6 galaxies are concerned, the metal-rich GCSs have very similar [Fe/H] distributions to their mother galaxies, while the metal-poor GCSs have the [Fe/H] distributions entirely shifted towards the lower metallicities and are nearly identical in all galaxies. This suggests that the metal-poor GCSs formed before the onset of major star formation in elliptical galaxies, and thus their metallicity distributions are independent of chemical enrichment of the mother galaxies. On the other hand, the metal-ich GCSs seem to have formed at the same time when the bulk of stars of the galaxy were born under an influence of chemical enrichment of the galaxy itself. This can be explained by a dissipational collapse picture of galaxy formation. The bimodal metallicity distribution of GCSs does not require the mergers of disc galaxies as the origin of ellipticals.

7. Conclusions

Spectroscopic measurements of metallic lines are rather difficult tasks to perform, and it is crucial to use reliable observational data alone to derive any global scaling relations.

The Mg_2 gradient distribution of elliptical galaxies is a single gaussian with a peak at Δ[Fe/H]$/\Delta \log x = -0.3$, and does not show any bimodial distribution.

The mean stellar metallicities of elliptical galaxies range from $\langle$[Fe/H]$\rangle \simeq -0.8$ to $+0.2$ and is not universal. There is a tight relationship between $\langle$[Fe/H]$\rangle$ and the central velocity dispersion $\log \sigma_0$, suggesting that the CM relation holds not only for stars at the galaxy centre, but also for all stars in a galaxy. The CM relation for global galaxy is systematically shifted to lower metallicity at any fixed luminosity by 0.3 dex. This would reduce the contribution of elliptical galaxies to the ICM enrichment.

The stellar metallicity distribution of elliptical galaxies has a single gaussian profile. The mean dispersion of [Fe/H] distribution is $\sigma_{[Fe/H]} \simeq 0.3$ and is almost independent of the central velocity dispersion. Clearly, giant elliptical galaxies lack the metal-deficient stars, thus the G-dwarf problem of elliptical galaxies exists.

The metal-poor GCSs of ellipticals are primary and formed before the onset of major star formation in their mother galaxy. The metal-rich GCSs show very similar [Fe/H] distributions to the galaxies and thus should be secondary and formed under the influence of chemical enrichment of ellipticals themselves. The bimodal [Fe/H] distribution of GCSs is consistent with the dissipative collapse picture of galaxy formation and the merger scenario is not necessary required to explain it.

References

Ajhar, E.A., Blakeslee, J.P. and Tonry, J.L.: 1994, *Astron. J.* **108**, 2087.

Arimoto, N., Matsushita, K., Ishimaru, Y., Ohashi, T. and Renzini, A.: 1997, *Astrophys. J.* **477**, 128.

Arimoto, N. and Yoshii, Y.: 1987, *Astron. Astrophys.* **173**, 23.

Bender, R. and Surma, P.: 1992, *Astron. Astrophys.* **258**, 250.

Binggeli, B. and Jerjen, H.: 1997, *astro-ph/9704027*.

Carlberg, R.G.: 1984, *Astrophys. J.* **286**, 403.

Carollo, C.M., Danziger, I.J. and Buson, L.: 1993, *Mon. Not. R. Astron. Soc.* **265**, 553.

Carollo, C.M. and Danziger, I.J.: 1994a, *Mon. Not. R. Astron. Soc.* **270**, 523.

Carollo, C.M. and Danziger, I.J.: 1994b, *Mon. Not. R. Astron. Soc.* **270**, 743.

Cole, S., Aragón-Salamanca, A., Frenk, C.S., Navarro, J.F. and Zepf, S.E.: 1994, *Mon. Not. R. Astron. Soc.* **271**, 781.

Davies, R.L., Burstein, D., Dressler, A., Faber, S.M., Lynden-Bell, D., Terlevich, R.J. and Wegner, G.: 1987, *Astrophys. J. Suppl.* **64**, 581.

Davies, R.L. and Sadler, E.M.: 1987, in: T. de Zeeuw (ed.), *Structure and Dynamics of Elliptical Galaxies, IAU Symp.* **127**, Reidel, Dordrecht, p. 441.

Davies, R.L., Sadler, E.M. and Peletier, R.F.: 1993, *Mon. Not. R. Astron. Soc.* **262**, 650.

Faber, S.M.: 1977, in: B.M. Tinsley and R.B. Larson (eds.), *The Evolution of Galaxies and Stellar Populations*, Yale Univ. Press, New Heaven, p. 157.

Fisher, D., Illingworth, G. and Franx, M.: 1995, *Astrophys. J.* **438**, 539.

Forbes, D.A., Brodie, J.P. and Grillmair, C.J.: 1997, *Astron. J.* **113**, 1652.

Geisler, D., Lee, M.G. and Kim, E.: 1996, *Astron. J.* **111**, 1529.

Gonzalez, J.J.: 1993, *PhD thesis*, Univ. of California.

Gonzalez, J.J. and Gorgas, J.: 1996, in: A. Buzzoni, A. Renzini and A. Serrano (eds.), *Fresh Views of Elliptical Galaxies, ASP Conference Series* **86**, p. 225.

Gorgas, J., Efstathiou, G. and Aragón-Salamanca, A.: 1990, *Mon. Not. R. Astron. Soc.* **245**, 217.

Greggio, L.: 1997, *Mon. Not. R. Astron. Soc.* **285**, 151.

Harris, W.E.: 1991, *Annu. Rev. Astron. Astrophys.* **29**, 543.

Hes, R. and Peletier, R.F.: 1993, *Astron. Astrophys.* **268**, 539.

Larson, R.B.: 1974, *Mon. Not. R. Astron. Soc.* **169**, 229.

Lee, M.G. and Geisler, D.: 1993, *Astron. J.* **106**, 493.

Pagel, B.E.J.: 1989, in: J.E. Beckman and B.E.J. Pagel (eds.), *Evolutionary Phenomena in Galaxies*, Cambridge University Press, p. 201.

Worthey, G.: 1994, *Astrophys. J. Suppl.* **95**, 107.

Worthey, G., Faber, S.M. and Gonzalez, J.J.: 1992, *Astrophys. J.* **398**, 69.

FIELD STARS AND STAR CLUSTERS IN M31

P. JABLONKA

DAEC-URA173, Observatoire de Paris-Meudon, Place Jules Janssen, F-92195 Meudon, France

Abstract. This review reports on recent results obtained from HST/WFPC2 high resolution observations of field stars and star clusters in the bulge and the halo of the nearby galaxy M31.

1. Introduction

Globular cluster systems are often used as tools to probe star formation histories in galaxies, since their high surface brightness make them distinguishable at larger distances than individual field stars, this particularly from the ground. There has been a long standing consensus, implying that globular clusters could not reach as high as a chemical enrichment as field stars, the latter being the ancestors of the former. A number of recent observations and analyses have now challenged this conventional view and open a new area where both field stars and star clusters are complementary in addressing the process of formation and evolution of galaxies.

The globular cluster system of M31 is one of the best studied systems in external galaxies. However, most of our knowledge comes mainly from cluster integrated properties, not from their individual stars. Only very few attempts have been made from the ground to get the Color-Magnitude Diagrams (hereafter CMDs) of the brightest of these globular clusters. Even with some of the best ground-based facilities, Heasley *et al.* (1988) and Christian and Heasley (1991) were only able to reach, at the CFHT, the upper part of the red giant branch. It is only recently, with the refurbished HST, that it has become possible to build CMDs of halo clusters, reaching 1 mag fainter than the horizontal branch (Ajhar *et al.*, 1996; Fusi Pecci *et al.*, 1994).

Having as a goal the study of the origin of the bulges of spiral galaxies, we obtained, during Cycles 5 and 6, Hubble Space Telescope WFPC2 images with the F555W and F814W filters. Our targets were three fields centered on super-metal-rich star clusters in the bulge of M31, namely G170, G177 and G198 (Jablonka *et al.*, 1992, 1998). Adopting 1 arcmin = 250 pc, their angular separations correspond to projected distances from the M31 center of about 1.55, 0.80, and 0.92 kpc, respectively. In order to secure the quality of our photometry and to establish a relative scale for our analysis, we took also, with the same filters, images of the halo globular cluster G1 ($\equiv$ Mayall II), located 39 kpc away from the center of M31 and equally previously observed by Rich *et al.* (1996). The metallicity of G1

Astrophysics and Space Science is the original source of publication of this article. It is recommended that this article is cited as: *Astrophysics and Space Science* **265**: 523–529, 1999.
© 1999 *Kluwer Academic Publishers.*

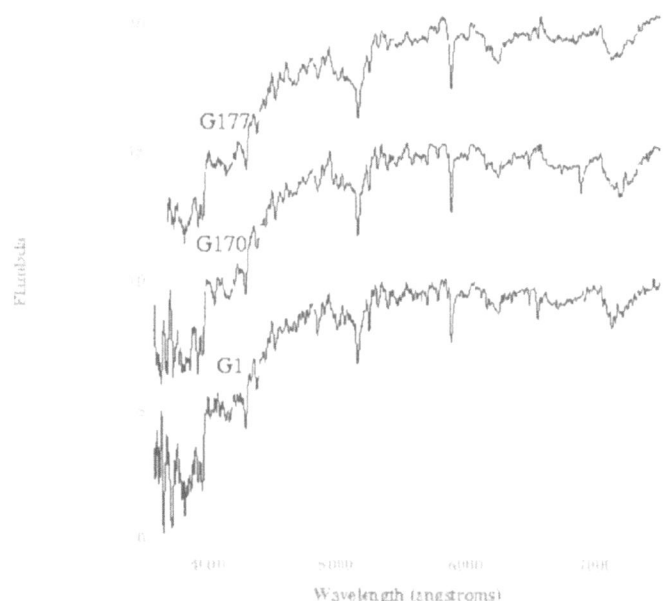

Figure 1. The ground-based integrated spectra of the three M31 globular clusters G1, G170, and G177.

falls in a non-conflictual regime (below solar, similar to the one of 47 Tuc) and its location, away from the M31 galaxy center, avoids any possible complication due to crowding.

We display in Figure 1, the integrated spectra of each of the globular clusters G1, G170, and G177, as obtained in a previous program (Jablonka *et al.*, 1992, 1998). These data show that G170 and G177 are definitely among the most metal-rich globular clusters known, with spectral features as deep as those of stars at the center of elliptical galaxies.

The CMDs deduced from our new HST/WFPC2 data allow us to address the stellar population properties not only of these clusters but also of the field bulge stars surrounding them.

2. Field Stars and Star Clusters in The Bulge

2.1. FIELD STARS

After standard recalibration of our HST data, we used the ALLFRAME software package for crowded-field photometry (Stetson, 1994), along with the PSFs kindly provided to us by the Cepheid-Distance-Scale HST key project. The F555W and F814W instrumental magnitudes were then converted to Johnson *V* and Cousins *I* magnitudes (Hughes *et al.*, 1998). See Jablonka *et al.* (1999) for further details

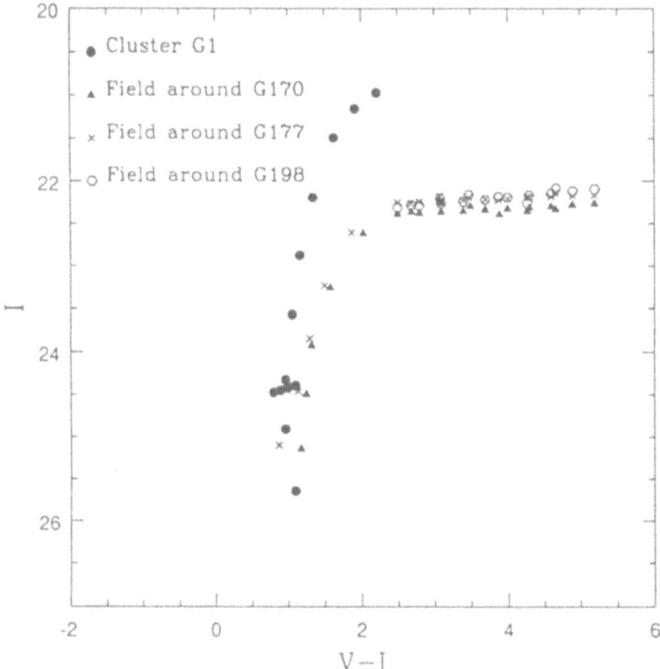

Figure 2. The mean locii of color-magnitude diagrams of the cluster G1 and of the field stars around G170, G177, and G198.

about our observations and data reduction. Between $\sim 45\,000$ and $\sim 65\,000$ bulge stars were detected in the PC frame only, depending on the cluster field.

Rich *et al.* (1996) have shown that G1 upper CMD sequences are perfectly well matched by 47 Tuc's ones. In Figure 2, the I vs. $V - I$ plan displays the mean locii of the CMDs for the three bulge fields compared to the mean locus of the CMD of G1, as observed with the PC. The level of the horizontal branch at the distance of M31 is indicated along the G1 mean locus.

Our goal here is to present, on a relative scale fixed by G1, the *mean* properties of the bulge stellar population at the location of the observed fields. Their Red Giant Branch (RGB) morphology, far more bended towards the lower right than for G1, is characteristic of an old and very metal-rich – approximately solar – population. As will be discussed in more details in a forthcoming paper, the dispersion around this mean metallicity is broad.

The high chemical enrichment in these three bulge stellar fields, likely reached within a rather short period of time, – bulge and halo stars have indistinguishable ages, as also reported by Ortolani *et al.* (1995) in the case of our Galaxy – certainly pleads for a spiral galaxy formation scenario where the bulge is formed at the very early stages.

Figure 3. Square regions of 64 by 64 pix centered on the G170 globular cluster (top) and on a control field away from the cluster. The corresponding deconvolved images with the new procedure are on the right.

2.2. STAR CLUSTERS

The acquisition of the photometry of the field bulge stellar population is already a challenge for any kind of photometric analysis software, due to the extremely high density of field stars. This is even more true for the cluster stars. The acquisition of the photometry of the cluster stars were done in parallel, using the above ALLFRAME analysis and a new image deconvolution procedure.

Image deconvolution is one of the techniques which has been most intensively studied and used for decontaminating images from blurring by the instrumental Point Spread Function. However, none of the algorithms proposed so far has led to fully satisfactory photometric results. They tended to produce the so-called 'ringing effects', i.e., artefacts which prevented to perform quantitative measurement on the deconvolved images. A new deconvolution algorithm has been recently proposed (Magain *et al.*, 1998). It is based on the recognition of the fact that the sampled data can not be fully deconvolved without violating the sampling theorem. With this new algorithm, the images are not deconvolved by the total PSF but rather by a narrower one, so that the deconvolved image is properly sampled. In the case of our M31 data, the choice of the spatial resolution is 2 pixels FWHM, where 1 pixel =

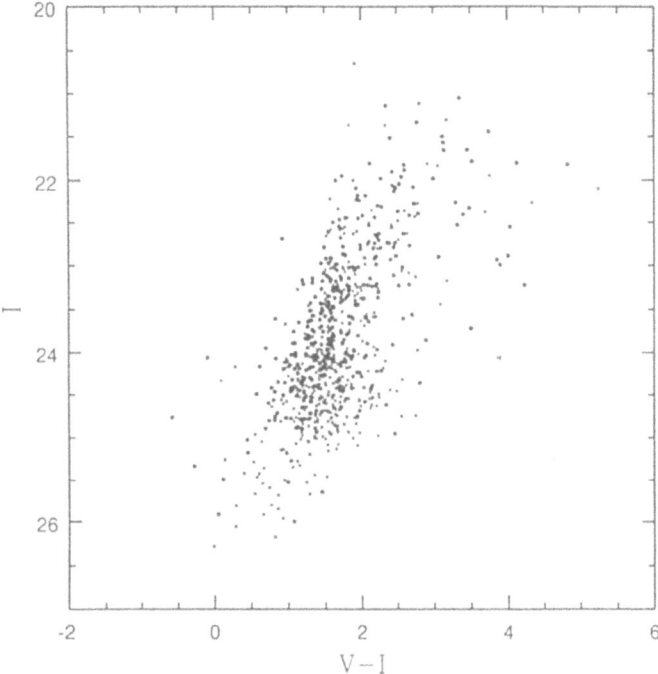

Figure 4. The color-magnitude diagrams for the deconvolved images. Dots represent the cluster stars, while crosses represent the field stars.

0.0455/2 arcsec = 0.0227 arcsec. Stars after deconvolution have a known gaussian shape. The background is smoothed at the scale of the deconvolution resolution. In Figure 3, we present square regions of 64 by 64 pixels, one centered on the cluster G170 (upper panels) and another, on a 'control' field located well outside the region of the cluster (lower panels). On the left are the original PC images and on the right the corresponding results after deconvolution.

Figure 4 displays the result of the photometry on the deconvolved images. Crosses represent the stars in the control field, while dots are the identified stars in the cluster. Clearly, within the precision of our photometry, one cannot distinguish between the CMDs from the field and from cluster stars, confirming the clusters as having metallicities similar to genuine bulge objects.

3. An Outstanding Cluster in the Halo

Our high-quality HST/WFPC2 data revealed that G1 is more than a interesting calibrator.

The KECK/HIRES spectrum of G1 permitted to derive its central velocity dispersion $\sigma_{obs} = 25.1$ km s^{-1} (Djorgovski *et al.*, 1997). By fitting G1 surface brightness profile from our HST/WFPC2 images, we obtained its core and tidal radii, 0.9 pc and 200 pc, respectively, leading to a $c = 2.35$ concentration.

Two simple estimates of the total mass of this cluster can be obtained: first, the total King-model mass $M = 14 \times 10^6 \, M_\odot$, second, the total Virial mass $M = 6.8 \times 10^6 \, M_\odot$. By using a King-Michie model fitted simultaneously to the surface brightness profile and the central velocity dispersion value, and recovering the total integrated absolute luminosity $M_V = -10.55$ mag, a grid of about 10 000 models was calculated for a wide range of values of each parameter. Total mass estimates range from $10 \times 10^6 \, M_\odot$ to $18 \times 10^6 \, M_\odot$. Clearly, all mass estimates give G1 more than twice as massive as ω Centauri, the most massive Galactic globular cluster. G1 is significantly more concentrated than ω Cen, which has $c = 1.24$, and 47 Tuc, another massive Galactic globular cluster, which has $c = 2.04$. Although G1 is the heaviest of the weighted clusters, it would be a hasty conclusion to claim that G1, even more than ω Cen, is a kind of transition step between globular clusters and dwarf elliptical galaxies. When considering the positions of G1 in the different diagrams defined by Kormendy (1985) (involving the parameters μ, r_c, M_V, and $\sigma_p(0)$), G1 always appears close to the sequences defined by the globulars, and away from the sequences defined by elliptical galaxies, bulges, or dwarf spheroidal galaxies. Consequently, although being very bright and very massive, G1 is a genuine globular cluster.

In its color-magnitude diagram, the RGB of G1 displays a potentially significant color width. The statistical significance of this width has been addressed. We conducted artificial star experiments in order to estimate the true measurement error. Then, assuming that the intrinsic width we calculated was due to a metallicity dispersion, we estimated this quantity by using the standard RGB sequences of Da Costa and Armandroff (1990), constructing a relation between $[Fe/H]$ and $(V - I)_0$. The resultant relation depends on the reddening we assume. We independently determined, $E(V - I) = 0.05$ and thus get a significant metallicity dispersion of $\sigma_{[Fe/H]} = \pm 0.39$ dex. A number of investigators (see Norris et al., 1997) have concluded that ω Cen exhibits a substantial metallicity range among its member stars. Some estimates place this range as high as 1.5 dex in [Fe/H]. This is thought to be the result of self-enrichment by Type-Ia supernovae whose ejecta were retained by ω Cen's substantial gravitational potential; this remains true even in the case of a complicated formation scenario with merger(s) of primordial protocluster subclumps. Since G1 is even more luminous than ω Cen, such a metallicity range can indeed be expected among its stars.

Acknowledgements

I warmly thank my collaborators, T. Bridges, F. Courbin, G. Meylan and A. Sarajedini.

References

Ajhar, E.A, *et al.*: 1996, *Astron. J.* **111**, 1110.

Christian, C.A. and Heasley, J.N.: 1991, *Astron. J.* **101**, 848.

Da Costa, G.S. and Armandroff, T.E.: 1990, *Astron. J.* **100**, 162.

Djorgovski, S.G., *et al.*: 1997, *Astrophys. J.* **474**, L19.

Fusi Pecci, F., *et al.*: 1994, *Astron. Astrophys.* **284**, 349.

Heasley, J.N, C.A, Friel, E.D. and Janes, K.A.: 1988, *Astron. J.* **94**, 1312.

Hughes, S.M.G., *et al.*: 1998, *Astrophys. J.* **501**, 32 (astro-ph/9802184).

Jablonka *et al.*: 1999, to be published in *Astrophys. J. Lett.*

Jablonka, P., Bica, E., Bonatto, C., Bridges, T.J., Langlois, M. and Carter, D.: 1998, *Astron. Astrophys.* **335**, 867.

Jablonka, P., Alloin, D. and Bica, E.: 1992, *Astron. Astrophys.* **260**, 97.

Kormendy, J.: 1985, *Astrophys. J.* **295**, 73.

Magain, P., Courbin, F. and Sohy S.: 1998, *Astrophys. J.* **494**, 472.

Norris, J.E., Freeman, K.C., Mayor, M. and Seitzer, P.: 1997, *Astrophys. J.* **487**, L187.

Ortolani, S., *et al.*: 1995, *Nature* **377**, 70.

Rich, R.M.: 1996, *Astron. J.* **111**, 768.

Stetson, P.: 1994, *Publ. Astron. Soc. Pacific* **106**, 250.

UV SPECTROSCOPIC DATING OF STARS AND GALAXIES

SARA HEAP, THOMAS BROWN and THIERRY LANZ
NASA Goddard Space Flight Center, Greenbelt, MD 20771, USA

SUKYOUNG YI
CalTech, Pasadena CA, USA

We report on new stellar models and spectral analyses carried out for the purpose of dating high-redshift galaxies from their rest-frame UV spectra. The principle behind UV spectrscopic dating is simple. For passively evolving stellar populations younger than about 7 Gyr, the hottest stars are at the main sequence turnoff (MSTO). These stars are thus the only source of significant flux in the mid-ultraviolet ($2000 - 3200$ Å) spectra of such galaxies; in fact, the spectra are remarkably similar to that of a single F-star. By determining the effective temperature of the MSTO stars, we can measure the age of the population.

This principle was first applied by Dunlop *et al.* (1996) and Spinrad *et al.* (1997) to LBDS 53W091, a very red galaxy at a redshift of $z = 1.55$. Observational evidence (colors, morphology, etc.) indicates that 53W091 is a normal giant elliptical galaxy, so its red color is likely due to an aged stellar population. Spinrad *et al.* (1997) used archival IUE low-dispersion spectra (6 Å) of F-type stars to calibrate the flux breaks at $\lambda 2640$ and $\lambda 2900$ in the spectrum of the galaxy in terms of age (T_{eff}). Their result, $t = 3.5$ Gyr since the last major episode of star formation, puts an upper limit on the Hubble constant, $H_o < 45$ km s^{-1} Mpc^{-1} for an Einstein-de Sitter universe. This value is lower than other modern estimates of H_o.

Shortly thereafter, Heap *et al.* (1998) derived a younger age for 53W091, $t = 2$ Gyr for a solar metallicity, based on a STIS high-resolution ($R = 30\,000$) spectrum of the F-type star, 9 Comae. The STIS observation is part of a HST program (ID = 7433) to observe and analyze spectra of F-type stars of known atmospheric properties (Edvardsson *et al.*, 1993) and distance (Hipparcos). At this time, about one-third of the observations have been completed.

Part of the difficulty in deriving reliable ages is the problem of separating temperature and metallicity. Strictly speaking, neither the Berkeley (Spinrad *et al.*) nor Goddard (Heap *et al.*) groups' age estimate can be correct, because both groups assumed a single metallicity for the stellar population. It is more likely that a giant elliptical has a range of metallicities. To pursue this point, Yi (1999) has calculated new evolutionary models allowing for a composite metallicity according to Kodama and Arimoto's (1997) infall model. Using Kurucz' (1993) $Z-$dependent SED's, he finds that both the mid-UV spectral breaks and the restframe optical

Astrophysics and Space Science is the original source of publication of this article. It is recommended that this article is cited as: *Astrophysics and Space Science* **265**: 531–532, 1999.
© 1999 *Kluwer Academic Publishers.*

colors point to an age, $t = 1.6 \pm 0.2$ Gyr. He also finds that most of the mid-UV flux comes from stars of sub-solar metallicity ($Z < 0.01$), in contrast to the optical flux ($\lambda > 4000$ Å) whose major contributors are high-metallicity stars ($Z > 0.03$).

We have also worked to improve our analyses of F-type spectra. Our spectral models clearly show that the prime age diagnostic, the spectral break at $\lambda 2640$, is degenerate: the higher the metallicity, the higher the temperature and the lower the age inferred from this spectral break. It is not possible to derive the population age from this diagnostic alone. Lanz *et al.* (1999), however, find other spectral indices that show great promise for breaking the $T_{eff} - Z$ degeneracy. The best set of mid-UV spectral indices appears to be the strength of the Mg I $\lambda 2852$ resonance line (becomes weaker with increasing temperature as Mg becomes ionized) and the mid-UV color, [2310]–[3040] (becomes 'bluer' with decreasing metallicity and increasing temperature). Our LTE models nicely match the observed Mg I and UV colors of the program stars if we assume Gray's (1992) formula for van der Waal's broadening when we compute the strength of Mg I $\lambda 2852$. We plan to take advantage of these new spectral indices to date LBDS 53W091 and other high-redshift galaxies.

References

Dunlop, J., Peacock, J., Spinrad, H., *et al.*: 1996, *Nature* **381**, 481.

Edvardsson, B., Andersen, J., Gustafsson, B., *et al.*: 1993, *Astron. Astrophys.* **275**, 101.

Gray, D.F.: 1992, *The Observation and Analysis of Stellar Photospheres*, 2nd ed., Cambridge University Press.

Heap, S., Brown, T., Hubeny, I., Landsman, W., Yi, S., *et al.*: 1998, *Astrophys. J.* **492**, L131.

Lanz, T., *et al.*: 1998, *Bull. Am. Astron. Soc.* **30**.

Kodama, T. and Arimoto, N.: 1997, *Astron. Astrophys.* **320**, 41.

Kurucz, R.L.: 1993, *CD-ROM series*, SAO, Cambridge.

Spinrad, H., Dey, A., Stern, D., *et al.*: 1997, *Astrophys. J.* **484**, 581.

Yi, S.: 1999, in: B. Guiderdoni *et al.* (eds.), *The Birth of Galaxies*, in press.

YOUNG STELLAR POPULATIONS AT HIGH Z: WHERE ARE THEY?

B. GUIDERDONI

Institut d'Astrophysique de Paris, CNRS, 98bis boulevard Arago, F-75014 Paris, France

Abstract. Our view on the deep universe is strongly biassed towards the starlight that directly escapes from high-redshift galaxies, since we know very little on the fraction of luminosity absorbed by dust. Attempts to correct for this effect directly from the slope of the UV spectra seem to suggest that a significant fraction of the UV flux is extinguished. New constraints are now set on the fraction of dust-enshrouded young stellar populations by the detection of the Cosmic Infrared Background, and faint galaxy counts at IR and submm wavelengths. We briefly review the observations and use a semi-analytic model of galaxy formation and evolution to predict number counts consistent with the background.

1. Introduction

Recent observational breakthroughs have made possible the measurement of the Star Formation Rate (SFR) history of the universe from rest-frame UV fluxes of moderate- and high-redshift galaxies (Madau *et al.*, 1996, 1999). The strong peak observed at $z \sim 1.5$ seems to be correlated with the decrease of the cold-gas comoving density in damped Lyman-α systems between $z = 2$ and $z = 0$ (Storrie-Lombardi *et al.*, 1996), and suggests that we could have seen the bulk of star formation in the universe. These results nicely fit in a view where star formation in bursts triggered by interaction/merging consumes and enriches the gas content of galaxies as time goes on. Indeed, such a scenario is qualitatively predicted within the paradigm of hierarchical growth of structures in which galaxy formation is a continuous process.

However, we have only a partial view on galaxy evolution since these observational data come from optical surveys which probe the rest-frame UV and visible emission of high-z galaxies. A still unknown fraction of star/galaxy formation is hidden by dust that absorbs UV/visible starlight and re-radiates at larger wavelengths. Could most of the young stellar populations at high z be buried in dust shrouds ? We hereafter briefly review the attempts to correct the UV fluxes for the effect of extinction, and recent observations of the deep universe at IR/submm wavelengths. These observations strongly suggest that a significant fraction of the young stellar populations is hidden by dust. Finally, we propose a semi-analytic modelling of galaxy formation and evolution in which the computation of dust extinction and emission is explicitly implemented.

 Astrophysics and Space Science is the original source of publication of this article. It is recommended that this article is cited as: *Astrophysics and Space Science* **265**: 533–540, 1999.
© 1999 *Kluwer Academic Publishers.*

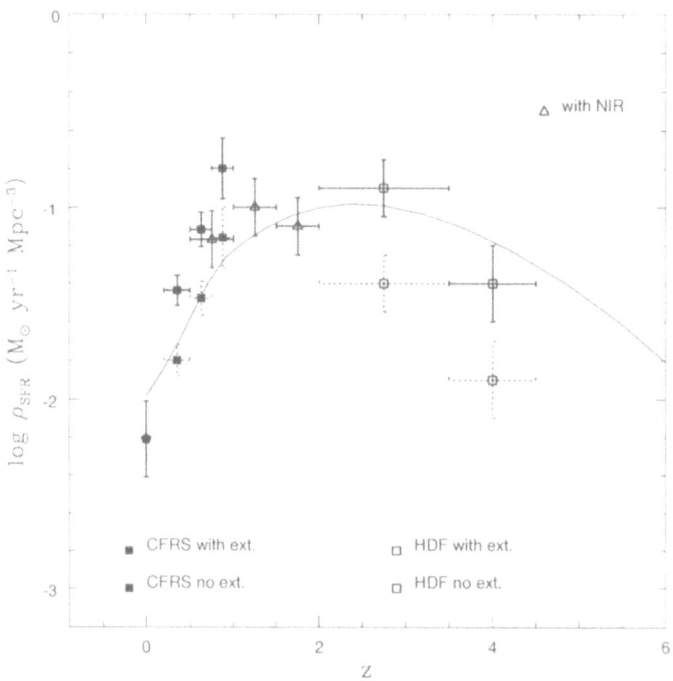

Figure 1. The evolution of the cosmic Star Formation Rate comoving density ρ_{SFR} with redshift z. For the Canada-France Redshift Survey ($z < 1$), the solid dots and error bars drawn with dotted lines and solid lines respectively give values uncorrected for extinction (Lilly *et al.*, 1996), and values estimated from a multi-wavelength analysis including IR, submm, and radio data (Flores *et al.*, 1999). For the Hubble Deep Field, the open squares and error bars drawn with dotted lines and solid lines respectively give values uncorrected for extinction (Madau *et al.*, 1996, 1998), and values corrected for an average $\langle E(B - V) \rangle = 0.09$ and the SMC extinction curve (Pettini *et al.*, 1997). The corrections derived by Meurer *et al.* (1997) would shift the corrected points upwards by ~ 0.5 dex. Finally, the open triangles show values determined from the HDF by a method with photometric redshifts involving visible and near-IR photometry (Connolly *et al.*, 1997). The rest-frame UV fluxes are converted into SFRs with the spectrophotometric model described in Guiderdoni *et al.* (1998). The solid line shows the best model in the latter paper (the so-called 'model E').

2. The Cosmic SFR Seen Through the Optical Window

The cosmic SFR comoving density at $z < 1$ is given by the Canada France Redshift Survey (CFRS). The optical observations probe the rest-frame UV flux that is readily converted into a SFR, *under an assumption on the IMF* (hereafter we take the standard Salpeter index $x = 1.35$). The study of this magnitude-limited sample leads to the measurement of the cosmic SFR density, once a correction for the low-luminosity end of the luminosity function is taken into account (Lilly *et al.*, 1996). Spectroscopic redshifts are usefully complemented by photometric redshifts at $z > 1$, and especially by the powerful UV drop-out technique (Steidel and Hamilton, 1993; Steidel *et al.*, 1996, 1999) that allows one to identify galaxies at redshift 3 and beyond. Madau *et al.* (1996, 1998) have applied this method to the galaxies of the Hubble Deep Field.

Since the early version of the reconstruction of the cosmic SFR density (the so-called 'Lilly-Madau' plot), much work has been done to address the luminosity budget of star-forming galaxies. The cosmic SFR density determined only from UV observations of the CFRS has been recenlty revisited with a multi-wavelength approach including IR, submm, and radio observations. The result is an upward correction of the previous values by an average factor 2.3 (Flores *et al.*, 1999). At higher redshift, various authors have attempted to estimate the extinction correction and to recover the fraction of UV starlight absorbed by dust (e.g. Meurer *et al.*, 1997; Pettini *et al.*, 1997). It turns out that the observed slope β of the UV spectral energy distribution $F_\lambda(\lambda) \propto \lambda^\beta$ (say, around 2200 Å) is flatter than the standard value computed from models of spectrophotometric evolution $\beta_0 \simeq -2.5$. The derived extinction corrections are large, and differ according to the method. Pettini *et al.* (1997) fit a typical extinction curve (the Small Magellanic Cloud one) to the observed colours and derive $\langle E(B - V) \rangle \simeq 0.09$ resulting in a factor 2.7 absorption at 1600 Å. Meurer *et al.* (1997) use an empirical relation between β and the IR to 2200 Å luminosity ratio in local starbursts. They derive $\langle E(B - V) \rangle \simeq 0.30$ resulting in a factor 10 absorption at 1600 Å. This discrepancy suggests something like a bimodal distribution of the young stellar populations: the first method would take into account the stars detected in the UV with relatively little reddening/extinction, while the second one would phenomenologically add the contribution of heavily-extinguished stars.

Figure 1 shows the cosmic SFR comoving density in the early version (no extinction), and after the work by Flores *et al.* (1999) at $z < 1$ and the extinction correction derived by Pettini *et al.* (1997) at higher redshift.

3. The Cosmic Infrared Background and the Submm Counts

In the submm range, Puget *et al.* (1996) have discovered an isotropic component in the COBE/FIRAS residuals between 200 μm and 2 mm. This measurement has been confirmed by subsequent work in the cleanest regions of the sky (Guiderdoni *et al.*, 1997), and by an independent determination (Fixsen *et al.*, 1998). The analysis of the COBE/DIRBE dark sky has also led to the detection of the isotropic background at 240 and 140 μm, and to upper limits at shorter wavelengths down to 2 μm (Hauser *et al.* 1998).

It appears very likely that this isotropic background is the long-sought 'Cosmic Infrared Background'. As shown in Figure 2, its level is about 5–10 times the no-evolution prediction based on the local IR luminosity function determined by IRAS. There is about twice as much flux in the CIRB than in the 'Cosmic Optical Background' (COB). If the dust that emits at IR/submm wavelengths is mainly heated by young stellar populations, the sum of the fluxes of the CIRB and COB

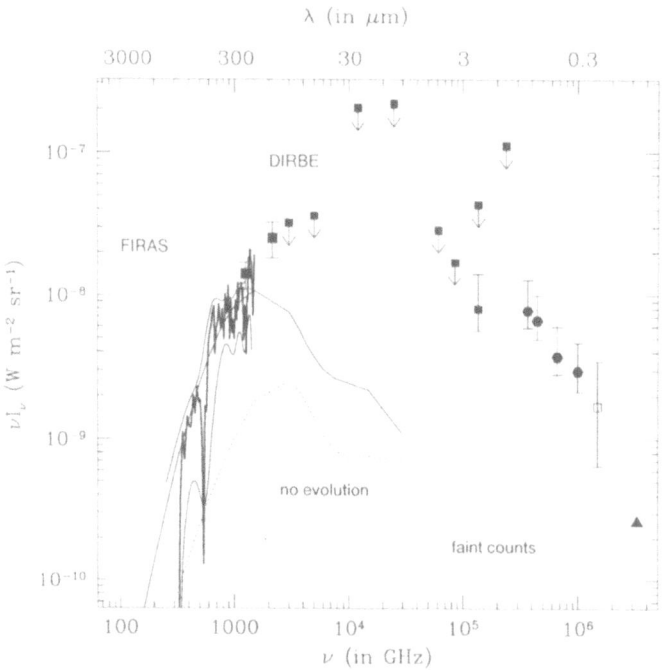

Figure 2. The cosmic optical and infrared backgrounds (respectively COB and CIRB). The COB (solid square, solid dots, and open square) is obtained from the faint counts and compiled by Pozzetti *et al.* (1998). The solid triangle gives an upper limit by Vogel *et al.* (1995). The thick solid lines show the CIRB extracted from FIRAS residuals in the cleanest regions of the sky (Puget *et al.*, 1996; Guiderdoni *et al.*, 1997). The thin lines bracket the range that is consistent with the error bars. The solid squares give the upper limits and detections (at 140 and 240 μm) from DIRBE (Hauser *et al.* 1998). The no-evolution curve (dotted line) is computed from the local IRAS luminosity function extrapolated to $z = 8$, for $H_0 = 50$ km s^{-1} Mpc^{-1} and $\Omega_0 = 1$. The solid line shows the best model (the so-called 'model E') in Guiderdoni *et al.* (1998).

gives the level of the Cosmic Background associated to stellar nucleosynthesis (Partridge and Peebles, 1967). As a matter of fact, the census of the local density of heavy elements $\rho_Z(z = 0) \sim 1 \times 10^7\ M_\odot$ Mpc^{-3} gives an expected bolometric intensity of the background $I_{bol} \simeq 50(1 + z_{eff})^{-1}$ nW m^{-2} sr^{-1} where z_{eff} is an effective redshift of star formation. This value is consistent with the observations.

Of course, it is not clear yet whether star formation is responsible for the bulk of dust heating, or there is a significant contribution of AGNs to the heating. At low z, the IRAS satellite has discovered 'luminous IR galaxies' (hereafter LIRGs), mostly interacting systems, and the spectacular 'ultraluminous IR galaxies' (hereafter ULIRGs), which are mergers and emit more than 95% of their energy in the IR (see e.g. the review by Sanders and Mirabel, 1996). It is very likely that the high-redshift counterparts of these local LIRGs and ULIRGs are responsible for the CIRB. According to Genzel *et al.* (1998), the starburst generally contributes

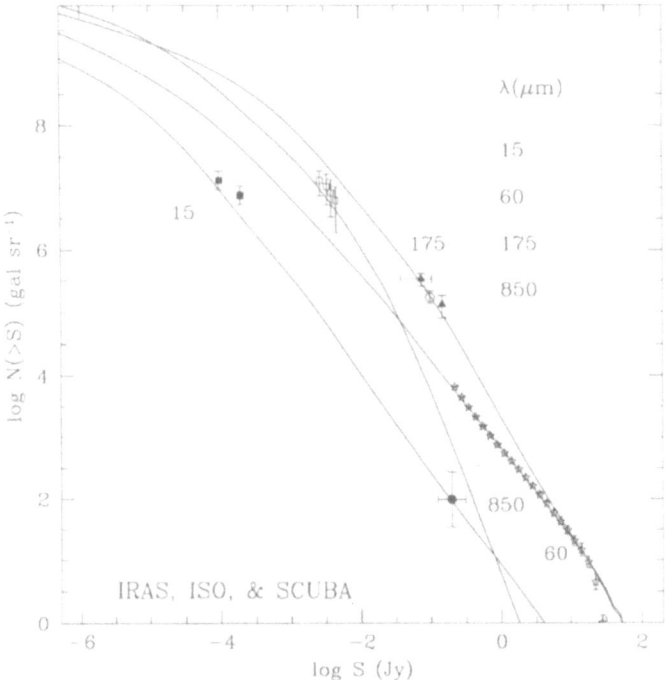

Figure 3. Predictions for faint counts at 15 μm, 60 μm, 175 μm, and 850 μm for model E. Open stars: Faint Source Survey (Lonsdale *et al.*, 1990). Solid hexagon: Rush *et al.* (1993). Solid squares: ISO-HDF (Oliver *et al.*, 1997) with ISOCAM at 15 μm. Open hexagon: Kawara *et al.* (1998) with ISOPHOT at 175 μm. Solid triangles: one of the Southern fields of the ISOPHOT FIRBACK survey (Puget *et al.*, 1998). Open squares: deep SCUBA survey (Smail *et al.*, 1997).

to 50–90% of the heating in local ULIRGs, but the evolution of this fraction with redshift is still unknown.

The FIRBACK programme is a deep survey of 4 deg^2 at 175 μm with the ISO-PHOT instrument aboard ISO. Preliminary results for one of the Southern fields unveil a surface density of sources (with $S_\nu > 100$ mJy) much larger than the no-evolution predictions based on the local IR luminosity function (Puget *et al.*, 1999). This strong evolution is confirmed by the other 175 μm deep survey by Kawara *et al.* (1998). Various deep surveys at 850 μm have been achieved with the SCUBA instrument on the JCMT (Smail *et al.*, 1997; Hughes *et al.*, 1998; Barger *et al.*, 1998; Eales *et al.*, 1999). They also unveil a surface density of sources (with $S_\nu > 2$ mJy) much larger than the no-evolution predictions. Figure 3 gives an account of the faint counts in the submm range.

4. Modelling Dust Spectra in a Semi-Analytic Framework

Guiderdoni *et al.* (1998) proposed a consistent modelling of IR/submm galaxy counts in the paradigm of hierarchical clustering. The extinction is consistently computed from chemical evolution, and standard assumptions about transfer in the so-called 'slab' geometry (where the star and dust components are homogeneously mixed with equal height scales). The spectral energy distribution of dust emission is computed by assuming a mix of various dust components. The contributions are fixed in order to reproduce the observational correlation of IRAS colours with total IR luminosity. These IR/submm spectra are implemented in a semi-analytic model of galaxy formation and evolution. This type of model has been very effective in computing the optical properties of galaxies in the paradigm of hierarchical clustering. We only extend this approach to the IR/submm range, and take the standard CDM case with $H_0 = 50$ km s^{-1} Mpc^{-1}, $\Omega_0 = 1$, $\Lambda = 0$, and $\sigma_8 = 0.67$. We assume a Star Formation Rate $SFR(t) = M_{gas}/t_*$, with $t_* \equiv \alpha t_{dyn}$. The efficiency parameter $1/\alpha = 0.01$ gives a nice fit of local spirals. The robust result of this type of modelling is a rather flat cosmic SFR history. As a phenomenological way of reproducing the steep rise of the cosmic SFR history from $z = 0$ to $z = 1$, we introduce a 'burst' mode of star formation involving a mass fraction that increases with z as $(1 + z)^4$, with ten times higher efficiencies $1/\alpha = 0.1$. The so-called 'model E' fairly reproduces the cosmic SFR and luminosity densities, as well as the CIRB (see Figures 1 and 2).

Figure 3 gives the predictions of number counts at 15, 60, 175, and 850 μm for this model. The agreement of the predictions with the data seems good enough to suggest that these counts do probe the evolving population contributing to the CIRB. The model shows that 15% and 60% of the CIRB respectively at 175 μm and 850 μm are built up by objects brighter than the current limits of ISOPHOT and SCUBA deep fields. The predicted median redshift of the ISO-HDF is $z \sim 0.8$. It increases to $z \sim 1.2$ for the deep ISOPHOT surveys, and to $z \geq 2$ for SCUBA, though the latter value seems to be very sensitive to the details of evolution. The optical follow-up of these sources is in progress.

5. Conclusions

1. There is now a strong evidence that high-redshift galaxies emit much more IR luminosity than predictions based on the local IR luminosity function, without evolution. The submm counts seem to unveil the bright end of the population that is responsible for the CIRB.
2. The issue of the relative contributions of the starbursts and AGNs to dust heating is still unsolved. Local ULIRGs seem to be dominated by starburst heating, but the behaviour at higher redshift is unknown.

3. It is difficult to correct for the influence of dust on the basis of the optical spectra alone. Multi-wavelength studies are clearly necessary to address the history of the cosmic SFR density.

4. Under the assumption that starburst heating is dominant, simple models in the paradigm of hierarchical clustering do reproduce the current IR/submm data.

5. The current studies on faint counts at submm wavelengths will guide models for the preparation of the observing strategies with forthcoming instruments : SIRTF, SOFIA, the PLANCK *High Frequency Instrument*, the FIRST *Spectral and Photometric Imaging REceiver*, and the MMA/LSA.

References

Barger, A.J., Cowie, L.L., Sanders, D.B., Fulton, E., Taniguchi, Y., Sato, Y., Kawara, K. and Okuda, H.: 1998, *Nature* **394**, 248.

Connolly, A.J., Szalay, A.S., Dickinson, M., SubbaRao, M.U. and Brunner, R.J.: 1997, *Astrophys. J.* **486**, L11.

Eales, S., Lilly, S., Gear, W., Dunne, L., Bond, J.R., Hammer, F., Le Fèvre, O. and Crampton, D.: 1999, *Astrophys. J.* **515**, 518.

Fixsen, D.J., *et al.*: 1998, *Astrophys. J.* **508**, 123.

Flores, H., Hammer, F., Thuan, T.X., Cesarsky, C., Désert, F.X., Omont, A., Lilly, S.J., Eales, S., Crampton, D. and Le Fèvre., O.: 1999, *Astrophys. J.* **517**, 148.

Genzel, R., Lutz, D., Sturm, E., Egami, E., Kunze, D., Moorwood, A.F.M., Rigopoulou, D., Spoon, H.W.W., Sternberg, A., Tacconi-Garman, L.E., Tacconi, L. and Thatte, N.: 1998, *Astrophys. J.* **498**, 579.

Guiderdoni, B., Bouchet, F.R., Puget, J.L., Lagache, G. and Hivon, E.: 1997, *Nature* **390**, 257.

Guiderdoni, B., Hivon, E., Bouchet, F.R. and Maffei, B.: 1998, *Mon. Not. R. Astron. Soc.* **295**, 877.

Hauser, M.G., *et al.*: 1998, *Astrophys. J.* **508**, 25.

Hughes, D., Serjeant, S., Dunlop, J., Rowan-Robinson, M., Blain, A., Mann, R.G., Ivison, R., Peacock, J., Efstathiou, A., Gear, W., Oliver, S., Lawrence, A., Longair, M., Goldschmidt, P. and Jenness, T.: 1998, *Nature* **394**, 241.

Kawara, K., Sato, Y., Matsuhara, H., *et al.*: 1998, *Astron. Astrophys.* **336**, L9.

Lilly, S.J., Le Fèvre, O., Hammer, F. and Crampton, D.: 1996, *Astrophys. J.* **460**, L1.

Lonsdale, C.J., Hacking, P.B., Conrow, T.P. and Rowan-Robinson, M.: 1990, *Astrophys. J.* **358**, 60.

Madau, P., Ferguson, H.C., Dickinson, M.E., Giavalisco, M., Steidel, C.C. and Fruchter, A.: 1996, *Mon. Not. R. Astron. Soc.* **283**, 1388.

Madau, P., Pozzetti, L. and Dickinson, M.E.: 1998, *Astrophys. J.* **498**, 106.

Meurer, G.R., Heckman, T.M., Lehnert, M.D., Leitherer, C. and Lowenthal, J.: 1997, *Astron. J.* **114**, 54.

Oliver, S.J., Goldschmidt, P., Franceschini, A., Serjeant, S.B.G., Efstathiou, A.N., *et al.*: 1997, *Mon. Not. R. Astron. Soc.* **289**, 471.

Partridge, B. and Peebles, P.J.E.: 1967, *Astrophys. J.* **148**, 377.

Pozzetti, L., Madau, P., Zamorani, G., Ferguson, H.C. and Bruzual, G.A.: 1998, *Mon. Not. R. Astron. Soc.* **298**, 1133.

Pettini, M., Steidel, C.C., Adelberger, K., Kellogg, M., Dickinson, M. and Giavalisco, M.: 1997, to appear in: J.M. Shull, C.E. Woodward and H. Thronson (eds.), *ORIGINS*, ASP Conference Series, astro-ph/9707200.

Puget, J.L., Abergel, A., Boulanger, F., Bernard, J.P. and Burton, W.B., *et al.*: 1996, *Astron. Astrophys.* **308**, L5.

Puget, J.L., Lagache, G., Clements, D.L., Reach, W.T., Aussel, H., Bouchet, F.R., Cesarsky, C., Désert, F.X., Dole, H., Elbaz, D., Franceschini, A., Guiderdoni, B. and Moorwood, A.F.M.: 1999, *Astron. Astrophys.* **345**, 29.

Rush, B., Malkan, M.A. and Spinoglio, L.: 1993, *Astrophys. J. Suppl. Ser.* **89**, 1.

Sanders, D.B. and Mirabel, I.F.: 1996, *Annu. Rev. Astron. Astrophys.* **34**, 749.

Smail, I., Ivison, R.J. and Blain, A.W.: 1997, *Astrophys. J.* **490**, L5.

Steidel, C.C. and Hamilton, D., 1993, *Astron. J.* **105**, 2017.

Steidel, C.C., Giavalisco, M., Pettini, M., Dickinson, M. and Adelberger, K.L.: 1996, *Astrophys. J.* **462**, L17.

Steidel, C.C., Adelberger, K.L., Giavalisco, M., Dickinson, M. and Pettini, M.: 1999, *Astrophys. J.* **519**, 1.

Storrie-Lombardi, L.J., McMahon, R.G. and Irwin, M.J.: 1996, *Mon. Not. R. Astron. Soc.* **283**, L79.

Vogel, S., Weymann, R., Rauch, M. and Hamilton, T.: 1995, *Astrophys. J.* **441**, 162.

SPECTROSCOPY OF GIANTS OF THE SAGITTARIUS DWARF GALAXY

P. BONIFACIO

Osservatorio Astronomico di Trieste, Via G.B. Tiepolo 11, I-34131 Trieste, Italy

L. PASQUINI

European Southern Observatory, K. Schwarzschild Strasse 2, D-85748 Garching bei Munchen, Germany

P. MOLARO

Osservatorio Astronomico di Trieste, Via G.B. Tiepolo 11, I-34131 Trieste, Italy

G. MARCONI

Osservatorio Astronomico di Roma, Via dell'Osservatorio 2, Monte Porzio Catone, I-00040 Roma, Italy

Abstract. In this paper we present the first results of the analysis of intermediate resolution ($\Delta\lambda \sim$ 3.5 Å) spectra of giants of the Sagittarius dwarf galaxy acquired using the ESO NTT telescope. From the deep CCD photometry of Marconi *et al.* (1998a) we have selected a sample of giants representative of the metallicity spread suggested by the comparison of the colour-magnitude diagram of Sagittarius with those of galactic globular clusters. The spectra have been used to measure radial velocities, to confirm the membership to Sagittarius, and to provide a metallicity estimate by using spectral synthesis techniques. The analyzed stars show a spread in metallicities in the range $-1.0 \leq$ [Fe/H] $\leq +0.7$, some 0.5 dex more metal-rich than the photometric estimates.

1. Introduction

The Sagittarius dwarf galaxy, discovered by Ibata *et al.* (1994, 1995), at a distance of only 25 Kpc, offers a unique opportunity to study the stellar populations of an external galaxy. Comparison of the color-magnitude diagram of Sagittarius with those of Galactic globular clusters has suggested a spread in metallicity in the range $-0.7 \leq$ [Fe/H] ≤ -1.6 (Marconi *et al.*, 1998a). This may be the sign of a complex star-formation history characterized by several bursts. Because Sagittarius is some 16 Kpc behind the Galactic Center there is confusion between Sagittarius and Bulge stars. Sagittarius revealed itself as a population with a mean radial heliocentric velocity around 140 km s^{-1} and a small velocity dispersion, thus probable membership may be ascribed on the basis of the radial velocity. Intermediate resolution spectra allow to determine radial velocities with a precision around 20 km s^{-1}, sufficient to confirm membership. Moreover such low resolution spectra may be used to obtain a crude estimate of the metallicity which may be compared

Astrophysics and Space Science is the original source of publication of this article. It is recommended that this article is cited as: *Astrophysics and Space Science* **265**: 541–544, 1999.
© 1999 *Kluwer Academic Publishers.*

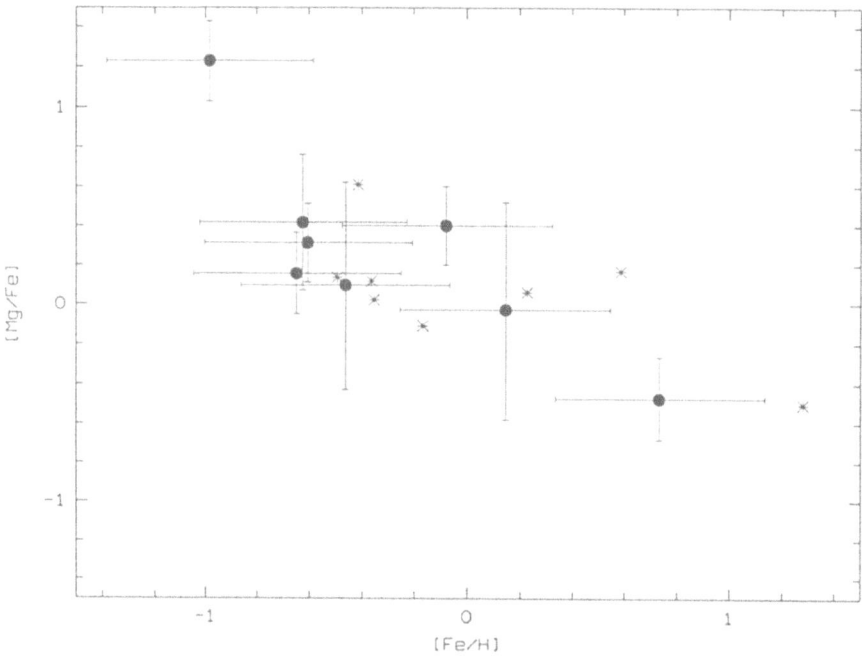

Figure 1. [Mg/Fe] ratios versus [Fe/H] for $\xi = 2.0$ km s^{-1} (filled symbols) and $\xi = 1.0$ km s^{-1}.

with the estimates based on the colour-magnitude diagrams. In this paper we report on radial velocities and abundances derived from grism spectra obtained with NTT+EMMI/MOS at ESO La Silla.

TABLE I

Abundances and atmospheric parameters

#	V_0	T_{eff} K	log g	[Fe/H] $\xi = 2$ km s^{-1}	[Mg/Fe] $\xi = 2$ km s^{-1}	[Fe/H] $\xi = 1$ km s^{-1}	[Mg/Fe] $\xi = 1$ km s^{-1}
105	17.55	5041	2.59	−0.5	+0.1	−0.17	−0.11
115	17.84	4953	2.36	−1.0	+1.2	−0.41	+0.61
124	17.00	4891	2.21	−0.6	+0.3	−0.35	+0.02
128	17.40	4778	1.93	+0.7	−0.5	+1.28	−0.50
139	17.78	4891	2.21	−0.1	+0.4	+0.23	−0.50
141	17.45	4977	2.42	−0.6	+0.4	−0.36	+0.12
142	17.55	5118	2.81	−0.7	+0.2	−0.49	+0.14
201	17.54	5003	2.49	+0.1	−0.0	+0.59	+0.17

2. Observations

We used the multi-object-spectroscopy (MOS) mode of the EMMI instrument on the NTT 3.5m telescope at ESO La Silla. The resolving power was about 1500, the usable spectral range was from about 480 nm to about 620 nm. We acquired spectra of 57 stars, the log of the observations and further details may be found in Marconi *et al.* (1998b).

3. Radial Velocities and Abundances

The range 480–530 nm was used to determine heliocentric radial velocities using cross-correlation with synthetic spectra as templates We considered the stars with heliocentric radial velocities in the range 100–180 km s^{-1} to be members of Sagittarius, following Ibata *et al.* (1997), this left us with a sample of 23 stars. In order to estimate abundances we defined six spectral indices which measure the Mgb triplet and some Fe and iron-peak elements features. We developed an iterative procedure which makes use of the SYNTHE code (Kurucz, 1993) to determine the abundances which best match the observed to the synthetic indices; details may be found in Marconi *et al.* (1998b). The procedure requires that the atmospheric parameters T_{eff}, log g and ξ be fixed for each star. The colour $(V - I)_0$ of Marconi *et al.* (1998a) was used to determine the effective temperatures using the calibration of Alonso *et al.* (1996); this calibration refers to dwarfs only, however it is theoretically known that the $(V - I)$ colour depends weakly on gravity; we expect the error introduced by neglecting the gravity dependence of $(V - I)$ to be on the order of 100 K. The isochrones of Straniero, Chieffi and Limongi (1997) for an age of 8 Gyr and [Fe/H] $= -0.5$ were used to estimate log g. Microturbulence cannot be determined from this intermediate-resolution data, we therefore performed the computations for both 1 km s^{-1} and 2 km s^{-1}, these values cover the range usually found in cool giants and allow to estimate the effect of microturbulence on the derived abundances. This work is still in progress; results for 8 out of the 23 stars with a radial velocity consistent with Sagittarius membership, are presented in Table I and displayed in Figure 1.

4. Discussion

Our stars have been selected in order to highlight the spread in metallicity derived from the colour-magnitude diagram and in fact among the stars analyzed we see a spread of over one dex. However the metallicities range from super-solar to [Fe/H] ≈ -1.0, about 0.5 dex more metal-rich than the range implied by the photometry. A systematic error in the spectroscopic or photometric (or both !) abundance estimates could reconcile the two results. However closer inspection of Figure 1 reveals that of the 8 analyzed stars 3 are solar or super-solar while the remaining 5 show little or no dispersion in metallicity. It is legitimate to ask whether the metal-rich stars belong in fact to the Bulge, in spite of their radial velocity. This hypothesis may not be ruled out, we note however that the [Mg/Fe] ratios appear to be solar or sub-solar. The large errors associated with our data preclude any firm conclusion, however, taken at face value, the [Mg/Fe] in our metal-rich stars appears to be different from the enhanced ratios displayed by the Bulge K giants of McWilliam and Rich (1994), Figure 20a. Our findings are in keeping with those of Smecker-Hane, McWilliam and Ibata (1998), who, on the basis of Keck HIRES spectra, found 2 out of 7 stars to be metal-rich and with α elements O and Ca under-abundant with respect to solar.

References

Alonso, A., Arribas, S. and Martinez-Roger, C.: 1996, *Astron. Astrophys. Suppl.* **117**, 227.

Ibata, R.A., Gilmore, G. and Irwin, M.J.: 1994, *Nature* **370**, 194.

Ibata, R.A., Gilmore, G. and Irwin, M.J.: 1995, *Mon. Not. R. Astron. Soc.* **277**, 781.

Ibata, R.A., Wyse, R.F.G., Gilmore, G., Irwin, M.J. and Suntzeff, N.B.: 1997, *Astron. J.* **113**, 634.

Kurucz, R.L.: 1993, *CD-ROM* 13, 18.

Marconi, G., Buonanno, R., Castellani, M., Iannicola, G., Molaro, P., Pasquini, L. and Pulone, L.: 1998a, *Astron. Astrophys.* **330**, 453.

Marconi, G., Bonifacio, P., Pasquini, L. and Molaro, P.: 1998b, in: P. Whitelock and R. Cannon (eds.), *The Stellar Content of Local Group Galaxies*, held in Cape Town, South Africa, 7–11 September 1998, to be published by ASP, p. 8.

McWilliam, A. and Rich, R.R: 1994, *Astrophys. J. Suppl.* **91**, 749.

Smecker-Hane, T., McWilliam, A. and Ibata, R.A.: 1998, *AAS Meeting* **192** 6613.

Straniero, O., Chieffi, A. and Limongi, M.: 1997, *Astrophys. J.* **490**, 425.

VIII – FUTURE PROSPECTS AND CONCLUSIONS

FUTURE LARGE-SCALE SURVEYS OF 'INTERESTING' STARS IN THE HALO AND THICK DISK OF THE GALAXY

T.C. BEERS

Department of Physics & Astronomy, Michigan State University, East Lansing, MI, USA

Abstract. The age of slow, methodical, star-by-star, single-slit spectroscopic observations of rare stars in the halo and thick disk of the Milky Way has come to an end. The use of new spectroscopic survey techniques which obtain large sky coverage to faint magnitudes will enable substantially better 'return on investment' in the near future. We review the present state of surveys for low metallicity, field horizontal-branch, and sdM stars in the Galaxy, and describe new lines of attack which should open the way to a more than hundredfold increase in the numbers of interesting stars with available spectroscopic information.

1. Introduction

How well do astronomers know the nature of the stellar populations of the thick disk and halo of the Galaxy? Despite five decades of research, the honest answer to this question is 'not very well'. Although most of the major classifications of luminous stars are probably recognized, only recently have sufficiently large samples of thick disk and halo stars been obtained to enable astronomers to finally test some of their most basic ideas about the formation and evolution of the Galaxy.

This limitation is imposed, to a great extent, by the difficulty of obtaining medium- and high-resolution spectroscopy of rare, and hence, widely spaced, field stars over the entire sky. Recent developments in instrumentation for wide-field Schmidt telescopes are now making it feasible to break through this barrier, and attack important questions of Galactic astronomy with samples of stars which are hundreds of times larger than was heretofore possible. In this brief review, we mention a number of samples of stars which should prove 'interesting' for detailed follow-up with these techniques in the near future.

2. 'Interesting' Stars

2.1. INTERMEDIATE- AND LOW-METALLICITY F, G, AND K-TYPE STARS

As is evident from the numerous discussions at this conference, and in the recent literature, stars which are depleted in their heavy metals by factors of 10 to 10 000 relative to the Sun contain precious chemical and kinematic information concerning the formation and evolution of the Galaxy, and the first generations of stars

Astrophysics and Space Science is the original source of publication of this article. It is recommended that this article is cited as: *Astrophysics and Space Science* **265**: 547–552, 1999.
© 1999 *Kluwer Academic Publishers.*

to have been born in the Universe. In the fifty years since such stars were first recognized, the total number of stars with abundances [Fe/H] ≤ -1.0 has grown from 2 (Chamberlain and Aller, 1951) to over 4 000 (Beers, 1999b).

Although the raw number appears impressive, it should be at once realized that the existing samples of metal-poor stars are still far from satisfactory in several respects. For example, most of the recognized low-metallicity stars are located within a few kpc of the Sun; very few low-metallicity stars are yet identified more than 10 kpc away. As a result, there is presently no means to consider possible changes in abundance distribution and/or kinematics of the metal-poor stellar populations as a function of distance. Furthermore, less than several hundred metal-poor stars have abundances measured for any element other than iron. Astronomers have only limited knowledge of the relative abundances of other important species, such as carbon, nitrogen, the alpha elements, and the neutron-capture elements. If there exist correlations between kinematics and elemental abundance variations in the stars of the thick disk and halo, we remain by-and-large ignorant of them.

2.2. FIELD HORIZONTAL-BRANCH AND BLUE METAL-POOR STARS

Stars in the horizontal-branch stage of evolution, collectively referred to here as Field Horizontal-Branch (FHB) stars, offer a number of advantages for probing the nature of stellar populations in the Galaxy. They are luminous and they are numerous. As such, the FHB stars can be used as kinematic tracers of the change in the velocity ellipsoid of the Galaxy from the solar neighborhood out to the distant halo (e.g., Sommer-Larsen *et al.*, 1997). The FHB stars provide one of the best populations for investigation of the proposed counter-rotating 'high halo', which apparently begins at 4–5 kpc from the Galactic plane (Wilhelm *et al.*, 1996, and references therein). If the so-called second-parameter problem is associated with differences in the ages of stellar populations, the resulting change in the colors of FHB stars provides information about the gradient in age as a function of Galacto-centric distance (e.g., Preston, Shectman and Beers, 1991). FHB stars also provide the means by which astronomers can make estimates of distance to clouds of high-velocity HI gas which may be associated with either the halo of the Galaxy or the Local Group, using the FHB stars as background and foreground illuminators (Wakker and van Woerden, 1997; Beers, 1999a).

Some 10 000 FHB/A star *candidates* have been identified from the HK survey of Beers and collaborators (Beers, Preston and Shectman, 1988; Beers *et al.*, 1996), but fewer than 1 000 of these stars have had the required medium-resolution spectroscopy obtained which enables measurements of their radial velocities, and estimates of their physical parameters (T_{eff}, log g, and [Fe/H]) to be made (Wilhelm *et al.*, 1999).

Preston, Beers and Shectman (1994) have argued that there exists a substantial population of apparently metal-poor, but young(ish) main-sequence stars in the Galaxy (referred to as Blue Metal-Poor, or BMP stars). These stars, previously described as halo blue stragglers, exhibit a distinctive set of kinematics (and perhaps abundances) which indicates that the BMPs may have been accreted during previous encounters of our Galaxy with dwarf galaxies such as Sagittarius.

Only on the order of 200 BMP stars have yet been confirmed on the basis of follow-up spectroscopy and photometry (Wilhelm, 1995).

2.3. SUBDWARF M STARS IN THE THICK DISK AND HALO

Low mass M-type stars potentially comprise the *dominant* stellar component of the thick disk and halo populations, yet only recently has their nature come under scrutiny. Because of their great numbers, and under the (testable!) assumption that the IMF does not change significantly at low abundances, the sdM stars provide excellent samples for examination of subtle variations in the halo and thick disk metallicity distribution functions.

At present, only a handful of sdMs are known with [Fe/H] $\leq$ -1.0, mostly from spectroscopic examination of high proper-motion catalogs (e.g., Gizis and Reid, 1996). However, Gizis has shown that sdMs can be readily recognized from inspection of various two-color planes (e.g., J–H vs. H–K), and further, that reliable estimates of abundance can be obtained from moderate-resolution spectroscopic analysis of their molecular bands (CaH, TiO) in the spectral region $6300 \leq \lambda \leq 7200$ Å (Gizis, 1997).

Gizis (1998) has estimated that on the order of 15,000 sdMs with [Fe/H] $\leq$ -1.0 can be identified from follow-up spectroscopy of candidates selected from the recently-initiated 2Mass survey.

3. Survey Work in Progress

In addition to the survey efforts which have already been mentioned, large numbers of 'interesting' stars are being identified in ongoing prism surveys of the thick disk and halo. The two we will discuss here are the 'digital' HK survey, and the Hamburg/ESO Stellar survey.

3.1. THE 'DIGITAL' HK SURVEY

As the name suggests, the digital HK survey is based on automatic scans of the HK objective-prism plates, using the APM in Cambridge (in collaboration with Mike Irwin). The APM scans result in a set of extracted prism spectra which are complete in the magnitude interval $11.0 < B < 15.0$ over the 7 000 square degrees of sky covered by the HK plates. The spectra are then inspected with a neural-network approach in order to objectively identify the stars most worthy of later spectroscopic follow-up. Rough color information is available from other direct sky surveys.

An objective selection criterion assures one that stars which were easily missed in the 'visual' HK survey are now captured. In particular, giants of low temperature (and colors $0.8 < B - V < 1.2$) whose CaII K lines, even at quite low metal abundance, appear strong. Preliminary results of this approach indicate that roughly 75% of the visually-identified MP candidates are recovered, as expected, but an additional 25–50 cooler MP candidates are also found per plate. We anticipate that, within two years, there will exist another 10 000 candidate MP stars from the HK survey which will be in need of spectroscopic and photometric follow-up.

3.2. THE HAMBURG/ESO STELLAR SURVEY PROJECT

Christlieb *et al.* (1999) describe this survey in more detail. Here we just touch on the salient points.

The Hamburg/ESO Stellar Survey (hereafter, HES) covers a larger area of the southern high-latitude sky than does the HK survey (8 000 vs. 4 500 square degrees). Because it is based on *un-widened* prism spectra, the HES obtains a limiting magnitude roughly two magnitudes fainter than the HK survey ($B \sim 17.5$ vs. $B \sim 15.5$). Furthermore, it has been demonstrated that rough ($\sigma(B - V) \sim 0.1$) color information can be obtained directly from the extracted prism spectra. Objective selection algorithms have been implemented, and the first spectroscopic follow-ups of HES MP candidates are now underway. It is presently estimated that some 20 000 MP candidates will result from the HES.

In addition to the MP stars, the HES also readily identifies FHB and other A-type stars. It is presently estimated that some 30 000 FHB/A stars will eventually come from the HES effort.

4. Multi-Fiber Wide-Field Spectroscopic Surveys

Single-slit spectroscopic follow-up of the $\sim 75\,000 - 100\,000$ newly-identified 'interesting' stars from the surveys described above would be a daunting observational task. Fortunately, a far more efficient follow-up technique is now feasible. The pioneering work of Watson and Parker has given the astronomical community the FLAIR II wide-field spectrograph, now in operation at the UK Schmidt in Australia. Recent approval of the AAT Board for preparation of a robotic fiber positioning system for the FLAIR II instrument will result in a new six-degree field (6DF) instrument to be in place in late 1999 (Watson, 1998). By early 2000, it is anticipated that the 6DF will be in routine operation. With the benefit of a CCD upgrade, and some 150 active fibers, acceptable integration times will yield ~ 2 Å resolution spectra of stellar sources down to $B = 16.5$.

One interesting challenge remains. Filling all the fibers on the 6DF with 'interesting' stars! Clearly, the optimal approach is to combine catalogues extracted from projects with similar wavelength and resolution requirements. The large numbers of stars to come from the digital HK survey, the HES, and other proper-motion selected stars, suggest themselves. The sdM stars, owing to their lack of flux in the blue, would probably have to be undertaken as a separate survey. Obviously, developing a similar instrument to the 6DF to be used on northern-hemisphere Schmidt telescopes should be given high priority as well.

5. High-Resolution Spectroscopic Surveys of MP Stars

One of the most important reasons to carry out large-scale medium-resolution spectroscopic surveys of 'interesting' stars is so that a subset of the MOST interesting stars might be identified for further study at high spectroscopic resolution. The instruments to carry out such investigations will shortly come online – Gemini, Hobby-Eberly, LBT, Magellan, and the VLT. There already exist a sufficient number of 'interesting' stars (e.g., the 1 000 stars from the HK survey with [Fe/H] $\leq$ −2.0) to provide a substantial candidate list. What is now required is high-resolution spectroscopy of this sample of very low-metallicity stars:

 – At High Resolution: R = 30 000 to 60 000
 – At High Signal-to-Noise: $S/N > 100/1$
 – With Complete Wavelength Coverage: $3100 \leq \lambda \leq 10000$ Å
 – With CONSISTENT Analysis.

   With such an effort, 'next-generation' astronomers (well, hopefully some of the present generation too) will be able to obtain:

 – Knowledge of the distribution of light elements in the early Universe
 – Knowledge of the heavy element abundances in the early Universe
 – Knowledge of REAL trends, exceptions to trends, and surprises, in the relative abundances of a host of elemental species in the early Universe.

   We have the telescopes, we have the stars, we have the analysis tools. All that we, as a community, require, is the desire and the cooperation to make this dream, one certainly shared by Roger and Giusa Cayrel, a reality.

Acknowledgements

TCB acknowledges support for this work received from grants AST 92-22326 and AST 95-29454 from the National Science Foundation.

References

Beers, T.C.: 1999a, High Velocity Clouds, in: B. Gibson and M. Putnam (eds.), *ASP Conference Series*, Astronomical Society of the Pacific, San Francisco, in press.
Beers, T.C.: 1999b, this volume.

Chamberlain, J.W. and Aller, L.H.: 1951, *Astrophys. J.* **114**, 52.

Christlieb, N., Wisotzki, L., Reimers, D., Gehren, T., Reetz, J., Gehren, T., Grasshoff, G. and Beers, T.C.: 1999, this volume.

Gizis, J.: 1997, *Astron. J.* **113**, 806.

Gizis, J.: 1998, private communication.

Gizis, J. and Reid, N.: 1996, *Astron. J.* **111**, 365.

Preston, G.W., Beers, T.C. and Shectman, S.A.: 1994, *Astron. J.* **108**, 538.

Preston, G.W., Shectman, S.A. and Beers, T.C.: 1991, *Astrophys. J.* **375**, 121.

Sommer-Larsen, J., Beers, T.C., Flynn, C., Wilhelm, R. and Christensen, P.-R.: 1997, *Astrophys. J.* **481**, 775.

Wakker, B.P. and van Woerden, H.: 1997, *Annu. Rev. Astron. Astrophys.* **35**, 217.

Watson, F.: 1998, *AAO Newsletter* **85**, 11.

Wilhelm, R.: 1995, *PhD Thesis*, Michigan State University.

Wilhelm, R., Beers, T.C., Kriessler, J.R., Pier, J.R., Sommer-Larsen, J. and Layden, A.C.: 1996, Formation of the Galactic Halo ... Inside and Out, in: H. Morrison and A. Sarajedini (eds.), *ASP Conference Series*, Astronomical Society of the Pacific, San Francisco, p. 171.

Wilhelm, R., Beers, T.C., Sommer-Larsen, J., Pier, J.R., Layden, A.C., Flynn, C., Rossi, S. and Christensen, P.-R.: 1999, *Astron. J.* submitted.

FUTURE PROSPECTS FOR SPECTROSCOPY WITH LARGE AND SMALL TELESCOPES

C.A. PILACHOWSKI and S.C. BARDEN

National Optical Astronomy Observatories*, *P.O. Box 26732, Tucson, AZ 85726-6732, USA*

Abstract. In the coming few years, more new telescopes with large aperture will become available for observations of stars in the Milky Way and in Local Group galaxies, and, increasingly, of stars in more distant galaxies. A wide range of new targets will come within reach not only from the increase of telescope aperture, but also from new technology which improves the performance goals of modern instrumentation. New technologies on the horizon will be explored to evaluate their impact on scientific programs in the future.

1. Introduction

This conference has brought together astronomers working in diverse aspects of the Milky Way Galaxy, from the halo to the bulge to the thick and thin disk, to the local solar neighborhood, with the goal to understand the big picture of the origin and evolution of our galaxy and how it ties to the origin and evolution of all galaxies. Just as our galaxy assembled from bits and pieces of intergalactic stuff, so our understanding of the whole picture assembles from the bits and pieces of our knowledge.

As we put the pieces together, larger questions emerge to guide our research using the new facilities now coming into being. What happened before the halo formed? What were the earliest structures formed during the pre-Galactic epoch? What remnants remain to tell us, and how do we find them? What role did the high velocity clouds play, if any? How are halos formed? Where to the stars come from? How typical is the Milky Way halo? Where does the Milky Way halo end, and where do Local Group, and intergalactic-scale structures begin? How do bulges form and how do they evolve dynamically over time? What are the relationships among the halo, the bulge, and the disk? Can we account for their diverse star formation histories of the dwarf spheroidals? How can star formation stop, and then restart billions of years later? And certainly, what initiated the burst of galaxy formation that created the major components of the local group?

---

* The National Optical Astronomy Observatories is operated by the Association of Universities for Research in Astronomy, Inc., under cooperative agreement with the National Science Foundation.

Astrophysics and Space Science is the original source of publication of this article. It is recommended that this article is cited as: *Astrophysics and Space Science* **265**: 553–559, 1999.

2. Critical Techniques

Some of these questions will be answered in the next decade using the new telescopes and instruments now becoming available. The Hipparcos satellite has greatly extended our reach for astrometric information about important stellar populations, and eventually SIM, GAIA, and other space-based interferometers will give us precise distances and kinematic information for countless stars throughout the galaxy. Imaging studies continue to be important to determine star formation histories of important stellar populations: dwarf galaxies, halo fields in the Milky Way and in other galaxies, and in bulges, both here and elsewhere. The Hubble Space Telescope as well as large and small ground-based telescopes will continue to play essential roles.

But spectroscopy is key to determine the kinematics and compositions of the stellar populations that make up galaxies. The large apertures of our new ground-based telescopes will allow us to reach fainter limiting magnitudes than ever before. Our high dispersion spectrographs will reach to Andromeda, and our low dispersion spectrographs can access stars in galaxies as far away as the Virgo Cluster. Large sample sizes and surveys are increasingly important to the study of stellar populations, and we will need multi-object spectrographs and fast single-object spectrographs to continue our work.

We add one more new technique that will become increasingly important to understanding the origin and history of stellar populations – asteroseismology. From the oscillation spectrum of stars we can determine precisely and independently ages and helium fractions of stars in the field, as well as in clusters. This critical information is often missing and unavailable, but as the technique of stellar seismology develops into a reliable and proven methodology, we will reap great benefit.

3. Instrumentation

New instrumentation will play a big role in helping us achieve our scientific goals. Not only are we gaining access to telescopes of very large aperture, but also our instruments are designed to be much more capable, and to meet higher standards, than the instruments of previous decades. In Table I we summarize the low resolution, multi-aperture, optical spectrographs that will soon be available. Every new large telescope sees such a spectrograph as a workhorse instrument, and every such telescope plans to have one. Their capabilities differ modestly, with some reaching further to the ultraviolet, some to higher resolution, and some to wider fields.

High dispersion, optical spectrographs will also be offered on most, if not all, new, large ground-based telescopes. Spectral resolutions span a wide range from R=16,000 up to perhaps as high as $R = 500\,000$, but most will work between $30\,000 < R < 60\,000$. Some will be equipped with optical fibers, and some, such

TABLE I

Multi-aperture spectrographs

Telescope	Instrument	Wavelength (μ)	Resolution	FOV (arcmin)
Gemini	GMOS	0.4–1.1	500–10 000	5.5
HET	LRS	0.4–1.0	600–3000	4
Keck	LRIS	0.31–1.0	300–1000	6 × 8
Keck	DEIMOS	0.39–0.9	1000–6000	5 × 16 (2)
Keck	ESI	0.39–1.1	300–9000	2 × 8
LBT	(Planned)			
Magellan	WF Camera	0.3–1.0	300–3000	10
Magellan	IMACS	0.36–1.05	1800–10 000	15, 27
MMT	MMT Spec.	0.32–1.0	300–3000	3
MMT	Binospec	0.39–1.0	1000–15 000	16 × 15
SALT	(Planned)			
Subaru	FOCAS	0.36–0.9	10–3000	6
VLT	FORS1/2	0.3–1.0	100–3000	2.5
	VIMOS	0.37–1.8	100–3000	14 × 16
GTC	ODIN	0.3–1.0	40 000	12

as the Hectoechelle at the new MMT Observatory will offer multi-fiber capability. The new high resolution spectrographs are summarized in Table II.

For many studies of galactic stellar populations, wide field, multi-fiber capability will be critical, but few of the new, large ground-based telescopes will offer such instruments. Table III summarizes the capabilities of planned wide-field, multi-fiber spectrographs on new, large telescopes. The Hectospec and Hectochelle instruments at the new MMT Observatory and the GIRAFFE on ESO's VLT are impressive instruments, with fields of view of 60′ and 25′, respectively, and with more than a hundred fibers each. Since these instruments are so important for

TABLE II

High Resolution Spectrographs

Telescope	Instrument	Wavelength (μ)	Resolution	Fiber?
Gemini	HROS	0.3–1.1	50 000–90 000	no
Gemini	USHiRS	0.38–1.0	120 K–500 K	yes
HET	HRS	0.42–1.1	30 000–120 000	yes
HET	MRS	0.39–1.8	3500–21 000	yes
Keck	HIRES	0.3–1.0	$\leq 60\,000$	no
LBT	(Planned)			
Magellan	Dual Ech.	0.32–1.0	30 000–60 000	no
MMT	MAESTRO	0.3–1.0	49 500	no
MMT	Hectochelle	0.35–1.0	32 000–40 000	255
MMT	AO Echelle	0.4–1.0	200 000	yes
SALT	(Planned)			
Subaru	HDS	0.3–1.0	$\leq 100\,000$	no
VLT	UVES	0.3–1.1	$\leq 100\,000$	no/8
VLT	AVES	0.56–1.1	16 000	no

studies of stellar populations, we must all work with our respective communities to see that more such instruments become available.

4. New Technologies

New technologies will also play an important role in new instrumentation. One area that has seen rapid development in recent years is spectroscopy with very high velocity precision. While this development is largely motivated by the search for extra-solar planets, it does enable new types of studies in stellar physics as well. In particular, stellar seismology, which requires velocity precision comparable to what is needed to detect extra-solar planets, will become feasible as stabile spectrographs are mated to large telescopes. Such spectrographs are usually fed with optical fiber,

TABLE III
Wide-Field Fiber Multi-Object Spectrographs

Telescope	Instrument	Wavelength (μ)	Resolution	FOV (arcmin)	Fiber?
Gemini	None				
HET	None				
Keck	None				
LBT	(Planned)				
Magellan	None				
MMT	Hectospec	0.35–1.0	1000–5000	60	300
MMT	Hectochelle	0.35–1.0	32 000–40 000	60	255
SALT	None				
Subaru	None				
VLT	GIRAFFE	0.37–0.9	7500–25 000	25	130

thermally stablized, and equipped with an absorption cell for velocity reference. Diego *et al.* (1998) describes a high resolution spectrograph envisioned for the Gemini telescope which meets these conditions.

The use of adaptive optics for spectroscopy is a second new technology which will become increasingly important in new instrumentation. The use of adaptive optics decouples the effective spectral resolution from the telescope aperture and allows higher spectral resolution (e.g. 200 000 to 500 000) without the need for very large instruments. The very sharp spatial profile achieved using adaptive optics will permit echelle grating orders to be closely packed on a detector, thereby providing excellent spectral coverage. The relatively small apertures will also reduce sky background, which is increasingly important for faint sources accessible with large aperture telescopes.

Two examples of existing or planned adaptive optics spectrographs are an echelle spectrograph planned for the new 6.5-m telescope of the MMT Observatory (Ge *et al.*, 1998) and a moderate resolution spectrograph (AVES) planned for the VLT (Pasquini *et al.*, 1998). The former instrument has already been tested at the Starfire Optical Range telescope in New Mexico. It provides a resolving power of $R = 200,000$, with a 10 μm, almost single-mode, optical fiber feed. Spectral coverage is almost complete from 0.4–1.0 μm. The VLT instrument will utilize a 0.3" aperture to attain a resolving power of $R \sim 16\,000$. It is expected to outperform a comparable non-adaptive optics spectrograph by more than 1 magnitude for sky or detector limited observations, and is intended for use for spectroscopy of solar-type stars in nearby galaxies.

Finally, we wish to highlight a new and promising grating technology which may change dramatically the design of spectrographs in the future – volume phase holographic (VPH) gratings (Barden *et al.*, 1998). VPH gratings consist of a layer of dichromated gelatin sandwiched between two layers of an optically thin substrate. The index of refraction of the gelatin is modulated to produce diffraction, and the devices can be used as either transmission gratings (the modulation is at right angles to the plane of the substrate) or as reflection gratings (the modulation is parallel to the plane of the substrate). Light is diffracted an angles according to classical diffraction and according to Bragg diffraction; the Bragg condition defines the spectral energy distribution. Line densities of 300 to 6000 l mm^{-1} are possible, and the gratings offer very high diffraction efficiencies (typically > 80%). Customization for specific instrument requirements is relatively easy, and complex grating structures, such as stacked gratings in a single element to allow cross dispersion, spectral multiplexing, and beam splitting are possible. VPH gratings are already available at modest sizes to work in the wavelength range $0.4 < \lambda < 1.0$ μm, and $0.3 < \lambda < 2.8$ μm may be possible. Grating sizes of up to 200 by 280 mm can be made, and it might be possible to make gratings as large as 500 by 700 mm.

An astronomical test of one VPH grating was performed using the 2.1-m telescope of the Kitt Peak National Observatory. The spectrograph, which consisted of simple achomatic lenses for the camera and collimator, combined with a VPH grating, was fed with a 200 μm optical fiber. The spectrograph achieved a 17% total efficiency including sky, telescope, fiber, optics, and detector in first and second order, with a 60% efficiency for the grating and two lenses alone. Further testing and development of VPH gratings for use in astronomical instrumentation, funded by the U.S. National Science Foundation, will continue at the National Optical Astronomy Observatories.

A VPH grating was successfully installed in the LDSS++ at the Anglo-Australian Observatory (Glazebrook, 1998a, 1998b) as part of the spectrograph upgrade. Simulations suggested that the use of the VPH grating would offer a significant increase in instrumental throughput compared to other alternatives. A grating was procured and installed in the spectrograph in October, 1998, and tests have confirmed the promise of high throughput. VPH gratings will almost assuredly see increased use in modern spectroscopic instrumentation.

5. In Gratitude

Finally, we wish to close with a few words of appreciation to Giusa and Roger Cayrel, who have inspired us to take joy in all there is to learn, and to do our best, both in our science and in our instruments.

References

Barden, S.C., Arns, J.A. and Colburn, W.S.: 1998, *SPIE* **3355**, 866.
Diego, F., Crawford, I.A. and Walker, D.D.: 1998, *SPIE* **3355**, 218.
Ge, J., Angel, R. and Shelton, C.: 1998, *SPIE* **3355**, 253.
Glazebrook, K.: 1998a, *Anglo-Australian Observatory Newsleter* **84**, 9.
Glazebrook, K.: 1998b, *Anglo-Australian Observatory Newsletter* **87**, 11.
Pasquini, L., Delabre, B., Avila, G. and Bonaccini, D.: 1998, *SPIE* **3355**, 105.

THE DETECTION OF EXTREMELY DISTANT GALAXIES

J. LEQUEUX

Observatoire de Paris, 61 Avenue de l'Observatoire, 75014 Paris, France

Abstract. Very recent observations have shown that there are dusty galaxies at redshifts larger that 5. Their integrated contribution to the far-IR/submillimeter sky background is known. These galaxies are very hard to detect in the visible and in the infrared, but they will be within reach of the NGST, although it will be very difficult to recognize them without further studies. However their far-infrared emission will be detectable by FIRST and more easily by the large submillimeter interferometer (ALMA) presently under study. ALMA, FIRST, SIRTF and the NGST will make an ideal set for investigating the formation and early evolution of galaxies.

1. Introduction

Very recent observations have shown that the formation of galaxies and the first generation of stars have occured much earlier than previously thought in the early evolution of the Universe. Roger Cayrel was a pioneer in this topic when he predicted that 'the low-metallicity stars (in the Galactic halo) were likely formed at an early cosmological epoch ($z > 5$ if $H_0 \simeq 65$ km s^{-1}), before the Galaxy had developed a disk' (Cayrel, 1996). A galaxy at $z = 5.34$ has been found with a probable internal reddening $A_V > 0.5$ mag. (Armus *et al.*, 1998). Dusty galaxies emit a fraction of their luminosity in the far-infrared, due to the thermal emission of dust heated by the UV and visible stellar radiation. As a consequence, the detection of such galaxies should be made in the far-IR/submillimeter wavelength range if they contain dust, while they are difficult to detect in the visible/near-IR due to extinction. Their detection is the subject of the present paper. In Section 2, I will briefly recall the very important contribution of distant galaxies to the far-IR/submillimeter background, and in Section 3 I will describe the strategies for detection of these galaxies, with a projection on future instruments.

2. Distant Galaxies and the Extragalactic Background

The recent progresses in our knowledge of the extragalactic background have been spectacular. In the visible, the Hubble deep field (Williams *et al.*, 1996) is so deep that essentially all the observable galaxies have been resolved. A simple integration over their images gives a good measure of the extragalactic background surface brightness. The situation is not so clear in the near-IR, because the galaxy counts

Astrophysics and Space Science is the original source of publication of this article. It is recommended that this article is cited as: *Astrophysics and Space Science* **265**: 561–566, 1999.
© 1999 *Kluwer Academic Publishers*.

still do not seem to converge at the faintest levels (J and $K \simeq 24$) reached with the Keck telescope (Bershady *et al.*, 1998). However it is likely that the observed galaxies already make most of the extragalactic background. In the far-IR and submillimeter range, the extragalactic background has been clearly measured by the COBE satellite (Puget *et al.*, 1996; Dwek *et al.*, 1998 and references herein; Fixsen *et al.*, 1998; Lagache, 1998). There is also a lower limit around 15 μm from galaxy counts with ISO (Aussel, 1999). An upper limit at similar wavelengths comes from the mere fact that we observe TeV-energy gamma rays from several distant active galactic nuclei (Biller *et al.*, 1995, 1998; unpublished observations with CAT; see also Stecker and de Jager, 1998): the TeV gamma-rays interact strongly with mid-IR photons to produce e^+/e^- pairs. These lower and upper limits almost coincide, so that the background near 15 μm is well constrained. This wavelength range corresponds to a minimum in the spectral energy distribution of the background and is a rough natural limit between the (redshifted) stellar radiation and the radiation by dust. Putting together all the data above one can conclude that only 1/3 of the energy of the extragalactic background is direct stellar radiation, while the other 2/3 are re-radiated by dust (Lagache, 1998).

It is very likely that most of the far-IR emission of the very distant galaxies corresponds to starbursts rather than to active galactic nuclei (AGNs). This can be seen by comparing the X-ray extragalactic background, which is dominated by AGNs, to the extragalactic background from the visible to the submillimeter. The spectral distribution of the radiation from AGNs is roughly flat in νI_ν units from the X to the submillimeter ranges. The corresponding predicted background from the visible to the submillimeter is fainter than the observed one by more than 2 orders of magnitude (Puget, 1999). Also, a simple calculation shows that the total energy density of the extragalactic background corresponds within a factor 2 to the total energy produced by nucleosynthesis until the present time, a satisfactory result given the uncertainties (Puget, 1999).

It seems that the submillimeter background is dominated by very luminous, high-redshift starburst galaxies. The galaxies detected with SCUBA at 850 μm are extremely dusty and have luminosities of $\simeq 10^{12} \ L_\odot$, comparable to that of Arp 220. They already make a substantial fraction of the background at this wavelength (Hughes *et al.*, 1998).

If the distant dusty galaxies have a spectral energy distribution in the far-IR similar to Arp 220, i.e. a relatively sharp peak near 60 μm, it is possible to deconvolve roughly the spectrum of the submillimeter extragalactic background and to obtain an idea of the redshift distribution of the contributing galaxies. This has been done by Puget (1999) with the result that there are galaxies up to a redshift of 10. The star formation rate per unit covolume peaks at a redshift of about 3, instead of 1–2 if one ignores the dust-enshrouded galaxies (Madau *et al.*, 1998).

3. Strategies for Detecting Very Distant Galaxies

Galaxies with little dust will be detected as it is done now by deep imaging in the visible and near-IR. Their redshift can be roughly determined by multi-filter photometry, or more accurately by spectroscopy when possible. At very high redshifts, the observed Lyman limit is in the red or even in the near-IR (at 1 μm for $z = 10$) so that the detection has to be made in the near-IR as well as the measurement of the spectral energy distribution. The Next Generation Space Telescope (NGST) will obviously be the ideal instrument for this.

For dusty galaxies, the detection is better made by observing the thermal dust emission in the far-IR/submillimeter range. Model spectral energy distributions of dusty 10^{12} $L_\odot$ galaxies have been published by Guiderdoni *et al.* (1998) as a function of redshift, for an Universe with $H_0 = 50$ km s^{-1} Mpc^{-2} and $\Omega_0 = 1$. Figure 1 adapted from Guiderdoni *et al.* (1999) with minor changes shows an updated version in which the spectral energy distribution is that of Arp 220. It also displays the wavelength range and the sensitivity limits (3 σ for 1 hour exposure) of a variety of present and future instruments.

It is clear from Figure 1 that the NGST is so sensitive that it will be able to detect even strongly dust-enshrouded galaxies. However these galaxies will be buried amongst a very large number of other galaxies and will not be recognizable without further studies. Moreover their total luminosity cannot be determined by the NGST alone, preventing a knowledge of the star formation rate at very high redshift if no complementary submillimeter observations are made.

A the other end of the spectrum, a remarquable fact is the degeneracy of the submillimeter continuum spectra of galaxies at different redshifts. This is due to a compensation between the steep slope of the spectrum and the effect of redshift. Thus a continuum survey at say 1 mm wavelength will be roughly sensitive to galaxies of the same luminosity whatever their redshift from 0.3 to 10. It will thus provide an unbiased view of the Universe at least for those galaxies which emit most of their energy in the far-IR. Large millimeter/submillimeter interferometers will be ideal instruments for this detection. The foreseen Atacama Large Millimeter Array (ALMA) resulting from the fusion of the american MilliMeter Array (MMA) with the european Large Southern Array (LSA) will have a collecting area of about 7000 m^2 and a continuum bandwidth of 4 GHz. It will be able to detect at 1 mm galaxies 50 times fainter than 10^{12} $L_\odot$ whatever their redshift. There will be no confusion due to the high angular resolution of the instrument and the position measurements will be extremely accurate, allowing unambiguous identifications of the objects. Redshift determinations will be possible using molecular lines if they are strong enough. The present observations of highly redshifted CO lines allows optimism in this respect. For galaxies at $z > 4$, it will also be possible to obtain a rough 'photometric' redshift by measuring the spectral energy distribution in the different accessible atmospheric windows, in particular at 450 and 350 μm. Complementary studies with the NGST will allow redshift measurements, etc.

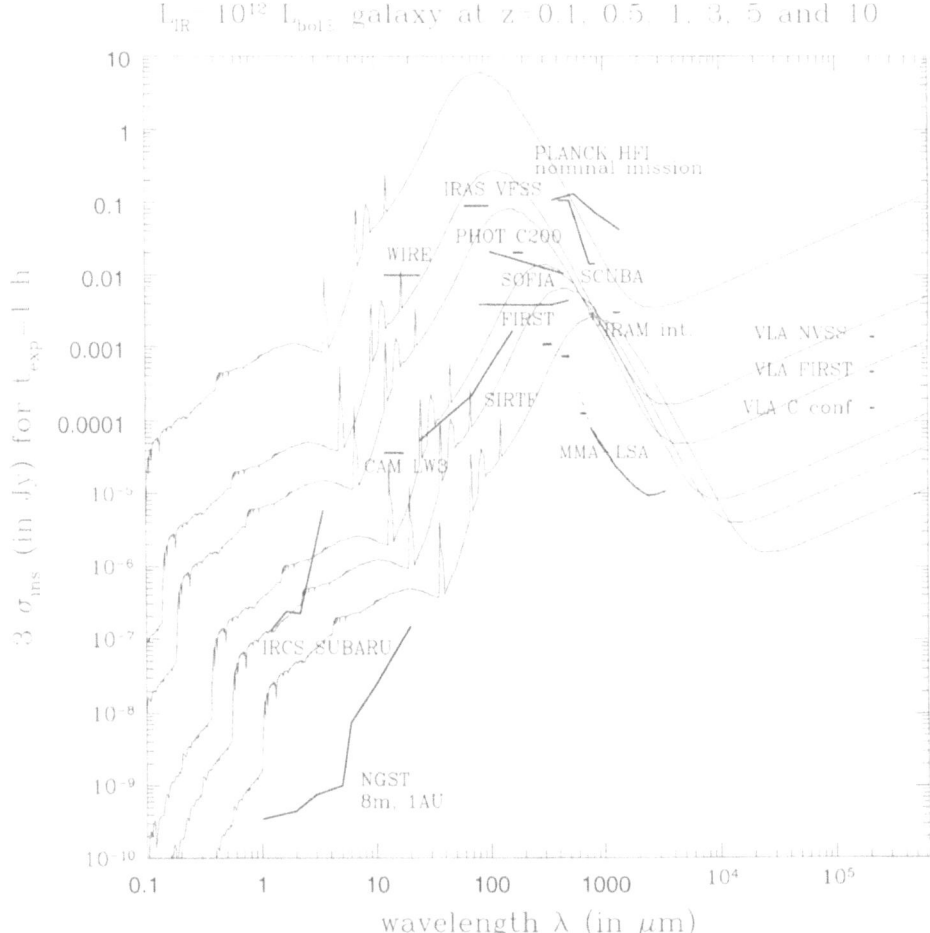

Figure 1. Sensitivity to point sources of various existing or planned instruments (3σ for 1 hour of on-source integration). The spectral energy distribution of a 10^{12} $L_\odot$ dusty starburst galaxy at different redshifts (0.1 to 10 from top to bottom) is indicated. The sensitivity of FIRST could be slightly better than indicated. The sensitivity of the LSA/MMA (now ALMA) corresponds to the median atmospheric conditions at the foreseen site of Chajnantor at 5000 m elevation in northern Chile, and a total antenna surface of 7000 m^2; note the atmospheric windows at 350 and 450 μm. 'CAM LW3' and 'PHOT C200' are for two surveys made with ISO. *Adapted from Guiderdoni et al. (1999).*

Dusty galaxies at lower redshifts will also be discovered either by the Far-Infrared and Submillimeter Space Telescope (FIRST) in the range 80–700 μm, or by the Space Infra-Red Telescope Facility (SIRTF) at shorter wavelengths with a large overlap with FIRST. These instruments are considerably less sensitive than ALMA. They are also limited by confusion due to their relatively poor angular resolution. But both have larger fields of view and hence are well adapted to large-scale shallow surveys. They will mainly detect bright galaxies with $z < 4$. They will also allow to obtain photometric redshifts for these galaxies and for the brightest galaxies with higher redshifts through multi-band observations of their spectral energy distribution.

4. Conclusions

I have shown that there are evidences that galaxies exist and contain dust hence heavy elements at very high redshifts, certainly larger than 5 and perhaps as high as 10. However we known nothing about their properties. In order to study these objects, and galaxy formation in general, a set of almost ideal facilities will become available approximately at the same time, 10 years from now. The NGST will be the prime instrument to detect and study relatively dust-free galaxies, and to make complementary studies of dusty galaxies preferably detected otherwise. ALMA will be ideal to make unbiased surveys of dusty galaxies at all redshifts and to obtain their redshifts. FIRST and SIRTF will survey bright dusty galaxies at lower redshifts and obtain their spectral energy distribution. These instruments will wonderfully complement each other. We are already living a great time for cosmology thanks to large optical telescopes, the Hubble Space telescope and infrared, submillimeter and millimeter instruments in space or on the ground. The end of the next decade will be even more exciting.

Acknowledgements

I wish to thank Jean-Loup Puget for enlightening discussions and Bruno Guider-doni for giving me the basis of Figure 1.

References

Armus, L., Matthews, K., Neugebauer, G. and Soifer, B.T.: 1998, Near-Infrared Observations of a Redshift 5.34 Galaxy: Further Evidence for Dust Absorption in the Early Universe, *Astrophys. J.* **506**, L89–L92.

Aussel, H.: 1999, Thèse, Université Paris XI.

Bershady, M.A., Lowental, J.D. and Koo, D.C.: 1998, Near-Infrared Galaxy Counts to J and K 24 as a Function of Image Size, *Astrophys. J.* **505**, 50–73.

Biller, S.D., Ackerlof, C.W., Buckley, J., *et al.*: 1995, An upper limit to the infrared background from observations of TeV gamma rays, *Astrophys. J.* **445**, 227–230.

Biller, S.D, Buckley, J., Burdett, A., *et al.*: 1998, New Limits to the IR Background: Bounds on Radiative Neutrino Decay and on VMO Contributions to the Dark Matter Problem, *Phys. Rev. Lett.* **80**, 2992–2995.

Cayrel, R.: 1996, The first generation of stars, *Astron. Astrophys. Rev.* **7**, 217–242.

Dwek, E., Arendt, R.G., Hauser, M.G., *et al.*: 1998, The COBE Diffuse Infrared Background Experiment Search for the Cosmic Infrared Background. IV. Cosmological Implications, *Astrophys. J.* **508**, 106–122.

Fixsen, D.J., Dwek, E., Mather, J.C., Bennett, C.L. and Shafer, R.A.: 1998, The Spectrum of the Extragalactic Far-Infrared Background from the COBE FIRAS Observations, *Astrophys. J.* **508**, 123–128.

Guiderdoni, B., Hivon, E., Bouchet, F.R. and Maffei, B.: 1998, Semi-analytic modelling of galaxy evolution in the IR/submm range, *Mon. Not. R. Astron. Soc.* **295**, 877–898.

Guiderdoni, B., Bouchet, F.R., Devrient, J.E.G., Hivon, E. and Puget, J.-L.: 1999, in: B. Guiderdoni *et al.* (eds.), *The Birth of Galaxies*, Editions Frontières, in press.

Hughes, D.H., Serjeant, S., Dunlop, J., *et al.*: 1998, High-redshift star formation in the Hubble Deep Field revealed by a submillimetre-wavelength survey, *Nature* **394**, 241–247.

Lagache, G.: 1998, Thèse, Université Paris XI.

Madau, P., Pozzetti, L. and Dickinson, M.: 1998, The Star Formation History of Field Galaxies, *Astrophys. J.* **498**, 106–116.

Puget, J.-L.: 1999, Cosmology with ISO, in: *Space Infrared Astronomy, Today and Tomorrow, Les Houches Lectures, August 1998*, to be published.

Puget, J.-L., Abergel, A., Bernard, J.-P., Boulanger, F., Burton, W.B., Désert, F.-X. and Hartmann, D.: 1996, Tentative detection of a cosmic far-infrared background with COBE, *Astron. Astrophys.* **308**, L5–L8.

Stecker, F.W. and de Jager, O.C.: 1998, Absorption of very high energy gamma-rays by intergalactic infared radiation: A new determination, *Astron. Astrophys.* **334**, L85–L8.

Williams, R.E., Blacker, B., Dickinson, M., *et al.*: 1996, The Hubble Deep Field: Observations, Data Reduction, and Galaxy Photometry, *Astron. J.* **112**, 1335–1389.

WHERE IS SMC1? A CAUTIONARY TALE: RECOMMENDATIONS ABOUT ASTRONOMICAL OBJECTS DESIGNATION

H.R. DICKEL

Univ. of Illinois, Astr. Dept./103 Astr. Bldg, 1002 W. Green Street, Urbana, IL 61801, USA

F. SPITE

Obs. de Paris, Section de Meudon, 92195 Meudon CEDEX, France

TASK GROUP ON DESIGNATIONS
IAU Commission 5

1. Introduction

A growing number of papers and (more recently) of data in machine-readable form are made available to the community. For a full use of this exploding astronomical information, some actions have to be undertaken. One of the first actions, initiated by J. Delhaye and R. Cayrel, was the gathering, star by star, of all the bibliographic references quoting this star. Made first on handwritten cards, then on punched cards, this venture took advantage of the progresses in data processing by computers, was extended to non-stellar objects, to a larger number of publications, becoming the well-known SIMBAD data base, operated by the Centre de Données de Strasbourg (CDS). The first Bulletin d'Information of the CDS was published in 1971 (Jung, 1971). Other astronomical data-bases are now available. One of the difficulties of the gathering of information about an astronomical object, is the variety of names attributed to a single object.

2. Designations

The designations found in the literature are sometimes non-optimally chosen. For example, SMC 1 is a planetary nebula in the LMC (Large Magellanic Cloud), designated from the initials of the authors Savage, Murdin and Clark (cf. Savage *et al.*, 1982). Moreover, authors sometimes choose, for a list of objects, designations which have already been adopted previously by other authors for other kind of objects. In order ot avoid confusion, some common rules, derived from common sense, have obviously to be followed.

Astrophysics and Space Science is the original source of publication of this article. It is recommended that this article is cited as: *Astrophysics and Space Science* **265:** 567–568, 1999.
© 1999 *Kluwer Academic Publishers.*

3. Commission 5 of the IAU

In an effort to promote clear and unambiguous identification of all astronomical objects outside the solar system (inside the solar system, rules have been observed for a long time), the IAU Task Group on Designations attempts to clarify existing astronomical designations, and the TG reviews, updates and advertises the IAU Recommendations for Nomenclature. The following documents on the Web are provided as a service to astronomers, to help them with the designation of astronomical sources of radiation outside the solar system:

How to refer to a source or designate a new one
http://cdsweb.u-strasbg.fr/how.html

IAU Recommendations for Nomenclature
http://cdsweb.u-strasbg.fr/iau-spec.html

Second Reference Dictionary of Nomenclature of Celestial Objects
http://cdsweb.u-strasbg.fr/cgi-bin/Dic

NEW (pre-)Registry of New acronyms
http://cdsweb.u-strasbg.fr/cgi-bin/DicForm

The Task Group, in collaboration with several editors of astronomical journals and managers of large data archives, is now studying the feasibility of an automated system to detect non conforming designations when an article and/or survey data are submitted for publication and/or to an electronic archive, in order to avoid any confusion.

4. Conclusion

The authors who have to name astronomical objects, are urged to look at the documents indicated hereabove. They will get additional help and advice by contacting members of the Task Group: names and adresses are listed in the documents of the two first WEB addresses provided hereabove.

References

Jung, J.: 1971, **1**, 1.
Savage, A., Murdin, P.G. and Clark, D.H.: 1982, *The Observatory* **102**, 229.

CONCLUDING REMARKS

L. WOLTJER

Observatoire de Haute Provence, Saint-Michel l'Observatoire, France
Osservatorio Astrofisico di Arcetri, Firenze, Italy

1. Introduction

Frequently meetings in honor of scientists risk being diffuse because the papers presented are unified by personal rather than scientific coherence. While it is certainly true that we have assembled here because of our admiration and friendship for Giusa and Roger Cayrel, at the same time this has been a very well focused meeting on the connection between the local history (mainly in our galaxy) as a function of time t and the aspect of the distant universe as a function of redshift z. To make this connection fully quantitative, we would have to know $z(t)$ and this involves Hubble constant, deceleration parameter, etc. The fact that the impreciseness of our knowledge in this respect has not very much bothered anyone, shows that we still have a long way to go before a fully quantitative correspondence can be established.

In any big bang cosmology the abundance of elements from carbon and up ('metals') started out at zero and so did the fraction of matter contained in stars. Subsequently, these increased in different places at different rates. However, the lowest metallicities we see in our Galaxy are well above zero. Among field stars these are at around –4 (logarithmically with respect to the sun), while, as Sargent reported, the Ly α clouds with low column densities, which presumably have undergone little processing, have upper limits in the range –3 to –3.5.

Globular clusters have higher metallicities with the lowest values around –2.5. It should be noted, however, that if the globulars had the same metallicity distribution as the field stars, no more than about one with a metallicity below –3 would be expected. It would be particularly interesting to see the metallicity distribution in metal poor dwarf galaxies, like the Sculptor and Fornax dwarfs.

Quasar absorption lines give information on galaxy haloes at redshifts around $z = 3$ and indicate metallicities of typically –2. However, the fraction of the metals locked up in dust remains uncertain and the results for the gas phase may be unduly low. An important part of the baryomic matter in the Universe is found as rather hot X-ray emitting gas in clusters of galaxies, and it is remarkable that here metallicities tend to be above –1.

Astrophysics and Space Science is the original source of publication of this article. It is recommended that this article is cited as: *Astrophysics and Space Science* **265**: 569–572, 1999.
© 1999 *Kluwer Academic Publishers. Printed in the Netherlands.*

Much higher metallicities occur in the super metal rich stars where values of up to +0.6 are reached even among stars some 12 Gyr old (as shown in Giusa Cayrel's paper). Metal production apparently took place rapidly. In a recent review paper Roger Cayrel asked the question 'will a star with metallicity below –4 be found by the year 2000?'. The odds don't seem to look very good; perhaps a few early massive stars polluted the Galactic gas before lower mass stars formed. Alternatively, as proposed by Suzy Collin, if black holes formed early – maybe before the galaxies – metals could have been synthesized in the surrounding accretion disks and have been ejected and dispersed into the primeval gas.

Talking about metallicity is still rather ambiguous, since different elements have different histories. Fe/H is a relatively clear indicator in that Fe has been synthesized exclusively in supernovae: small amounts in SN II (and possibly SN Ib,c) relatively instantaneously after the formation of massive stars and larger amounts in SN Ia. In the latter case, long waiting times are necessary at least if these involve mass transfer from a one to three solar mass star onto a white dwarf. If so, no SN Ia should be seen at high redshifts, the more so since recent models seem to indicate that the SN Ia channel becomes relatively less probable at low metallicities. It should, however, be noted that SN Ia might also occasionally result from the evolution of massive close binaries. More in general, it would be important in population synthesis and chemical evolution models to take binaries more explicitly into account than done until now.

The interpretation of stellar metallicities in Galactic evolution depends on the belief that what is observed now is the same as what went into the star at the time of formation. To ensure this, it is important (except for Fe/H) to utilize unevolved stars. Unfortunately, these tend to be faint, but with the new large telescopes the situation should improve. Furthermore, one should be sure that rotational circulation, sedimentation, surface effects, etc. have not affected the results. While this may well be true, some lingering doubts appear not unjustified.

Determinations of abundances in stars also have suffered from the classical problems of non LTE effects, uncertain f-values, opacities, etc. Here the situation has fortunately much improved. In the gas phase of interstellar matter greater difficulties persist: in the cold gas dust formation may change the composition, while in the hot gas temperature inhomogeneities may create problems.

The very light elements have had a different history from the others. $^7Li/H$ remains at a constant level at very low metallicities (the 'Spite plateau'). Variations in this lithium abundance appear to be less than 10%. On the one hand, this gives support to the belief that it was produced in the Big Bang, on the other, it also indicates that during the stellar lifetime there has not been much change owing to the effects enumerated before, at least in those stars where the convective envelope is sufficiently shallow to avoid lithium burning.

The constancy of the lithium abundance contrasts with the claimed variations in the deuterium to hydrogen ratio. In the case of quasar absorption results the apparent variation may well be due to the difficulties of determining the velocity profile of the hydrogen lines. Even in the solar neighborhood along different lines of sight different values have been claimed. Typical determinations of the D/H ratio now cluster around D/H 2×10^{-5}, compatible with traditional Big Bang parameters.

2. Galactic Populations

Various components of the Galaxy are now generally recognized: halo, thick disk, thin disk, bulge and center. There has been much discussion as to whether the thick disk is a discrete unit or whether all these components have a more gradual transition. The fact that we have heard here terminology like superhalo, halo A and B, thin-thick disk, tail or extension of thick disk would seem to support the latter view. Nevertheless, it is also clear that these components represent different star formation histories in different parts of the galaxy, with the thin disk having an extended star formation history and that of the other components mainly concentrated in the early history of the Galaxy, before about 10 Gyr ago. Halo and bulge are made up of low angular momentum stars, while the thick and thin disks are mainly rotationally supported, the former with somewhat larger random motions. Whether the bulge formed before, after or at the same time as the thick disk is still being debated, but its metallicity appears to be relatively high – not very different from solar, with evidence for metallicity gradients. Claims for the presence of a young bulge population have not found much support, but very close (pc scale) to the Galactic center infrared observations have shown the presence of very massive stars.

Metallicities differ in different populations, but there is much scatter in all. This may be related to different places of origin, to inadequate mixing processes in the gas and to accretionary events. Differences in the ratios between different metals also appear to be related to kinematical details.

Of course, we would like to connect the metallicity differences to different places and times of origin of the stars in the Galaxy. However, orbital diffusion and especially effects associated with non axial symmetry of the Galactic gravitational potential may wash out evidence about the original orbits and thereby account for part of the scatter.

3. The Future

Some new instruments will become available in the coming decade which will qualitatively advance the subject of our conference.

The spatially resolved composition of the hot gas in distant clusters of galaxies should be observed with XMM, the large X-ray collector to be launched by ESA early in 2000.

The cool gas and the dust at high redshifts will be observed with FIRST, the far infrared (80–800 m) 3.5 m ESA space telescope to be launched in 2006 or

2007. Unfortunately, the angular resolution (20 arcsec at mid-range) is poor, and therefore it is to be expected that the Large Submm Array or the MMA or a combination of the two will be constructed by the middle of next decade. While the array size of 10 km will give sub-arcsec resolution, its wavelength coverage will be much more limited, even at a very high mountain site. The combination of FIRST and LSA/MMA should be particularly powerful for the analysis of high redshift chemistry.

The optical/near IR analysis of high redshift galaxies, Ly α clouds and faint, old, unevolved stars in our Galaxy should be advanced by the future 8-m class telescopes: the Next Generation Space Telescope (2007?) and the thirteen 8-m class telescopes that should be available on the ground by the middle of the next decade. With so much large telescope time, it will be important to devote significant blocks of time to specific problems: what are the metallicities in the low column density Ly α clouds or what are the lowest metal abundances found in Local Group dwarf galaxies?

For the understanding of the populations of our own Galaxy future astrometric satellites, which should be able to reach 10 μas precision and to observe up to 10^9 stars, will be invaluable but are unlikely to become available already in the coming decade. Combined with radial velocities, these should enable us to place huge numbers of stars in half of the Galaxy in phase space with 10% accuracy in all coordinates. While analyzing the astrometric observations of 10^9 stars may pose daunting problems, determining their radial velocities may be equally difficult to achieve.

All in all it is clear that the coming two decades will be an exciting time for all those who study our Galaxy, its chemical and dynamical evolution and the connections thereof with the high redshift universe. For Giusa and Roger Cayrel and for all of us there remains a lot to do.

Author Index

574

The manufacturer's authorised representative in the EU is Springer
Nature Customer Service Centre GmbH, Europaplatz 3, 69115 Heidelberg,
Germany. If you have any concerns regarding our products, please
contact ProductSafety@springernature.com

Printed and bound by CPI Group (UK) Ltd, Croydon, CR0 4YY

29/04/2026

02099472-0012